Atomic Physics
with Positrons

NATO ASI Series

Advanced Science Institutes Series

A series presenting the results of activities sponsored by the NATO Science Committee, which aims at the dissemination of advanced scientific and technological knowledge, with a view to strengthening links between scientific communities.

The series is published by an international board of publishers in conjunction with the NATO Scientific Affairs Division

A	**Life Sciences**	Plenum Publishing Corporation
B	**Physics**	New York and London
C	**Mathematical and Physical Sciences**	D. Reidel Publishing Company Dordrecht, Boston, and Lancaster
D	**Behavioral and Social Sciences**	Martinus Nijhoff Publishers
E	**Engineering and Materials Sciences**	The Hague, Boston, Dordrecht, and Lancaster
F	**Computer and Systems Sciences**	Springer-Verlag
G	**Ecological Sciences**	Berlin, Heidelberg, New York, London,
H	**Cell Biology**	Paris, and Tokyo

Recent Volumes in this Series

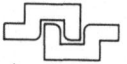

Series B: Physics

Atomic Physics with Positrons

Edited by

J. W. Humberston

University College London
London, England

and

E. A. G. Armour

University of Nottingham
Nottingham, England

Plenum Press
New York and London
Published in cooperation with NATO Scientific Affairs Division

Proceedings of a NATO Advanced Research Workshop on
Atomic Physics with Positrons,
held July 15–18, 1987,
in London, England

Library of Congress Cataloging in Publication Data

NATO Advanced Research Workshop on Atomic Physics with Positrons (1987:
London, England)
 Atomic physics with positrons.

 (NATO ASI series. Series B, Physics; vol. 169)
 "Published in cooperation with NATO Scientific Affairs Division."
 "Proceedings of a NATO Advanced Research Workshop on Atomic Physics
with Positrons, held July 15–18, 1987, in London, England"—T.p. verso.
 Bibliography: p.
 Includes index.
 1. Positrons—Scattering—Congresses. 2. Collisions (Nuclear physics)—
Congresses. 3. Cross sections (Nuclear physics)—Congresses. I. Humberston,
John W. II. Armour, E. A. G. III. North Atlantic Treaty Organization. Scientific Af-
fairs Division. IV. Title. V. Series: NATO ASI series. Series B, Physics; v. 169.
QC793.5.P628N375   1987   539.7′214   88-2336
ISBN-13:978-1-4612-8267-9   e-ISBN-13:978-1-4613-0963-5
DOI:10.1007/978-1-4613-0963-5

INTERNATIONAL ORGANISING COMMITTEE

E.A.G. Armour, University of Nottingham, England
E. Ficocelli Varracchio, University of Bari, Italy
J.W. Humberston, University College London, England
C.J. Joachain, Free University of Brussels, Belgium
R.P. McEachran, York University, Toronto, Canada
W. Raith, University of Bielefeld, Germany
T.S. Stein, Wayne State University, Detroit, U.S.A.

LOCAL ORGANISING COMMITTEE

E.A.G. Armour, University of Nottingham
C.D. Beling, University College London
M. Charlton, University College London
J.W. Darewych, University College London
T.C. Griffith, University College London
J.W. Humberston, University College London

PREFACE

    The NATO Advanced Research Workshop on Atomic Physics with Positrons, which was held at University College London during 15-18 July 1987, was the fourth meeting in a series devoted to the general theme of positron colli- sions in gases. Previous meetings have been held at York University, Toronto (1981); Royal Holloway College, Egham (1983) and Wayne State Uni- versity, Detroit (1985).

    Recent very significant improvements in positron beam currents, due to the development of more efficient moderators and the use of more intense positron sources, are making possible an increasingly sophisticated range of experiments in atomic collision physics. Whereas a few years ago only total scattering cross sections could be determined, measurements can now be made of various partial and differential cross sections. Intense positron beams are also being used to produce positronium beams and already, as reported here, preliminary investigations have been made of collisions of positronium with several target systems.

    These experimental developments have stimulated, and been stimulated by, steady, if somewhat less spectacular, progress in associated theoretical studies. Both aspects of the field are well represented in these Proceed- ings.

    The final session of the Workshop was devoted to such 'exotic' topics as the formation of antihydrogen and the behaviour of positive muons in gases. Considerable interest is currently being shown in the formation of antimatter and various possible methods of synthesising antihydrogen were discussed. The justification for including positive muons in a workshop ostensibly devoted to positrons is that in some respects the positive muon can be thought of as a heavy 'positron' (or a light 'proton'), and interes- ting comparisons between the behaviour of muons and positrons can be made.

    One afternoon of the Workshop was designated as a poster session, and more than thirty posters were displayed. Abbreviated versions of most of the posters are included in these Proceedings in addition to the contribu- tions from the invited speakers.

    Many readers will already know of the tragic illness which has afflic- ted Coulter McDowell and brought his scientific career to an end. He made many contributions to theoretical atomic collision physics and he was closely involved with this series of workshops, having been the Director of the Royal Holloway College meeting in 1983. A moving account of his distingui- shed career was presented by his friend and former colleague, Brian Bransden.

    The Organising Committee gratefully acknowledge the generous financial support provided by the NATO Science Committee and also the financial and administrative assistance provided by University College London. In addition they wish to thank the Royal Society for an interest free loan. Thanks are due to Ms. Siranee Pollakorn for her most helpful secretarial asistance in the preparation of the Workshop.

The editors wish to thank the invited speakers for their co-operation in submitting the final versions of their manuscripts on time and, also, Mrs. Anne Davis for so ably retyping many of the contributions to the Proceedings.

The continuing rapid growth in the field of positron collisions has made it appropriate to hold meetings such as this one every few years in order to review progress and speculate on the direction of future developments. The next Workshop will be held at the Goddard Space Flight Center, Greenbelt, Maryland, U.S.A. in 1989.

J. W. Humberston

E. A. G. Armour

5 October 1987

CONTENTS

CONTRIBUTED PAPERS

POSITRON-IMPACT IONIZATION AND POSITRONIUM FORMATION

Wilhelm Raith

Fakultät für Physik
Universität Bielefeld
D-4800 Bielefeld, Fed. Rep. of Germany

INTRODUCTION

Energy losses of charged particles in gases are mainly determined by impact ionization. At low energies, below the range of validity of the First Born Approximation (FBA), our theoretical understanding of impact ionization is not yet satisfactory. With data on impact ionization by positrons (as well as antiprotons) becoming available we can now compare particles of equal mass and different charge and vice versa. The study of positron-impact ionization will contribute significantly to the theory of electron-impact ionization; the absence of the exchange interaction with the atomic electrons makes positron-impact ionization a much simpler process theoretically. The Wannier theory of threshold behaviour, which has proven to be compatible with all electron tests performed so far, will be put to yet another test with positrons for which a different power-law exponent has been predicted. Positronium (Ps) formation on atoms is a unique quantum mechanical rearrangement problem; of interest here is a comparison with electron transfer in proton-atom collisions.

All data obtained thus far come from gas-target experiments and consist of angle-integrated cross sections on single ionization and Ps-formation. (The differential Ps-formation in the forward direction and Ps-formation by electron capture from solids, utilized in Ps-beam creation, are covered in other invited papers of this Workshop). Double ionization has not yet been observed. Cross sections for Ps-formation usually include all the Ps bound states but not the Ps continuum states.

Atomic-beam experiments aimed at integral as well as differential cross sections are under way, including measurements on atomic hydrogen[1] which are desired for comparison with fundamental scattering theory[2] and for interpretation of the Ps-annihilation radiation from the galactic center[3]. While waiting for atomic-hydrogen data we can already gain some valuable insights by comparing theory and experiment for helium.

THE SIGNATURES

Here we consider the ionizing interactions of a positron with an atom A or a molecule. Each signature listed below has already been utilized by the research group mentioned in parentheses.

1

Impact ionization:

$$e^+ + A \rightarrow A^+ + e^- + e^+ \quad \text{with } \Delta E(e^+) > E_{ion}$$

Unique signatures are

- the positron energy loss (Tokyo[4])
- the emergence of a free electron (Arlington[5])
- the production of an ion <u>in conjunction with the survival of the positron</u> (Bielefeld[6])

Ps formation:

$$e^+ + A \rightarrow A^+ + Ps$$
$$\hookrightarrow 2\gamma \text{ or } 3\gamma \ .$$

Unique signatures are

- the short-lived Ps atom (London[7])
- the emission of 2γ or 3γ annihilation radiation (London[8])
- the vanishing of the primary positron (Arlington[9])
- the production of an ion <u>in conjunction with the non-survival of the positron</u> (Bielefeld[6]),

All these different methods were applied to helium. The Bielefeld measurements with the ion-detection method are the most recent ones.[6] The positron impact-ionization cross section was found to merge with the electron cross section above 600 eV, at intermediate energies the positron cross section is larger but at about 35 eV the two cross section curves appear to cross over.[10] The energy-loss method of Sueoka[4] yielded data up to 120 eV which show merging of positron and electron cross sections already above 40 eV with σ_{ion} + < σ_{ion} - below 40 eV. The electron-detection method of Diana et al.[5] was used for measurements up to 200 eV; but the scattered data points show only that σ_{ion} + > σ_{ion} -. The Ps-formation cross section measured with the ion-detection method[6] rises steeply from threshold, has a maximum of 0.44×10^{-16} cm^2 near 40 eV, and falls off rather slowly with increasing energy. The Ps-detection method[7] has been used for differential measurements in the forward direction but not for angle-integrated cross sections. The 3γ-detection method was employed by Charlton et al.[8] in the very first measurement of Ps formation cross sections. Their cross sections are almost a factor 1/4 lower at the maximum and approach zero more rapidly with increasing energy. The data obtained with the vanishing-positron method[9] have a maximum value of 0.5×10^{-16} cm^2 (statistically consistent with the results of the ion-detection method), but on the higher-energy side they exhibit peculiar oscillations.

<u>Conclusion</u>: At present no two methods give consistent results over a wide energy range.

The following discussions will be based on the results obtained with the ion-detection method for helium and molecular hydrogen.

THE ION-DETECTION METHOD (Bielefeld)

Extraction and detection of ions provides a relative measure of the sum of the impact-ionization and Ps-formation cross sections $\sigma_{ion}^+ + \sigma_{Ps}$. Counting only those ions which are time-correlated with an outgoing positron leads to a relative measure of the impact-ionization cross section. In both cases the relative cross sections can be normalized by performing ananlogous electron measurements and comparing with electron literature values. The Ps-formation cross section is then obtained by subtracting the ionization cross section from the sum of both cross sections.

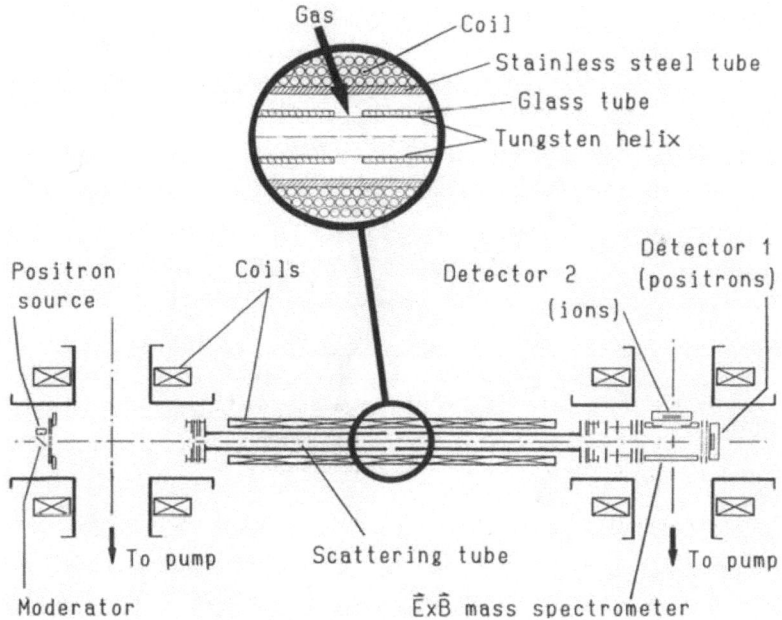

Fig. 1.   Experimental arrangement for measurement of σ_{ion}^+ and σ_{Ps} by the ion-detection method (from Fromme et al.[6]).

In order to facilitate the extraction of scattered positrons and produced ions the whole apparatus (Fig. 1) is surrounded by coils providing a longitudinal magnetic guiding field. Inside the long, thin scattering tube a current flows through the tungsten helix which lines the inside wall of the glass tube. This produces a longitudinal electric field for ion extraction towards the $\vec{E} \times \vec{B}$ mass spectrometer, which separates positrons from ions and directs them onto the detectors 1 and 2.

Measurements are made of the counts in both detectors and also the time between events registered by detectors 1 and 2. The time-correlation spectrum shows a flat continuum due to uncorrelated ions from Ps-formation (and background events) and a pronounced peak due to positron-correlated ions from impact ionization. The peak area is taken as a relative measure of the impact-ionization cross section.

Data on Ps-formation in molecular hydrogen obtained by this method (Fig. 2 top) agree reasonably well with data obtained by the Arlington group (triangles), utilizing the vanishing positron-method, but disagree significantly with the London data (bars) obtained by 3γ detection. Qualitatively, this assessment equals that for helium. The agreement with theoretical predictions (Fig. 2 bottom) is unsatisfactory, which is also true for helium.[6]

Up to now the electric extraction field did not allow for the restriction of the energy width of the cross section measurement to less than 4 eV. The experimental arrangement is being modified to increase the energy resolution. This involves increasing the longitudinal magnetic field strength decreasing the electric field strength and installing an $\vec{E} \times \vec{B}$ positron monochromator between the moderator and the scattering tube.

In an atomic-beam experiment a modified version of the ion-detection method will be utilized to measure both cross sections for atomic hydrogen.[1]

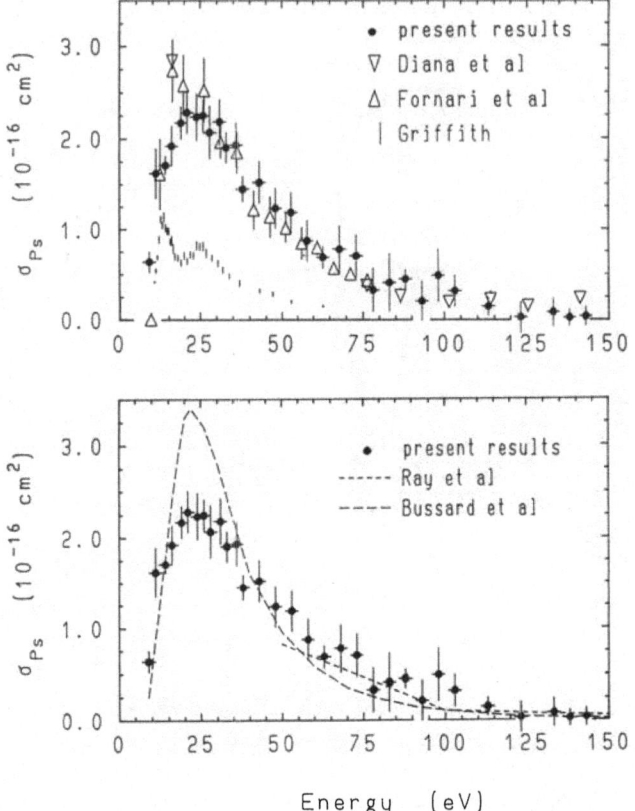

Fig. 2.   Positronium-formation cross section of molecular hydrogen (from Fromme et al.[12]). Top: comparison with other experimental results.[13,9,14] Bottom: comparison with theoretical results.[15,16]

CROSS SECTIONS NEAR THRESHOLD

For impact ionization the Wannier theory predicts that the angle-integrated cross section obeys

$$\sigma_{ion}^{\pm} (E) \ \alpha (E - E_{ion})^{n^{\pm}}$$

with $n^- = 1.127$ and $n^+ = 2.651$.[17] This implies $\sigma_{ion}^+ < \sigma_{ion}^-$ near threshold whereas measurements show that $\sigma_{ion}^+ > \sigma_{ion}^-$ at intermediate energies up to the convergence region.

The experimental results for helium have already indicated a cross-over of $\sigma_{ion}^+(E)$ and $\sigma_{ion}^-(E)$ near $E = 35$ eV[10] and the theoretical results of Campeanu et al. support this contention.[18] A similar behaviour is exhibited by the results for molecular hydrogen (Fig. 3). Threshold studies with higher energy resolution are needed for a determination of the power-law exponent.

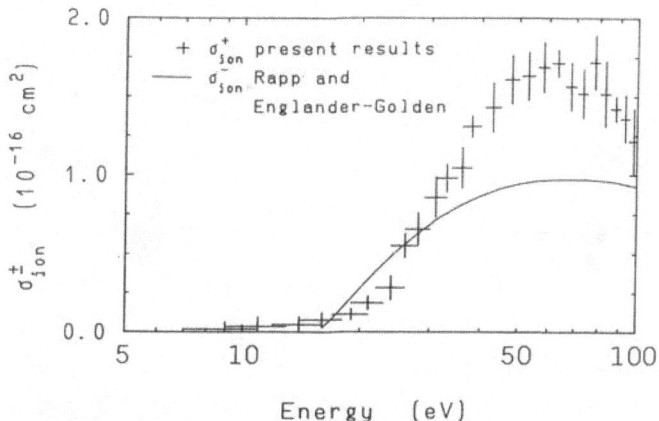

Fig. 3. Positron-impact ionization of molecular hydrogen compared
with electron-impact ionization[19] near threshold (from
Kruse[20]).

Very close to threshold the Ps-formation cross section is given by the
first partial-wave contribution for which

$$\sigma_{Ps}(E) \; \alpha (E - E_{Ps})^{\frac{1}{2}} \; .$$

The available data can be fitted and extrapolated to $\sigma_{Ps} = 0$ at $E = E_{Ps}$ by
a steep straight line. In the energy range between E_{Ps} and the first
excitation threshold the deduction of $\sigma_{Ps}(E)$ from experimental values of the
total cross section, σ_t, gives the angle-integrated elastic cross section.
This procedure produced the first experimental evidence for a threshold
anomaly (cusp) in the elastic cross section (Fig. 4). Recent measurements
for molecular hydrogen combined with the total cross section data of
Hoffman et al.[21], also gave evidence of a pronounced threshold anomaly
(Fig. 5).

COMPARISON WITH OTHER PARTIAL CROSS SECTIONS

At its maximum, the ionizing cross sections, $\sigma_{ion}^+ + \sigma_{Ps}$, account for
more than 60% of the total cross section, σ_t, in the case of helium. In
view of the fact that total and ionizing cross sections can now be measured
rather accurately but not yet calculated reliably, whereas elastic and
excitation cross sections can be calculated but not yet measured, it is
worthwhile to define a "total cross section without the ionizing channels"
for comparison of theory and experiment at the present state of knowledge:

$$\sigma_{no\ ion}^+ = \sigma_t^+ - \sigma_{ion}^+ - \sigma_{Ps}$$

$$= \sigma_{el}^+ + \sum \sigma_{exc}^+$$

5

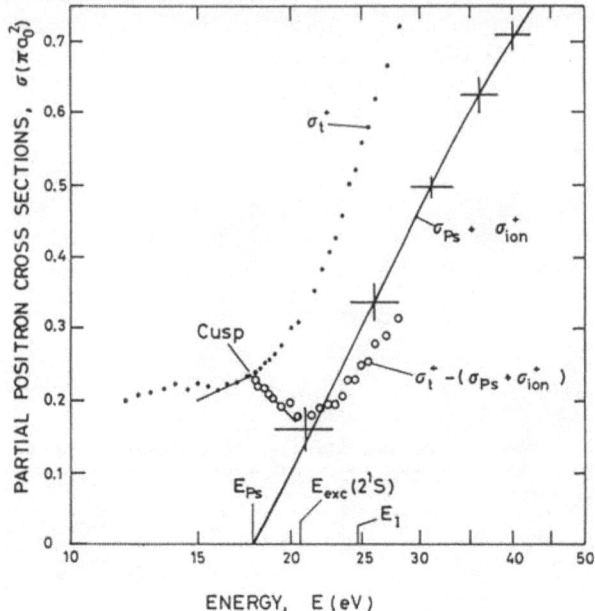

Fig. 4. Cross section anomaly in the angle-integrated elastic cross section at the Ps-formation threshold for helium (from Campeanu et al.[10]).

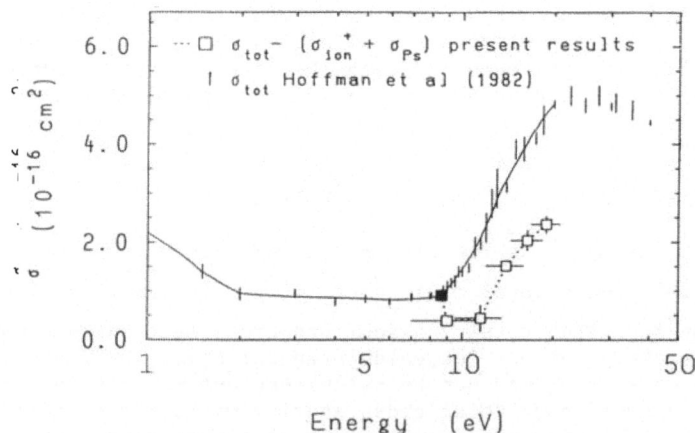

Fig. 5. Cross section anomaly in the elastic cross section at the Ps-formation threshold for H_2 (from Kruse[20]). Here the difference curve represents the cross section for elastic scattering as well as rotational and vibrational excitation.

The same expression with supercript "−" and $\sigma_{Ps} = 0$ applies to electron scattering. For helium such a comparison (Fig. 6) shows the deviation of the calculated excitation cross sections at low energies and absence of such calculations at high energies, but also a rather satisfactory agreement at intermediate energies of 100 to 300 eV.

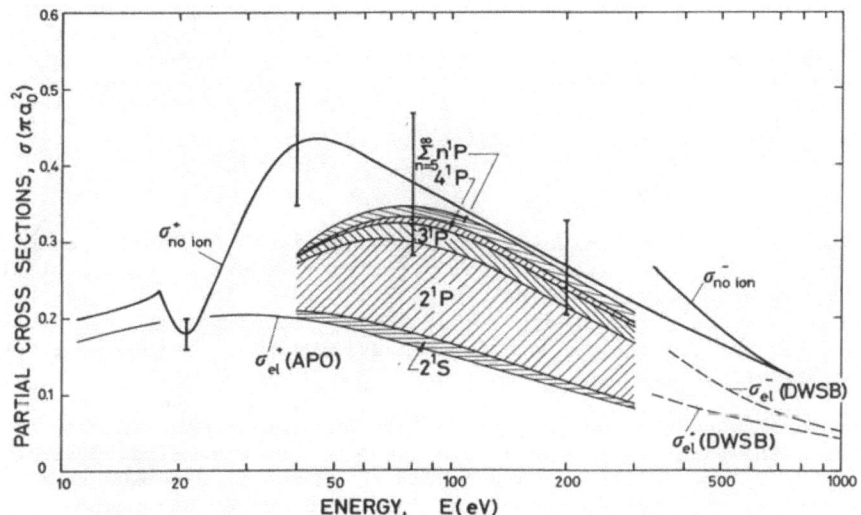

Fig. 6. Partitioning of the helium cross section (adapted from Campeanu et al.[10]). Fat solid curves $\sigma_{no\ ion}^{+}$ and $\sigma_{no\ ion}^{-}$ are from experimental results. The elastic cross sections with specifications APO and DWSB are calculations based on "adiabatic polarized orbital" and "distorted-wave second Born" approximation, respectively. The hatched areas represent distorted-wave excitation cross sections.

THE CONVERGENCE REGION

The merging of corresponding positron and electron cross sections at high energies is of great interest as it indicates that the FBA becomes valid, provided that the convergence is not caused by compensation due to counteracting non-FBA effects. The convergence in the case of helium has been an intriguing question ever since the total cross section measurements of Kauppila et al.[22] showed an early merging at 200 eV whereas the elaborate distorted-wave second Born approximation calculation of Dewangan and Walters[23] predict for the elastic cross sections $\sigma_{el}^{-} > \sigma_{el}^{+}$ even above 2000 eV. Regarding both results as valid requires that the partial cross sections for excitation and ionization must provide for a rather exact compensation of the difference in electron and positron elastic scattering over a wide energy range.

The ionization cross sections merge above 600 eV. Between 200 and 600 eV the excess of electron over positron elastic scattering is mainly compensated by $\sigma_{Ps} + \sigma_{ion}^{+} > \sigma_{ion}^{-}$. But above 600 eV the compensation can only come from the sum over all excitation cross sections, which ought to produce a positron excess up to energies of 2000 - 3000 eV. Theoretical studies on positron and electron excitation at higher energies could elucidate this problem.

COMPARISON OF POSITRON AND PROTON COLLISIONS

Before positron and antiproton cross sections could be measured comparisons were restricted to electron and proton cross sections which could not distinguish between effects related to the differences in charge-sign or mass-value. It has been customary to make such comparisons for "equal velocities", not equal kinetic energies, because the magnitude of the cross section depends on the duration of the interaction which is inversely proportional to the velocity.

A comparison of impact-ionization cross sections for positrons, protons, and electrons on helium versus initial velocity v_i shows a merging of all cross sections at about 15×10^6m/s (Fig. 7). At lower energies the proton cross section is much larger than the positron and electron cross sections. For antiprotons the single-ionization cross sections were found to be the same as the protons.[27]

Ps-formation can be compared with electron transfer in a proton collision. Since the particle produced is an H-atom the cross section symbol σ_H is appropriate. For helium as target atom the comparison versus equal initial velocity (Fig. 8) does not exhibit the expected convergence at higher energies. Instead the H-formation cross section, σ_H, which largely exceeds σ_{Ps} at low velocities, is much smaller at higher energies. The cross-over lies at about 4×10^6m/s.

The two processes differ somewhat in threshold energy:

$$E_{Ps}(He) = E_{ion}(He) - E_{ion}(Ps) = 17.8 \text{ eV}$$

$$E_H(He) = E_{ion}(He) - E_{ion}(H) = 11.0 \text{ eV}$$

The main difference, however, is that the energy needed for the acceleration of the captured electron to the outgoing speed amounts to about half of the incoming positron's energy, but is only a tiny fraction of the energy of the massive incoming proton. Consequently, for the proton collision initial and final velocities are almost equal whereas for the positron collision the final velocity is about $1/\sqrt{2}$ of the initial velocity (except very close to threshold where it is even less). This significant difference in final velocities raises the question of which velocity is more appropriate for comparison. Both processes have one charged particle and a neutral atom in the initial as well as the final state. Thus the long-range pre- and post-collision interaction is determined by the polarizability of the atom which for Ps and H is 25.6 and 3.2 times that of He, respectively.[30] This fact strongly suggests that a comparison for "equal final velocities" might indeed be more meaningful. This new way of plotting (Fig. 9) demonstrates that the final velocity is the relevant parameter for comparing electron transfer to the positron and the proton. Now, instead of crossing over, the two curves exhibit the expected convergence behaviour at high energies. (For impact ionization plotting versus final rather than initial velocity does not change the convergence behaviour because in that case $v_i \approx v_f$ holds for $E \gg E_{ion}$).

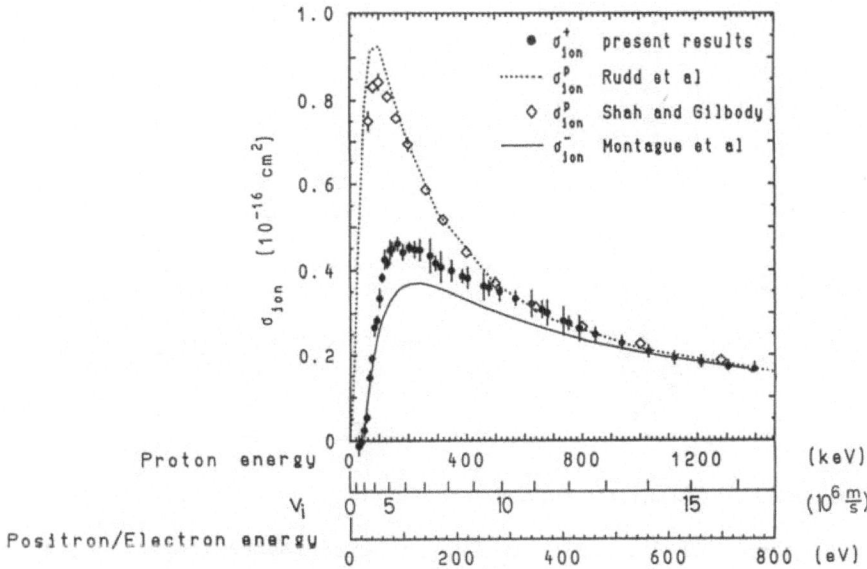

Fig. 7. Comparison of impact-ionization cross sections for
positrons,[6] protons,[24,25] and electrons[26] on helium
(from Fromme et al.[28]).

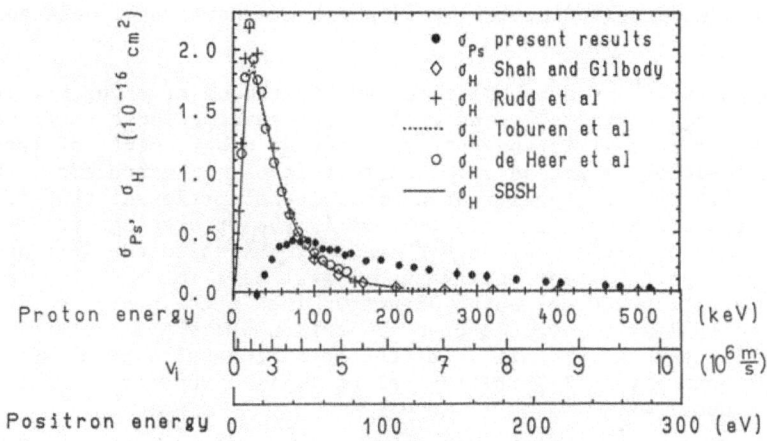

Fig. 8. Comparison of electron-transfer cross sections for
positron[6] and proton[25,29] collisions leading to
formation of Ps (σ_{Ps}) and H(σ_H), respectively,
plotted versus initial velocities (from Fromme et
al.[28]).

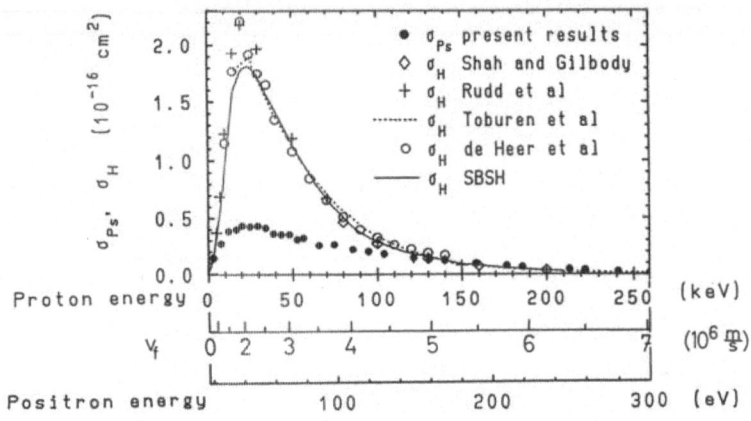

Fig. 9. Same as in Fig. 8 but now plotted versus **final**
velocity of the outgoing atom Ps or H, respectively
(from Fromme et al.[28]).

TESTING THE IMPACT-IONIZATION THEORY

Comparing the experimental data for positron-impact ionization of
helium with theoretical results (Fig. 10) leads to a surprising conclusion:
The four theoretical results, obtained with rather different methods, all
agree fairly well with the experiment. Furthermore, two of them are calculations for electron-impact ionization with exchange excluded. Peach and
McDowell[31] used a half-range FBA, and McGuire[32] a Glauber approximation;
both methods do not distinguish between the positive and negative charge
on the incoming particle.

In addition, the curve labelled Basu et al, gives the results of
their distorted-wave approximation "DW2" which are nearly the same as their
FBA results.[33]

Campeanu et al.[18] used bound-state wave functions of high quality
and took into account distortion as well as screening effects which were
found to be important. This suggests that the good agreements of less
elaborate calculations might be somewhat fortuitous. Nevertheless, the
comparison of Fig. 10 indicates that a significant portion of the difference
between positron and electron cross sections is due to exchange which
reduces the electron cross section. Therefore, the absence of exchange
makes positron-impact ionization a valuable test case for approximations
describing distortion and screening effects. Then these approximations
can be applied analogously to electron ionization where the presence of
exchange complicates the comparison of theoretical results with experimental data. Such a parallel theoretical analysis of positron and electron
ionization is under way.[35]

OPEN QUESTIONS

What model describes the differences between positron and electron
interactions with atoms? The consideration of static and dynamic Coulomb
interactions (referring to the interaction of the positron/electron with
the screened nuclear charge and the induced atomic dipole moment, respectively) which add in electron scattering but subtract in positron scattering,
has been used to explain the general differences in total cross sections.

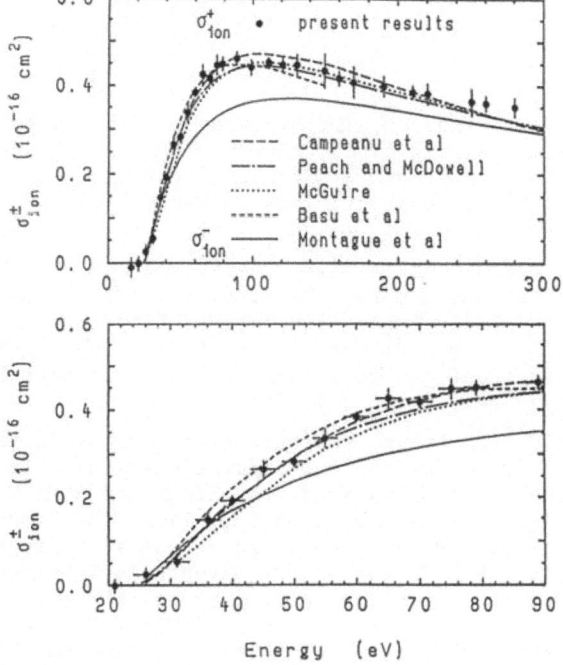

Fig. 10. Positron-impact ionization of helium, Experimental cross sections[6] compared with several theories[18,31-33] and with experimental electron cross sections[26] (from Fromme[34]).

But this seems to hold only for elastic scattering. Apparently, the interaction with the induced dipole moment is insignificant for ionization. Why? And what is the explanation for $\sigma_{exc}^+ > \sigma_{exc}^-$ above 600 eV in case of helium, which follows from the convergence behaviour of σ_t, σ_{el}, and σ_{ion}? Is it the attractive (repulsive) two-particle interaction of the beam positron (electron) with one of the atomic electrons which favours inelastic processes - excitation as well as ionization - for positron scattering?

Are Ps-formation and positron-impact ionization really two entirely different processes, or can both be described by one theory? The experimental values of $(\sigma_{Ps} + \sigma_{ion}^+)$ do not exhibit any structure at $E = E_{ion}$, not even a change of curvature. As $\sigma_{ion}(E)$ starts from zero at threshold the still rising curve of $\sigma_{Ps}(E)$ seems to reduce its growth rate accordingly. Is this observation already an answer to the question raised? Electron capture to continuum states of Ps, characterized by nearly zero relative velocity between the outgoing electron and positron, is expected to lead to a peak in the electron energy distribution[36] of which Charlton et al.[37] observed a first indication. How much does such continuum-state Ps-formation contribute to σ_{ion}^+? Could it affect the threshold behaviour?

ACKNOWLEDGEMENTS

The author gratefully acknowledges valuable discussions with his co-workers at Bielefeld, Dr. G. Sinapius, Dr. D. Fromme and G. Kruse, as well as the theorists at York University, Toronto, Prof. R.P. McEachran, Prof. A.D. Stauffer and Dr. R.I. Campeanu. The experimental work at Bielefeld has

11

been supported by the Deutsche Forschungsgemeinschaft.

REFERENCES

1. M.S. Lubell and G. Sinapius, The Brookhaven positron-hydrogen scatter-
    ing experiment, invited paper, this Workshop; G. Spicher, A. Glasker,
    W. Raith, G. Sinapius, and W. Sperber, Ionisation of Atomic Hydrogen
    by Positron Impact, Abstract, this Workshop.

2. J.W. Humberston, Positronium Formation, in "Positron (Electron)-Gas
    Scattering", W.E. Kauppila, T.S. Stein, and J.M. Wadehra, eds.
    World Scientific, Singapore (1986), p. 35; A.S. Gosh, P.S. Majundar
    and M. Basu, Positron-Impact Ionization of Hydrogen Atoms, Can. J.
    Phys. 63:621 (1985); A.E. Wetmore and R.E. Olson, Ionisation of H
    and He$^+$ by Electrons and Positrons Colliding at Near-Threshold
    Energies, Phys. Rev. A: 2822 (1986).

3. M. Leventhal and B.L. Brown, Positron Astrophysics in the Galactic
    Center Region: Observations and Simulations, in: "Positron (Electron)-
    Gas Scattering", W.E. Kauppila, T.S. Stein and J.M. Wadehra, eds.,
    World Scientific, Singapore (1986), p. 140; B.L. Brown, A possible
    galactic positron annihilation medium: neutral hydrogen, Astrophys.J.
    292:L67 (1985); B.L. Brown and M. Leventhal, Laboratory Simulation
    of Direct Positron Annihilation in a Neutral-Hydrogen Galactic
    Environment, Phys. Rev. Lett. 57:1651 (1986); W.R. Webber, V. Schön-
    felder, and R. Diehl, Is there a common origin for the cosmic γ ray
    lines at 0.51 and 1.81 MeV near the galactic center?, Nature 323:692
    (1986).

4. O Sueoka, Excitation and Ionization of He Atom by Positron Impact, J.
    Phys. Soc. Japan 51:3757 (1982); Y Katayama, O Sueoka and S. Mori,
    Inelastic cross section measurements for slow positron-O_2 collisions,
    J. Phys. B. 20:1645 (1987).

5. L.M. Diana, L.S. Fornari, S.C. Sharma, P.K. Pendleton and P.G. Coleman,
    Measurements of total ionization cross sections for positrons, in:
    "Positron Annihilation", P.C. Jain, R.M. Singru and K.P. Gopinathan,
    eds., World Scientific, Singapore (1985), p. 342.

6. D. Fromme, G. Kruse, W. Raith and G. Sinapius, Partial-Cross-Section
    Measurements for Ionization of Helium by Positron Impact, Phys. Rev.
    Lett. 57:3031 (1986).

7. G. Laricchia, M. Charlton, T.C. Griffith and F.M. Jacobsen, Prelimi-
    nary Results on the Angular Dependence of Ps Emission in e$^+$-Gas
    Collisions, in: "Positron(Electron)-Gas Scattering", W.E. Kauppila,
    T.S. Stein and J.M. Wadehra, eds. World Scientific, Singapore (1986)
    p. 303.

8. M.Charlton, G. Clark, T.C. Griffith and G.R. Heyland, Positronium
    formation cross sections in the inert gases, J. Phys. B. 16:L465
    (1983).

9. L.S. Fornari, L.M. Diana and P.G. Coleman, Positronium Formation in
    Collisions of Positrons with He. Ar, and H_2. Phys. Rev. Lett. 51:
    2276 (1983).

10. R.I. Campeanu, D. Fromme, G. Kruse, R.P. McEachran, L.A. Parcell,
    W. Raith, G. Sinapius and A.D. Stauffer, Partitioning of the posi-
    tron-helium total scattering cross section, J. Phys. B. 20:3557
    (1987).

11. G. Sinapius, D. Fromme and W. Raith, Positron Impact Ionization Cross
    Section of Helium, in: "Positron (Electron)-Gas Scattering", W.E.
    Kauppila, T.S. Stein and J.M. Wadehra, eds. World Scientific,
    Singapore (1986), p. 61

12. D. Fromme, G. Kruse, W. Raith and G. Sinapius, Measurement of the impact ionization and positronium formation cross sections for positron scattering on molecular hydrogen, Abstract, this Workshop.

13. L.M. Diana, P.G. Coleman, D.L. Brooks, P.K. Pendleton and D.N. Norman, Positronium formation cross sections in He and H_2 at intermediate energies, Phys. Rev. A 34:2731 (1986).

14. T.C. Griffith, Positronium Formation Cross-Sections in Various Gases, in: "Positron Scattering in Gases", J.W. Humberston and M.R.C. McDowell, eds., Plenum, New York (1984), p. 53.

15. A. Ray, P.P. Ray and B.C. Saha, Positronium formation in positron-hydrogen-molecule collisions, J. Phys. B 13:4509 (1980).

16. R.W. Bussard, R. Ramaty and R.J. Drachman, The annihilation of galactic positrons, Astrophys. J. 228:928 (1979).

17. H. Klar, Threshold ionization of atoms by positrons, J. Phys. B 14:4165 (1981)

18. R.I. Campeanu, R.P. McEachran and A.D. Stauffer, Screening and distortion effects in positron impact ionization of helium, J. Phys. B 20:1635 (1987).

19. D. Rapp and P. Englander-Golden, Total Cross Sections for Ionization and Attachment in Gases by Electron Impact. I. Positive Ionization, J. Chem. Phys. 43:1464 (1965)

20. G. Kruse, Diplomarbeit, Universität Bielefeld, 1987.

21. K.R. Hoffman, M.S. Dababneh, Y.-F. Hsieh, W.E. Kauppila, V. Pol, J.H. Smart and T.S. Stein, Total-cross-section measurements for positrons and electrons collding with H_2, N_2 and CO_2, Phys. Rev. A 25:1393 (1982)

22. W.E. Kauppila, J.P. Downing and V. Pol, Measurements of total scattering cross sections for intermediate-energy positrons and electrons colliding with helium, neon and argon, Phys. Rev. A 24:725 (1981).

23. D.P. Dewangan and H.R.J. Walters, The elastic scattering of electrons and positrons by helium and neon: the distorted-wave second Born approximation, J. Phys. B 10:637 (1977).

24. M.E. Rudd, Y.-K. Kim, D.H. Madison and J.W. Gallagher, Electron production in proton collisions: total cross sections, Rev. Mod. Phys. 57:965 (1985).

25. M.B. Shah and H.B. Gilbody, Single and double ionization of helium by H^+, He^{2+}, and Li^{3+} ions, J. Phys. B 18:899 (1985)

26. R.G. Montague, M.F.A. Harrison and A.C.H. Smith, A measurement of the cross section for ionisation of helium by electron impact using a fast crossed-beam technique, J. Phys. B 17:3295 (1984).

27. L.H. Anderson, P. Hvelplund, H. Knudsen, S.P. Møller, A.H. Sørensen, K. Elsener, K.-G. Rensfelt and E. Uggerhøj, Multiple Ionization of He, Ne, and Ar by Fast Protons and Antiprotons, CERN-EP/87-72 (1987). The new data on single ionzation supersede those published earlier (Phys, Rev. Lett. 57:2147 (1986)).

28. D. Fromme, G. Kruse, W. Raith and G. Sinapius, Comparison of impact-ionisation and charge-transfer cross sections for positron and proton scattering on helium, Abstract, this Workshop.

29. D. Fromme, G. Kruse, W. Raith and G. Sinapius, Comparison of impact-ionisation and charge-transfer cross sections oor positron and proton scattering on helium, Abstract, this Workshop.

29. M.R. Rudd, R.D. Dubois, L.H. Toburen, C.A. Ratcliffe, and T.V. Goffe, Cross Sections for ionization of gases by 5-4000-keV protons and for electron capture by 5-150-keV protons, Phys. Rev. A 28:3244 (1983);

L.H. Toburen, M.Y. Nakai and R.A. Langley, Measurements of High-Energy Charge-Transfer Cross Sections for Incident Protons and Atomic Hydrogen in Various Gases, Phys. Rev. 171:114 (1968); F.J. de Heer, J. Schutten and H. Moustafa, Ionization and electron capture cross sections for protons incident on noble and diatomic gases between 10 and 140 eV, Physica 32:1766 (1966); P.M. Stier and C.F. Barnett, Charge Exchange Cross Sections of Hydrogen Ions in Gases, Phys. Rev. 103:896 (1956); J.B.H. Stedeford and J.B. Hasted, Further investigations of charge exchange and electron detachment. Proc. Roy. Soc. (London) A 227:466 (1955).

30. H. Stuart, Molekularrefraktion und elektrische Polarisier-barkeit von Atomen und Ionen (auch Molekülionen),in: Landolt-Börnstein, Vol. I/1, A Eucken, ed., Springer, Berlin (1950), p. 399; J.W. Humberston, Theoretical aspects of positron collisions in gases, Adv. Atomic. Molec, Phys. 15:101 (1979).

31. G. Peach and M.R.C. McDowell, private communcation.

32. J.H. McGuire, private communication, plotted are the "GM2" results of J.E. Golden and J.H. McGuire, Phys. Rev. A 13:1012 (1976).

33. M. Basu, P.S. Mazumdar and A.S. Ghosh, Ionisation cross sections in positron-helium scattering J. Phys. B 18:369 (1985).

34. D. Fromme, Ionisierung von Helium durch Positronen, Dissertation, Universität Bielefeld, 1987.

35. R.I. Campeanu, R.P. McEachran and A.D. Stauffer, Electron and positron impact ionisation of helium at intermediate energies, to be published.

36. M. Brauner and J.S. Briggs, Ionisation to the projectile continuum by positron and electron collisions with neutral atoms, J. Phys. B 19:L325 (1986).

37. M. Charlton, G. Laricchia, N. Zafar and F.M. Jacobsen, Inelastic positron collisions in gases, invited paper, this Workshop.

INELASTIC POSITRON COLLISIONS IN GASES

M. Charlton, G. Laricchia, N. Zafar and F.M. Jacobsen[*]

Department of Physics and Astronomy
University College London
Gower Street, London WC1E 6BT, U.K.

A number of positron (e^+) inelastic scattering effects are discussed. These include the e^+-CO_2 system at low energies, the energy spectra of electrons emitted following e^+ impact ionisation and positronium (Ps) formation. A new method of investigating Ps formation in gases is described and preliminary data are described. Some discussion is devoted to methods of measuring the angular dependence of the Ps formation cross section σ_{Ps}.

1.   POSITRON - CO_2 SCATTERING

This system has attracted interest since independent determinations of the total scattering cross section, σ_T (1,2) have, as shown in Figure 1, found pronounced structure in the vicinity of the thresholds for positronium formation in both ground and excited states. It has also been noted that σ_T for N_2O has similar features at the relevant energies (3). CO_2 was investigated at U.C.L. in an experiment designed to detect n = 2 Ps (Ps^*) and, though large signals were found, these could not be attributed to Ps^* (4,5).

The Ps^* experiment was similar to that of Canter and co-workers (6,7) and consisted of an ultra-violet sensitive phototube viewing the scattering chamber and placed in coincidence with a NaI(Tl) γ-ray detector and incorporating the usual TAC-MCA arrangments. Ps^* manifested itself as a long delayed exponential component at analyser times, $t_A > 0$, ($h\nu$ used as start signal; γ-ray as stop) which could be eliminated by inserting a borosilicate glass disc between the scattering chamber and the u-v phototube. Performing this check on the signals observed in CO_2 failed to remove the $t_A > 0$ component, though some attenuation did occur. This indicates that the energy of the photon responsible for this component was less than $\simeq 3.5 - 4$ eV (8) (see Figure 4). In addition large signals were also observed at $t_A < 0$. A typical $h\nu - \gamma$-ray spectrum for CO_2 is shown in Figure 2 (upper curve) for 14 eV positrons. The presence of these components was unexpected. As a starting point for analysis we assume that none of the spectra can be assigned to Ps^*. This will be valid at $t_A < 0$ but may not be strictly true at $t_A > 0$. We first consider $t_A < 0$.

---

[*] Present address: The Physics Institute, Aarhus University,
                     Aarhus-C, DK-8000, Denmark.

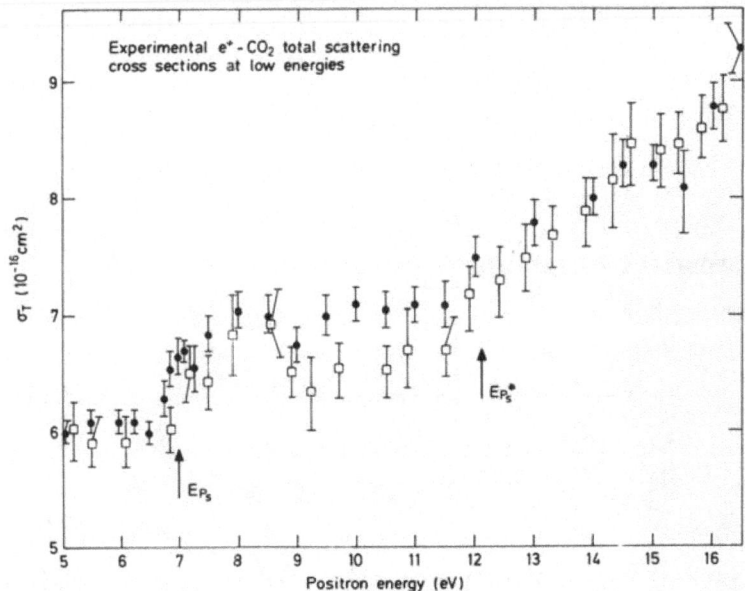

Fig. 1. The U.C.L. (□) (1) and Detroit (●) (2) e^+-CO_2 total scattering cross section at low energy. The thresholds for Ps formation in the ground and first excited states are shown.

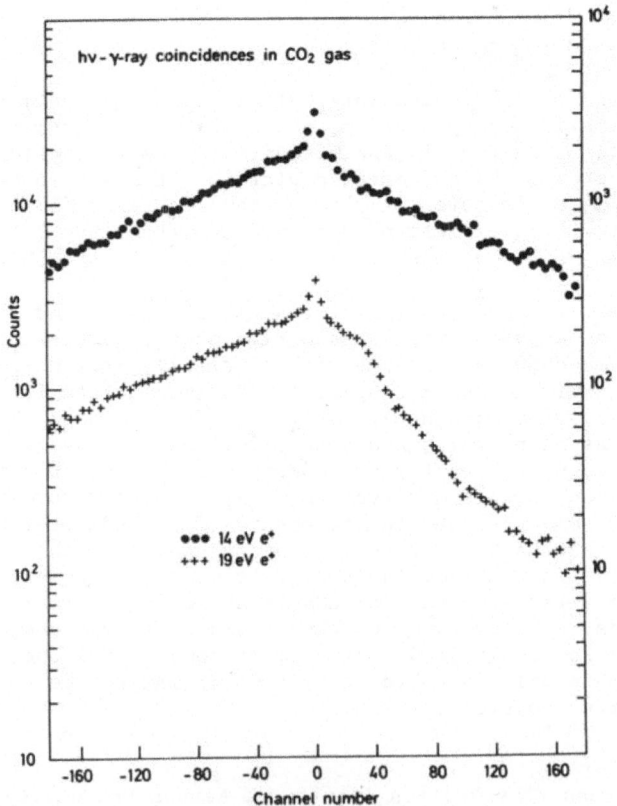

Fig. 2. $h\nu$-γ-ray coincidences in CO_2 gas for 14 eV and 19 eV e^+. There are 0.75 ns per channel.

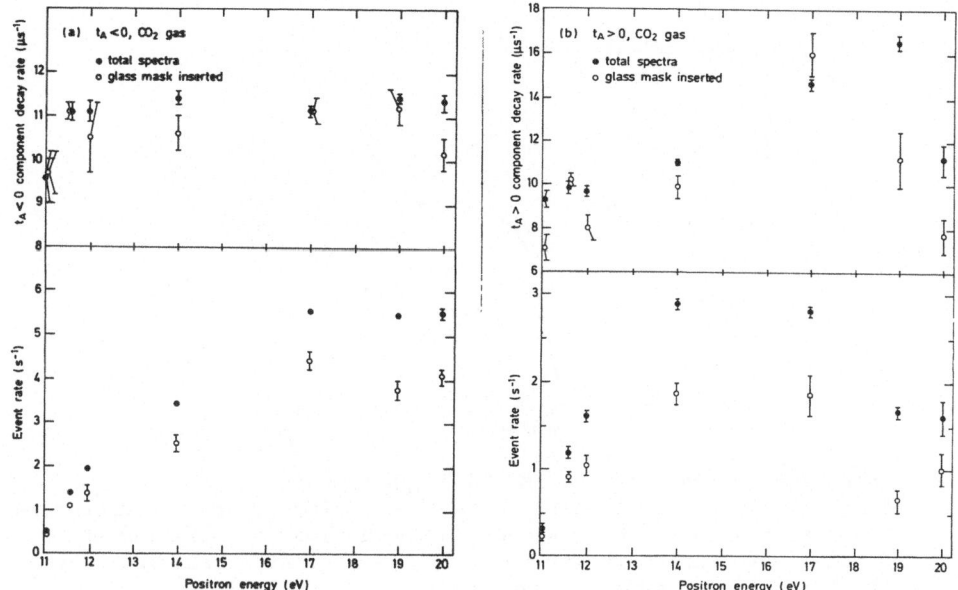

Fig. 3. The measured decay rates and yields obtained using the hν-γ-ray coincidence method in CO_2 gas. The results, ●, are for total spectrum whilst those, o, are with the borosilicate glass mask inserted. a) $t_A < 0$ b) $t_A > 0$.

Figure 3(a) show the measured decay rates and yields (total measured counts s^{-1}) for positrons in the energy range 11 - 20 eV. Note that the average energy of the beam is given with the energy spread for annealed polycrystaline W ≃ 2 - 3 eV and further that no signal was visible at 10.5 eV for both $t_A < 0$ and $t_A > 0$. The main points to note are as follows:-

i)   the decay rates are roughly constant (both total and with glass inserted) at ≃ 11 $μs^{-1}$,

ii) the event rate rises rapidly at an energy close to that observed in the $σ_T$ measurements (see Figure 1),

iii) the yield with the borosilicate mask inserted is always lower than the total yield with the average difference being 24 ± 3% and

iv) the existence of the $t_A < 0$ component means that the annihilation γ-ray preceeded in time the emission of the u-v photon.

These factors can give clues regarding the physical origins of the observed effects. Point (i), also illustrated by comparing the upper and lower curves of Figure 2, probably means that the source of the u-v photon was slowly moving - i.e. not Ps. Point (ii) gives us the threshold whilst point(iii) gives some information regarding the energy of the u-v photon. The response curve of the phototube with and without the borosilicate mask are sketched qualitatively in Figure 4. Referring to this, since some of the light is prevented from reaching the detector when the glass mask is inserted, the energy of the photon must be on the rapidly-varying portion of the curve (full line) at around 3.5 eV.

Point (iv) is also relevant. The sharpness of the exponential compo-nent, even close to threshold, suggests that the γ-ray does not, as postu-

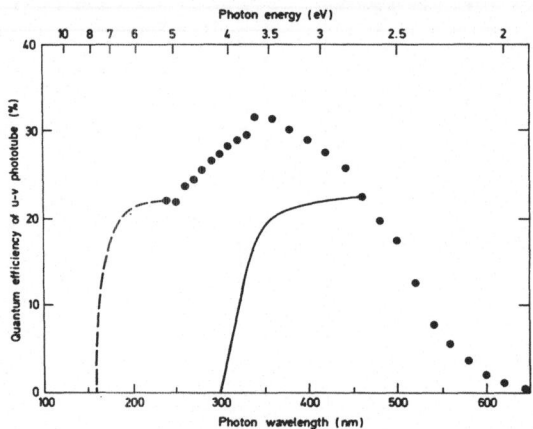

Fig. 4. Quantum efficiency of the EMI 9829Q photo-tube at various
incident photon wavelengths. The points, (ooo), are manufacturers
measurements, (---) corresponds to the estimated variation at
shorter wavelengths whilst (—) is the estimated behaviour when
the borosilicate mask is used (8).

lated previously (5), arise from e^+ which strike the gas cell apertures
but are due to genuine e^+-CO_2 events. Such prompt γ-ray signals can arise
from annihilation of p-Ps so the $h\nu$-γ-ray coincidence may be due to the
reaction

$$e^+ + CO_2 \rightarrow p\text{-Ps} \quad\quad + CO_2^{+*}$$

$$0.125\text{ns} \downarrow \text{prompt} \quad\quad \downarrow \text{delayed} \quad\quad (1)$$

$$2 \quad \gamma\text{-rays} \quad\quad h\nu + CO_2^+$$

in which the e^- is captured from one of the more tightly bound levels and
the photon is emitted in the subsequent rearrangement of the molecular ion.
It is worth noting that data taken at various gas pressures has confirmed
that the $h\nu$-γ-ray coincidences arise from single positron scatters and are
not, for example, due to excitation and Ps formation in two separate collisions. Reaction (1) satisfies all of the points (i) - (iv); however it is
necessary to examine the excited state spectra of CO_2^+ to determine whether
the mechanism is allowed as observed.

A summary of the energies and configurations of the first four excited
states of CO_2^+ is given in Table 1. The only reference which could be
found relating to the lifetimes of these states was Malakhov et al (9) (in
Russian) who find the a $^2\Sigma_u^+$ - $^2\Pi g$ lifetime of $\simeq 110$ ns at a CO_2 gas pressure
of 5×10^{-3} torr. This is similar to the pressure used in these studies and
the lifetime is close to the inverse of the rates shown in Figure 3(a).

Turning to the events at $t_A > 0$ it is clear that a process similar to
(1) must be responsible. The decay rates and yields for the spectra are
shown in Figure 3(b) and are more widely varying than for those at $t_A < 0$.
The measured decay rates change markedly above 14 eV and the yields rise
and then fall. Also, as shown for 19 eV (Figure 2 (lower curve)), the
spectra are now more complicated with, at these higher energies, 'shoulder'
type regions.

Firstly note that here the sequence of events is $h\nu$ as start signal
followed by a delayed γ-ray. Such an annihilation photon can only rise
from o-Ps with the u-v photon emanating from a gas decay. If this gas decay

Table 1. The first 4 states of CO_2^+ (10)

state	energy (eV)	Relevant Ps formation threshold (eV)
$^2\Pi g$	13.8	7.0
$^2\Pi u$	17.3	10.5
$^2\Sigma u$	18.1	11.3
$^2\Sigma g$	19.4	12.6

is prompt the spectrum will consist of a single exponential. This will not be the case if the emission of the u-v photon is delayed or if there is the possibility of detecting either photon from a cascade leading to the ground state. The presence of o-Ps suggests that the observed process is

$$e^+ + CO_2 \rightarrow o\text{-}Ps + CO_2^{+*}$$
$$142ns \quad \downarrow delayed \qquad \downarrow prompt$$
$$3 \; \gamma\text{-rays} \qquad\qquad h\nu + CO_2^+$$

(2)

where again the captured electron is from an inner level. Further evidence that the events at $t_A > 0$ and $t_A < 0$ are due to similar processes is that both signals appear at the same mean impact energy ($\simeq$ 11 eV). The non-exponential behaviour observed in the spectra and the variations in the decay rates and yields at $t_A > 0$ may also be complicated by the rapidly moving o-Ps.

Although it is not currently possible to unambigously assign the observed signal events to processes (1) and (2) the explanations above are plausible. Some of the major unanswered questions concern the origin of the non-exponential features in the spectra and why, if reactions (1) and (2) are correct, the ratio of the event rates at $t_A > 0$ and $t_A < 0$ were not closer to 3:1? Direct measurements of σ_{Ps} in this energy range would be helpful. It should also be noted finally that several other molecules were studied using this method to search for other effects of this kind. These were only found in N_2O gas which displayed similar characteristics to CO_2. As mentioned above the similarity of the behaviour of positrons in these gases has also been noted in the total scattering cross sections (3).

2. SECONDARY ELECTRONS FROM POSITRON IMPACT IONISATION

Positron impact ionisation,

$$e^+ + A \rightarrow e^+ + e^- + A^+$$

(3)

with a threshold at a positron impact energy, E_I, can lead to electron emission over a wide range of energies. The longitudinal energy spectra of electrons emitted in collisions of this type can be studied by measuring the retarding field integral profile using an axial magnetic field system. The results of such a study for e^+-He collisions were given by Coleman (11) and are reproduced in Figure 5 showing that the upper cut-off for electron energies is approximately 30 eV.

We have performed a similar study using a more intense beam of slow positrons. The gas studied was Ne since this was the simplest target atom which could be introduced to our short scattering chamber (20 mm length)

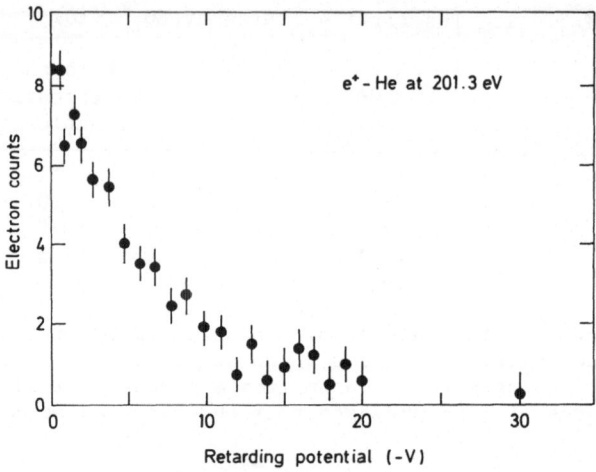

Fig. 5. The electron count versus retarding voltage plot obtained by the Arlington group (11) for e$^+$ impact ionisation.

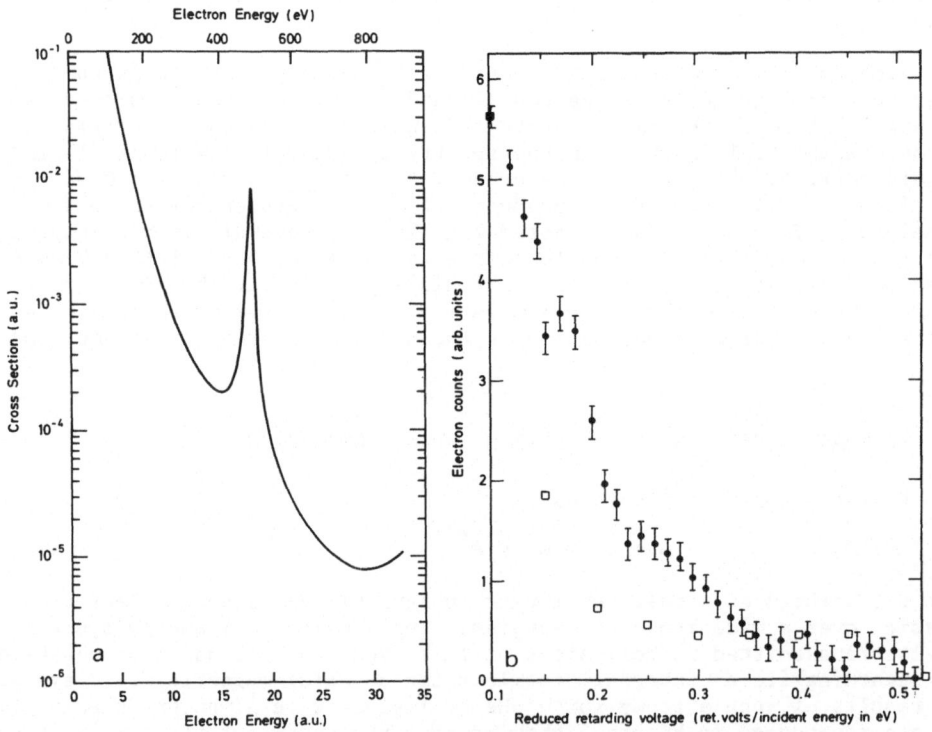

Fig. 6. a) Results of Brauner and Briggs (12) for emission of electrons in the beam direction following e$^+$ impact ionisation at 1 keV.
      b) □; Results of Brauner and Briggs (12) converted into an integral retarding spectrum and plotted at reduced retarding voltages (retarding volts/incident energy in eV). Φ; experimental points for Ne gas at 200 eV e$^+$ impact.

in sufficient quantities to attenuate the e^+ beam by 10%. The experiment was motivated by the recent calculation of Brauner and Briggs (12) who have pointed out that the presence of a positron/electron pair in the final state of (3) (rather than 2 electrons for e^- impact) results in a process known as ionisation to the projectile (positron) continuum in which electrons may be emitted with very much higher energies than was found in the Texas experiment (11). This process has long been known in positive ion-atom collisions (13) but is of interest here due to the small mass of the positron. The signature of this type of collision is the emission of electrons with a velocity of the projectile. The electrons are, effectively, dragged along in the continuum of the projectile resulting, in the special case of positron impact at an energy E_e^+, a peak corresponding to a sharing of the residual kinetic energy, $(E_e^+ - E_I)$.

Brauner and Briggs (12) have illustrated this process by performing a first Born approximation calculation for positron-atom collisions and their results at 1 keV for electron emission in the direction of the incident beam are shown in Figure 6(a). The experiment measures an integral retarding profile and the data of (12) have been converted into such a plot in Figure 6(b) and rescaled for any impact energy. The notable feature of this integral spectrum is the flat portion of the curve which falls rapidly to zero at an energy close to $(E_e^+ - E_I)/2$. The flat region corresponds to the peak in Figure 6(a).

In reality however the electrons may be emitted over a wide range of angles and thus, in our experiment, this feature will be obscured. The maximum measured e^- energy should still be around $(E_e^+ - E_I)/2$. The experimental spectrum for Ne gas at 200 eV incident positron energy is shown in Figure 6(b). These points have had background subtracted and were taken by using a digital voltage ramp to sweep the potential applied to a retarding grid with the counts taken in 10 second intervals stored in a multi-channel analyser used in a multi-channel scalar mode. The ramp unit also provided a channel advance pulse at the end of each timing interval.

The experimental and theoretical data have both been plotted versus reduced retarding voltage with the expected 'cut-off' close to $(E_e^+ - E_i)/2 \simeq E_e^+/2$. The two sets of data (experimental taken at 200 eV impact, theoretical data for 1 keV) have been normalised at the reduced voltage of 0.1. Both curves have a similar form with the electron counts dropping steeply as the retarding voltage is raised. The plateau region observed in the theoretical plot is also, to some extent, present in the experimental data though it is less distinct. The experiment also supports the fact that electrons are emitted with energies close to half the energy of the incident beam. Better accord with theory could not be expected bearing in mind the use of the axial magnetic field and the crudity of the calculation. Further studies of these interesting phenomena using electrostatically controlled positron beams are highly desirable. Note also that similar features to those found experimentally at 200 eV were also observed in data taken at 100 and 400 eV.

3. POSITRONIUM FORMATION

In this section we describe a new attempt to measure σ_{P_s}. In addition two proposed methods of measuring the differential Ps formation cross section $d\sigma_{P_s}/d\Omega$, are discussed.

The first determination of σ_{P_s} by the U.C.L. group using a triple coincidence technique to measure o-Ps decays have been found to be too low (15,16). This can be attributed to loss of o-Ps by annihilation on impact with the walls of the scattering chamber and motion of the o-Ps atoms out

of the detection region. Other methods of σ_{Ps} determination developed by
the Bielefeld (14) and Texas (15,16) groups have given values for He gas in
reasonable accord with one another and with some of the more recent calculations (eg. 17).

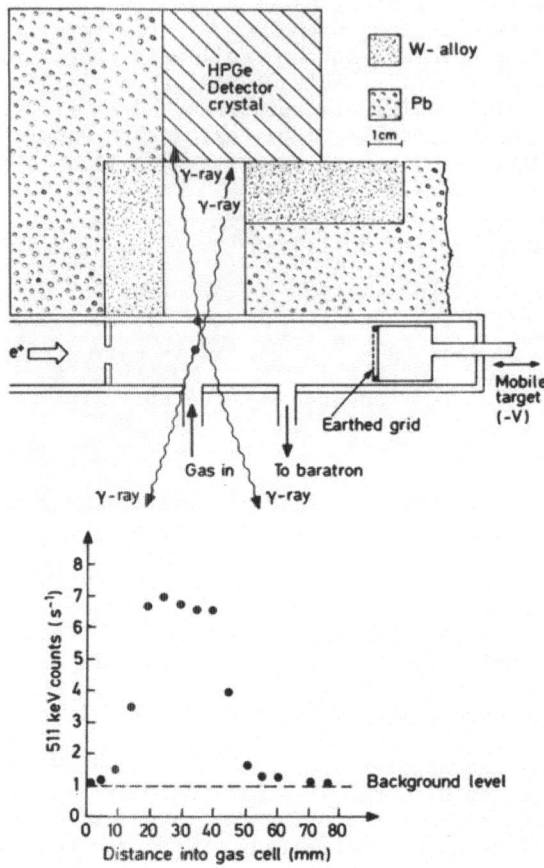

Fig. 7.   The upper drawing shows apparatus currently used at U.C.L. to
investigate Ps formation.   The 20 mm wide,~ 80 mm long scattering
chamber is partially viewed using an intrinsic Ge detector.   The
lower drawing shows the beam count rate (511 keV peak) obtained
by moving the target to various positions within the cell.

Our first attempt at remeasuring σ_{Ps} involved detecting the Doppler
shifted γ-rays emitted by p-Ps annihilating in flight. With a lifetime of
only 125 ps it will annihilate without collision even at the highest e^+
impact energies.   An intrinsic Ge γ-ray detector with a resolution of
1.5 keV F.W.H.M. at 511 keV was used to monitor the annihilations in a
scattering chamber with and without gas present. The slow e^+ were reflected using a 2 grid system with the detector located on the beam axis.
Although the Doppler shifted p-Ps signal was clearly visible in the spectra
it proved to be difficult to extract the shifted peaks (blue and red) from
the unshifted background annihilations which occured on one of the reflecting grids.   Alternative methods of deflecting the e^+ which have traversed
the scattering cell into a region which was shielded from the detector can
be contemplated though these were impractical here due to the low beam
strengths ($< 5 \times 10^3$ s^{-1}).

Accordingly a new experiment was devised which, as shown in Figure 7, consists of a narrow scattering cell which terminates at a mobile target which can be negatively biassed. The Ge detector views a restricted section of the chamber defined by the W-alloy and Pb γ-ray shielding and monitors the 511 keV signal with and without gas in the cell. These signal levels differ by $\simeq$ 10% with the vacuum count rate being $\simeq$ 1.0 s^{-1}. The target can be moved along the axis up to the defining aperture and Figure 7 also contains a plot of the 511 keV rates recorded at various distances along the cell. This can be used to generate the relative detection efficiency at each position.

Helium gas has been investigated between threshold and 125 eV. After normalizing our relative value to that of the Bielefeld group (14) at 40 eV, good agreement was obtained over most of the energy range. At present the statistical error on our data are at the $\pm$ 10% level. Further investigations will be carried out to ascertain whether this method can yield accurate relative/absolute values for σ_{Ps}. The major uncertainty arises due to the unknown fraction of the o-Ps which will be detected. In the present arrangement most of the o-Ps will strike, and annihilate at, the chamber walls. If the detected fraction is independent of energy then this method will be reliable.

Finally, we describe two methods which have been proposed to study the differential Ps formation cross section, $d\sigma_{Ps}/d\Omega$. The importance of making measurements of this quantity has recently been discussed by Charlton and Jacobsen (18) and McGuire (19) and these are, briefly, to offer more stringent tests of approximate scattering theory calculations of this process and also, at higher energies, to search for the evidence of double scattering (Thomas scattering) capture mechanisms.

The first of the $d\sigma_{Ps}/d\Omega$ systems (18) is shown in Figure 8 and comprises a scattering chamber located in an axial magnetic field in which the gaseous target is formed using a capillary array (20) thus allowing the o-Ps formed over a wide range of angles to travel freely to the chamber walls. o-Ps which does not decay in flight may strike the wall and annihilate there in view of a pair of NaI(Tl) or similar γ-ray counters. The angular resolution of the system will be fixed by the properties of the e$^+$ beam and the chamber dimensions but can probably be kept at 10° or below. In assessing the feasibility of using this system it is necessary to estimate the coincidence rate. This must be done by using theoretical data for $d\sigma_{Ps}/d\Omega$ and allowing for solid angles, detection efficiency and in-flight annihilation. At low energies where σ_{Ps} is relatively large estimated yields vary from $\simeq$ 10^{-5} (0-10°) to $\simeq$ 10^{-8} (close to 90°) per incident e$^+$. This type of experiment needs a high flux and is ideally suited for operation with pulsed LINAC-based beams since here the machine pulse can be used to veto unwanted cosmic-ray coincidences. At higher energies where Thomas scattering is expected to be important the calculations of McGuire et al (21) suggest that the expected yields will be only 10^{-9} (0-10°) and 10^{-10} (90°) per incident e$^+$. Thus, higher energy experiments of this type are probably not feasible on any existing machine.

The second $d\sigma_{Ps}/d\Omega$ system, shown in Figure 9, circumvents the low rate problem by accepting all of the o-Ps emitted over a narrow range of angles at different positions along a scattering cell. The fast o-Ps which passes through the 1 mm wide annulus can be detected using a C.E.M.A. which has an X-Y position sensitive readout. It is this feature which will allow the determination of $d\sigma_{Ps}/d\Omega$ since the position at which the o-Ps strikes the C.E.M.A. will be determined by its angle of emission and its point of scatter along the cell.

This design was originated by E. Horsdaal-Pedersen (22) and will be

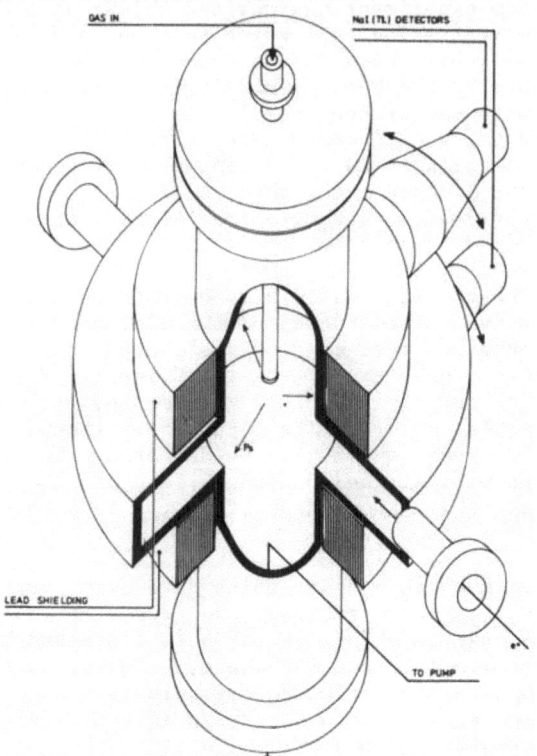

Fig. 8. Proposed system for the measurement of $d\sigma_{Ps}/d\Omega$ (19). The scattering region is located in the centre of the chamber and is formed by gas emanating from a differentially pumped capillary array. The Pb shielding is to reduce unwanted coincidences. Mobile internal shutters can also be used to restrict the volume of the chamber viewed by the detectors.

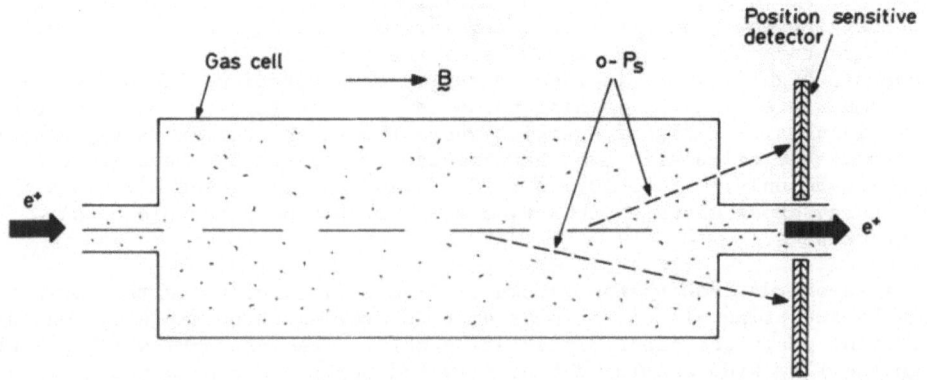

Fig. 9. $d\sigma_{Ps}/d\Omega$ system proposed by E. Horsdaal-Pedersen (21). The fast o-Ps produced at various points along the scattering cell may pass through the 1 mm annular aperture to the position sensitive detector.

most useful at the higher impact energies since its detection efficiency can easily be shown to be peaked around 40-60° (the Thomas angle for e^+ is 45°) with an ideal resolution of $\simeq$ 3° (22). Estimates based upon solid angle give rates of > 0.1 s^{-1} at 45° and 500 eV impact energy when using LLNL e^+ beam parameters (23). Due to the variation of the solid angle with which the o-Ps can be detected this method cannot be used near 0° and at angles close to 90° or larger. Thus to measure $d\sigma_{Ps}/d\Omega$ over a wide range of angles and impact energies it will probably be necessary to use both types of apparatus.

ACKNOWLEDGEMENTS

Thanks are due to the other members of the e^+ physics group at U.C.L. for helpful discussions. Dr. Erik Horsdaal-Pedersen, Aarhus, is thanked for allowing the inclusion of his proposed $d\sigma_{Ps}/d\Omega$ system. M.C. is grateful to the Royal Society for the provision of a 1983 University Research Fellowship.

REFERENCES

1. M. Charlton, T.C. Griffith, G.R. Heyland and G.L. Wright J. Phys. B. 16:323 (1983).
2. K.R. Hoffman, M.S. Dababneh, Y-F. Hsieh, W.E. Kauppila, V. Pol, J.H. Smart and T.S. Stein Phys. Rev. A. 25:1393 (1982).
3. Ch. K. Kwan, Y-F. Hsieh, W.E. Kauppila, S.J. Smith, T.S. Stein, M.N. Uddin and M.S. Dababneh Phys. Rev. Lett. 52:417 (1984).
4. G. Laricchia, M. Charlton, G. Clark and T.C. Griffith Phys. Lett. 109A:97 (1985).
5. M. Charlton and G. Laricchia in 'Positron (Electron) - Gas Scattering' eds. W.E. Kauppila, T.S. Stein and J.M. Wadehra, World Scientific Singapore (1986) pp. 73-84.
6. K.F. Canter, A.P. Mills, Jr. and S. Berko Phys. Rev. Lett. 34:177 (1975).
7. D.C. Schoepf, S. Berko, K.F. Canter and A.H. Weiss. pp. 165-7 in "Positron Annihilation - Proc. 6th Int. Conf. on Positron Annihilation", eds. P.G. Coleman, S.C. Sharma and L.M. Diana, North Holland New York (1982).
8. eg. Thorn-EMI "Photomultipliers" (1982).
9. Malakhov and Basier Opts. Spek (Russian) 30:421 (1971).
10. D.L. Judge, G.S. B oom and A.L. Morse Can. J. Phys. 47:489 (1969).
11. P.G. Coleman, pp 25-34 in: as Ref. 5 (1986).
12. M. Brauner and J.S. Briggs, J. Phys. B. 19:L325 (1986).
13. eg. M.W. Lucas, K.F. Man and W. Steckelmacher p1 in: Lecture Notes in Physics Vol. 213. K.O. Greeneveld, W. Meckbach, I. Sellin eds. Springer: Berlin (1984).
14. D. Fromme, G. Kruse, W. Raith and G. Sinapius Phys. Rev. Lett. 57:3031 (1986).
15. L.S. Fornari, L.M. Diana and P.G. Coleman Phys. Rev. Lett. 51:2276 (1983).
16. L.M. Diana, P.G. Coleman, D.L. Brooks, P.K. Pendleton and D.M. Norman Phys. Rev. A 34:2731 (1986).
17. P. Khan and A.S. Ghosh, Phys. Rev. A. 28:2181 (1983).
18. M. Charlton and F.M. Jacobsen, Appl. Phys. A43:235 (1987).
19. J.H. McGuire pp 222-31 in: as Ref. 5 (1986).
20. K.F. Man, W. Steckelmacher and M.L. Lucas, J. Phys.B. 19:4171 (1986).
21. J.H. McGuire, N.C. Sil and N.C. Deb Phys. Rev. A. 34:685 (1986).
22. E. Horsdaal-Pedersen. Commun. to M.C., C.L. Cocke and R.H. Howell (1986).
23. R.H. Howell, R.A. Alvarez, K.A. Woodle, S. Dhawan, P.O. Egan, V.A. Hughes and M.W. Ritter I.E.E.E. Trans.Nuc.Sci. 30:1438 (1983).

POSITRON–ATOM DIFFERENTIAL SCATTERING MEASUREMENTS

W.E. Kauppila and T.S. Stein

Department of Physics and Astronomy
Wayne State University
Detroit, Michigan 48202 USA

INTRODUCTION

It is well known that measurements of differential cross sections
(DCS's) for a specific scattering channel can provide a more sensitive
test of a scattering theory than the measurement of the total cross
section (TCS) for that channel. To illustrate this, the DCS for elastic
scattering of a positron (of momentum k) by an atom for impact energies
below the first inelastic threshold is given by (using the method of
partial wave analysis for a central potential)

$$DCS(k,\theta) = \left| f(k,\theta) \right|^2 = \frac{1}{k^2} \sum_{\ell,m} (2\ell+1)(2m+1)\sin\delta_\ell \sin\delta_m \cos(\delta_\ell - \delta_m) P_\ell P_m \quad (1)$$

while

$$TCS(k) = \int DCS(k,\theta)\,d\Omega = \frac{4\pi}{k^2} \sum_\ell (2\ell+1)\sin^2\delta_\ell \qquad (2)$$

where $f(k,\theta)$ is the scattering amplitude, $\delta_{\ell,m}(k)$ are energy-dependent and
real scattering phase shifts for each relative angular momentum $\ell\hbar$ of the
system, and $P_{\ell,m}(\cos\theta)$ are Legendre polynomials. At a given energy of
investigation, DCS measurements made as a function of the projectile
scattering angle θ could, in principle, provide sufficient information to
determine the individual scattering phase shifts, while this is not
possible with TCS measurements.

Up to the present time, there have been two groups that have reported
experimental DCS investigations[1,2] for positron scattering, in both
cases, for elastic scattering by argon atoms. The methods employed in
these two experiments are quite different with a time–of–flight (TOF)
approach being used at the University of Texas at Arlington[1,3] and a
crossed–beam technique used by our group at Wayne State.[2] An earlier
investigation, which will not be discussed further here, was made by
Jaduszliwer and Paul[4] to extract individual phase shifts for low energy
positron–helium scattering by analyzing the varying transmission of a
positron beam through a gas cell as a function of an axial magnetic field.

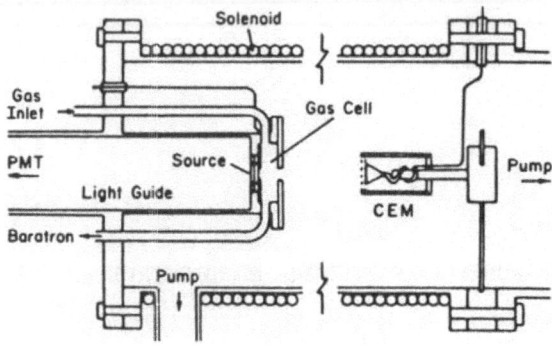

Fig. 1. TOF spectrometer used by Coleman et al. (Refs. 1 and 3).

TOF METHOD

        The first DCS measurements were made by Coleman and McNutt[1] using a
time-of-flight approach. A schematic diagram of their TOF spectrometer[3]
is shown in Fig. 1. In their experiment slow positrons coming from a
source pass through a 1 cm long gas cell, travel 25 cm in an evacuated
straight flight tube (possessing a strong axial magnetic field), and then
are detected by a channeltron electron multiplier (CEM). The TOF of each
positron detected by the CEM as it travels from the source to the CEM is
determined. The axial magnetic field preferentially guides positrons
scattered in the forward direction to the CEM. By alternately admitting
gas into and then evacuating the gas cell, TOF histograms for detected
positrons were obtained as shown in Fig. 2. The TOF spectrum obtained
when gas is in the scattering cell is found to have a "tail" on the
long-time side which relates to positrons that have experienced forward
elastic scattering. An appropriately adjusted "vacuum" TOF spectrum is
subtracted from the "gas" spectrum in order to obtain a "difference"
spectrum. The signal in this "difference" spectrum is then correlated to
various angles of forward elastic scattering in order to obtain absolute
values for the DCS. The angular range for which it was felt that the best
DCS results were obtained in this experiment ranged from 20 - 60°, which
is a direct consequence of this TOF approach.

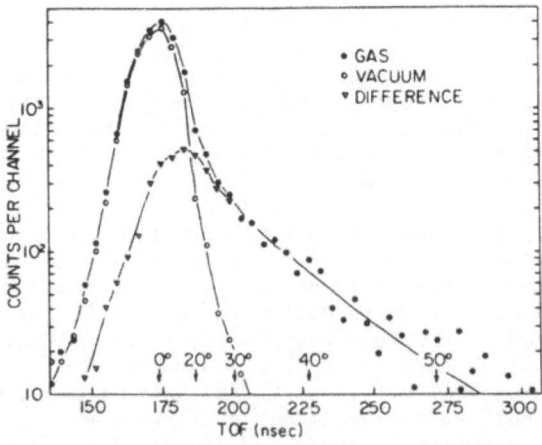

Fig. 2. TOF histograms obtained by Coleman and McNutt (Ref. 1) for
        positron-Ar scattering at 6.7 eV for a run time of 55,000 s.

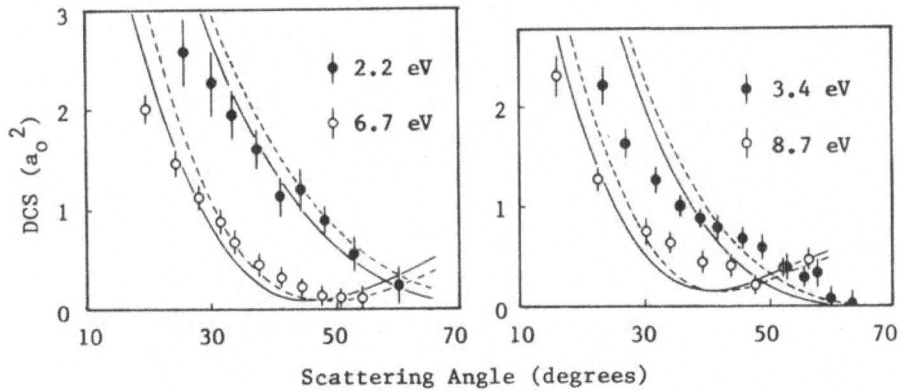

Fig. 3. Positron-Ar elastic DCS measurements of Coleman et al. (Ref. 1)
compared with theoretical calculations.

The DCS measurements obtained by Coleman and McNutt[1] are shown in
Fig. 3 where they are compared with the results of a semiempirical
calculation by Schrader[5] (solid lines) and some "scaled-down" results of
a polarized orbital calculation by McEachran et al.[6] (dashed lines).
The measurements are in reasonable agreement with the theoretical
calculations.

CROSSED-BEAM METHOD

The first application of a crossed-beam experiment for positron-gas
scattering was made by our group with our DCS measurements for positron
elastic scattering by argon atoms.[2] The experimental setup for this
experiment is shown in Fig. 4. A positron beam obtained from a
tungsten-moderated sodium-22 source is crossed with an atom beam effusing
from a multichannel capillary array oriented perpendicular to the
projectile (positron or electron) beam. CEM's are used to detect both the
primary (#1) and scattered (#2) projectiles with the position of CEM#2
being variable between 30 - 135° with respect to the primary beam

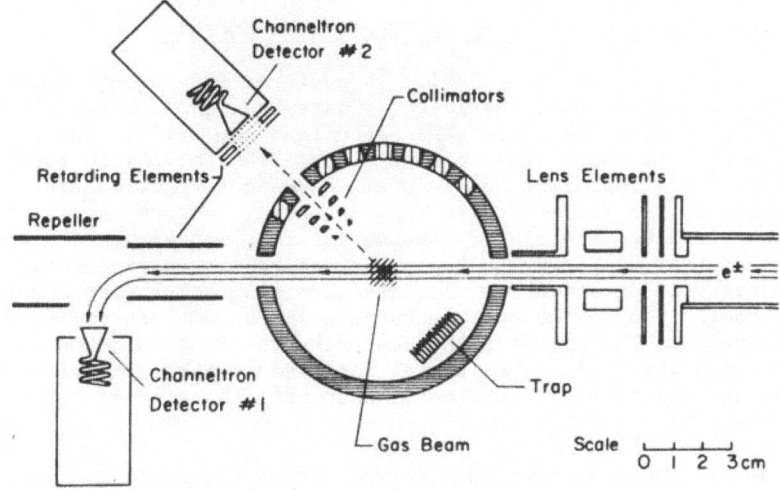

Fig. 4. Crossed-beam setup used at Wayne State.
(From Hyder et al., Ref. 2)

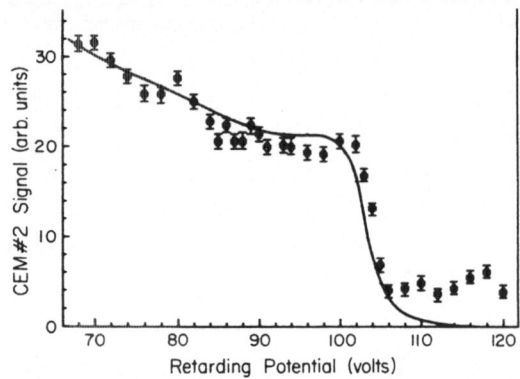

Fig. 5. Retarding-potential curves for positrons (•) and electrons
(——) scattered at 30° from argon and detected by CEM#2
when ±100 V is applied to the projectile source.
(From Hyder et al., Ref. 2)

direction. The angular acceptance of CEM#2 is estimated to be about
±8°. A retarding element preceding CEM#2 is used for separation
of the elastic differential scattering signal from noise and inelastically
scattered projectiles. For example, in the retarding potential curves
shown in Fig. 5, which displays a difference signal (with and without the
gas beam on) detected with CEM#2 at 30° when +100 or −100 V is applied
to the projectile source for positrons and electrons, respectively, there
is a significant drop in the signal when the retarding potential rejects
the elastically scattered projectiles. This occurs within a few volts of
the applied voltage. Inelastically scattered projectiles will have a
minimum energy loss of 11.5 eV for argon, where the retarding curve begins
trending upward with decreasing retarding potential. The elastic DCS
signal is the difference between the CEM#2 signal with the retarding
potential set a few volts below and a few volts above the applied voltage.
It is to be noted that with this crossed-beam method only the relative
shape of DCS values vs angle are determined at a given energy and it is
the shape of the DCS curve that is compared with the results of
theoretical calculations or, in the case of electrons, prior experiments.

$e^{\pm}$-Ar Elastic DCS Measurements at Intermediate Energies

The results of our first published positron DCS measurements[2] are
shown in Fig. 6 where the elastic DCS's have been measured for 100, 200,
and 300 eV positrons and electrons scattering from argon. Our results are
compared with some theoretical calculations[7-9] and also with prior
measurements[10,11] for electrons. In all cases our relative DCS
measurements are normalized to the results being compared to at 90°.
Our reason for doing the electron measurements is to serve as a test of
our experiment by comparing to the prior well-known measurements, for
which rather good agreement is obtained, and also to provide a good
relative comparison to the corresponding positron DCS measurements made in
our same system. The positron DCS measurements at 100 and 200 eV are in
good agreement with the shapes of the polarized orbital calculations of
McEachran and Stauffer[7] and the model potential calculations of Nahar
and Wadehra.[8]

At 300 eV the positron and electron DCS measurements are compared
with each other and with the optical model calculations of Joachain et
al.[9] One of the motivations for these positron and electron comparison
measurements derives from the total (elastic plus inelastic) cross section

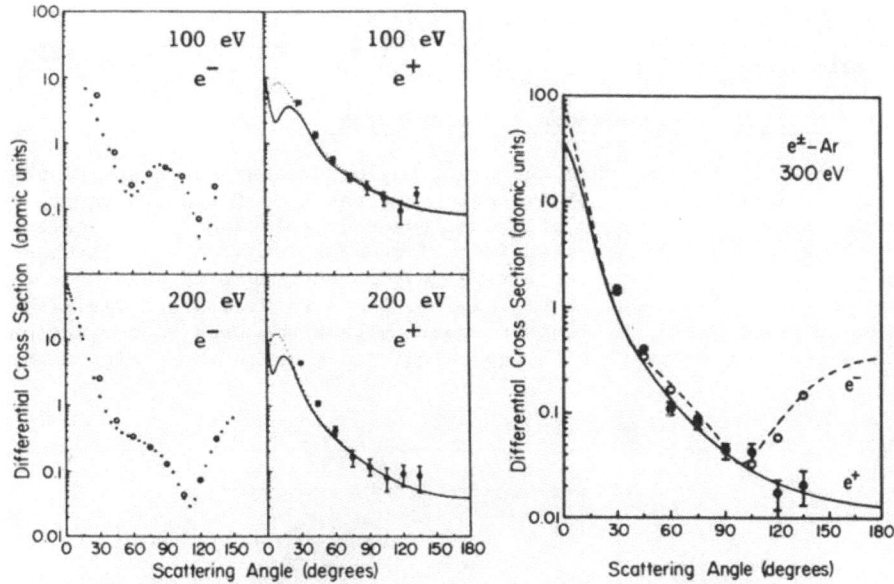

Fig. 6. Elastic DCS measurements of Ref. 2 for positrons (●) and
electrons (o) scattered from Ar. These results are compared
for electrons with measurements of Ref. 10 (···) at 100 eV
and Ref. 11 (···) at 200 eV; for positrons with calculations
of Ref. 7 (– –) and Ref. 8 (···) at 100 and 200 eV; and
at 300 eV with the calculations of Ref. 9 for positrons (——)
and electrons (– –). (From Hyder et al., Ref. 2)

(Q_t) measurements made in our laboratory[12] which have shown that the
respective Q_t curves are tending towards merging at these energies with
the positron Q_t values being 18% and 13% lower than the electron Q_t
values at 100 and 300 eV, respectively. Furthermore, the optical model
calculations of Joachain et al.[9] for the corresponding elastic DCS's for
positrons and electrons at 300 eV also predict a near merging of the
respective curves for scattering angles less than 100°. Our relative
measurements at 300 eV are seen to be in very good agreement with the
calculations of Joachain et al.

It is interesting to compare the relative shapes of the positron and
electron DCS curves in that the electron DCS curves exhibit considerable
structure for angles >30°, while the positron curves are monotonically
decreasing with larger angles. The double minima for electrons at 100 eV,
which also persists for lower energies, arises in a partial wave analysis
treatment from a significant contribution from the d-wave phase shift,[10]
which if it was present alone would produce zeros in the DCS at 55 and
125°. It is curious that the number of elastic DCS minima at 100 eV for
electron scattering from various inert gases (# minima)[13] is He (0), Ne
(1), Ar (2), Kr (3), and Xe (4), indicating that as the atoms get larger,
the higher the ℓ-values of the phase shifts that dominate the structure in
the DCS. Predictions of the elastic DCS for positron scattering at
similar energies from He and Ne by Byron and Joachain,[14] using an
optical model formalism, and Dewangen and Walters,[15] using the
distorted-wave second Born approximation, also indicate monotonically
decreasing DCS's with increasing angle, as is the case for Ar. It is to
be noted that the first Born approximation[14,16], which does not
distinguish between a positron or electron, indicates a DCS that decreases

monotonically with larger angles. The only suggestions of possible
structure for positron scattering at these intermediate energies are the
theoretical predictions[7,8] for angles <30°.

e+-Ar Elastic DCS Measurements at Lower Energies

In order to investigate the small angle structure at higher energies,
we are now in the process of extending our positron DCS measurements to
lower energies because both of the relevant calculations[7,8] indicate
that the structural features shift to larger angles as the positron energy
is decreased. Some initial relative measurements that we have made[17]
for Ar at 20 eV are shown in Fig. 7, where our results are normalized to
the calculations at 75°. Another approach that was used to compare our
experimental measurements with the calculations is shown in Fig. 8 where

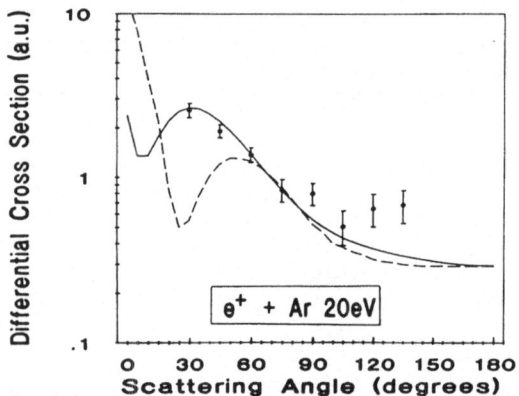

Fig. 7. Elastic DCS measurements of Smith et al. (Ref. 17) compared with
calculations of McEachran and Stauffer (Ref. 7) (- -), and Nahar
and Wadehra (Ref. 8) (——).

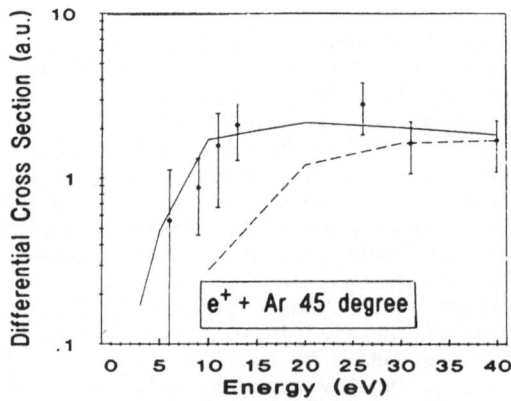

Fig. 8. Elastic DCS measurements of Smith et al. (Ref. 17) compared with
calculations of McEachran and Stauffer (Ref. 7) (- -), and Nahar
and Wadehra (Ref. 8) (——). Each theoretical "curve" consists
merely of straight line segments joining discrete points.

DCS measurements were made at a fixed angle of 45° while varying the positron energy from 6 to 40 eV with our results normalized to the calculation of McEachran and Stauffer at 40 eV. It is seen that our measurements tend to be in better agreement with the calculation of Nahar and Wadehra.

FUTURE DIRECTIONS

A continuation of positron DCS measurements for various gases at various energies is certainly warranted in order to obtain more specific information about positron scattering, which in addition to being of interest on its own merit, could help to obtain a deeper understanding of electron scattering processes. In the case of electron scattering the effects of the static and polarization interactions is that they are additive (both attractive), while for positrons there is a tendency toward cancellation since the static interaction is repulsive and polarization is attractive. Two additional differences between electron and positron scattering is the presence of exchange for electrons and the positronium formation channel for positrons. A few possible investigations of special interest will be discussed next.

Helium at Intermediate Energies

A particularly interesting investigation would be to perform elastic DCS measurements for positrons and electrons scattering from helium especially for energies at and above 200 eV, which is where it has been observed[12] by our group that the total cross sections (Q_t's) merge to within 2% for energies between 200 and 600 eV (our highest measured energy). Meanwhile, the distorted-wave second Born approximation calculation of Dewangen and Walters[15] indicates that this merging of the Q_t's should not occur until about 2000 eV, and also indicates that the elastic scattering cross section for 200 eV electrons, which agrees to within 12% of the experimental values of Register et al.,[18] should be 238% larger than the positron elastic scattering cross section at 200 eV. The main limitation on performing these measurements now is that the predicted DCS's for positron scattering by helium[14] are typically an order of magnitude lower than for argon.[9]

Critical Points and DCS Minima

It has already been mentioned that the elastic DCS's for electrons scattering from the heavier inert gases exhibit several minima (e.g., see Fig. 6), which can be attributed to electron diffraction effects. In general there will be certain points, called "critical points", where the DCS reaches a very deep minimum at a given electron energy and scattering angle, such that if either the electron energy or scattering angle is changed by a small amount the DCS may increase by more than several orders of magnitude. Such critical points have been observed[19] in electron scattering from Ne, Ar, Kr, and Xe for electron energies ranging from 49 to 772 eV and scattering angles ranging from 52 to 143°. One reason for studying critical points, given by Buhring,[20] is that they should provide a more sensitive test of the atomic potentials contributing to scattering than measurements of DCS's away from critical points. It has also been pointed out by Buhring[21] that electrons elastically scattered near a critical point can be totally polarized.

In the case of positron differential elastic scattering, no minima in the DCS curves have yet been definitively observed, although the 20 to 60° measurements of Coleman and McNutt seem to suggest a minimum at about 50° for positron energies of 6.7 and 8.7 eV. Theoretically, a single minimum has been predicted in the elastic DCS curves for positrons scattering at small forward angles from He, Ne, and Ar.[7] Wadehra et al.[22] have used the phase shifts of the first seven partial waves for elastic scattering of positrons by Ar, Kr, and Xe (calculated by McEachran et al.[6,23]), and the Born approximation with known polarization potentials to obtain the phase shifts for higher partial waves,[24] in order to predict the existence of critical points for positron elastic scattering by Ar, Kr, and Xe. A 3-dimensional plot of the DCS for positrons elastically scattered from Ar as a function of scattering angle and positron momentum is shown in Fig. 9, where it can be seen how the single minimum in the DCS curve at a fixed energy develops into a critical point. The 3-D DCS plots are similar for Kr and Xe. Two-dimensional plots of the DCS vs positron energy and vs scattering angle, in both cases through the critical points, shown in Fig. 10 demonstrate the depth and narrowness of the DCS in the vicinity of these critical points. Nakanishi and Schrader[25] have predicted the existence of critical points for positron scattering by H and He based on their semiempirical calculations. A summary of the locations of the predicted critical points is given in Table 1. It is interesting to note, as pointed out by Wadehra et al.,[22] that if only the first two terms (ℓ = 0, 1) in the expansion in Eq. 1 are important (as would be the case at sufficiently low energies), then a critical point would occur when the s-wave phase shift is equal to the p-wave phase shift -- these energies, E(s=p), have been estimated from available phase shift curves or tabulations and are given in Table 1. In each case E(s=p) is found to be close to the critical point energy. Furthermore, it was shown by Wadehra et al.[22] that if only the first two partial waves are important then the critical angle should be 109.5°, which is indeed close to the predicted critical angles. The presence of higher partial waves causes the predicted critical points to differ slightly from E(s=p) and 109.5°.

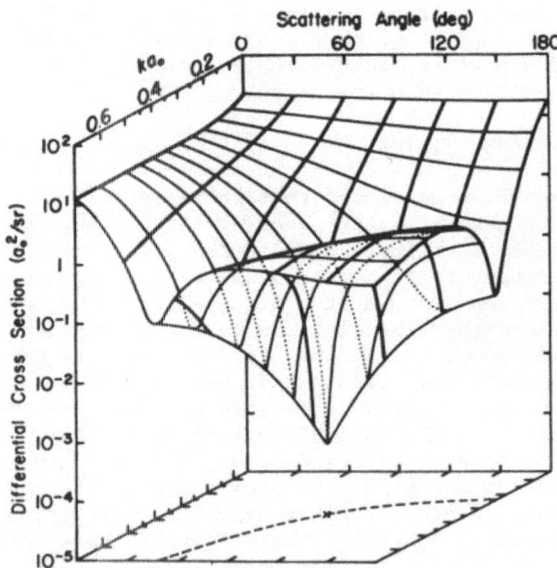

Fig. 9. Three-dimensionsal perspective of the predicted elastic DCS for positron-Ar scattering. The dashed curve is the locus of the projections of the DCS minima onto the ka_0 vs scattering-angle plane with the "x" being the critical point. (From Wadehra et al., Ref. 22)

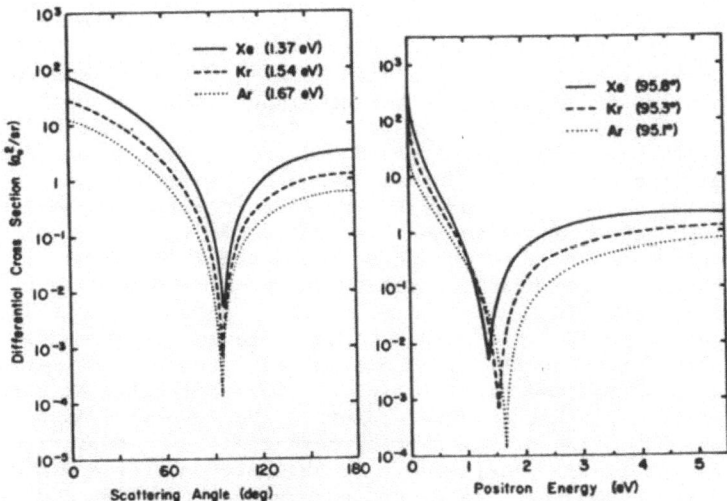

Fig. 10. Predicted elastic DCS's through the critical points for
positrons scattering from Ar, Kr, and Xe.
(From Wadehra et al., Ref. 22)

An interesting general thought concerning critical points is that for
electron scattering many have been observed at intermediate energies,[19]
while for positrons they are predicted to occur only for low energies. It
has already been mentioned that the static and polarization interaction
potentials for electron scattering by atoms are both attractive. In the
case of positron scattering only the polarization interaction is
attractive, while the static interaction which dominates at higher

Table 1. Predicted critical points (energy and angle) for positron
scattering by various gases, along with energies where the
s-wave phase shift is equal to the p-wave phase shift and
the s-wave phase shift passes through zero, and the energy
location of observed Ramsauer-Townsend minima in Q_t.
All energies expressed in eV and angles in degrees.

Gas	Critical Point E	Angle	E(s=p)	E(s=o)	E(RT)
H	2.75[a]	105.0[a]	2.9[a]	5.4[a]	
He	1.67[a]	100.0[a]	1.7[a]	2.8[a]	2[b]
Ne	–	–	0.7[c]	1.1[c]	0.6[b]
Ar	1.67[d]	95.1[d]	1.4[e]	2.7[e]	2[f]
Kr	1.54[d]	95.3[d]	1.3[g]	2.7[g]	
Xe	1.37[g]	95.8[d]	1.1[g]	2.7[g]	

a: Nakanishi and Schrader (Ref. 25)
b: Stein et al. (Ref. 26)
c: McEachran et al. (Ref. 27)
d: Wadehra et al. (Ref. 22)
e: McEachran et al. (Ref. 6)
f. Kauppila et al. (Ref. 28)
g: McEachran et al. (Ref. 23)

energies is repulsive. It is known that a Ramsauer-Townsend effect (a minimum in Q_t at low energy) occurs for positron scattering by He,[26] Ne,[26] and possibly Ar,[28] and it is also known[29] that in order for a Ramsauer-Townsend effect to occur the net interaction potential must be attractive. In general, it is found that Ramsauer-Townsend effects are associated with the s-wave phase shift passing through zero (or an integral multiple of π). In Table 1, we have also estimated from available phase shifts the energy E(s=o) where the s-wave phase shift passes through zero and listed the energy locations of the observed Ramsauer-Townsend effects for positron scattering. From Table 1 it can be seen that the predicted critical point energies are less than E(s=o) for each listed gas suggesting that these critical points for positron scattering are expected only where the net interaction potential is expected to be attractive (i.e., where the attractive polarization interaction dominates the repulsive static interaction). As a result, it seems rather curious that critical points, like Ramsauer-Townsend effects, may occur only when the net scattering potentials are attractive. Furthermore, the minima predicted in lower energy positron differential elastic scattering by inert gases are each associated with a critical point (as seen in Fig. 9), and most, if not all, of the DCS minima for electron scattering by inert gases are also linked to critical points. Therefore, it may be expected that positron DCS curves, in general, may be for the most part monotonically decreasing with increasing scattering angle, except perhaps at lower energies where the net interaction potential becomes attractive and critical points may exist, or perhaps in the vicinity of cusps (discussed in the next section). Since DCS minima seem to shift to smaller angles for higher energies, the remnants of these critical points may be small scattering angle minima at somewhat higher energies. It is interesting to speculate that perhaps for positron differential elastic scattering by alkali metal atoms, which have very large polarizabilites and should have correspondingly large attractive polarization interaction potentials for positron scattering, that critical points and associated DCS minima may exist for somewhat higher energies than the atoms listed in Table 1.

Cusps

Another interesting investigation is a search for possible "cusps" in the elastic scattering cross section when the threshold for Ps formation is crossed. The basic idea of a cusp refers to a general and fundamental type of phenomenon discovered in theoretical calculations by Wigner[30] who pointed out that "...all reaction and scattering cross sections, which are possible below the threshold energy of the new mode, show a cusp at the threshold energy for the new mode", which is to say that the opening of a new scattering channel as the projectile energy is increased, influences the cross section in the "old" channel that was already open prior to crossing the threshold for the new channel. Although such effects were clearly observed in the realm of nuclear physics in the 1950's, it was not until the 1970's that such effects were observed in the realm of atomic physics. Eyb and Hoffman[31] clearly observed cusps in the energy dependences of the elastic DCS's at many different scattering angles for electron-Na and electron-K collisions, and pointed out that "The excitation of the first excited state is expected to influence the elastic scattering from alkali metals much more strongly because of the great ease with which the valence electron goes from the ground ^2S to the first excited ^2P state..."

In the case of positron scattering it is well known that for many target gases Ps formation appears to make a very significant contribution to Q_t as one crosses the Ps formation threshold, as is shown in Fig. 11 for the inert gases.[26,28,32] As a result, the situation for positron scattering is quite similar to the alkali metals for electron scattering where excitation of the ^2P state becomes an important contributor to Q_t as one crosses that threshold, which is where very pronounced cusps have been observed. The first indication that the opening up of the Ps formation channel may have an effect on the elastic scattering channel was provided by Humberston,[33] whose calculations for positron-atomic hydrogen scattering suggest that there is an appreciable change in the slope of the total elastic scattering cross section curve when the positron energy crosses the Ps formation threshold energy. More recently, Raith[34] has provided indirect evidence that a cusp exists in the total elastic scattering cross section for positron scattering from helium at the Ps formation threshold by subtracting recent Ps formation measurements made by the Bielefeld group from prior Q_t measurements.[26] Similarly, we have subtracted the low energy Ps formation cross sections measured[35-7] at the University of Texas at Arlington from the corresponding Q_t curves in Fig. 11 to obtain the "x" points through which we have drawn smooth "dashed" curves starting at the respective Ps formation thresholds. The dashed curves can be considered to be estimates of the behavior of the total elastic cross sections immediately above the Ps formation thresholds and it is seen that there are suggestions of cusps at these thresholds. A typical cusp search would take the form of the investigation shown in Fig. 8, but with more experimental points taken at smaller energy intervals, where if a cusp were present you would expect to

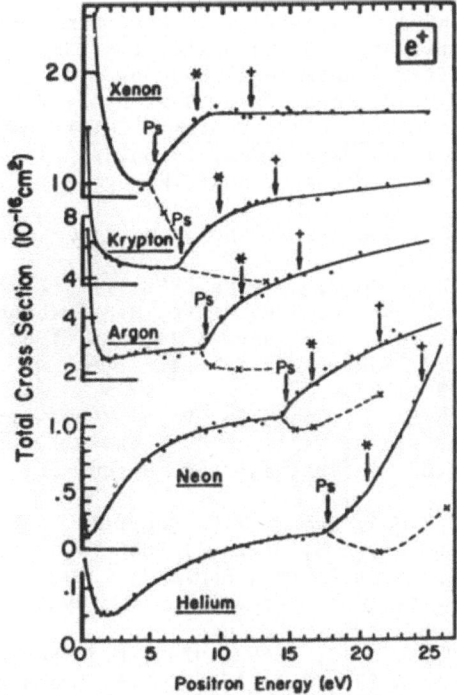

Fig. 11. Measured total scattering cross sections (——) for positrons colliding with He, Ne, Ar, Kr, and Xe (Refs. 26, 28, and 32). The arrows indicate the threshold energies for Ps formation, atomic excitation (*), and atomic ionization (+). The dashed curve is drawn smoothly from the Q_t curve at the Ps formation threshold through the x's, which represent the cross section remaining after measured Ps formation cross sections (Refs. 35-7) are subtracted from the corresponding energy Q_t values.

find some structural feature appearing at the Ps formation threshold, which for Ar is 9.0 eV. It is to be noted that neither theoretical "curve" in Fig. 8 takes Ps formation into account.

ACKNOWLEDGMENTS

We wish to acknowledge helpful discussions with our colleague Dr. J.M. Wadehra. The Wayne State positron (electron)-gas scattering group is supported by the National Science Foundation (Grant #PHY83-11705).

REFERENCES

1. P.G. Coleman and J.D. McNutt, Phys. Rev. Letters 42, 1130, 1979.
2. G.M.A. Hyder, M.S. Dababneh, Y.-F. Hsieh, W.E. Kauppila, C.K. Kwan, M. Mahdavi-Hezaveh, and T.S. Stein, Phys. Rev. Lett. 57, 2252 (1986).
3. P.G. Coleman, J.D. McNutt, J.T. Hutton, L.M. Diana, and J.L. Fry, Rev. Sci. Instrum. 51, 935 (1980).
4. B. Jaduszliwer and D.A.L. Paul, Can. J. Phys. 51, 1565 (1973); Can. J. Phys. 52, 1047 (1974).
5. D.M. Schrader, Phys. Rev. A 20, 918 (1979).
6. R.P. McEachran, A.G. Ryman, and A.D. Stauffer, J. Phys. B 12, 1031 (1979).
7. R.P. McEachran and A.D. Stauffer, in Positron (Electron)-Gas Scattering, edited by W.E. Kauppila, T.S. Stein and J.M. Wadehra (World Scientific, Singapore, 1986), p. 122.
8. S.N. Nahar and J.M. Wadehra, Phys. Rev. A 35, 2051 (1987).
9. C.J. Joachain, in Electronic and Atomic Collisions, edited by G. Watel (North-Holland, Amsterdam, 1978), p. 71; C.J. Joachain, R. Vanderpoorten, K.H. Winters, and F.W. Byron, Jr., J. Phys. B 10, 227 (1977).
10. S.K. Srivastava, H. Tanaka, A. Chutjian, and S. Trajmar, Phys. Rev. A 23, 2156 (1981).
11. R.D. Dubois and M.E. Rudd, J. Phys. B 8, 1474 (1975).
12. W.E. Kauppila, T.S. Stein, J.H. Smart, M.S. Dababneh, Y.K. Ho, J.P. Downing, and V. Pol, Phys. Rev. A 24, 725 (1981).
13. I.E. McCarthy, C.J. Noble, B.A. Phillips, and A.D. Turnbull, Phys. Rev. A 15, 2173 (1977).
14. F.W. Byron, Jr. and C.J. Joachain, Phys. Rev. A 15, 128 (1977).
15. D.P. Dewangen and H.R.J. Walters, J. Phys. B 10, 637 (1977).
16. C.J. Joachain, K.H. Winters, and F.W. Byron, Jr, J. Phys. B 8, L289 (1975).
17. Steven J. Smith, G.M.A. Hyder, W.E. Kauppila, C.K. Kwan, M. Mahdavi-Hezaveh, and T.S. Stein, ICPEAC XV -- Abstracts of Contributed Papers (Brighton, England, 1987).
18. D.F. Register, S. Trajmar, and S.K. Srivastava, Phys. Rev. A 21, 1134 (1980).
19. C.B. Lucas, J. Phys. B 12, 1549 (1979); K.J. Kollath and C.B. Lucas, Z. Physik A 292, 215 (1979); C.B. Lucas and J. Liedtke, in Abstracts of Papers of ICPEAC IX, edited by J.S. Risley and R. Geballe, (Univ. of Washington Press, Seattle, 1975), p. 460.
20. W. Buhring, in Proceedings of the 4th ICAP - Abstracts of Contributed Papers, edited by J. Kowalski and H.G. Weber (Heidelberg Univ. Press, Heidelberg, Germany, 1974), p. 417.
21. W. Buhring, Z. Physik 208, 286 (1968).
22. J.M. Wadehra, T.S. Stein, and W.E. Kauppila, Phys. Rev. A 29, 2912 (1984).
23. R.P. McEachran, A.D. Stauffer, and L.E.M. Campbell, J. Phys. B 13, 1281 (1980).

24. T.F. O'Malley, L. Spruch, and L. Rosenberg, J. Math Phys. $\underline{2}$, 491 (1961).
25. H. Nakanishi and D.M. Schrader, Phys. Rev. A $\underline{34}$, 1810 (1986).
26. T.S. Stein, W.E. Kauppila, V. Pol, J.H. Smart, and G. Jesion, Phys. Rev. A $\underline{17}$, 1600 (1978).
27. R.P. McEachran, A.G. Ryman, and A.D. Stauffer, J. Phys. B $\underline{11}$, 551 (1978).
28. W.E. Kauppila, T.S. Stein, and G. Jesion, Phys. Rev. Lett. $\underline{36}$, 580 (1976).
29. H.S.W. Massey, Physics Today $\underline{29}$, 42 (March 1976).
30. E.P. Wigner, Phys. Rev. $\underline{73}$, 1002 (1948).
31. M. Eyb and H. Hofmann, J. Phys. B $\underline{8}$, 1095 (1975).
32. M.S. Dababneh, W.E. Kauppila, J.P. Downing, F. Laperriere, V. Pol, J.H. Smart, and T.S. Stein, Phys. Rev. A $\underline{22}$, 1872 (1980).
33. J.W. Humberston, in Positron (Electron)-Gas Scattering, edited by W.E. Kauppila, T.S. Stein, and J.M. Wadehra (World Scientific, Singapore, 1986), p. 35.
34. W. Raith, private communication.
35. L.S. Fornari, L.M. Diana, and P.G. Coleman, Phys. Rev. Lett. $\underline{51}$, 2276 (1983).
36. L.M. Diana, S.C. Sharma, L.S. Fornari, P.G. Coleman, P.K. Pendleton, D.L. Brooks, and B.E. Seay, in Positron Annihilation, edited by P.C. Jain, R.M. Singru, and K.P. Gopinathan (World Scientific, Singapore, 1985), p. 428.
37. L.M. Diana, D.L. Brooks, P.C. Coleman, P.K. Pendleton, D.M. Norman, B.E. Seay, and S.C. Sharma, in Positron (Electron)-Gas Scattering, edited by W.E. Kauppila, T.S. Stein, and J.M. Wadehra (World Scientific, Singapore, 1986), p. 293.

TOTAL AND INELASTIC POSITRON SCATTERING CROSS SECTIONS

Osamu Sueoka

Institute of Physics, College of Arts and Sciences
University of Tokyo, 3-8-1 Komba
Meguro-ku, Tokyo, 153 Japan

1. INTRODUCTION

Values are presented of the total cross sections (TCS) for positrons
and electrons colliding with the hydrocarbons CH_4, C_2H_2, C_2H_4, C_2H_6 and
C_6H_6, the polar molecules H_2O and NH_3 and also CF_4, CCl_4, SiH_4 and O_2,which
have recently been measured in our laboratory. A simple method for measu-
ring the cross section for positronium (Ps) formation is demonstrated for
these molecules. In the next subsection a brief comment on TCS is added.
Considerable interest is nowadays concentrated on inelastic collision studies.
We treat experimentally the inelastic cross section for the Schumann-Runge
excitation in O_2 by positron and electron impact, and inelastic collisions
with other molecules. Methods for narrowing the beam energy were tried and
demonstrations were performed for several inelastic collisions. Theoretical
calculations of total cross sections (elastic and absorption) for CH_4, SiH_4,
H_2O and NH_3 have been made by Jain[1] and are reported elsewhere in these
Proceedings.

Recent Measurements of Total Cross Sections

Measurements of total cross sections (TCS) for positrons colliding with
atoms and molecules have been performed, and experimental methods and the
accuracy of the measurements have been discussed. The main results have
been described in the review of Charlton[2] and in other reviews. However,
not all problems have been investigated. The main topics in the study of
TCS are as follows.

I. TCS measurements of high accuracy

Precise measurements of TCS are necessary for high grade approximations
in polarisation potential calculations in e^+-gas collisions. It is not
difficult to carry out measurements with statistical errors less than 1%,
but larger uncertainties arise due to forward scattering. A new precise
measurement for positron collisions with He has recently been carried out
by Mizogawa et al[3] for the purpose of searching for the existence of a
positron resonance, corresponding to the resonance at 19.36 eV in electron
collisions. In spite of very careful and precise measurements, no sharp
variation was found in the TCS-curve. They also demonstrated that, within
a few percent overall error, the TCS in the elastic collision region are in
good agreement with the theoretical values of Campeanu and Humberston[4].

II. Measurements for special atoms or molecules

The order in which TCS measurements have been made for target gases has not always been the same as the order of importance in the study of this field. Since positron beams are weak in intensity and low in brightness, measurements for special gases such as atomic hydrogen and metal gases[5] are not easily made. A TCS measurement for positrons colliding with ions has not yet been performed.

III. Measurements in the low energy range

The lowest positron beam energy for TCS experiments is 0.5 eV[6]. The beam of the Wayne State University group is monochromatic of 0.1 eV breadth. Our measurements were made with broad energy beams and the lowest energy was 0.7 eV, disregarding the energy perpendicular to the flight path. Lower energy positron beams are necessary for the study of the effects of rotations and vibrations in molecules, which will open a new face of atomic physics with positrons.

IV. Comparison of TCS for e^+ and e^- in the intermediate energy range

The comparison of TCS for e^+ and e^- in the intermediate energy is an important subject. Measurements are performed under almost the same experimental conditions. The data of the Wayne State University group (inert gases, H_2[6], N_2[6], CO[8], CO_2[8]), Bielefeld University group (9 species of hydrocarbons[9]) and our group (N_2[10], CO[10], CO_2[10], CH_4[7], C_2H_2[11], C_2H_4[7], C_2H_6[7], H_2O[12], NH_3[13], C_6H_6, CF_4[14] and SiH_4[14]) do not lead to a simple and definite conclusion.

2. TOTAL CROSS SECTION MEASUREMENTS

2.1 Outline of the Experimental Method and Procedure[7,10]

Total cross sections (TCS) were measured using a beam transmission technique. The TCS is obtained from the equation

$$I_g = I_v \exp(-Q_t n l) \tag{1}$$

where I_v and I_g are net counts obtained after subtracting the accidental coincidences in the vacuum and the gas runs, n is the target-gas number density in the collision cell and l is the effective length of the cell.

We prepared a combined apparatus for inelastic collision experiments and for TCS measurements. The schematic diagram of the experimental setup is shown in Figure 1. As the length of the collision cell (geometrical length of 54 mm) is rather short, it is difficult to obtain absolute values of TCS, and the values were obtained by a normalisation to TCS in e^+-N_2 of Hoffman et al[6]. Solenoid coils around the flight path and the collision cell were used for e^+ (or e^-) beam-transportation. For measurements of TCS, the field around the collision cell must be weak. The intensity of slow positrons passing along the flight path depends on the strength of the field. The field around the cell was 3-9 G and 50 G elsewhere. Even for measurements in which the scattering is sharply forward, the TCS curve in the weak magnetic field shows the saturation of TCS values for the variation of the field. The measurements are performed under a weaker field than the saturation field for electron measurements. But for positron measurements, there were also measurements in a stronger field than that for saturation because

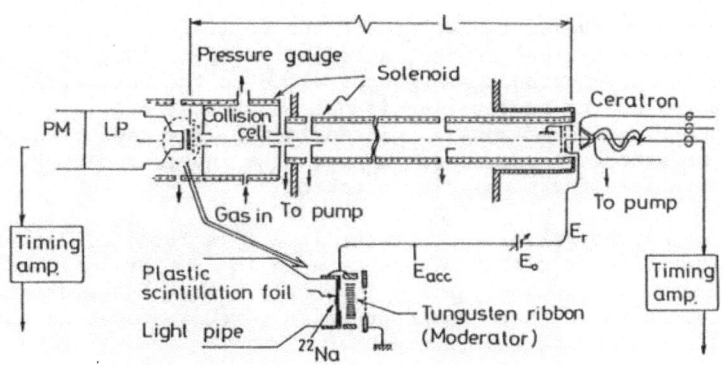

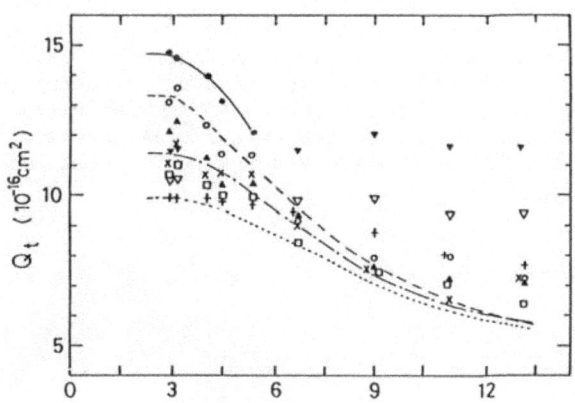

The diameter of the exit aperture of the collision cell is rather large, 8 mm. A positron (electron) source, a radio-isotope, ^{22}Na, of 50 μCi, and a moderator of tungsten-ribbon are placed close to the collision cell. So the source of low energy projectiles is not in vacuum, but in gas. The number of slow positrons (electrons) produced is not affected by the surface state of the moderator in vacuum or in gas. The independence of TCS on gas pressure and species of gas was also checked.

The typical number of slow positrons is 1 e$^+$/s for 3 G, 2 e$^+$/s for 4.5 G at the acceleration energy of 0 eV, and 6 e$^+$/s for 9 G at intermediate energies. In the next subsections, 2.2 - 2.6, only parts of the data are shown in the figures. Detailed data and discussion were presented in the original papers or will be given in the future.

2.2 TCS for e$^+$ and e$^-$ Colliding with CH$_4$, C$_2$H$_4$, C$_2$H$_6$ and Other Hydrocarbons

Floeder et al[9] measured the TCS for 5 - 400 eV positrons and electrons colliding with CH$_4$, C$_2$H$_4$, C$_2$H$_6$ and six other sorts of alkanes, C$_n$H$_{2n+2}$, alkanes C$_n$H$_{2n}$ and isocarbonic molecules. They derived from the results that the TCS for e$^+$ and e$^-$ were proportional to the number of molecular electrons. The TCS for 1 - 400 eV positrons and electrons with CH$_4$, C$_2$H$_4$ and C$_2$H$_6$ were measured in our laboratory. The results for e$^+$-CH$_4$ are shown together with the experimental data of Charlton et al[15], Kauppila et al[16] and Floeder et al[9] and the theoretical elastic scattering data of Jain and Thompson [17] in Figure 3. The TCS for positrons colliding with C$_2$H$_6$, C$_2$H$_4$, C$_2$H$_2$ and CH$_4$ in the energy range below 10 eV are given in Figure 4. The data of these gases except C$_2$H$_2$ were recently obtained and are more accurate but differ from the previous work! The TCS for e$^+$ and e$^-$ in the intermediate energy range (100 - 400 eV) are shown with the data of Floeder et al[9] in Figure 5. The same plots for N$_2$, CO and CO$_2$ are also given with the data of Hoffman et al[16] in Figure 6.

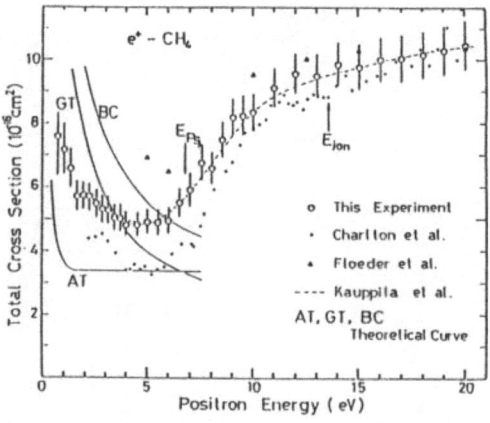

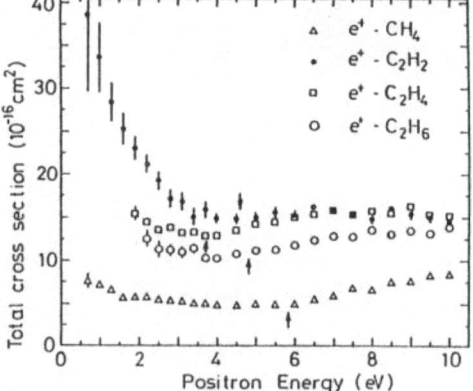

Fig. 3. TCS for e$^+$ colliding with CH$_4$. Experimental points: O, Sueoka and Mori[7]; ●, Charlton et al[15]; Δ, Floeder et al[9]; ----, Kauppila et al al[16], and with theoretical curves ———, AT, GT and BC from Jain and Thompson[17]. Error bars show uncertainties in the present results except the errors due to forward scattering.

Fig. 4. TCS for e$^+$ colliding with C$_2$H$_2$, C$_2$H$_4$, C$_2$H$_6$ and CH$_4$. Thresholds for Ps formation are indicated by arrows.

The TCS measurements for e^+ and e^- colliding with C_2H_2 (acetylene)[11] were made under the same conditions in the range 1 – 400 eV. The results for e^+ and e^- in the range 1 – 20 eV are shown in Figure 7. It was deduced from the measurements of the magnetic field dependence that the scattering of positrons by C_2H_2 at low energies is strongly forward peaked. The effect on the scattering of double or triple electronic bonds in hydrocarbons is not explicitly revealed in TCS. As shown in Figure 4, however, the magnitude of TCS at lower energies for these hydrocarbons does not depend on the number of hydrogen atoms. On the other hand, at higher energies TCS is proportional to the number of hydrogen atoms.

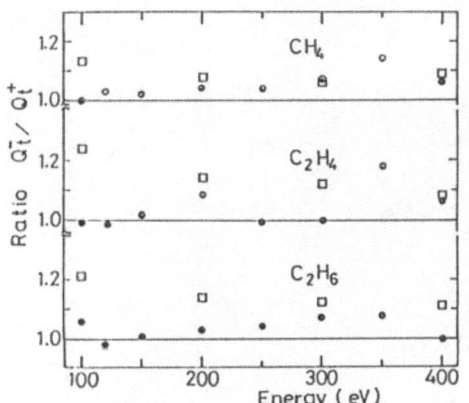

Fig. 5. Ratio of TCS for e^- to that for e^+, Q^-_t/Q^+_t, plotted against impact energy for CH_4, C_2H_4 and C_2H_6. ●, Sueoka and Mori[7]; □, Floeder et al[9].

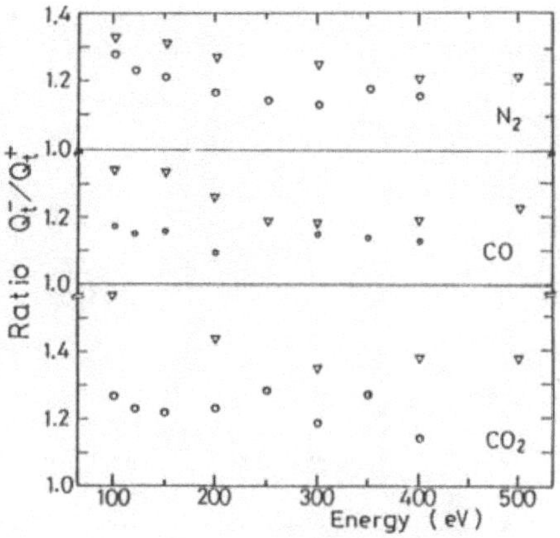

Fig. 6. Q^-_t/Q^+_t plotted against impact energy for N_2, CO and CO_2. o, Sueoka and Mori[10]; ∇, the data of Wayne State University group[6,8].

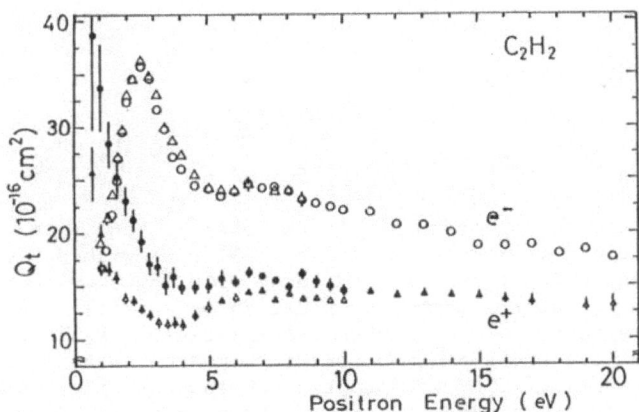

Fig. 7. TCS for e^+ and e^- with C_2H_2 in the range below 20 eV. e^+ data, Δ, 9 G; ●, 4.5 G. e^- data, o, 4.5 G; Δ, 3.6 G.

2.3 TCS for e^+ and e^- colliding with C_6H_6 (benzene)

The TCS for 0.7 – 400 eV positrons and 1 – 400 eV electrons colliding with C_6H_6 (benzene) molecules have recently been measured in our laboratory in order to compare with the TCS for chain-hydrocarbons. From the dependence on the magnetic field, the scattering of positrons is deduced to be more forward peaked than that of e^- in the low energy range. The TCS data for e^+ and e^- in the range below 20 eV are presented in Figure 8. The TCS for positrons are very high, such as those in e^+-K[5], and higher than those in e^+-H_2O and e^+-NH_3. The TCS for e^+ and e^- extending to intermediate energies are shown in Figure 9.

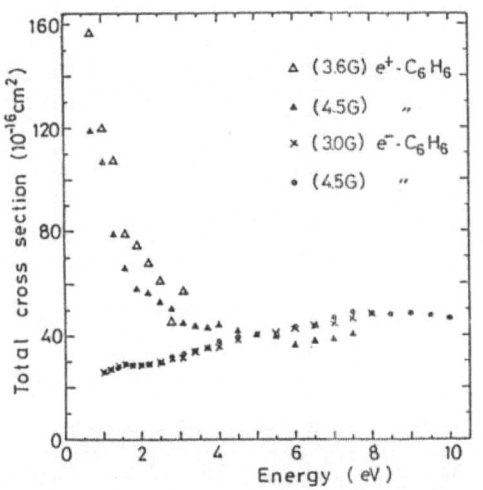

Fig. 8. TCS for e^+ and e^- colliding with C_6H_6 in the range below 10 eV.

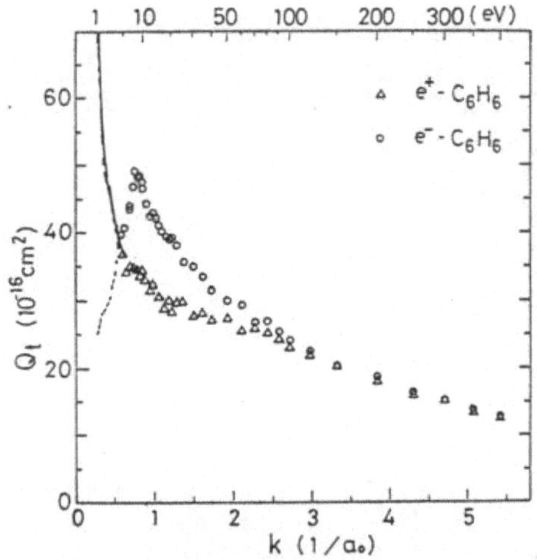

Fig. 9. TCS for e^+ and e^- colliding with C_6H_6 extending to interdiate energies.

2.4 TCS with H_2O[12] and NH_3[13]

We have presented the experimental data of the total cross sections for e^+ and e^- colliding with the polar molecules H_2O and NH_3. As shown in Figure 10, in the low energy region TCS for e^+ and e^- are high, but 1/4 or 1/3 of the theoretical values of Jain[18]. This large discrepancy with the theoretical data is not understood. In the energy range below 2 eV and above 200 eV, the TCS values for e^+ and e^- colliding with H_2O coincide with each other, as shown in Figure 11, and those with NH_3 show the same tendency. These results show that the Born approximation is a fairly good approximation for these regions. One of the characteristics of scattering by polar molecules is that the cross section for Ps formation is very low. In the ordinary measurements of TCS, Ps formation cross sections cannot be observed because of high background counts. We have obtained these values just above the threshold energy from the TCS curve under stronger magnetic fields for the beam-guide, as shown in Figure 12. This simple method is useful for molecules with low cross sections for Ps formation or for molecules with low threshold energies for Ps formation, such as C_2H_4 and C_2H_6. It was demonstrated by applying this method that the increase of TCS in the range above the threshold of Ps formation is mainly due to Ps formation. The low probability of Ps formation in polar molecules must be checked in lifetime measurements.

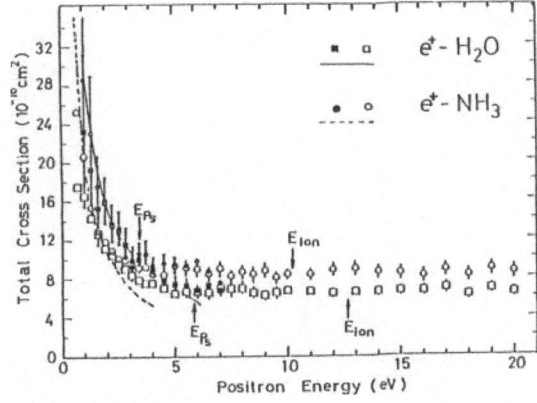

Fig. 10. TCS for positrons colliding with H_2O and NH_3 in the range below
20 eV, ■, ●, at 4.5 G; □, o, at 9 G for
NH_3 and H_2O. The solid line and the
broken line are the theoretical data
for H_2O and NH_3 of Jain[18] reduced by a
factor of 5.

Fig. 11. TCS for e^+ and e^-
colliding with H_2O in the range
1 – 400 eV.

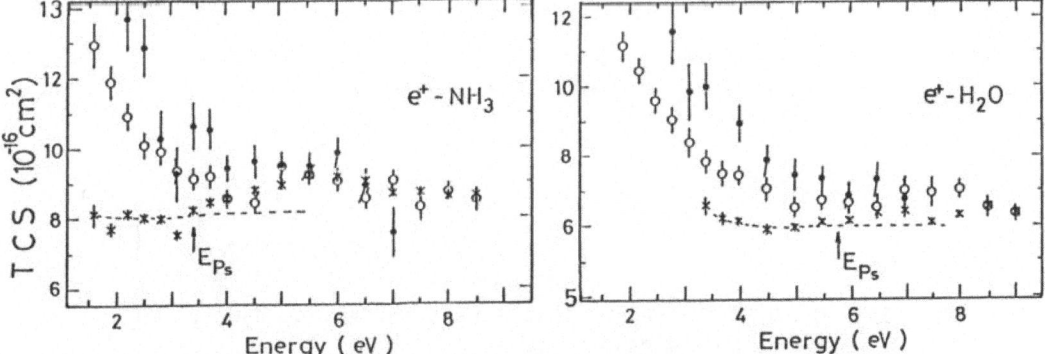

Fig. 12. Enlarged plot of Figure 10 with additional data (x) at 23 G.
the broken curves are the extrapolated cross section curves
without the contribution of Ps formation. ●, 4.5 G; o, 9 G
for each case in H_2O and NH_3. The error bars show statistical errors only.

2.5 TCS for CH_4, CF_4, CCl_4 and SiH_4

The main purpose was to measure the TCS for positrons and electrons
colliding with monosilane molecules, SiH_4, but the TCS for molecules with
similar structure, CH_4, CF_4 and CCl_4 were also obtained. As mono-silane
molecules are inflammable, careful gas-handling is needed. The TCS for
positrons in the range 1 – 50 eV and those for electrons in the same range
are given in Figures 13 and 14 respectively. The theory for e^+-SiH_4 and
e^--SiH_4 has been developed by Jain[1] in detail. As shown in the figures, the
results for CH_4 and SiH_4 are similar to each other for both e^+ and e^- collisions. But the collisions with CF_4 and CCl_4 may be quite different from
those with CH_4. The ratio of TCS for e^- to that for e^+ in the intermediate
energy range 100 – 400 eV is plotted in Figure 15 for H_2O, NH_3, SiH_4,C_2H_2
and H_2 molecules.

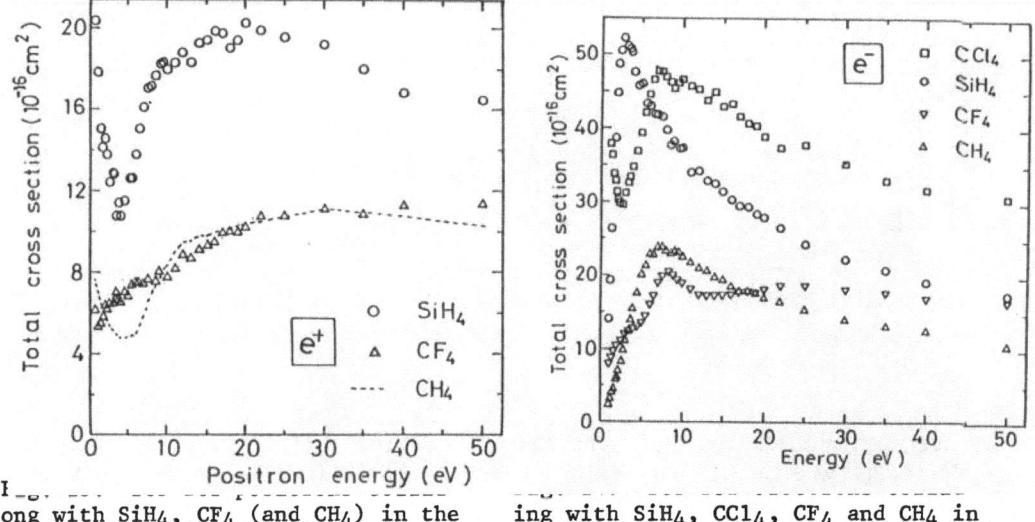

ong with SiH$_4$, CF$_4$ (and CH$_4$) in the range 1 - 50 eV.

ing with SiH$_4$, CCl$_4$, CF$_4$ and CH$_4$ in the range 1 - 50 eV.

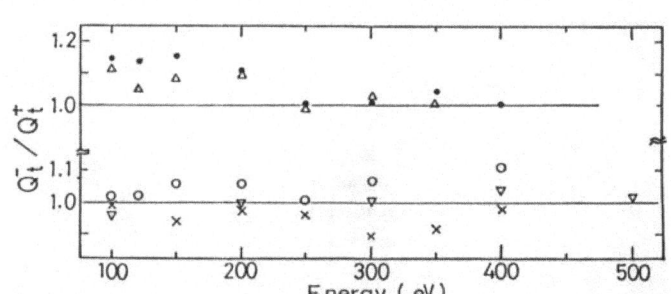

Fig. 15. Q^-_t/Q^+_t plotted against impact energy. Δ, H$_2$O; •, NH$_3$, ∇, H$_2^6$; o, SiH$_4$, ✕, C$_2$H$_2$.

2.6 TCS with O$_2$ Molecules

In the range 1 - 400 eV, the TCS for e$^+$ and e$^-$ colliding with O$_2$ molecules were measured for the purpose of using them in the determination of the inelastic cross section of O$_2$ by the relative method.

3. INELASTIC CROSS SECTIONS OF e$^+$ AND e$^-$ WITH MOLECULES

3.1 Schumann-Runge Excitation of O$_2$ by Positron Impact[19]

Considerable difficulties arise in the experimental measurement of inelastic cross sections due to the broad energy spread and low intensity of the positron beam. The same beam is used for positron inelastic colli- sion experiments and for TCS experiments. If the energy spread perpendicu- lar to the flight path is neglected on TCS-measurements, the experimental accuracy of TCS does not depend on the energy width of the beam. For inelas- tic-collision experiments, however, the energy width of the primary beam makes the resolution worse as also does the length of the collision cell. For these reasons, inelastic collision experiments have been attempted only for the inert gases He, Ne and Ar in which the energy levels are fairly wide apart.

The inelastic collision experiment in O_2 molecules, the Schumann-Runge excitation, which is a strong absorption band in the ultra-violet region for O_2, has recently been measured for positron impact. The measurement was carried out using the apparatus shown in Figure 1 with a strong magnetic field of 50 G for beam-transportation in the flight tube, contrary to that used for TCS measurements, with a weaker magnetic field in the vicinity of the collision cell. Thus, a major fraction of the forward scattered positrons was collected on the detector by a strong magnetic field. In inelastic measurements, of course, a retarding potential is not applied.

The cross sections of the Schumann-Runge excitation, Q_{SR}, are not integrated over all angles, but integrated within a limited angle (the critical angle) which depends on the energy of scattered particles and the magnetic field. So, the positron data of Katayama et al[19] is not directly comparable with the electron inelastic collision data of Wakiya[20] which were obtained by the conventional electron-beam method. In order to compare the positron data with the electron data for Q_{SR}, it is necessary to perform electron inelastic-collision measurements under the same experimental conditions. The same method was applied to the electron-beam experiment, but the energy resolution was not sufficiently good. The number of positrons collected at the detector is determined by the critical angle which depends on the energy of the beam.

The cross section for Schumann-Runge excitation in the forward cone, Q_{SR}, was derived from the energy loss spectra by the relative value of the total cross section Q_t, namely

$$Q_{SR} = I_{SR}Q_T/I_t = I_{SR}Q_t/I(1 - \exp(Q_t n)) \tag{2}$$

where Q_t is the total cross section and n is the gas number density derived from the gas pressure and gas temperature. The length of the collision cell was determined as described in subsection 2.1. I_t and I are the intensities of the scattered and incident positrons, respectively. I_{SR} is the intensity derived from the summation of the positron count in the SR energy loss region.

The original data in the time spectrum are transferred to the energy loss spectrum. Examples of the spectra of positrons colliding with O_2 molecules are shown for several incident energies in Figure 16. The Schumann-Runge excitation intensities, I_{SR}, were obtained by summation of the counts in a range wider than the SR range (7.1 - 9.7 eV) in consideration of the energy resolution. Examples of the summation-range are shown in Figure 16 with the broken curves. The SR inelastic cross sections calculated by equation (2) are plotted together with the SR cross sections of Wakiya[20] for electron impact and the cross sections of Griffith[21] in Figure 17. In the energy region of the SR excitation, the Ps formation channel opens.

3.2 Inelastic Positron Collisions with H_2O and NH_3

In the energy loss spectra of forward scattered electrons for H_2O and NH_3 molecules obtained by Lassetre et al[23], fairly strong and broad peaks of the first excitation for H_2O and NH_3 are shown at 7.42 eV for H_2O and 6.38 eV for NH_3, respectively. The measurement of positron energy-loss spectra for these structures was tried using the same method as in the measurement of Schumann-Runge excitation in O_2. The inelastic cross sections for the structure were obtained from the intensities derived from the

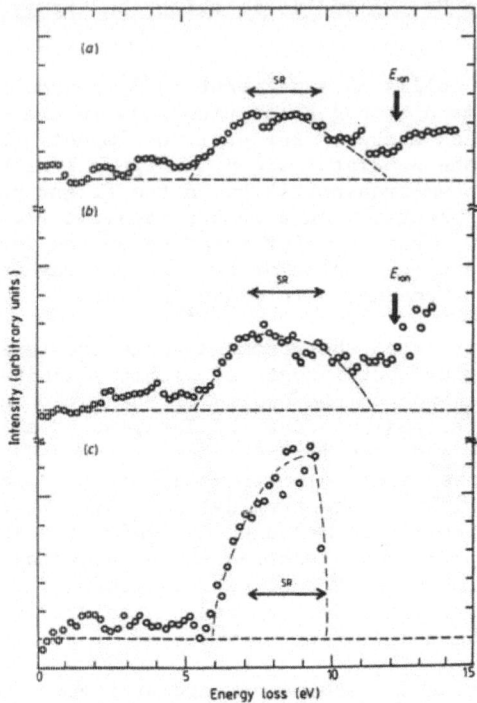

Fig. 16. Energy loss spectra in O_2 from slow positron impact at (a) 22.3, (b) 14.3 and (c). 10.5 eV. The SR region and the threshold for the first ionisation are indicated by horizontal and vertical arrows respectively. The areas used to determine I_{SR} are shown by broken curves.

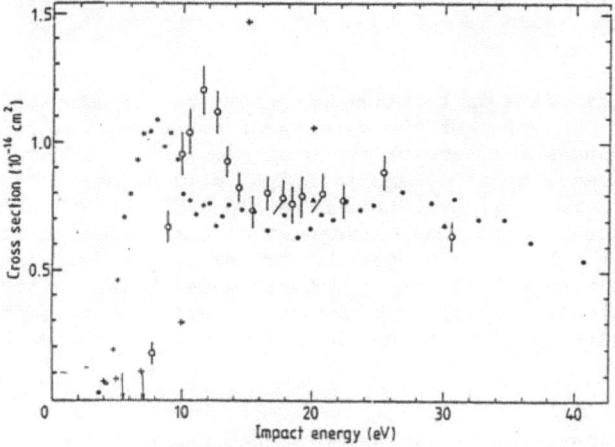

Fig. 17. o, SR cross sections for e^+-O_2 collision between 7.8 and 30.6 eV; •, Ps formation cross section data of Griffith[21]; ▲, SR cross section data of Wakiya[20] for electrons. +, total electronic excitation data of Trajmar[22]. The error bars show the summation of the statistical errors and errors in Q_t, n and ℓ in equation (2). The arrows indicate the threshold energy of Ps formation (5.3 eV) and that of the SR continuum (7.1 eV).

summation of positron counts in the energy loss ranges 6.8 - 8.3 eV for
H_2O and 5.7 - 7.3 eV for NH_3 as a function of impact energy. The loss
spectra for positrons have more complicated features than those deduced from
the electron loss data. Of course, measurements by narrow beams and corresponding electron-beams are necessary.

3.3 Beam-Narrowing Method

In order to prepare beams with a narrow energy width, novel TOF methods
A and B were tried. The schematic diagram of these systems is shown in
Figure 18.

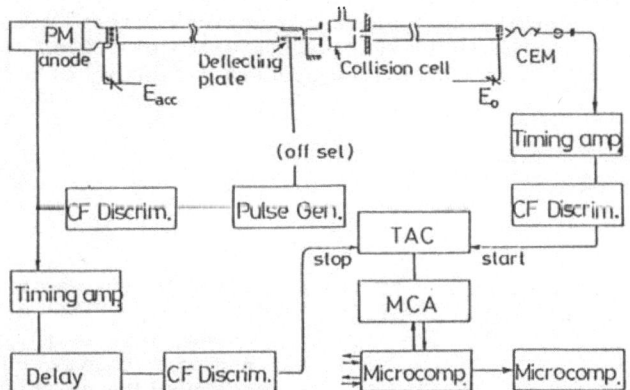

Fig. 18. Schematic diagram of the inelastic experiments.

Method A. The electron (positron) beam passed through a flight path of
0.56 m length and was deflected for only a very short time by a set of
beam deflecting plates 50 mm in length and 6 mm apart, in order to remove
the higher energy parts of the beam produced at the moderator. Only the
higher energy part of the beam was deflected by applying a 10 volt-pulse,
triggered by the PM signal, to the deflecting plates placed in front of the
collision cell. Since the effect of the beam-narrowing depends critically
on the width of the pulse applied on the deflecting plates, the optimum
value of the pulse width must be chosen. The starting time of the pulse for
the deflection is delayed about 60 ns from a PM-signal. The time lag is due
mainly to the propagation time in the pulse generator and the CF discrima-
tor. The electron-beam in the case of a 1.5 eV acceleration potential at
the moderator was measured with pulses of various widths applied to the
deflecting plates. Spectra of electron beams taken with various delayed
pulses are shown in Figure 19 (a) and those for a positron beam with 10 V
accelerating potential are shown in Figure 19 (b). For beams of 10 - 15 eV,
the energy width was reduced to about half, but for beams higher than 25 eV,
the width was only slightly improved because of insufficient time resolution
in the deflecting system. To accomplish better beam resolution a faster
pulse generator with higher output voltage and improved geometry of the
deflecting plates are necessary.

Method B. For the positron beam used in inelastic experiments, the narrowing method was modified as follows. After the beam has been accelerated
with a low potential at the moderator it passes through the flight path on

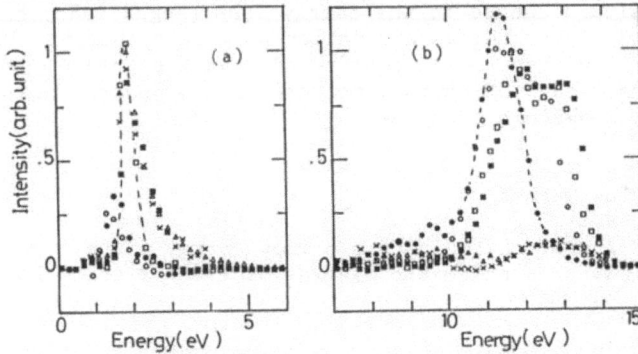

Figs. 19(a) and 19(b). Energy spectra of e^- beam with 1.5 eV acceleration (a) and e^+ beam with 10 eV acceleration (b) for various
pulse widths. For the best conditions for e^- beam of 1.5 eV
E_{acc}, pulses for the deflection are applied with a delay of
400 ns from the PM signal, and with a delay of 200 ns for e^+
beam of 10 eV E_{acc}. The narrow spectra are shown as broken
curves.

the PM –side. It is then shaped by the deflector and is again accelerated
to the required energy by the potential between two parallel nets placed in
front of the collision cell. The potential on the left flight–path of the
deflecting plates shown in Figure 18 is earthed in method A, but is floated
by the off-set potential in method B. The energy width of the beam in the
collision cell is almost the same as that of the low acceleration–beam in
the case of method A. In practice in TOF experiments, however, the beam is
distributed over many channels in the MVA. Which is the better method in
inelastic experiments for positrons and electrons depends on the conditions
of the measurements. For beams higher than 20 eV method B is useful.

3.4   Schumann–Runge Excitation Cross Section by Electron Impact

Under the same conditions as for the positron–energy–loss experiment
mentioned in 3.1, the peak by SR-loss in the electron energy loss spectrum
was not observed. Then, an electron beam, with its high energy tail removed
by the method described in the previous subsection, was used for the cross
section measurement for the Schumann–Runge band excitation in O_2 molecules.
As demonstrated in Figure 20, the distinctness of the peak due to the SR-loss
depends on the magnetic field, namely, a strong magnetic field collects
electrons scattered in the forward direction as a result of low energy
excitations ($b^1\Sigma_g^+$ and $a'\Delta_g$) as well as scattering contributions from the
apertures and/or the walls of the flight path. The smearing effect on the
peak of the SR excitation due to various scattering mechanisms is much
larger for electron beams than for positron beams.

The comparison of the SR excitation cross section for positrons with
that for electrons has been performed by method A. The preliminary data for
the ratio Q^+_{SR}/Q^-_{SR} are as follows: 1.4, 1.6 and 1.4 for 16.4, 19.3 and
22.2 eV (mean energy of the incidence), respectively. The TOF method for
inelastic collisions was performed by method A for less than 20 eV impact
energy and method B for higher than 20 eV. The experimental procedure, data

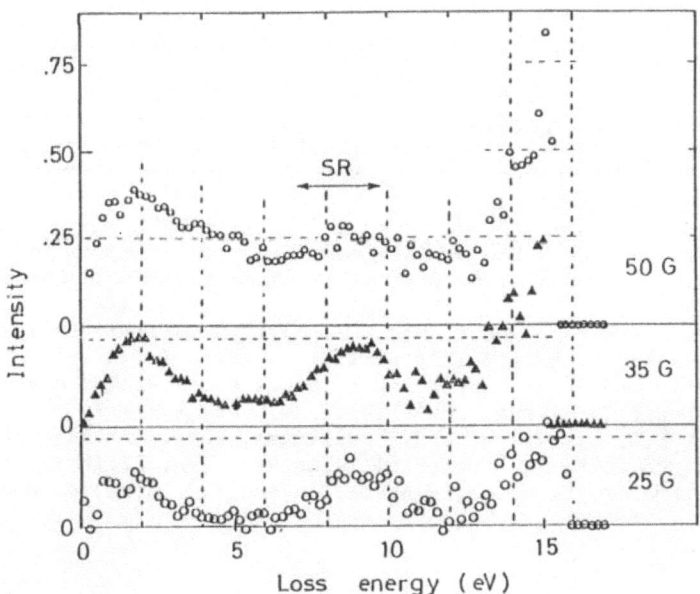

Fig. 20. Energy loss spectra of electrons with 15.0 eV E_{acc} in O_2
under various magnetic fields using the method A.

taking and data analysis are the same as in the positron impact measurements
of Katayama et al[19].

ACKNOWLEDGEMENTS

The work of the author's group has been performed in collaboration:
with S. Mori and Y. Katayama in the TCS—measurements and in O_2 inelastic
collisions; with M. Yamazaki in the new beam method.

REFERENCES

1. A. Jain, Atomic Physics with Positrons ed. J.W. Humberston and E.A.G.
       Armour, (Plenum, New York), (1988).
2. M. Charlton, Rep. Prog. Phys. 48:737 (1985).
3. T. Mizogawa, Y. Nakayama, T. Kawaratani and M. Tosaki, Phys. Rev. A
       31:2171 (1985).
4. R.I. Campeanu and J.W. Humberston, J. Phys. B 10:L153 (1977).
5. T.S. Stein, R.D. Gomez, Y.-F. Hsieh, W.E. Kauppila, C.K. Kwan and
       Y.T. Wan, Phys. Rev. Lett. 55:488 (1985).
6. K.R. Hoffman, M.S. Dababneh, Y.-F. Hsieh, W.E. Kauppila, V. Pol, J.H.
       Smart and T.S. Stein, Phys. Rev. A 25:1393 (1982).
7. O. Sueoka and S. Mori, J. Phys. B -9:4035 (1986).
8. Ch. K. Kwan, Y.-F. Hsieh, W.E. Kauppila, S.J. Smith, T.S. Stein and
       M.N. Uddin, Phys. Rev. A 27:1328 (1983).
9. K. Floeder, D. Fromme, W. Raith, A. Schwab and G. Sinapius, J. Phys.
       B 18:3347 (1985).
10. O. Sueoka and S. Mori, J. Phys. Soc. Jpn. 53:2491 (1984).
11. O. Sueoka and S. Mori, Atomic Collision Res. Jpn. 12:16 (1986).
12. O Sueoka, S. Mori and Y. Katayama, J. Phys. B 19:L373 (1986).
13. O. Sueoka, S. Mori and Y. Katayama, J. Phys. 20:3237 (1987).
14. S. Mori, Y. Katayama and O. Sueoka, Atomic Collisions Res. Jpn. 11:19

(1985).

15. M. Charlton, T.C. Griffith, G.R. Heyland and G.L. Wright, J. Phys. B 16:323 (1983).

16. W.E. Kauppila, M.S. Dababneh,Y.-F. Hsieh, Ch. K. Kwan, S.J. Smith, T.S. Stein and M.N. Uddin, Proc. 13th Int. Conf. on the Physics of Electronic and Atomic Collisions (Berlin) ed. J. Eichler, W. Fritsch, V. Hertel, N. Stolterfort and U. Wille, (Amsterdam: North-Holland), Abst. p. 303. (1983).

17. A. Jain and D.G. Thompson, J. Phys. B 16:1113 (1983).

18. A. Jain, 3rd Int. Workshop on Positron (Electron)-Gas Scattering, ed. by W.E. Kauppila, T.S. Stein, J.M. Wadehra, (World Scientific) p. 283 (1986).

19. Y. Katayama, O. Sueoka and S. Mori, J. Phys. B 20:1645 (1987).

20. K. Wakiya, J. Phys. B 11:3913 (1978).

21. T.C. Griffith, Positron Scattering in Gases ed. J.W. Humberston and M.R.C.McDowell, (Plenum, New York), p. 53-63 (1984).

22. S. Trajmar, D.C. Cartwright and W. Williams, Phys. Rev. A 4:1482 (1971).

23. O. Sueoka, and S. Mori, Atomic Collision Res. Jpn. 12:16 (1986).

24. E.N. Lassettre, A. Skerbele, M.A. Dillon and K.J. Ross, J. Chem Phys. 48:5066. (1968).

STUDIES OF INELASTIC POSITRON SCATTERING USING 2.3 AND 3 M SPECTROMETERS

L. M. Diana, P. G. Coleman,[*] D. L. Brooks,
and R. L. Chaplin

Department of Physics
University of Texas at Arlington
Arlington, Texas, U.S.A.

[*]School of Mathematics and Physics
University of East Anglia
Norwich, U.K.

We report total positronium formation cross sections for
krypton with beam energies from the threshold region to
352.2 eV. The primary maximum at about 16 eV is followed
by a small secondary maximum in the vicinity of 107 eV.
There is, also, an increase in the rate of decline of the
cross section with energy between 325 and 352 eV. We
present estimates of total elastic scattering cross
sections for argon and neon obtained by subtracting inelastic from total cross sections. In both cases the
elastic cross section exhibits a dip just after the
positronium formation threshold and a recovery in the
neighborhood of the ionization threshold. We have
tabulated positronium formation and proton electron
capture cross sections and the ratios of the former to
the latter for molecular hydrogen, helium, and krypton.
In all cases the ratios climb steeply as particle
velocity increases, the sharp rise beginning at appreciably lower velocities for hydrogen and krypton than
for helium. We discuss initial work with the 3 m
spectrometer and future measurements with it and the
2.3 m spectrometer.

INTRODUCTION

The primary goal in our laboratory is the measurement of cross sections
for inelastic interactions of positrons with gases. The astrophysicist can
use these numbers for estimation of the fraction of positrons that annihilate in flight and the fraction that thermalize, and thus improve his understanding of positron annihilation in solar flares.[1] Additionally, such
measurements permit the partitioning of the total cross section (Q_{tot}) for

positrons into the elastic and inelastic channels, contribute to the comparison of positron and electron total cross sections, and, in the case of total positronium formation cross section (Q_{Ps}), shed light on the anomalously low positronium formation fractions obtained for Kr and Xe from positron lifetime spectra.

In the recent past we have centered our attention on Q_{Ps}, the cross section unique to positrons, and we present in this work new measurements of Q_{Ps} for Kr. We include, also, estimates of total elastic cross sections (Q_{el}) for Ar and Ne obtained by subtraction of inelastic from total cross sections, comparisons of Q_{Ps} and cross sections for electron capture by protons (Q_{ec}), and descriptions of activities to be undertaken in the near future with the 2.3 m and the 3 m spectrometers.

TOTAL CROSS SECTION FOR POSITRONIUM FORMATION IN KRYPTON

We are currently using our 2.3 m spectrometer, a schematic diagram of which comprises Fig. 1, to measure Q_{Ps} for Kr. The number of determinations made to date is nearly three times as many as were reported earlier.[2] The techniques being employed, the possible sources of systematic error, and the correction for double scattering are discussed in detail in Ref. 3. As stated there $Q_{Ps} = fQ_{tot}/F$, where f is the fraction of the incident positrons that form positronium, and F is the fraction that scatter into all available channels. We have scaled Q_{Ps} to the Q_{tot} published by the Wayne State University (WSU) group.[4,5]

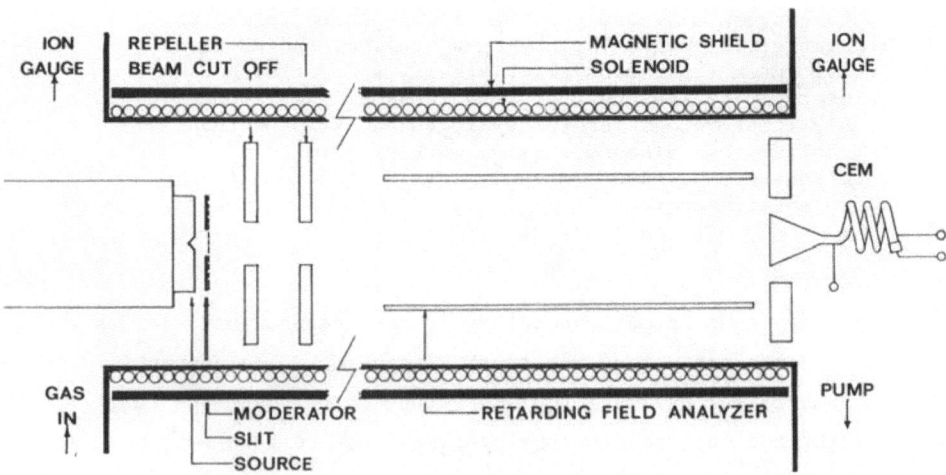

Fig. 1.   Schematic diagram of apparatus used to measure Q_{Ps}.
Flight path is 2.3 m.

Table 1. Total Positronium Formation Cross Sections in Kr (preliminary values, statistical uncertainties in parentheses)

Positron Energy (eV)	Q_{Ps} (πa_0^2)	Positron Energy (eV)	Q_{Ps} (πa_0^2)
7.7	1.2 (0.8)	141.3	2.19(0.32)
9.9	3.5 (0.5)	151.5	1.73(0.18)[b]
13.0	5.8 (0.8)	163.1	1.79(0.21)
16.1	6.4 (0.6)	171.3	1.42(0.22)
26.2	5.9 (0.4)[a]	182.0	1.78(0.25)
51.2	3.8 (1.2)	201.5	1.58(0.23)[b]
52.3	3.8 (0.6)	214.0	1.66(0.18)
76.3	2.3 (0.6)	225.9	1.51(0.20)
86.3	2.43(0.32)	240.9	1.49(0.20)
93.9	2.5 (0.4)[b]	255.9	1.55(0.18)
101.3	2.48(0.21)[a]	275.1	1.67(0.25)
107.0	2.7 (0.5)	300.2	1.42(0.23)
112.4	2.61(0.29)[b]	325.1	1.56(0.22)
126.3	1.96(0.31)	352.2	0.83(0.19)[b]

a,b Weighted mean of three or two measurements respectively.

All of the cross sections in Table 1 have been corrected for double scattering, and those at positron energies of 7.7 and 9.9 eV have been corrected, also, for imperfect reflection of the backscattered projectiles by the moderator. This correction is important because determining f requires the detection of all positrons that do not form positronium. The median beam energy which yielded the results at 7.7 eV was 6.6 eV; 7.7 eV is the median energy of the beam positrons capable of forming positronium. To correct for imperfect reflection, the maximum scattering angle for a positron of median beam energy at which reflection would occur was computed. Then the angular distribution of differential elastic scattering cross sections for 8.70 eV positrons calculated by Massey et al.[6] was used to estimate the fraction of the scattered positrons that would not be reflected. The results were 0.152 for the 6.6 eV beam and 0.116 for 9.9 eV. A similar calculation with the angular distribution for 30.6 eV, the highest energy addressed in Ref. 6, yielded 0.008, and it is assumed that the correction is negligible at higher beam energies, also. Using angular distributions for differential elastic scattering is appropriate for 6.6 eV positrons

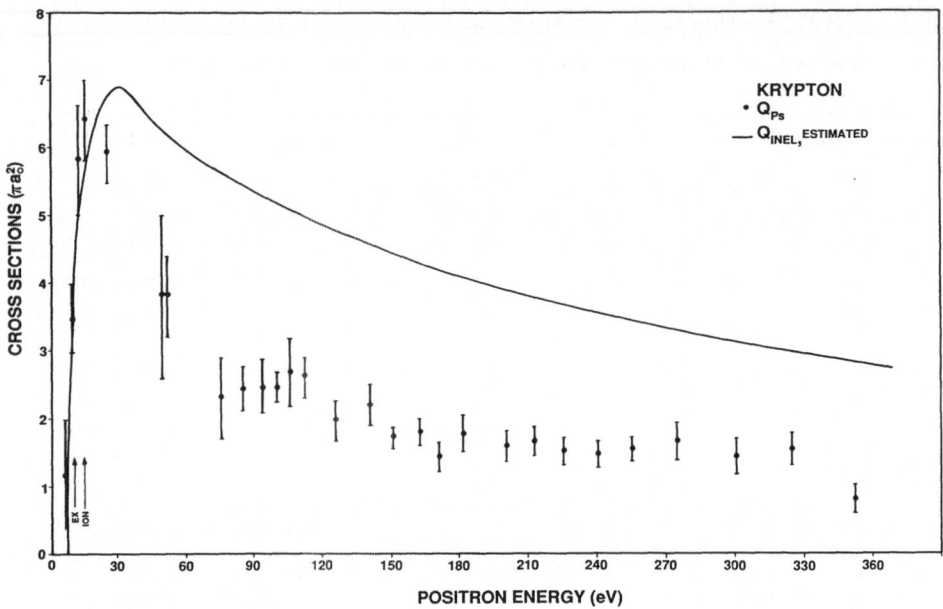

Fig. 2. Total Ps formation cross sections in Kr. Solid line is estimated Q_{inel} (see text).

since the only other open channel is positronium formation and not grossly inappropriate for 9.9 and 30.6 eV. Very few of the positrons in a beam of 9.9 eV median energy are expected to cause excitation, and, as Fig. 2 indicates, only a small fraction of a 30 eV beam would engage in excitation and ionization.

The curve in Fig. 2, representing the sum of all total inelastic cross sections (Q_{inel}), was obtained by subtracting estimated total elastic scattering cross sections Q_{el} from a curve smoothly drawn through Q_{tot} values obtained from Refs. 4 and 5. A smooth curve through the Q_{tot} of Ref. 4 just below the positronium formation threshold energy E_{Ps}, extrapolated to E_{Ps} yielded an estimate of 5.5 πa_0^2 for Q_{el} at E_{Ps}. We took Q_{inel} = smoothed Q_{tot} - 5.5 πa_0^2 up to 30 eV and at higher energies Q_{inel} = 0.56 (smooth Q_{tot}) because at 30 eV 5.5 πa_0^2,is about 0.44 Q_{tot}. An arbitrary choice was necessary because we were unable to locate any theoretical calculations above 23 eV and because Q_{inel} = smooth Q_{tot} - 5.5 πa_0^2 leads to Q_{inel} approximately equal to Q_{Ps} at about 200 eV. Q_{Ps} at 7.7 and 9.9 eV agree well with Q_{inel}. The Q_{Ps} at 13.0 and 16.1 eV indicate that Q_{el} at these energies dips below 5.5 πa_0^2 perhaps forming something similar to the cusp at E_{Ps} noted by Campeanu et al.[7] for Q_{el} of He. A small secondary maximum is discernible in the neighborhood of

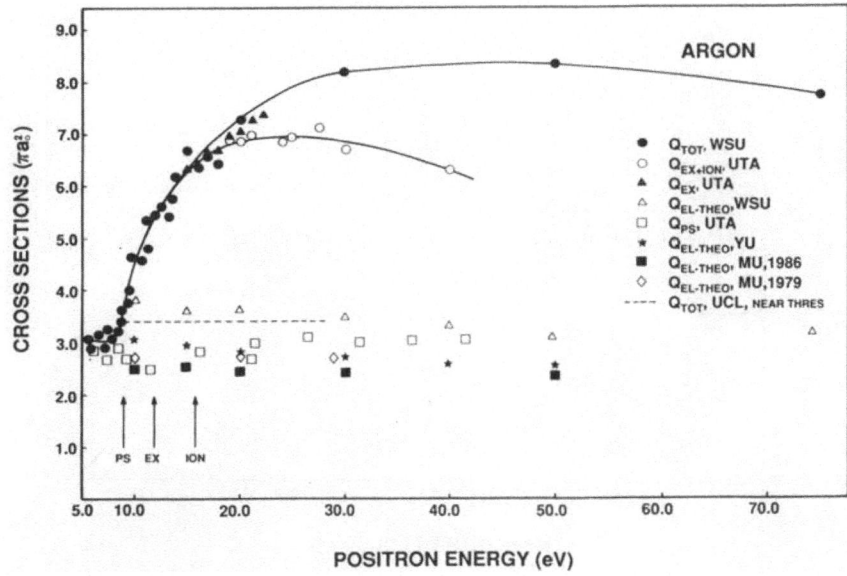

Fig. 3. Cross sections for positron scattering from Ar. See explanation in text.

107 eV; the energy required to remove the first S electron from the N shell is 111.0 eV.[8] Also worthy of note is the increased steepness of the negative slope of the Q_{PS} vs. energy curve between 325 and 352 eV. We plan additional measurements between and beyond these energies and between 16 and 150 eV.

ESTIMATED ELASTIC CROSS SECTIONS FOR ARGON AND NEON

Our Q_{PS} measurements for Ar[9,10] and Ne[11] for positron energies up to about 50 eV are used together with the excitation, Q_{ex}, and excitation plus ionization, Q_{ex+ion}, cross sections obtained earlier in our (UTA) laboratories[12] to estimate Q_{el}. The results are shown in Figs. 3 and 4. Smooth curves are drawn through Q_{tot} reported by the WSU laboratories.[13,14,15] From these are subtracted Q_{ex} and Q_{ex+ion} and smooth curves drawn through the remainders. Finally, from these latter curves the Q_{PS} are subtracted, and their positions represent our estimates of the elastic cross section. The dashed lines represent estimates of Q_{tot} near the E_{PS} obtained from the values reported by Charlton et al.

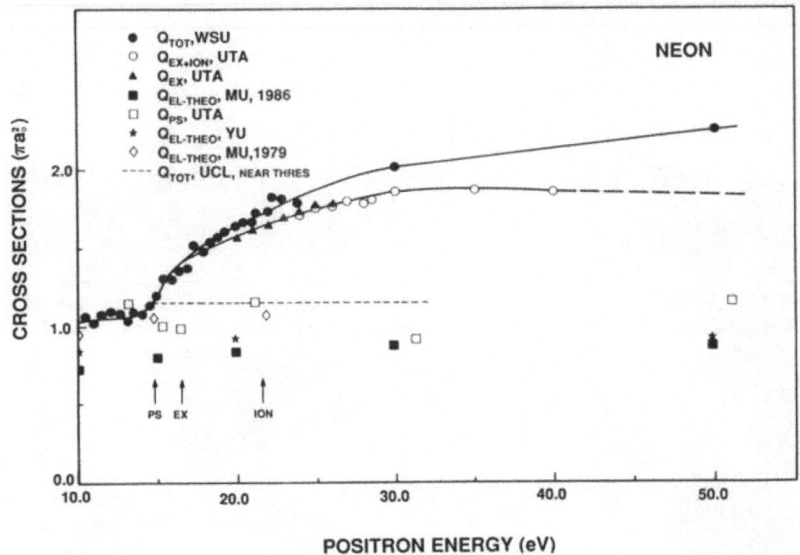

Fig. 4. Cross sections for positron scattering from Ne. See explanation in text.

.of University College London (UCL).[16] On the plot for Ar are shown the Q_{el} calculated by Nahar and Wadehra[17] (WSU), by theorists in York University[18] (YU), and by Nakanishi and Schrader[19] and Schrader[20] of Marquette University (MU). The theoretical Q_{el} values on the Ne plot were taken from Refs. 19 and 20 and from the calculations of McEachran and Stauffer[21,22] (YU).

Our estimates of Q_{el} for Ar and Ne behave similarly. They exhibit a decline in value near E_{Ps}, reach a minimum at about the excitation threshold energy (E_{ex}), and return to the level near E_{Ps} in the vicinity of the ionization threshold energy (E_{ion}). It appears that competition from opening channels brings about a decline in Q_{el} until there is sufficient energy for all scattering processes.

ELECTRON CAPTURE BY POSITRONS AND PROTONS

The work of J. H. McGuire and his colleagues[23] has created interest in comparisons of Q_{Ps} and the cross section for electron capture by protons, Q_{ec}, obtained for projectiles with equal velocities. We list in Table 2

Q_{Ps} measured in our laboratories for H_2, He, and Kr (from Refs. 3 and 9 and this article) and Q_{ec} reported by several laboratories (from Refs. 24 through 31). The beam velocities and energies at which the Q_{Ps} were determined comprise the first two columns. Beside each positron beam energy is the proton energy required for equal projectile velocity, and proton energies not too dissimilar from this ideal are tabulated above and/or below the ideal.

The ratio of Q_{Ps} to Q_{ec} (R) increases with velocity for all three gases beginning at one order of magnitude or more below unity. R becomes approximately unity for H_2 and Kr at a velocity of about 3×10^6 m/s and at about 4×10^6 m/s for He. When the positron beam energy is slightly above 200 eV, R is several tens for He, about 100 for Kr, and several hundred for H_2.

Plots of R vs. velocity exhibit reasonably smooth variation with some irregularities apparent as a result of the involvement of several laboratories, the uncertainties in cross sections and energies, comparisons at positron and proton velocities that are excessively disparate, or some combination of these or of these and other causes.

3 M SPECTROMETER

The UTA 3 m time-of-flight (TOF) spectrometer is shown in Fig. 5. Until recently we have been conducting exploratory measurements with He and Ar which have led to design improvements and to some interesting initial results,the confirmation of which we are now undertaking.

TOF spectra are generated by signals from the photomultiplier tube (PMT) and the channel electron multiplier (CEM). Some of the positrons from 250 μCi of ^{22}NaCl pass through a 0.25 mm thick KL236 fast plastic scintillator and cause light flashes which, after passing through the plexiglas light guide and being collected at the photocathode of the Amperex XP1021 photomultiplier, produce fast electrical pulses suitable for timing. A fraction of these positrons are intercepted by a 50% transmission annealed tungsten moderator with consequent reemission with mean energy of approximately 1.3 eV.

The desired accelerating potential is applied to the moderator. The accelerated positrons pass through a 3 mm diameter aperture into a 4.3 cm diameter, 2.4 cm deep gas cell with 7 mm diameter exit port. Two sets of plates for establishing electric fields transverse to the axial magnetic (B) field reduce background and permit fine alignment of source and detector. The external coil which generates the B field extends without discontinuity from the source area to the CEM detector. The coil is

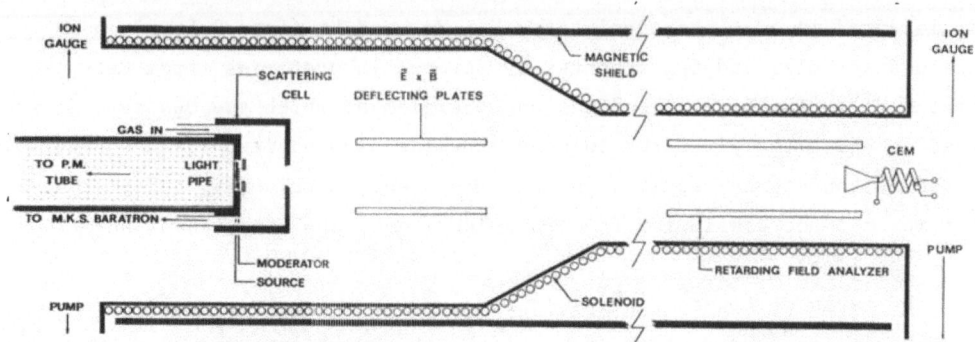

Fig. 5. UTA 3 m time-of-flight spectrometer with cylindrical retarding
field analyzer. There are four $\vec{E}$ x $\vec{B}$ deflecting plates.

shielded from the earth's field.   The B field constrains positrons
with transverse momentum components to helical paths of sufficiently small
diameter to insure their exiting the scattering chamber and striking the
CEM, the pulses from which start a time-to-amplitude converter (TAC).   The
delayed photomultiplier pulses stop the TAC, the output pulses of which
are stored by a multichannel analyzer.   This reversal of the order of the
timing signals in the production of TOF spectra prevents swamping of the
circuitry by the profusion of PM pulses.

The useful energy range of this high resolution spectrometer is
extended by the cylindrical retarding field' analyzer which permits extraction of information from unresolved peaks by the method of Sueoka.[32]

We calculate cross sections using $Q_{scat} = N_{scat} Q_{tot}/N_{tot}$ where the
subscript scat represents excitation, ionization, or excitation plus
ionization in cases where these events fail to be resolved. N_{scat}, the
number of positrons scattered into the channel or channels being considered,
is the sum of the counts in the appropriate peak in the TOF spectrum after
the signal-restoration-background-subtraction procedures reported by
Coleman et al.[33] and by Coleman[34] have been applied. N_{tot}, the number of
positrons scattered into all channels, is calculated from
$N_{tot} = N_0[1 - \exp(-Q_{tot}nL)]$. N_0, the total number of incident positrons,
is measured with the gas cell evacuated, nL, the gas density-path length
product, is determined with the assistance of an MKS Baratron pressure
gauge, and Q_{tot} is obtained from the literature.

A restored TOF spectrum, obtained with 70 eV positrons incident on Ar,
is shown in Fig. 6.

Table 2. Cross Sections and Ratios at Similar e^+ and H^+ Velocities

Particle Velocity (10^6 m/s)	Particle e^+ (eV)	Energy H^+ (keV)	H2 Q_{Ps} Q_{ec} (πa_0^2)	R	He Q_{Ps} Q_{ec} (πa_0^2)	R	Kr Q_{Ps} Q_{ec} (πa_0^2)	R
1.66	7.7	14.3					1.2(0.8)	0.082
		14.4					14.6[a]	
		16.0	3.88[b]					
1.81	9.3	17.1	0.01(0.05)	0.003				
1.86	9.9	18.1					3.5(0.5)	0.29
		18.2					11.9[a]	
2.08	12.3	22.6	1.83(0.44)	0.64				
		25.0	2.88[c]					
		23.7					10.5[d]	
2.14	13.0	23.9					5.8(0.8)	0.55
2.38	16.1	29.6					6.4(0.6)	0.78
		30.0					8.24[e]	
2.39	16.2	29.8	3.23(0.10)	0.72				
		30.0	4.51[e]					
		35.0	3.69[e]					
2.64	19.8	36.4	2.93(0.37)	0.79				
2.74	21.3	39.1			0.16(0.04)	0.10		
		40.0			1.53[e]			
3.04	26.2	48.1					5.9(0.4)	1.0
		50.0					5.89[a]	
3.04	26.3	48.3	2.87(0.39)	1.4	0.37(0.06)	0.30		
		50.0	2.02[e]		1.25[f]			
3.31	31.3	57.5	2.23(0.20)	1.6	0.47(0.09)	0.48		
		60.0	1.38[e]		0.97[e]			
3.57	36.3	66.7	2.10(0.34)	2.2	0.50(0.09)	0.67		
		70.0	0.94[e]		0.75[e]			
3.81	41.3	75.8	1.38(0.24)	2.0	0.53(0.08)	0.91		
		80.0	0.69[e]		0.58[e]			
		90.0			0.45[e]		1.84[e]	
4.24	51.2	94.0			0.52(0.03)	1.2	3.8(1.2)	2.1
		90.0	0.53[e]					
4.25	51.3	94.2	1.15(0.18)	2.2				
4.29	52.3	96.0					3.8(0.6)	2.6
		100.0	0.38[e]		0.38[e]		1.47[e]	

(continued)

Table 2. Continued.

Particle Velocity (10^6 m/s)	Particle e^+ (eV)	Energy H^+ (keV)	H2 Q_{Ps} (πa_0^2)	Q_{ec}	R	He Q_{Ps} (πa_0^2)	Q_{ec}	R	Kr Q_{Ps} (πa_0^2)	Q_{ec}	R
4.45	56.3	103.4	0.96(0.20)		2.5	0.44(0.08)		1.2			
		110.0					0.30[e]				
4.63	61.0	112.0				0.414(0.027)		1.4			
		110.0		0.19[e]							
4.64	61.3	112.6	0.90(0.09)		4.7						
		120.0		0.14[e]			0.26[e]				
4.83	66.3	121.7	0.64(0.08)		4.6	0.33(0.07)		1.3			
		130.0		0.13[e]			0.22[e]				
5.01	71.3	130.9	0.57(0.07)		4.4	0.30(0.06)		1.4			
		140.0		0.07[e]			0.19[e]			0.44[e]	
51.8	76.3	140.1	0.49(0.06)		7.0	0.28(0.07)		1.5	2.3(0.06)		5.2
5.51	86.3	158.5							2.43(0.32)		5.9
		160.0								0.41[d]	
5.51	86.4	158.6	0.30(0.09)		12.8						
		168.6		0.024[b]							
		180.0					0.23[d]				
5.97	101.3	186.0	0.23(0.07)		15.0	0.2(0.025)		0.9			
		188.0		0.015[b]							
		165.0								0.41[d]	
5.97	101.3	186.0							2.48(0.21)		6.0
		200.0								0.097[g]	
6.29	112.4	206.3							2.61(0.29)		26.9
		200.0		0.0102[c]							
6.32	113.7	208.8	0.26(0.10)		25.5						
		180.0					0.23[d]				
6.33	113.8	208.9				0.23(0.04)		1.0			
6.64	125.3	230.1	0.18(0.07)		85.7						
		242.0		0.0021[c]							
7.57	163.1	299.4							1.79(0.21)		63.9
		300.0								0.0028[g]	
7.76	171.3	314.6	0.03(0.05)		52.6						
		337.0		0.00057[c]							
		363.0					0.0053[c]				

Table 2. Continued.

Particle Velocity (10^6 m/s)	Particle e^+ (eV)	Energy H^+ (keV)	H_2 Q_{Ps} (πa_0^2)	Q_{ec}	R	He Q_{Ps} (πa_0^2)	Q_{ec}	R	Kr Q_{Ps} (πa_0^2)	Q_{ec}	R
8.41	201.3	369.6				0.041	(0.023)	7.7			
8.68	214.0	393.0							1.66	(0.18)	104
		400.0		0.00018[h]						0.016[g]	
8.92	226.3	415.5	0.07	(0.08)	389						
		437.0					0.0020[c]				
9.17	239.1	439.0				0.06	(0.04)	30.0			
9.40	251.4	461.6				0.017	(0.019)	15.5			
		500.0					0.0011[c]				
		550.0								0.0097[g]	
10.28	300.2	551.2							1.42	(0.23)	146

[a] Stedeford and Hasted, Ref. 24.

[b] Stier and Barnett, Ref. 25.

[c] Barnett and Reynolds, Ref. 26.

[d] Afrosimov, Il'in, and Solov'ev, Ref. 27.

[e] DeHeer, Schutten, and Moustafa, Ref. 28.

[f] Williams and Dunbar, Ref. 29.

[g] Toburen, Nakai, and Langley, Ref. 30.

[h] Williams, Ref. 31.

THE IMMEDIATE FUTURE

After Q_{Ps} measurements with the noble gases are completed, we will use the 2.3 m spectrometer to determine total ionization cross sections (Q_{ion}) for these same gases. An important question to be answered by the Q_{ion} is whether they complement Q_{Ps} in the energy regions where it forms maxima and minima to account for a smoothly varying energy dependence of Q_{tot}. Additionally, values obtained at intermediate energies will assist in determining Q_{ex} from the TOF spectra of the 3 m spectrometer with incompletely resolved excitation and ionization peaks.

On the other hand, the low energy Q_{ion} results from the 3 m spectrometer will serve to corroborate the measurements by the 2.3 m spectrometer.

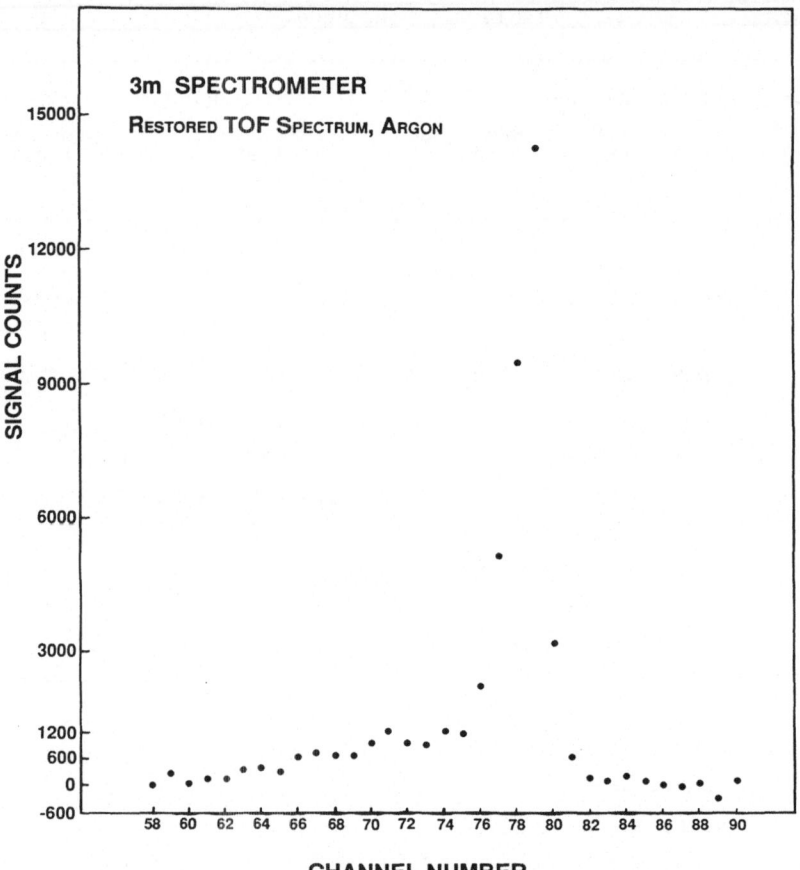

Fig. 6. Spectrum obtained with 70 eV positrons.

The principal first numbers from the 3 m instrument are to be, however, Q_{ex}, Q_{ion}, and differential elastic scattering cross sections for the noble gases.

ACKNOWLEDGMENTS

We acknowledge with pleasure and are grateful for the assistance given us by J. K. Chu, J. P. Howell, and D. M. Norman in designing, building, and repairing apparatus, in writing and executing computer programs, and in taking data; by P. K. Pendleton and B. E. Seay in conscientious data taking and careful equipment repair; by Dr. L. S. Fornari in contributing importantly to the design and construction of the spectrometers; by W. Liu and J. T. Adams in thorough library work, data taking, and computations; by Dr. M. S. Dababneh in making computations; by M. W. Lutes and D. Miller in precise execution of numerous construction projects; and by Dr. R. N. Claytor and D. W. Coyne in expert design and repair of electronic equipment.

ACKNOWLEDGEMENT

This work was supported by National Science Foundation Grant PHY-8506933.

REFERENCES

1. John M. McKinley, Positrons in Astrophysics, in: "Proceedings of the Third International Workshop on Positron (Electron) – Gas Scattering," Walter E. Kauppila, Talbert S. Stein, and Jogindra M. Wadehra, ed., World Scientific, Singapore (1986).
2. L. M. Diana, D. L. Brooks, P. G. Coleman, P. K. Pendleton, D. M. Norman, B. E. Seay, and S. C. Sharma, Total Cross Sections for Positronium Formation in Molecular Hydrogen, Krypton, and Xenon, in: Ref. 1.
3. L. M. Diana, P. G. Coleman, D. L. Brooks, P. K. Pendleton, and D. M. Norman, Positronium formation cross sections in He and H_2 at intermediate energies, Phys. Rev. A 34: 2731 (1986).
4. M. S. Dababneh, W. E. Kauppila, J. P. Downing, F. Laperriere, V. Pol, J. H. Smart, and T. S. Stein, Measurements of total scattering cross sections for low-energy positrons and electrons colliding with krypton and xenon, Phys. Rev. A 22:1872 (1980).
5. M. S. Dababneh, Y.-F. Hsieh, W. E. Kauppila, V. Pol, and T. S. Stein, Total-scattering cross-section measurements for intermediate-energy positrons and electrons colliding with Kr and Xe, Phys. Rev. A 26:1252 (1982).
6. H. S. W. Massey, J. Lawson, and D. G. Thompson, Collisions of Slow Positrons with Atoms, in: "Quantum Theory of Atoms, Molecules, and the Solid State," Per-Olov Lowdin, ed., Academic Press, New York (1966).
7. R. I. Campeanu, D. Fromme, G. Kruse, R. P. McEachran, L. A. Parcell, W. Raith, G. Sinapius, and A. D. Stauffer, Partitioning of the positron-helium total scattering cross section, submitted for publication.
8. C. E. Moore, in: "Analyses of Optical Spectra," NSRDS–NBS 34, Office of Standard Reference Data, National Bureau of Standards, Washington, listed in: "Handbook of Chemistry and Physics, 66th Ed., 1985–86," Robert C. Weast, ed., CRC Press, Boca Raton (1985).
9. L. S. Fornari, L. M. Diana, and P. G. Coleman, Positronium Formation in Collisions of Positrons with He, Ar, and H_2, Phys. Rev. Lett. 51: 2276 (1983).
10. L. M. Diana, P. G. Coleman, D. L. Brooks, and R. L. Chaplin, in preparation.
11. L. M. Diana, S. C. Sharma, L. S. Fornari, P. G. Coleman, P. K. Pendleton, D. L. Brooks, and B. E. Seay, Total positronium formation cross sections for neon from the threshold region to intermediate energies, in: "Positron Annihilation, Proceedings of the Seventh International Conference on Positron Annihilation," P. C. Jain, R. M. Singru, and K. P. Gopinathan, ed., World Scientific, Singapore (1985).

12. P. G. Coleman, J. T. Hutton, D. R. Cook and C. A. Chandler, Inelastic scattering of slow positrons by helium, neon, and argon atoms, Can. J. Phys. 60: 584 (1982).

13. W. E. Kauppila, T. S. Stein, and G. Jesion, Direct Observation of a Ramsauer-Townsend Effect in Positron-Argon Collisions, Phys. Rev. Lett. 36: 580 (1976).

14. W. E. Kauppila, T. S. Stein, J. H. Smart, M. S. Dababneh, Y. K. Ho, J. P. Downing, and V. Pol, Measurements of total scattering cross sections for intermediate-energy positrons and electrons colliding with helium, neon, and argon, Phys. Rev. A 24: 725 (1981).

15. T. S. Stein, W. E. Kauppila, V. Pol, J. H. Smart, and G. Jesion, Measurements of total scattering cross sections for low-energy positrons and electrons colliding with helium and neon atoms, Phys. Rev. A 17: 1600 (1978).

16. M. Charlton, G. Laricchia, T. C. Griffith, G. L. Wright, and G. R. Heyland, Measurements of low-energy $e^{\pm}$-Ne and e^{+}-Ar total scattering cross sections, J. Phys. B 17: 4945 (1984).

17. Sultana N. Nahar and J. M. Wadehra, Elastic scattering of positrons and electrons by argon, Phys. Rev. A 35: 2051 (1987).

18. A. D. Stauffer, personal communication.

19. Hiroshi Nakanishi and D. M. Schrader, Simple but accurate calculations on the elastic scattering of electrons and positrons from neon and argon, Phys. Rev. A 34: 1823 (1986).

20. David M. Schrader, Semiempirical polarization potential for low-energy positron-atom and positron-atomic-ion interactions. I. Theory: Hydrogen and the noble gases, Phys. Rev. A 20: 918 (1979).

21. R. P. McEachran and A. D. Stauffer, Differential Cross-Sections for Positron Noble Gas Collisions, in: "Proceedings of the Third International Workshop on Positron (Electron)-Gas Scattering," Walter E. Kauppila, Talbert S. Stein, and Jogindra M. Wadehra, ed., World Scientific, Singapore (1986).

22. R. P. McEachran and A. D. Stauffer, Electron Scattering from Neon, Phys. Lett. A 107: 397 (1985).

23. J. H. McGuire, N. C. Sil, and N. C. Deb, Capture of atomic electrons by high-velocity positrons, Phys. Rev. A 34: 685 (1986).

24. J. B. H. Stedeford and J. B. Hasted, Further Investigations of Charge Exchange and Electron Detachment. I. Ion Energies 3 to 40 keV, by J. B. H. Stedeford. II. Ion Energies 100 to 4,000 keV, by J. B. Hasted, Proc. Roy. Soc. A 227: 466 (1955).

25. P. M. Stier and C. F. Barnett, Charge Exchange Cross Sections of Hydrogen Ions in Gases, Phys. Rev. 103: 896 (1956).

26. C. F. Barnett and H. K. Reynolds, Charge Exchange Cross Sections of Hydrogen Particles in Gases at High Energies, Phys. Rev. 109: 355 (1958).

27. V. V. Afrosimov, R. N. Il'in, and E. S. Solov'ev, Electron Capture by Protons in Inert Gases, Soviet Physics J.T.P. 5: 661 (1960).

28. F. J. deHeer, J. Schutten, and H. Moustafa, Ionization and Electron Capture Cross Sections for Protons Incident on Noble and Diatomic Gases between 10 and 140 keV, Physica 32: 1766 (1966).

29. J. F. Williams and D. N. F. Dunbar, Charge Exchange and Dissociation Cross Sections for H_1^+, H_2^+, and H_3^+ Ions of 2-to 50-keV Energy Incident Upon Hydrogen and the Inert Gases, Phys. Rev. 149: 62 (1966).

30. L. H. Toburen, M. Y. Nakai, and R. A. Langley, Measurement of High-Energy Charge-Transfer Cross Sections for Incident Protons and Atomic Hydrogen in Various Gases, Phys. Rev. 171: 114 (1968).

31. J. F. Williams, Measurement of Charge-Transfer Cross Sections for 0.25-to 2.5-MeV Protons and Hydrogen Atoms Incident upon Hydrogen and Helium Gases, Phys. Rev. 157: 97 (1967).

32. Osamu Sueoka, Retarding Potential-TOF Method for e^{+}-Gas Collisions, J. Phys. Soc. Jpn. 51: 2381 (1982).

33. P. G. Coleman, T. C. Griffith, and G. R. Heyland, The Analysis of Data Obtained with Time to Amplitude Converter and Multichannel Analyser Systems, _Appl. Phys._ 5: 223 (1974).
34. P. G. Coleman, The distortion of TAC–MCA spectra by the measuring process, _J. Phys._ E 12: 590 (1979).

RECENT DEVELOPMENTS IN THE THEORY OF FAST POSITRON-ATOM COLLISIONS

C.J. Joachain

Physique Théorique, Université Libre de Bruxelles, Belgium
and Institut de Physique Corpusculaire, Université de
Louvain, Louvain-La-Neuve, Belgium

1. INTRODUCTION

The field of positron-atom collisions has attracted an increasing
amount of interest during the last few years. The present survey concen-
trates on recent theoretical work dealing with positron-atom collisions,
for incident positron energies well above the ionization energy of the
target, but non-relativistic. It is an up-dated version of the review I
gave at the previous conference (Joachain, 1984), which contains references
to earlier work. Atomic units (a.u.) will be used.

2. ELASTIC SCATTERING AND TOTAL CROSS SECTIONS

Let us start by analyzing elastic scattering. At sufficiently large
values of the incident positron wave number k, the first Born approximation
- which yields identical cross sections for incident electrons and
positrons - is accurate. We are interested here in the "intermediate
energy" region, such that k is decreased while remaining somewhat larger
than unity. In this region higher order methods must be used to obtain
reliable scattering amplitudes and to investigate the differences between
electron and positron scattering. We shall briefly describe the most
sophisticated methods which have been proposed.

We begin by considering the Eikonal-Born series (EBS) theory of
Byron and Joachain (1973, 1977a). It is based on a detailed analysis of
the leading behaviour of the terms of the Born series, $\bar{f}_{Bn}$, in the large
k limit, and on a comparison of these terms with the corresponding terms
$\bar{f}_{Gn}$ of the Glauber series, obtained by expanding the Glauber amplitude
f_G in powers of the projectile-target interaction V. This analysis leads
to the EBS amplitude

$$f_{EBS} = \bar{f}_{B1} + \bar{f}_{B2} + \bar{f}_{G3} \qquad (1)$$

which is consistent through order k^{-2}. We remark that the EBS' amplitude
(Byron and Joachain, 1975)

$$f_{EBS'} = f_G - \bar{f}_{G2} + \bar{f}_{B2} \qquad (2)$$

- also called the modified Glauber amplitude - is also consistent through
order k^{-2}.

The EBS method is very successful for light atoms, at high incident energies and small momentum transfers, where perturbation theory converges rapidly. However, when the Born series is more slowly convergent, improvements are required. These can be achieved by using methods which include terms from all orders of perturbation theory in order to ensure unitarity. For example, Byron, Joachain and Potvliege (1981, 1982, 1985) have proposed to unitarize the EBS method in the following way. First, the potential scattering amplitude of Wallace (1973), which corrects in a systematic way the eikonal amplitude, is generalized to the case of $e^{\pm}$-atom scattering. The resulting multi-particle amplitude f_W includes terms from all orders of perturbation theory, but it still suffers from important small-angle deficiencies in second order due to the fact that f_W is a zero-excitation energy approximation. The amplitude f_W is therefore corrected in second order to yield the Unitarized Eikonal Born series (UEBS) amplitude

$$f_{UEBS} = f_W - \overline{f}_{W2} + \overline{f}_{B2} \tag{3}$$

where $\overline{f}_{W2}$ is the second order term in the expansion of f_W in powers of the interaction potential V. At small values of the momentum transfer Δ the UEBS amplitude (3), expanded through order k^{-2}, reduces to the EBS amplitude (1). At large Δ, f_{UEBS} differs negligibly from the multi-particle Wallace amplitude f_W, which contains the two leading terms (in powers of k^{-1}) of each order of perturbation theory summed to all orders. Thus the UEBS amplitude (3) combines the advantages of the EBS method at small momentum transfers with those of the many-particle Wallace amplitude f_W at large Δ.

Another way of unitarizing the EBS method is to transform it into an optical model approximation (Byron and Joachain, 1974, 1977b, 1981, Joachain et al., 1977). The basic idea is to convert the lowest order terms of perturbation theory, calculated by using the EBS approach, into an optical potential V_{opt}. For e^+-atom collisions, the optical potential is given by

$$V_{opt} = V^{(1)} + V^{(2)} + V^{(3)} + \ldots \tag{4}$$

where $V^{(1)} = \langle i|V|i \rangle$ is the static potential, $|i\rangle$ being the initial target state and V the positron-atom interaction potential. The second order part $V^{(2)}$ of the optical potential reads

$$V^{(2)} = \sum_{n \neq 0} \frac{\langle i|V|n \rangle \langle n|V|i \rangle}{k^2/2 - K - (w_n - w_i) + i\varepsilon} \quad , \quad \varepsilon \to 0^+ \tag{5}$$

where K is the kinetic energy operator of the projectile, $|n\rangle$ an intermediate target state of energy w_n, and w_i the energy of the initial state $|i\rangle$. This complicated (non-local, complex and energy-dependent) expression of $V^{(2)}$ can be approximated at sufficiently high energies by the local potential

$$V^{(2)} = V_{pol} + i V_{abs} \tag{6}$$

where V_{pol} and V_{abs} are real, central and energy-dependent. The term V_{pol} (which falls off like r^{-4} at large r) accounts for dynamic polarization effects and i V_{abs} for absorption effects due to loss of flux from the incident channel. The leading contribution of the third order potential $V^{(3)}$ has also been obtained for $e^{\pm}$-H scattering (Byron and Joachain, 1981). The full optical potential V_{opt} given by equation (4) is then treated in a unitary manner by using the method of partial waves. This retains all the advantages of the EBS approach at small momentum transfers, but includes approximations to higher orders of perturbation theory, which are important at large momentum transfers, where the (singular) positron-nucleus Coulomb potential governs the scattering.

Closely related to the optical potential approach is the second order potential (SOP) method developed by Bransden et al. (see for example Winters et al., 1974; Scott and Bransden, 1981), in which the close-coupling equations are modified by the addition of a potential matrix accounting approximately for the states omitted from the set of M states retained in the close-coupling expansion. To lowest order in the projectile-atom   interaction V this potential matrix reads

$$V_{fi}^{(2)} = \sum_{n \geqslant M+1} \frac{\langle f|V|n \rangle \langle n|V|i \rangle}{k^2/2 - K - (w_n - w_i) + i\varepsilon} \quad , \quad \varepsilon \to 0^+.$$
                                                                                    (7)

This is the expected generalization of the second order part $V^{(2)}$ of the optical potential, given by equation (5) for elastic scattering. As in the Byron-Joachain optical model approach, approximations must be made to handle the non-local, complex, energy-dependent potentials given by equation (7). For example, Mukherjee and Sural (1982) have investigated e^+-H and e^+-He elastic scattering in the single channel version of the SOP method, using the closure approximation to evaluate the second order optical potential. The same model has also been investigated recently for e^+-H elastic scattering by Makowski et al. (1986b), using the method of continued fractions (Horacek and Sasakawa, 1983, 1984, Makowski et al. 1985, 1986a, Znojil, 1984) to obtain the (complex) phase shifts. A multi-channel version of the SOP method (referred to as coupled-channel optical model, CCOM) has been investigated for e^+-H collisions by Bransden, McCarthy and Stelbovics (1985). In their calculations an equivalent local approximation to $V_{fi}^{(2)}$ is made, and the resulting close-coupling problem is then treated by solving coupled Lippmann-Schwinger integral equations in momentum space.

Finally, we mention distorted-wave treatments, which are characterised by the fact that the interaction is broken in two parts, one which is treated exactly and the other which is handled by perturbation theory. An example is given by the distorted-wave second Born approximation (DWSBA) of Dewangan and Walters (1977).

Let us now discuss some of the results which have been obtained, beginning with the simplest case : positron-atomic hydrogen elastic scattering. In Table 1 the differential cross sections calculated by using the UEBS, third order optical model (OM), second order optical model with the continued fraction approach (SOPCF) and coupled channel optical model (CCOM) methods are compared at an incident positron energy of 200 eV. The agreement between the predictions of these four sets of sophisticated calculations is seen to be very good. It should be noted that the SOPCF calculations of Makowski et al.(1986b) confirm the UEBS and OM results and disagree with the earlier one-channel SOP calculations of Mukherjee and Sural (1982). Also shown in Table 1 are the values of the differential cross sections for incident electrons of the same energy (200 eV), obtained by using the third order optical model theory. The differences between the positron and electron results are seen to be significant, the positron values being always smaller than the corresponding electron ones. In particular, the strong forward peak present in the case of electron scattering is significantly reduced for positron scattering. This is mainly due to the fact that for positrons the static potential $V^{(1)}$ is opposite in sign to the polarization potential V_{pol}.

Table 1. Differential cross sections (in a.u.) for e^+-H(1s) elastic
scattering at an incident energy of 200 eV. UEBS, OM, SOPCF and
CCOM refer to methods discussed in the text. Also shown in the
last column are the third order optical model results for incident
electrons. The numbers in parentheses indicate powers of ten.

θ (deg)	UEBS (1)	OM (2)	SOPCF (3)	CCOM (4)	OM(e^-) (2)
0	1.6	1.6	1.6	1.85	5.7
10	7.1(-1)	7.0(-1)	6.8(-1)	6.3(-1)	1.1
20	2.8(-1)	2.8(-1)	2.7(-1)	2.6(-1)	3.9(-1)
30	1.1(-1)	1.1(-1)	1.0(-1)	1.0(-1)	1.5(-1)
40	4.5(-2)	4.6(-2)	4.4(-1)	4.5(-2)	6.2(-2)
60	1.2(-2)	1.2(-2)	1.2(-2)	1.1(-2)	1.6(-2)
80	4.5(-3)	4.7(-3)	4.6(-3)	4.3(-3)	6.3(-3)
100	2.3(-3)	2.4(-3)	2.3(-3)	2.2(-3)	3.2(-3)
120	1.4(-3)	1.5(-3)	1.5(-3)	1.4(-3)	2.0(-3)
140	1.0(-3)	1.1(-3)	1.1(-3)	1.0(-3)	1.4(-3)
160	8.6(-4)	9.0(-4)	8.8(-4)	8.2(-4)	1.2(-3)
180	8.1(-4)	8.5(-4)	8.3(-4)	-	1.1(-3)

(1) Byron, Joachain and Potvliege, 1985
(2) Byron and Joachain, 1981
(3) Makowski, Raczynski and Staszewska, 1986b
(4) Bransden, McCarthy and Stelbovics, 1985

The angular distributions for elastic scattering of fast positrons
by noble gases (helium, neon and argon) have been discussed by Byron and
Joachain (1977a,b) and Joachain (1978). As a first example, we see in
Fig.1 the theoretical small – angle positron – helium differential cross
section at an incident energy of 500 eV, as obtained from a second order
optical model calculation (Byron and Joachain, 1977b). Also shown for
comparison are the theoretical values for electron impact, calculated from
various elaborate theories, which are seen to be in excellent agreement
with each other and with the absolute electron impact experimental data.
This feature gives additional confidence in the accuracy of the theoretical
predictions for positron impact.

As a second example, I shall now discuss the angular distribution for
positron-argon elastic scattering at intermediate energies, following
the analysis of Joachain and Potvliege (1987). This process is of particular interest because Hyder et al. (1986) have recently been able to
measure relative differential cross sections for $e^\pm$-Ar collisions. In
Fig.2 their data are compared with the second order optical model
predictions (Joachain et al., 1977, Joachain, 1978) at an incident energy
of 300 eV. The agreement between theory and experiment is seen to be
excellent.

The experiments of Hyder et al. constitute an important first step
in providing severe tests for the theoretical models of positron-atom
collisions. However, since these measurements are relative and do not
cover the angular region $\theta \lesssim 30°$, the test they provide is not yet stringent
enough to make an unambiguous selection among conflicting theoretical
models. For example, Hyder et al. have shown that their data at 100 eV and
200 eV are compatible with the polarized orbital calculations of McEachran
and Stauffer (1986) and with a closely related "static + polarization"
model potential calculation carried out by Nahar and Wadhera (1987). On
the other hand, these data also agree with the second order optical model
calculations of Joachain et al. (1977) at these energies, as illustrated
in Fig.3 for an incident energy of 100 eV. Now, as seen from Fig.3, the
full optical model calculations (solid line) which include the effects of

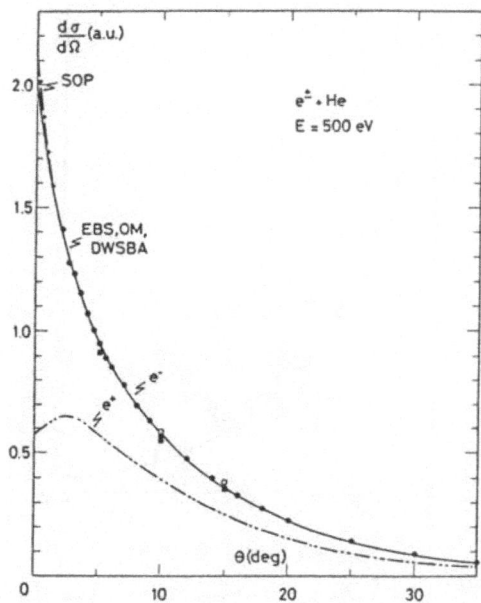

Fig. 1. Differential cross section (in a.u.) for the elastic scattering
of positrons and electrons by helium at an incident energy of 500
eV. Theoretical curves : —..— Second order optical model calculation for positrons (Byron and Joachain, 1977b), —— : EBS (Byron
and Joachain, 1973); SOP : second order potential (Winters et al.,
1974), OM : second order optical model (Byron and Joachain, 1977b)
and DWSBA : distorted wave second Born approximation (Dewangan and
Walters) calculations for electrons. Experimental data : ○ Oda et
al. (1972); ● Bromberg (1974); ■ Jansen et al. (1976).

the static, the polarization and the absorption potentials give markedly
different results from those obtained either by using only the static
potential (dotted line) or a "static + polarization" potential (dashed
line). Unfortunately, since these differences concern the small angle
region and the absolute magnitude of the cross sections, the experiments
of Hyder et al. can be accounted for either by the full optical model
calculations or by the "static + polarization" calculations, which omit
the absorption potential. On theoretical grounds, however, it is obvious
that the neglect of absorption effects (i.e. of the loss of flux due to
non-elastic scattering) is not legitimate at the energies considered here.
This strongly suggests that the full optical model calculations, which
properly account for absorption, should be preferred.

Let us now consider total cross sections. In Table 2 are given some
recent results concerning total (integrated) cross sections for elastic
e^+-H(1s) scattering. The agreement between the four higher order methods
is very good, except at 100 eV where the CCOM result is quite lower than
the others. We note that as the energy increases the cross sections slowly
tend (from below) to the first Born values.

The total (complete) cross sections σ_{tot}^+ for e^+-H(1s) scattering,
obtained from the forward elastic amplitude $f_{el}^+(\theta=0)$ by using the optical
theorem, $\sigma_{tot}^+=(4\pi/k)\,\mathrm{Im}\,f_{el}^+(\theta=0)$, are given in Table 3. Also shown in
Table 3 are the UEBS values for σ_{tot}^-, the corresponding total (complete)
electron cross section. Now, in terms of the elastic direct e^--H(1s)
amplitude f_{el}^- and exchange amplitude g_{el} one has $\sigma_{tot}^-=(4\pi/k)\,\mathrm{Im}\,[\,f_{el}^-(\theta=0)$
$-(1/2)\,g_{el}(\theta=0)]$. Since in the UEBS method the two quantities

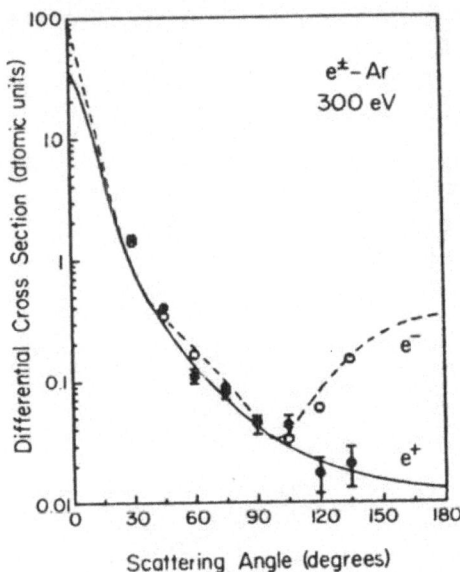

Fig. 2. Differential cross section (in a.u.) for the elastic scattering
of positrons and electrons by argon at an incident energy of 300
eV. The theoretical curves correspond to the second order optical
model calculations of Joachain et al. (1977) for positrons (———)
and for electrons (---). The experimental data are those of Hyder
et al. (1986) for positrons (●) and for electrons (○) normalized
at 90° to the theoretical results. From Hyder et al. (1986).

$Im\, f_{el}^{-}(\theta=0)$ and $Im\, f_{el}^{+}(\theta=0)$ are equal for $e^{\pm}$-H scattering, the difference
between σ_{tot}^{+} and σ_{tot}^{-} is entirely due to exchange effects and is seen
from Table 3 to remain quite small even at relatively low energies. We
also remark that since $\sigma_{tot}^{+} \simeq \sigma_{tot}^{-}$ while the total (integrated) elastic
cross sections are such that $\sigma_{el}^{+} \lesssim \sigma_{el}^{-}$ it follows that the non-elastic
cross sections must be such that $\sigma_{ne}^{+} > \sigma_{ne}^{-}$ in the intermediate energy
region considered here.

The total cross sections for fast positrons colliding with helium,
neon and argon have been discussed by Byron and Joachain (1977a) and
Stein and Kauppila (1982). Recent calculations of total cross sections
have been performed for e^{+}-He collisions by Khare and Lata (1985) and for
e^{+}-Ar scattering by Khare, Kumar and Lata (1986), using less sophisticated
methods than the optical model calculations of Byron and Joachain (1974)
and Joachain et al. (1977). We also remark that Gien (1986) has
calculated total cross sections for $e^{\pm}$-alkali atom (Li, Na, K) collisions,
using the EBS' approximation of Byron and Joachain (1975) and a model
potential approach to include core-interaction effects. His results at
40 and 50 eV for e^{+}-K collisions are in very poor agreement with the
experimental data of Stein et al. (1985), but these energies are probably
too low to draw meaningful conclusions. Finally we note that Khare, Kumar
and Vijayshri (1985) have calculated differential and total (integrated)
cross sections for the elastic scattering of positrons by Ca atoms, using
a model potential which does not take into account absorption effects.

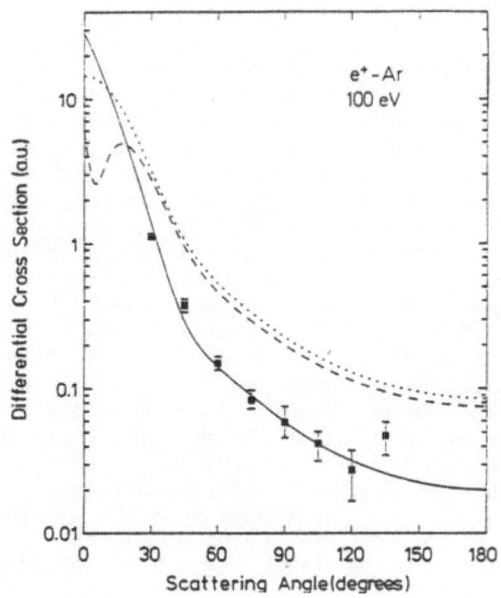

Fig. 3. Differential cross section (in a.u.) for the elastic scattering
of positrons by argon at an incident energy of 100 eV. Theoretical
curves : ___ : full second order optical model calculations
(Joachain et al., 1977); ... static potential only; --- static +
polarization potential; ■ Experimental data of Hyder et al (1986),
normalized to the optical model calculations at $\theta=90°$. From
Joachain and Potvliege (1987).

Table 2. Total (integrated) elastic cross sections (in units of πa_0^2) for
e^+-H(1s) scattering. FBA corresponds to the first Born approximation and UEBS, OM, SOPCF and CCOM to methods discussed in the
text. The references (1)-(4) are the same as in Table 1

E (eV)	FBA	UEBS (1)	OM (2)	SOPCF (3)	CCOM (4)
100	2.99(-1)	2.20(-1)	2.23(-1)	2.12(-1)	1.66(-1)
200	1.54(-1)	1.31(-1)	1.31(-1)	1.27(-1)	1.21(-1)
300	1.04(-1)	9.29(-1)	9.29(-1)	9.10(-2)	-
400	7.83(-2)	7.19(-1)	7.19(-1)	-	-

Table 3. Total (complete) cross sections (in units of πa_0^2) for positron
and electron-atomic hydrogen collisions. UEBS, OM, SOPCF and CCOM
refer to methods discussed in the text. The last column gives the
UEBS results for incident electrons. The references (1)-(4) are
the same as in Table 1.

E (eV)	UEBS (1)	OM (2)	SOPCF (3)	CCOM (4)	UEBS(e^-) (1)
100	2.18	2.15	2.13	2.01	2.24
200	1.34	1.32	1.27	1.28	1.35
300	9.78(-1)	9.64(-1)	9.23(-1)	-	9.84(-1)
400	7.77(-1)	7.70(-1)	-	-	7.80(-1)

3. EXCITATION AND IONIZATION

Let us now consider excitation collisions. In this case the first Born amplitude $\bar{f}_{B1}$ falls off rapidly at large momentum transfers Δ, so that large angle scattering is governed by the second Born term $\bar{f}_{B2}$ at high energies. This may be understood as follows. Large momentum transfer collisions can only take place if the projectile positron (or electron) collides with the much heavier nucleus. Now, for inelastic scattering the orthogonality of the target initial and final states $\langle f|i\rangle = 0$, removes the projectile-nucleus interaction term from the first Born amplitude $\bar{f}_{B1}$. As a result, the first Born approximation differential cross section for high-energy large angle inelastic scattering is orders of magnitude too small. The fact that $\bar{f}_{B2}$ falls off more slowly than $\bar{f}_{B1}$ for inelastic collisions at large Δ is due to the possibility of off-shell elastic scattering in the intermediate states $|i\rangle$ and $|f\rangle$ - the initial and final target states of the inelastic transition - where the projectile can experience the Coulomb potential of the nucleus. It is worth noting that since the values $\Delta \lesssim 1$ correspond to angles $\theta \lesssim k^{-1}$, the angular region in which the first Born approximation is valid shrinks with increasing energy. However, because the dominant contribution to the integrated cross section comes from the region $\Delta \lesssim 1$, the first Born values for integrated inelastic cross sections are reliable at sufficiently high energies.

In Table 4 are given some recent results concerning two basic excitation processes : the excitation of the 2s and 2p states of atomic

Table 4. Differential cross sections (in a.u.) for the excitation of the 2s and 2p states of atomic hydrogen by positron impact, at an incident energy of 200 eV, as obtained from the UEBS and CCOM methods. Also shown are the corresponding UEBS results for incident electrons.

	1s-2s transition			1s-2p transition		
θ (deg)	UEBS (1)	CCOM (2)	UEBS(e^-) (1)	UEBS (1)	CCOM (2)	UEBS(e^-) (1)
0	2.8	2.5	1.2	210	220	210
10	2.7(-1)	3.6(-1)	2.5(-1)	1.5	1.7	1.3
20	2.2(-2)	2.9(-2)	2.0(-2)	5.1(-2)	5.0(-2)	2.8(-2)
30	3.2(-3)	5.1(-3)	3.2(-3)	6.1(-3)	5.1(-3)	2.2(-3)
40	1.1(-3)	1.9(-3)	1.3(-3)	1.8(-3)	1.4(-3)	6.7(-4)
60	2.9(-4)	5.1(-4)	3.4(-4)	4.1(-4)	3.3(-4)	1.8(-4)
80	1.1(-4)	1.8(-4)	1.2(-4)	1.7(-4)	1.3(-4)	9.3(-5)
100	5.5(-5)	8.0(-5)	6.0(-5)	1.0(-4)	6.3(-5)	6.4(-5)
140	2.4(-5)	3.2(-5)	2.6(-5)	5.7(-5)	2.8(-5)	4.3(-5)
180	1.9(-5)	-	2.0(-5)	4.9(-5)	-	3.8(-5)

(1) Byron, Joachain and Potvliege, 1985
(2) Bransden, McCarthy and Stelbovics, 1985.

hydrogen by positron impact. Although the UEBS and CCOM angular distributions are qualitatively similar, the detailed agreement between the two sets of results is relatively poor outside the small angle region. Also given in Table 4 are the corresponding UEBS results for electrons, which are in good agreement with the optical model results of Bransden et al. (1982). We note that for both the 1s-2s and 1s-2p transitions the small angle positron differential cross sections are higher than the corresponding electron ones, so that the total (integrated) positron cross sections for these excitation processes are bigger for positron than for electron impact, in agreement with the discussion of Section 2.

Differential and integrated cross sections for five transitions
(1s-ns, n = 2,3,4 and 2s-ns, n = 3,4) of atomic hydrogen by positron
impact have been calculated by Saxena, Gupta and Mathur (1984), using
various distorted wave treatments. Khan and Ghosh (1986) have obtained the
n = 2 excitation cross sections of He^+ by using the close-coupling method.
Kumar, Srivastava and Tripathi (1985) have performed distorted wave calculations for the 1^1S-2^1S and 1^1S-2^1P transitions in helium at incident
positron energies of 100 eV and 200 eV, and have compared their results
with previous theoretical calculations (Willis et al, 1981, Parcell, Mc
Eachran and Stauffer, 1983 and Madison and Winters, 1983). The agreement
between the various theoretical angular distributions is satisfactory at
200 eV, but poor at the lower energy of 100 eV. We also note that Saxena
and Mathur (1986) have studied the angular correlation parameters for the
2s-2p transition in lithium by positron impact. Their results show a
variation similar to that found earlier in positron-helium collisions
(Willis et al., 1981) and positron-atomic hydrogen collisions (Fargher and
Roberts, 1984).

Let us now turn to ionizing collisions. Most of the detailed knowledge
concerning ionization reactions has been obtained during the recent years
by analyzing $(e^-,2e^-)$ coïncidence experiments (for a review, see Joachain
and Piraux, 1986). It is clear that in the same way (e^+,e^+e^-) experiments
would yield very useful information about ionization processes. As an
illustration, the triple differential cross sections (TDCS) for $e^\pm$-H(1s)
ionization were compared in my previous review (Joachain, 1984) for the
case of a coplanar asymmetric kinematics, using a second Born treatment.
Another interesting situation has been analyzed by Mandal, Ray and Sil
(1986), who have used the Faddeev formalism to study two competing processes
involved in the positron impact ionization of atomic hydrogen, namely i)
direct ionization and ii) electron capture (or positronium formation) in
the continuum. They showed that the second process is responsible for the
appearance of a cusp in the predicted differential cross section when the
positron and ejected electron have low relative momentum in the final
channel. The presence of this cusp is to be expected since, as pointed
out by Brauner and Briggs (1986), the singular structure in the differential
cross section for electron ejection observed for heavy-ion impact ionization
should also be present for positron impact ionization.

Unfortunately, differential measurements concerning positron impact
ionization are not yet available, so that comparison between theory and
experiment must still be made at the level of total (integrated) ionization
cross sections, σ^+_{ion}. In Fig. 4 are shown the values of σ^+_{ion} recently
measured by Fromme et al.(1986) for e^+-He ionization. These values differ
significantly from previous measurements of Sueoka (1982) and Diana et al.
(1985). It is seen from Fig.4 that the data of Fromme et al. are in very
good agreement with the calculations of Basu, Mazumdar and Ghosh (1985),
McGuire (1986), Peach and McDowell (1986) and Campeanu, McEachran and
Stauffer (1986), performed by using quite different theoretical models.
Thus these calculations give support to the data of Fromme et al., but at
the same time we see once again that integrated cross section measurements
are not a stringent test of the theory. Also shown in Fig. 4 is the
corresponding electron impact ionization cross section of helium, measured
by Montague, Harrison and Smith (1984); it is seen that the electron impact
ionization cross section is always lower than the corresponding positron
impact cross section.

4. POSITRONIUM FORMATION

Positronium (Ps) formation is a rearrangement process, similar to
charge exchange, in which an incident positron captures a bound electron
to form the bound system (e^+e^-). Several theoretical estimates of Ps

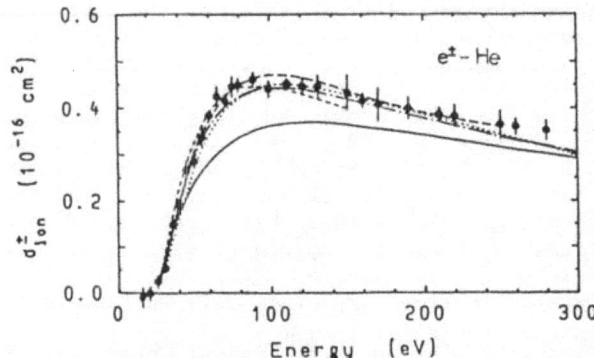

Fig. 4. The total cross section for positron impact ionization of helium.
Theoretical curves : --- Basu et al.(1985), ... McGuire (1986),
-·- Peach and McDowell (1986), — — — Campeanu et al. (1987), •,
Experimental data of Fromme et al.(1986). The solid curve (——)
shows the electron impact ionization cross section of Montague
et al.(1984). From Fromme et al.(1986).

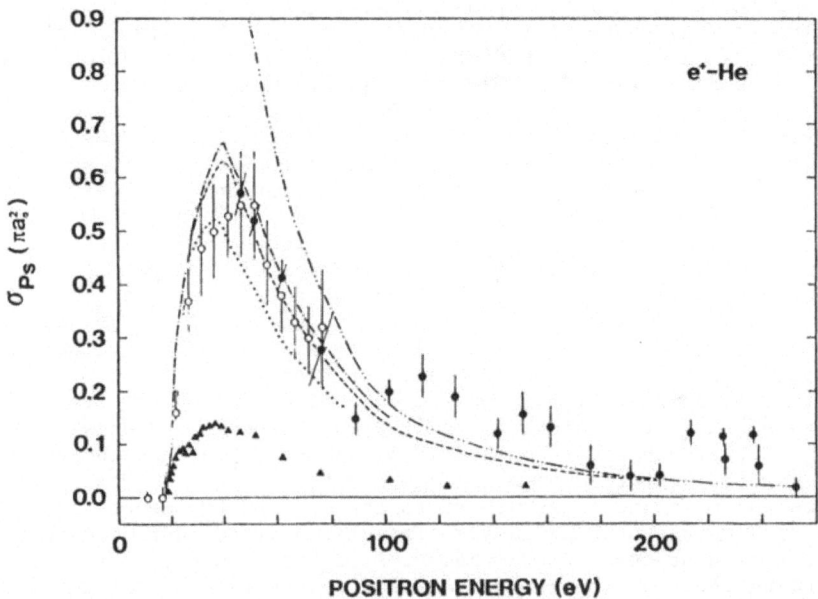

Fig. 5. The total cross section for positronium formation in helium.
Theoretical curves : -·- first Born approximation (Mandal et al.,
1980); -·- distorted wave approximation x 1.20, assuming an n^{-3}
rule for Ps formation in excited states (Mandal et al., 1979);
--- first order exchange approximation (Mandal et al., 1980);
... distorted wave polarised orbital approximation (Khan et al.,
1985). Experimental data : ▲, Charlton et al.(1983), o Fornani
et al.(1983), •, Diana et al. (1986).

formation from atomic hydrogen at intermediate and high energies have been discussed in my previous review (Joachain, 1984). In the case of Ps formation from helium, total cross sections have been measured by Charlton et al.(1983), Fornari et al. (1983) and more recently by Diana et al. (1986) and Fromme et al. (1986). As seen from Fig. 5, there is an important disagreement between the data of Charlton et al. and those obtained at the University of Texas (Fornari et al., Diana et al.). The recent results of Fromme et al. are close to those of the University of Texas group, except that they do not exhibit the structure found by Diana et al. on the high-energy side of the maximum. We also see from Fig. 5 that the first Born approximation is in reasonable agreement with the data of Diana et al. (1986) down to an energy of about 80 eV, and that the first-order exchange approximation of Mandal, Guha and Sil (1980) improves this agreement on the low-energy side. On the other hand, Fromme et al. (1986) find that their data exceed all theoretical results in the high-energy region, and that this discrepancy is unlikely to be due to neglect or incomplete consideration of positronium formation in excited states. More work is clearly required in order to understand the reasons for these differences.

Finally, we note that recently Choudhury, Mukherjee and Sural (1986) have studied the formation of positronium from H$^-$, and that Mazumdar and Ghosh (1986) have obtained Ps formation cross sections in e$^+$-Li scattering by using a distorted wave approximation.

REFERENCES

Basu M, Mazumdar PS and Ghosh AS 1985 J. Phys. B : At. Mol. Phys. 18 369
Bransden BH, Scott T, Shingal R and Raychoudhyry RK 1982 J. Phys. B : At. Mol. Phys. 15 4605
Bransden BH, McCarthy IE and Stelbovics AT 1985 J. Phys. B : At. Mol. Phys. 18 823
Brauner M and Briggs JS 1986 J. Phys. B : At. Mol. Phys. 19 L325
Bromberg JP 1974 J. Chem. Phys. 61 963
Byron FW Jr and Joachain CJ 1973 Phys. Rev. A 8 1267
_____ 1974 Phys. Lett. 49A 306
_____ 1975 J. Phys. B : Atom. Mol. Phys. 8 L284
_____ 1977 a Phys. Rep. 34 233
_____ 1977 b Phys. Rev. A15 128
_____ 1981 J. Phys. B : Atom. Mol. Phys. 14 2429
Byron FW Jr, Joachain CJ and Potvliege R 1981 J. Phys. B : At. Mol. Phys. 14 L609
_____ 1982 J. Phys. B : At. Mol. Phys. 15 3915
_____ 1985 J. Phys. B : At. Mol. Phys. 18 1637
Campeanu RI, McEachran RP and Stauffer AD 1987 J. Phys. B : At. Mol. Phys. (to be published)
Charlton M, Clarck G, Griffith TC and Heyland GR 1983 J. Phys. B : At. Mol. Phys. 16 L465
Choudhury KB, Mukherjee A and Sural DP 1986 Phys. Rev. A 33 2358
Dewangan DP and Walters HRJ 1977 J. Phys. B : Atom. Mol. Phys. 10 637
Diana LM, Fornari LS, Sharma SC, Pendleton PK and Coleman PG 1985 in Positron Annihilation, edited by Jain PC, Singru RM and Gopinathan KP (World Scientific, Singapore) p.342
Diana LM, Coleman PG, Brooks DL, Pendleton PK and Norman DM 1986 Phys. Rev. A 34 2731
Fargher HE and Roberts MJ 1984 J. Phys. B : At. Mol. Phys. 17 L587
Fornari LS, Diana LM and Coleman PG 1983 Phys. Rev. Lett. 51 2276
Fromme D, Kruse G, Raith W and Sinapius G 1986 Phys. Rev. Lett. 57 3031
Gien TT 1987 Phys. Rev. A 35 2026
Horacek J and Sasakawa T 1983 Phys. Rev. A 28 2151
_____ 1984 Phys. Rev. A 30 2274

Hyder GMA, Dababneh MS, Hsieh YF; Kauppila WE, Kwan CK, Mahdavi-Hezaveh M and Stein TS 1986 Phys. Rev. Lett. 57 2252

Jansen RHJ, de Heer FJ, Luyken HJ, Van Wingerden B and Blaauw HJ 1976 J. Phys. B : Atom. Mol. Phys. 9 185

Joachain CJ 1978 in Electronic and Atomic Collisions, edited by Watel G (North-Holland, Amsterdam) p.71

———— 1984 in Positron Scattering in Gases, edited by Humberston JW and McDowell MRC (Plenum, New York) p.39

Joachain CJ, Vanderpoorten R, Winters KH and Byron FW Jr 1977 J. Phys. B : Atom. Molec. Phys. 10 227

Joachain CJ and Piraux B 1986 Comments in At. and Mol. Phys. 17 261

Joachain CJ and Potvliege RM 1987 Phys. Rev. (to be published)

Khan P and Ghosh AS 1986 Phys. Rev. A 33 729

Khan P, Mazumdar PS and Ghosh AS 1985 Phys. Rev. A 31 1405

Khare SP and Lata K 1985 J. Phys. B : Atom. Mol. Phys. 14 2941

Khare SP, Kumar A and Lata K 1986 Phys. Rev. A 33 2795

Khare SP, Kumar A and Vijayshri 1985 J. Phys. B : Atom. Mol. Phys. 18 1827

Kumar M, Srivastava R and Tripathi AN 1985 J. Phys. B : Atom. Mol. Phys. 18 4169

McEachran PR and Stauffer AD 1986 in Positron (Electron)-Gas Scattering, edited by Kauppila WE, Stein TS and Wadhera JM (World Scientific, Singapore)

McGuire JH 1986, quoted in Fromme et al. (1986)

Madison DH and Winters KH 1983 J. Phys. B : At. Mol. Phys. 16 4437

Makowski AJ, Raczynski and Staszewska 1985 Chem. Phys. Lett. 114 325

———— 1986 a Phys. Rev. A 33 733

———— 1986 b J. Phys. B : Atom. Mol. Phys. 19 3367

Mandal P, Guha S and Sil NC 1979 J. Phys. B : Atom. Mol. Phys. 12 2913

———— 1980 Phys. Rev. A 22 2623

Mandal P, Ray K and Sil NC 1986 Phys. Rev. A 33 756

Mazumdar PS and Ghosh AS 1986 Phys. Rev. A 34 4433

Montague RG, Harrison MFA and Smith ACH 1984 J. Phys. B : Atom. Mol. Phys. 17 3295

Mukherjee A and Sural DP 1982 J. Phys. B : Atom. Mol. Phys. 15 1121

Nahar SN and Wadhera JM 1987 Phys. Rev. A 35 2051

Oda N, Nishimura F and Tahira S 1972 J. Phys. Soc. Japan 33 462

Parcell LA, McEachran RP and Stauffer AD 1983 J. Phys. B : Atom. Mol. Phys. 16 4249

Peach G and McDowell MRC 1986 quoted in Fromme et al. (1986)

Saxena S, Gupta GP and Mathur KC 1984 J. Phys. B : Atom. Mol. Phys. 18 3743

Saxena S and Mathur KC 1986 J. Phys. B : Atom. Mol. Phys. 19 3181

Scott T and Bransden BH 1981 J. Phys. B : Atom. Mol. Phys. 14 2277

Stein TS and Kauppila WE 1982 Adv. Atom. Mol. Phys. 18 53

Stein TS, Gomez RD, Hsieh YE, Kauppila WE, Kwan CK and Wan YJ 1985 Phys. Rev. Lett. 55 488

Sueoka O 1982 J. Phys. Soc. Japan 51 3757

Wallace SJ 1973 Ann. Phys. (NY) 78 190

Willis SL, Hata J, McDowell MRC, Joachain CJ and Byron FW Jr 1981 J. Phys. B : Atom. Mol. Phys. 14 2687

Winters KH, Clark CD, Bransden BH and Coleman JP 1974 J. Phys. B : At. Mol. Phys. 7 788

Znojil M 1984 Phys. Rev. A 30, 2080.

NEW UNDERSTANDING OF ATOMIC PHYSICS USING HIGH VELOCITY POSITRONS

J.H. McGuire and N.C.Deb

Department of Physics
Kansas State University
Manhattan, KS 66506 USA

ABSTRACT: We consider various high velocity scattering processes of
atoms and positrons that may lead to new understanding of atomic
collisions. In electron capture by positrons interference of second
order singularities may be observable in the differential cross sections,
while in total cross sections there are qualitative differences in
theoretical predictions of the ratio of capture by positrons and protons
with the same velocity. In double excitation, ionization and capture
plus ionization there is evidence from heavy ion data and analysis that
correlation is playing a significant and observable role. We suggest
that similar studies with positrons may yield further insight into the
role of correlation with few and many electron targets.

INTRODUCTION

What good are positrons for doing atomic physics? The fact that
positrons are a form of antimatter exhibiting properties that differ from
normal matter is virtually irrevelant to most current research in atomic
physics. For current atomic and molecular studies positrons are simply
particles of the opposite charge and same mass as the common electron,
and of the same charge but different mass as the common proton. Hence,
if one wishes to study something new in atomic physics using positrons as
projectiles, one is constrained to look for effects that depend on both
the mass and charge of the positron.

In this paper we discuss some scattering phenomena which depend on
both the charge and mass of the positron and therefore will give information about atomic collisions that cannot be obtained using either
electrons or protons. The first part of this paper deals with electron
capture by positrons. We emphasize comparisons of total capture cross
sections by positrons and protons at the same velocity since we expect
that these cross sections will differ by about an order of magnitude at
high velocities. We also emphasize studies of differential cross
sections for electron capture by positrons near a scattering angle of 45°
where a quantum interference of two singular second order amplitudes has
been predicted. In both cases some experiments are suggested which may
lead to new insights into understanding of atomic collisions. We also
consider electron capture to the continuum and particle capture in
systems with various mass ratios where observable second order
singularities vary with the mass ratio.

In the second part of this paper we consider single and double excitation and ionization and capture by various projectiles. In helium recent observations of single and double ionization by high velocity protons and antiprotons indicate that where the single ionization cross sections are the same for protons and antiprotons the double ionization cross sections differ by a factor of two. Antiproton data are similar to electron data. Data for positrons, which would help to complete the picture, are not yet available. The factor of two effect in double ionization has been interpreted as a correlation effect, although the physical nature of the correlation mechanism is not yet clear. Understanding this effect would give insight into a many body effect in atomic collisions. Various possible related studies are considered including double excitation by positrons and electrons, simultaneous capture and ionization in helium by positron impact and the importance of correlation for two continuum electrons. The thrust of these studies is to understand the role of correlation in atomic scattering, and hence to begin many body studies that may bear on understanding of our human environment.

SINGLE ELECTRON CAPTURE

Second Order Singularities

Electron capture at high velocity has been largely understood [1-6] as a two step or second order process, although some interest in regarding capture as a first order process continues. [7-8] The argument that capture is largely a second order process dates to the seminal work[9] of L.H. Thomas in 1927 who pointed out that the simplest way to capture a particle is via two binary collisions since capture in a single binary collision is forbidden classically by conservation of energy and momentum. Such a two step process is constrained classically to occur at a unique scattering angle[10] by conservation of energy and momentum. This may be understood from Figure 1. For capture to occur particles 1 and 3 must have the same velocity and this constrains all vectors in the collision to be coplanar. We regard the incident velocity $\vec{v}$ to be much larger than the orbital velocity of the captured electron which is considered to be at rest initially. Assuming that the incident velocity, $\vec{v}$, is known there are six unknowns, namely $\vec{v}''$, $\vec{u}$, and $\vec{v}'$. However, conservation of energy and momentum give three equations of constraint at each vertex so that $\vec{v}''$, $\vec{u}$, and $\vec{v}'$ are uniquely determined. Hence the scattering angle, i.e., the direction of $\vec{v}'$, is uniquely determined in a two step process.

For electron capture from atoms by heavy projectiles, e.g. protons, the classical angle at which capture occurs, called the Thomas angle, is given by

$$\sin(\Theta_T) = m/M_P \; \sin(60°)$$

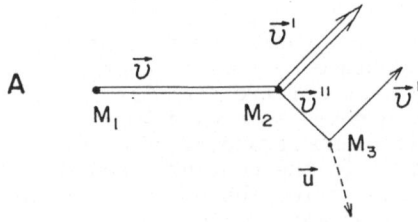

Fig. 1

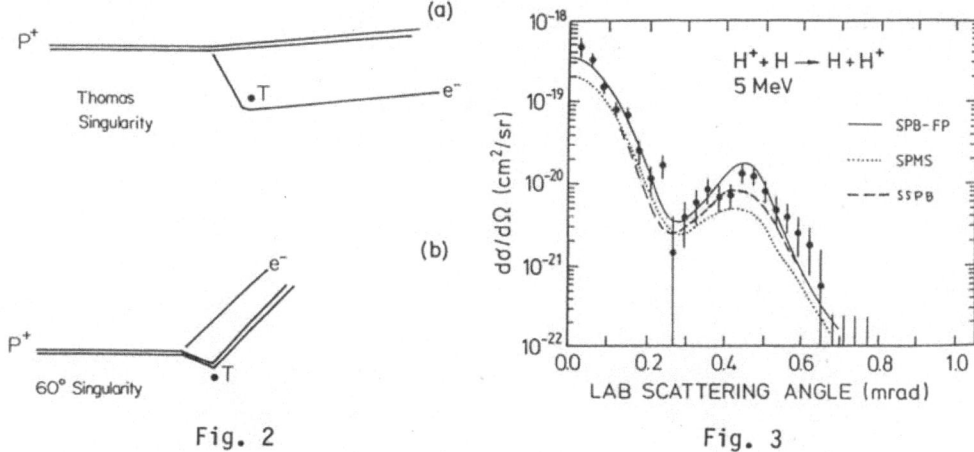

Fig. 2

Fig. 3

and the intermediate electron is scattered into an angle of 60° with
respect to the beam direction, and then simply redirected by the heavy
target nucleus as illustrated in Figure 2a. This result may be easily
confirmed by elementary calculations on a single piece of paper. If the
projectile is a positron then the Thomas angle moves to 45°[11] and the
intermediate scattering occurs at 45°, which is also fairly easy to
confirm. Experimental proof of this picture was achieved with the
observation of the Thomas peak[12,13] in ion-atom scattering. The recent
results of Schuch's group[13] are shown in Figure 3.

For heavy particle impact most of the total cross section is
confined to forward angles not much larger than the Thomas angle. And at
sufficiently high velocities total capture cross sections are expected to
be dominated by the contributions from the Thomas peak. Exact second
Born quantum calculations[3] and all other full second order theories[3,4-6]
confirm this picture of electron capture, namely that at sufficiently
high velocities electron capture is predominately a second order process
largely confined to the Thomas angle by energy and momentum conservation.
A second Thomas like singularity at 60°, as illustrated in Figure 2b is
in principle possible, but does not contribute significantly to total
capture cross sections and does not fundamentally change the second order
nature of the process for heavy particle impact. Some experimental
evidence for this 60° second order singularity has been obtained. Hence
a composite picture for electron capture by heavy particles is sketched
in Figure 4.

Electron capture by positrons is more complicated than capture by
protons because the mass of the positron is the same as the electron
mass. Specifically, the second order singularities which dominate in the
proton case come together, interfere, and cancel in the positron
case.[11,14,15] This phenomenon is sketched in Figure 3. For protons
there are two second order singularities. As the mass of the projectile
decreases these peaks move together, as is again easily verified by a
simple calculation. When the mass of the projectile is equal to the mass
of the captured electron both peaks occur at 45°. They cancel for the
dominant 1s-1s capture channel as may be seen from Figure 5 where the two
VGV second order terms have opposite signs since the second vertex is of
opposite sign in the two diagrams. For odd parity transitions, e.g.
1s-2p, the interference is constructive,[11,14] however, since there is a
relative sign change in the matrix elements of $(-1)^{\ell}$. (This requires a
little thought to understand conceptually.)

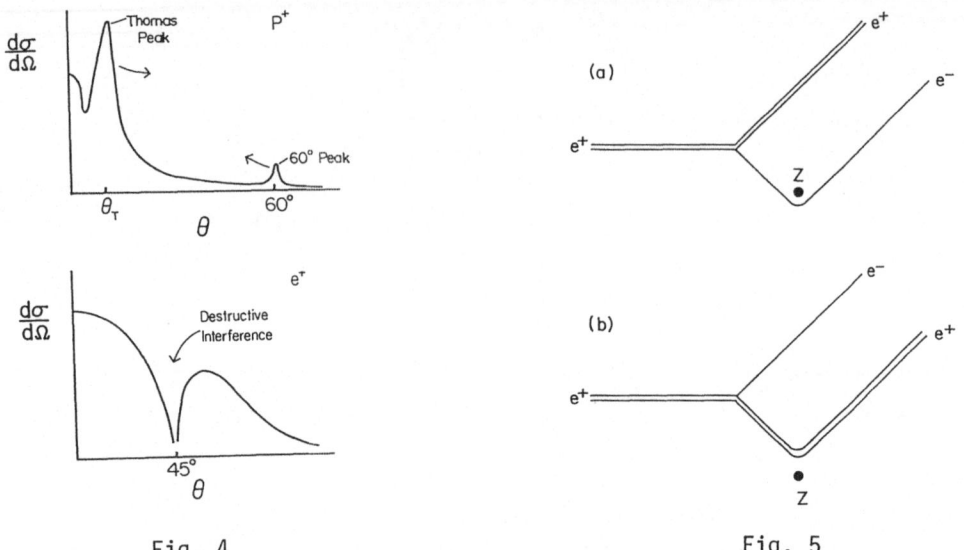

Fig. 4   Fig. 5

In short, second order terms may be dominant in electron capture at high collision velocities, and may not be ignored. In theory a calculation should be complete through second order in the Born series to be complete at high velocities. The question of higher order Born terms will be considered further below, but here we point out that the cancellation of second order terms on capture by positrons raises an interesting question about the possible significance of third order effects in positronium formation at high velocities. Unlike first Born theories, higher Born theories enable one to study the intermediate states of the system, i.e. understand something about what is happening during the collision process itself. Assuming that the potentials and asymptotic states are known then complete knowledge of the intermediate states of the system (e.g. via the exact Green's function) gives an exact representation of the full scattering event.

Theory

We have developed[14-16] a method to evaluate electron capture by positrons which is complete through second order and contains partial contributions from all higher order Born terms. We shall not try to explain here all the mathematical details which are given elsewhere,[12] but we will present the general approach and some of the significant features.

$$T_{if} = \langle \psi_f | \bar{V}_f | \psi \rangle = \langle \psi_f | \bar{V}_f (1 + G^+ \bar{V}_i) | \Psi_i \rangle$$

$$= \langle \Psi_f | \left(\frac{Z}{R} - \frac{Z}{r} \right) \left[1 + G^+ \left(\frac{-1}{\rho} \right) \right] | \Psi_i \rangle$$

$$+ \langle \psi_f | \left(\frac{Z}{R} - \frac{Z}{r} \right) G^+ \left(\frac{Z}{R} \right) | \Psi_i \rangle = T_{a+b} + T_d$$

Here Z/R, $-Z/r$, and $-1/\rho$ are positron-nucleus, electron nucleus, and positron- electron interactions respectively. The first amplitude, T_{a+b}, contains both of the singular second order terms illustrated in Figure 5.

86

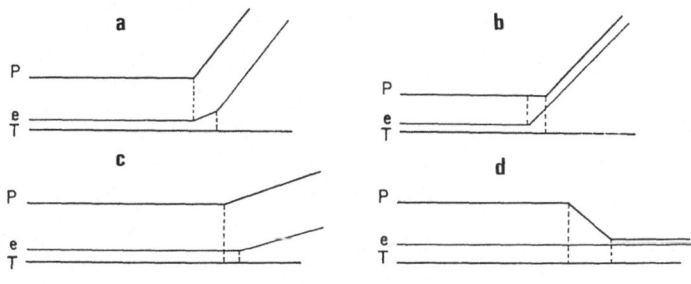

Fig. 6

The second amplitude, T_d, contains two more second order amplitudes which, however, are non-singular. These are called second order distortion amplitudes and are represented schematically in Figure 6. For electron capture by positrons it may be easily shown[16] that only four second order amplitudes exist, namely the four that we consider here. For capture by heavy particles only one amplitude, corresponding to the Z/r term in T_{a+b}, is normally considered.

We now introduce two approximations: (i) The interaction, $Z/R - Z/r$, is ignored in the Green's function, G^+, in the first term, thus giving a Coulomb Green's function, G_c^+, and (ii) the interaction $-1/\rho$ is ignored in G^+ in the second term. In the previous work of Shakeshaft and Wadehra,[11] plane-wave Green's functions were used in both parts of the T_{a+b} amplitude of Eq. (1) and T_d. We also include some selected higher-order terms via Coulomb Green's functions, so that our intermediate states contain Coulomb distortions. However, different Coulomb intermediate states are used for T_{a+b} and for T_d. In the first term there are positronium intermediate states. H-atom intermediate states are used in the second term. In both terms the electron propagates in the field of a positive charge. Nevertheless, the ultimate validiy of this procedure will <u>rest</u> on comparison with appropriate experimental data or a more complete theory.

Using the approximate off-energy-shell Coulomb wave function of Macek and co-workers,[5] and following the technique of the resulting amplitude may be reduced[15] to closed form. Here only two leading-order terms in a (Z/v) expansion have been retained. Hence, the error is of order (Z/v) (with a large coefficient), and this expression is accurate for systems of arbitrary charges at sufficiently high velocities. The second-order distortion term T_d may be expressed in terms of a one-dimensional integral.

Calculations

Differential cross sections for 1s-1s cpature are shown in Figs. 7-9. Three calculations are given in Fig. 7. In the curve labeled T_a only the amplitude for the Thomas peak, corresponding to Fig. 1(a), is included. This is the amplitude used for capture by heavy projectiles. The large Thomas peak at about 45° dominates the total cross section for T_a at very high velocities. The T_{a+b} curve in Fig. 7 includes contributions from Figs. 1(a) and 1 (b), but ignores the second-degree distortion, T_d in Eq. (1). The cancellation of the Thomas peak, as predicted by Shakeshaft and Wadehra[11] is quite evident near 45°. Full first Born calculation given a sailor dip. The dip at 23° is due to the interference between first and second Born amplitudes. The curve T_{a+b+d} represents our most complete calculation including the second-order distortion term in addition to contributions from Figs. 1(b) and 1(c). At 45° the amplitude T_{a+b} changes sign while the distortion amplitude T_d is finite and smoothly varying. The structure seen about 45° occurs because of interference due to all of the four second-order terms.

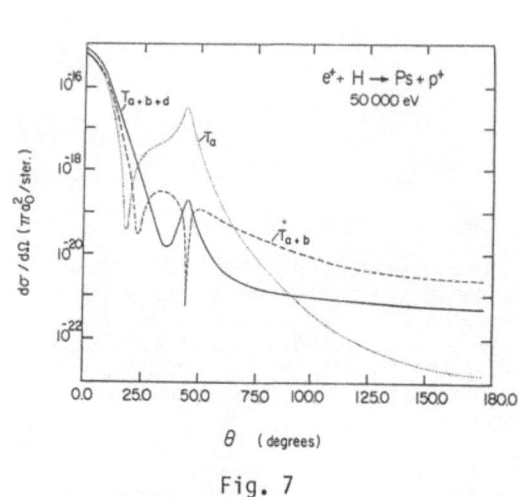

Fig. 7

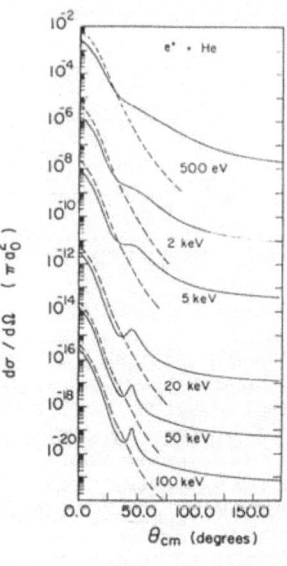

Fig. 8

In Figure 8 we present differential cross sections for 1s–1s capture of electrons from helium by positrons. There is a significant difference between the first-order Brinkman-Kramers (BK) and our second-order calculation at the larger scattering angles. Scattering from the target nucleus, which is ignored in the BK calculation, is probably significant at the larger angles. At energies about 1 keV, where experimental observation may be possible, there is a slight shoulder, due interference of higher order terms, in our calculation. This interference grows more pronounced as the projectile energy increases.

In Figure 9 differential cross sections are shown at a collision energy of 10 keV for various targets for 1s–1s capture. As the charge of the target increases, the interference structure tends to disappear. The velocity scale for the collision is set, at least in part, by the orbit velocity of the target electron. Since the orbit velocity increases linearly with Z, 10 keV positrons have a lower scaled velocity for large Z targets than for small Z targets. In other words, for large Z targets it is necessary to use higher projectile energies to attain the same ratio of projectile to target orbital velocity which tends to set the scale for physical phenomena.

Figure 10 shows a comparison of our results with experimental data[17,18] in helium. Below 170 eV both our DMS result and the BK result are much too high. This is not surprising since both theories are based on an expansion in Z_T/v which is not small at low energies. Furthermore, our calculation may not give correct cross sections at low energies due to our use of the peaking approximation valid at high velocities, $v \gg 1$. Above 170 eV our results are consistent with observation.[17,18] In this experimentally difficult region the uncertainties in observed results are large. However, we do note that a rough fit between E=120 and 200 eV to the data[17,18] varies as E^{-1}, while the theory varies as E^{-5}. There is general agreement among theorists that at asymptotically high energies the 1s–1s cross section varies as E^{-6}. In our opinion 250 eV is a high energy for helium, i.e., $v/Z_T = 2.5 > 1$, but not quite asymptotic. Here $\vec{v}$ is the projectile velocity in atomic units. And cross sections for electron capture by protons in helium do fall off much faster than E^{-1} at these velocities. If further observation confirms an E^{-1} dependence, then this result is likely to lead through to new insight about capture by positrons at high velocity.

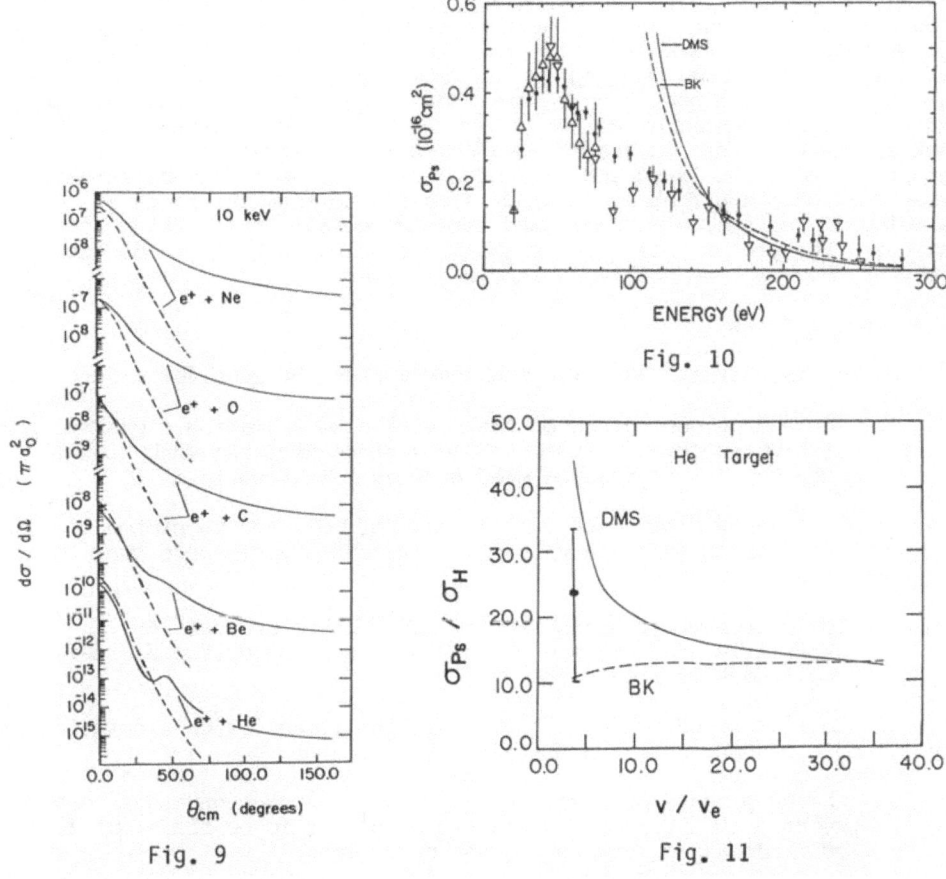

Fig. 9

Fig. 10

Fig. 11

In Figure 11 we plot the ratio σ_{Ps}/σ_H of total cross sections for Ps formation in e^+ + He collision to the corresponding results of H-atom formation in p + He collision to the corresponding results of H-atom formation in p + He collision as a function of v/Z_T. Both our DMS result and BK result shows that Ps formation cross sections are higher than H-atom formation cross sections at least by an order of magnitude. Physically $\sigma_{Ps} > \sigma_H$ because the positron slows down during the collision to share its kinetic energy with the electron to conserve overall energy. The proton does not slow down significantly since it is relatively massive. Because the capture cross section increases rapidly as v decreases, σ_{Ps} is larger than σ_H. For DMS the ratio σ_{Ps}/σ_H falls rapidly with v/Z_T up to about 15 and then continues to fall very slowly. The corresponding BK ratio however, rises slowly with increasing energy. These results both indicate that capture cross sections may not scale with projectile mass in a simple way, i.e., independent of projectile velocity as suggested by the results of Ma et al.[19] Olson et. al[20] have calculated the ratio of cross sections for electron capture by positrons to protons in helium form a few to 200 eV and find a curve which is monotonically increasing with projectile energy up to a value of about 30 at 200 eV. Hence three different calculations give qualitatively different shapes for the energy dependence of this cross section ratio, providing motivation for further experimental observation.

Theoretical Questions

In our opinion the theory for electron capture by positrons is not as well understood as the theory for capture by protons. While there are reasons to believe that for capture by positrons (a) the Thomas peak probably cancels, and (b) the total cross section is probably larger than for protons of the same velocity, the theory is not quantitatively reliable at this time. An especially interesting question is about the role of third and higher Born terms. How large are they? Are the intermediate states identifiable? And what are experimental tests possible? Another question that is easier to address is the effect of constructive interference in the 1s-2p and all odd Δ transitions in the differential cross section. And finally it will be useful to know the range of validity of the approximations used here, especially the peaking approximations. For proton impact the peaking approximations we use do not apply at the lower velocities which we consider.

Experimental Tests

We recommend that the following observations be considered further:

1. The energy dependence of the positronium formation cross section at high energies. Present observations decrease much slower than the E^{-5} dependence predicted by most theories.

2. The energy dependence of the ratio of positron to proton cross sections at the same velocity. Various theories are qualitatively different.

3. The differential cross section for capture by positrons, especially near 45° where interference between second order amplitudes is predicted.

In addition we note the following possibilities for consideration in the future (in some cases the distant future):

1. The cusp shape of electron capture to the continuum has been related to the Thomas mechanism. This could be different for positrons than for protons and heavy ions.

2. For relativistic positrons one could vary the mass of the projectile and the two second order singularities could be separated from the interference minimum as the energy of the projectile increases.

3. For capture from a diatomic molecule such as H_2 scattering from the two atomic centers could produce an interference pattern in the differential cross section analogous to two slit scattering.

4. Differences in capture from Ps by positrons and electrons could provide a test of violations of CPT invariance.

SINGLE AND DOUBLE EXCITATION, IONIZATION AND CAPTURE

Double Ionization

For single excitation and ionization one expects the first Born approximation to be valid at high velocities and consequently the cross section for impact by particles and antiparticles to be the same. As shown recently at CERN by a group from Aarhus,[24] single ionization cross sections in helium are in fact indentical within experimental errors of about 10% for protons and antiprotons above 6 atomic units of velocity. However, at around 10 units of velocity the double ionization cross sections differ by a factor of about two. This is shown in Figure 12. The antiproton data lies close to the electron data at the higher velocities shown in Figure 13, where a ratio of single to double ionization cross sections is shown. Data for the double ionization of

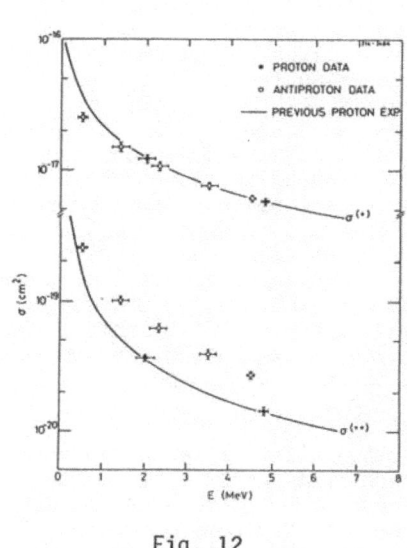

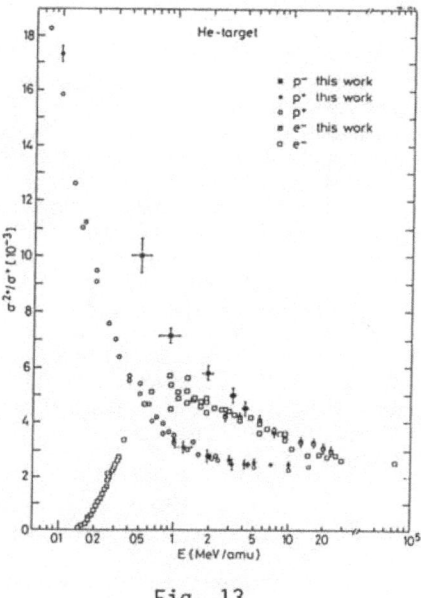

Fig. 12   Fig. 13

helium by positrons at energies about 1360 eV would be helpful in completing this picture.

Physical Interpretations

It is generally agreed[22-26] that the cause of the factor of 2 difference in double, but not single, ionization between projectiles of opposite charge is due to the electron-electron correlation interaction. However, different authors have different physical pictures of how correlation affects the double ionization cross section. Our interpretation[23] is that at the higher energies considered in Figures 12 and 13 there are two contributions to the quantum amplitude—a rearrangement amplitude arising from the correlation interaction linear in the projectile charge, Z_p, and a direct amplitude from the independent electron approximation quadratic in the projectile charge. The difference between projectiles of opposite charge comes from the Z_p^3 obtained from the square of the amplitude. Reading and Ford[24] have much more complete calculations, termed the forced impulse method, which also contain a Z_p^3 effect. They point out that this effect may arise when the projectile interacts twice with one target electron polarizing the target in the first step. The polarization effect differs for projectiles of opposite sign. Classical Trajectory Monte Carlo calculations including correlation, done[25] by Olson et. al., also give a Z_p^3 effect and are consistent with the intepretation of Reading and Ford. Olson also emphasizes the importance of electrons on a potential ridge between the target and projectile for positive projectiles at moderately high velocities. Sorensen has proposed[22] a two step mechanism where one electron, after collision with the projectile, collides with the second electron while leaving the collision region. What combination of these pictures, if any, is correct is not yet clear.

Sorting out the nature and importance of these correlation collision mechanisms[26] will require understanding of the few body problem in atomic collisions. This understanding could be useful in addressing many body problems in other areas of physics, chemistry, and perhaps biology. In short, the problem may be central to an understanding of the environment in which we human beings live.

Transfer Ionization

Another process of some interest is simultaneous transfer and ionization,[22-29] simply called transfer ionization or (TI). This process has been studied for impact of various projectiles on helium. We shall concentrate on high velocity collisions where analysis has been done[28] using the direct and rearrangement mechanisms. This analysis is in agreement with observation over a wider range of energies for $Z_p > 1$ than for protons, which we regard as somewhat anamolous. How do positrons compare with protons and with our analysis as well as classical scaling laws? At very high collision velocities it is expected that the ratio of transfer ionization to single capture cross sections goes to a value independent on the projectile velocity. It has been suggested[27] that this ratio will go to the same value as for double to single ionization by photons (but not protons, electrons or other charged particles) where final state correlation is absent. Some data supports[27-29] this simple sensible picture for proton impact. However, as we have discussed above, there are reasons to expect that capture for protons and positrons may be rather different.

It is possible that a second order singularity, like the Thomas mechanism discussed above, may contribute significantly[27-29] to this ratio at high velocities. The physical mechanism for this second order singularity is essentially the same as the one proposed by Sorensen (above) for double ionization, namely scattering of one electron by a second target electron. This second order mechanism depends on the mass of the projectile and may be different for positrons than for protons. In particular, a peak in the differential cross section, or ratio of cross sections, if present, will occur at forward angles for protons and larger angles for positrons.

Further Studies

We recommend that the following studies be given further consideration:

1. Double and single ionization of helium and other targets including H_2 by positrons. This will complete the picture for single and double ionization by electrons, protons, antiprotons and photons. Studies of differential cross sections could yield effects that are larger than those observed in total cross sections.

2. Double excitation by particles and antiparticles. Theory here is easier since one may avoid the problem of dealing with two continuum electrons which may be strongly correlated. Hence the tests may be more stringent.

3. Studies of transfer ionization. Are there differences between the ratio of transfer ionization to single capture for protons and positrons of the same velocity? Is a Thomas singularity observable in either ratios of total or differential cross sections and how do positron and proton results compare?

The following studies may also eventually be worthy of consideration:

1. Where are the continuum electrons after ionization? Are there many ridge electrons contributing at intermediate velocities to total cross sections for ionization for positrons and are they similar to ionization by protons? Do these electrons disappear for ionization by electrons?

2. Threshold ionization energy dependence. What is the coefficient of the exponent of the energy dependence for ionization by positrons at threshold? Some theories predict that this coefficient is different for electrons and positrons.

SUMMARY

In this paper we have tried to emphasize studies of atomic collisions with positrons that will lead to new understanding of physical mechanisms. Some emphasis was given to electron capture by positrons where, because the positron and electron masses are the same, destructive interference of second-order singularities occur. The residue of this interference may be experimentally observable in differential cross sections at high velocity. The ratio of positron to proton capture cross sections at the same time velocity is expected to be greater than one, as has been recently confirmed experimentally. However, the energy dependence of this ratio has not yet been observed at high velocity where various theories give qualitatively different results. The role and interpretation of higher Born terms has yet to be understood in electron capture by positrons. Emphasis is also given to studies of double excitation, double ionization and transfer ionization in high velocity collisions of positrons with targets containing two electrons. Recent observations and analysis of double ionization by protons and antiprotons suggest that correlation may play a significant and observable role in these processes, so that new insights into the few body and many body problem in atomic scattering may emerge. A number of further experimental as well as theoretical studies have been suggested.

This work has been supported by the U.S. Department of Energy, Office of Basic Energy Sciences, Division of Chemical Sciences.

REFERENCES

1. R. Shakeshaft and L. Spruch, Rev. Mod. Phys. 51, 369 (1979).
2. J.S. Briggs, J. Phys. B10, 3075 (1977).
3. J.H. McGuire, J. Eichler and P.R. Simony, Phys. Rev. A28, 2104 (1983).
4. J.H. Macek and K. Taulbjerg, Phys. Rev. Lett. 46, 170 (1980).
5. J. Macek and S. Alston, Phys. Rev. A26, 250 (1982).
6. K. Taulbjerg and J.S. Briggs, J. Phys. B16, 381 (1983).
7. D.P. Dewangan and J. Eichler, private communication.
8. D. Belkic, R. Gayet, J. Hanssen and A. Salin, private communication.
9. L.H. Thomas, Proc. R. Soc. London, Ser A 114, SGI (1927).
10. M. Lieber, private communication.
11. R. Shakeshaft and R.M. Wadehra, Phys. Rev. A22, 968 (1980).
12. E. Horsdal-Pedersen, C.L. Cocke and M. Stockli, Phys. Rev. Letter 50, 1910 (1983).
13. H. Vogt, W. Schwab, R. Schuch, M. Schulz and E. Junstiniano, Phys. Rev. Lett. 57, 2256 (1986).
14. J.H. McGuire, N.C. Sil and N.C. Deb, Phys. Rev. A34, 685 (1986).
15. N.C. Deb, J.H. McGuire and N.S. Sil, Phys. Rev. A (in press).
16. N.C. Deb, J.H. McGuire and N.C. Sil, Phys. Rev. A (in press).
17. D. Fromme, G. Kruse, W. Raith and G. Sinapius, Phys. Rev. Letters 57, 3031 (1987).
18. L.M. Diana, P.G. Coleman, P.L. Brooks, P.V. Pendelton and D. Norman, Phys. Rev. A34, 2731 (1986).
19. Q.C. Ma, X.X. Cheng, A.H. Liu, and T. Watanabe, Phys. Rev. A32, 2645 (1985).
20. J.H. McGuire, Phys. Rev. Lett. 49, 1153 (1982); Proceedings of the Second U.S.-Mexico Atomic Physics Symposium on Two Electron Phenomenon, Cocoyoc, Mexico 1986, to be published.

21. L.H. Andersen, P. Hvelplund, H. Knudsen, H.P. Moller, K. Elsner, K.G. Rensfelt and U. Uggerhoj, Phys. Rev. Lett. 57, 2147 (1986).

22. L.H. Andersen, P. Hvelplund, H. Knudsen, H.P. Moller, A.H. Sorensen, K. Elsnar, K.G. Rensfelt and E. Uggerhoj, preprint.

23. R.E. Olson, B.A.P.S. 32, (1987); private communication.

24. J.F. Reading and A.L. Ford, Phys. Rev. Lett. 58, 543 (1987); J. Phys. B (in press).

25. R.E. Olson, private communciation; A.E. Wetmore and R.E. Olson, Phys. Rev. A34, 2822 (1986); R.E. Olson, Phys. Rev. A33, 4397 (1986).

26. J.H. McGuire, Phys. Rev. A (in press) 1987.

27. E. Horsdel-Pedersen and J. Larsen, J. Phys. B12, 4085 (1985).

28. J.H. McGuire, E. Salzborn and A. Muller, Phys. Rev. A35 3265 (1987).

29. H. Knudsen, L.H. Andersen, P. Hvelplund, J. Sorensen and B. Ciric, J. Phys. B (in press) 1987.

CALCULATIONS OF SCATTERING CROSS SECTIONS AND ANNIHILATION RATES IN LOW ENERGY COLLISIONS OF POSITRONS WITH MOLECULAR HYDROGEN

Edward A. G. Armour

Department of Mathematics
University of Nottingham
Nottingham NG7 2RD
England

ABSTRACT

An analysis of the requirements for an accurate calculation of Z_{eff}, the effective number of electrons per molecule available to the positron for annihilation, for low energy e^+H_2 scattering using the Kohn method shows the importance of including basis functions which contain the positron-electron distance as a linear factor, i.e. Hylleraas-type funct-ions. Such functions are very complicated to include as they are not sep-arable, i.e. they cannot be expressed as a finite expansion of one partic-le functions. However, I have been able to extend the calculation of the lowest partial wave for low energy e^+H_2 scattering using the Kohn method and basis sets involving only separable functions, which I described at the last Workshop at Detroit in 1985, to include Hylleraas-type basis functions. The results for Z_{eff} at very low energies are much closer to the experimental value than any that have been obtained previously. The inclusion of Hylleraas-type functions also has a very significant effect on the low energy phase shift and total cross section, bringing the cross section into agreement with experiment for incident positron energies up to about 2 eV. As far as I am aware, this is the first time that Hylleraas-type functions have been used in a molecular scattering calc-ulation.

INTRODUCTION

At the last Workshop on positron-gas scattering in Detroit in 1985 I presented an application of the Kohn variational method to low energy positron-hydrogen-molecule scattering[1,2,3]. The calculations were car-ried out using trial functions involving only separable functions, i.e. functions which can be expressed as a finite expansion of products of one particle functions. The use of such functions makes the evaluation of the matrix elements required in a variational calculation comparatively easy. For this reason they have been extensively used both in atomic and molecular bound state calculations and in electron-atom and molecule scat-tering calculations.

In the calculation I carried out on the lowest partial wave, i.e. the lowest partial wave of $\sum_g^+$ symmetry, I was able to obtain a good qualitat-ive description of the behaviour of the phase shift by including sufficient separable functions in the trial function. Similar results have also been obtained by Tennyson[4] using the R-matrix method[5] with separable basis functions. A description of his calculation appears elsewhere in these proceedings.

In the work[1,3] I presented at Detroit, I drew attention to the inadequate description that the resulting wavefunction gave of the process of positron annihilation. Thus the calculated value of $Z_{eff}(k)$, the effective number of electrons per hydrogen molecule available to the positron for annihilation, at very low energies was found to be very much less than the experimental value[6].

The reason for this discrepancy is not hard to find. $Z_{eff}(k)$ measures the value of the wavefunction when the positron and an electron coincide[7,8]. Suppose the electron concerned is electron 1. As a consequence of the Coulombic attraction between the positron and the electron, the exact wavefunction, ψ, at points of coincidence must satisfy the cusp condition[8,9]

$$\left(\frac{\partial \hat{\psi}}{\partial r_{13}}\right)_{r_{13}=0} = -\tfrac{1}{2}(\psi)_{r_{13}=0} \tag{1}$$

where

$$\underset{\sim}{r}_{13} = \underset{\sim}{r}_3 - \underset{\sim}{r}_1 \tag{2}$$

and $\underset{\sim}{r}_1$ and $\underset{\sim}{r}_3$ are the positron vectors of the electron and positron, respectively. $\hat{\psi}$ is the average value of ψ taken over the sphere $r_{13} =$ constant, with variables other than $\underset{\sim}{r}_{13}$ fixed.

The Kohn trial function can only be made sufficiently flexible so that it can take this condition into account adequately by including basis functions containing r_{13} and r_{23} as linear factors. These factors are not separable functions and their inclusion makes the calculation very much more complicated.

The inclusion of linear factors of the inter-particle distance in the trial function in variational calculations has a long and interesting history. As early as 1929, Hylleraas[10] showed that their inclusion brought about rapid convergence in variational calculations of the ground state energy and wavefunction of helium. Soon afterwards, James and Coolidge[11] showed that this was also the case for the hydrogen molecule. The rapid convergence is due to the increased flexibility of trial functions containing such functions. This enables them to take into account adequately the cusp condition similar to equation (1) when the two electrons coincide[8,9]. In what follows I shall refer to basis functions which contain the inter-particle distance as a linear factor as Hylleraas-type functions.

Schwartz[12] showed in 1961 that the Kohn variational method with a basis set made up of functions containing linear and higher powers of the inter-particle distance gave accurate results for low energy positron-hydrogen-atom scattering. Basis sets of this type have subsequently been employed in Kohn and other variational calculations on low energy positron-hydrogen-atom and positron-helium scattering. For details, see the review article by Humberston[13]. It has not so far been possible to compare the results obtained for atomic hydrogen with experiment but the results obtained for helium are in good agreement with experiment, both in the case of the total cross section and $Z_{eff}(k)$.

The spherical symmetry of atomic targets made it possible to incorporate the inter-particle distance directly into the variables of integration, as was done by Hylleraas[10]. This made it possible to evaluate all

96

the necessary matrix elements exactly either analytically or by Gaussian quadrature.

The absence of spherical symmetry makes the use of Hylleraas-type functions in calculations involving molecules very difficult. Nevertheless, Clary[14] has succeeded in carrying out very accurate variational calculations of the energy of He_2^+ and He_2 for a given internuclear distance using basis sets containing Hylleraas-type functions.

For the reasons described above, I considered[3] that the key to an accurate calculation of $Z_{eff}(k)$ for low energy positron-hydrogen molecule scattering using the Kohn variation method would be the inclusion in the basis set of Hylleraas-type functions containing the positron-electron distance. It has proved possible to include such functions, using a method similar to the one used by Clary. The results[15] for $Z_{eff}(k)$ at very low energies are much closer to the experimental value than any that have been obtained previously. The phase shift of the lowest partial wave and the associated contribution to the cross section are very significantly increased[16]. The contribution from the lowest partial wave is large enough to account for the experimental value of the total cross section up to incident positron energies of about 2 eV. This is much higher than in the case of the calculation using only separable functions[1,2].

As far as I am aware, this is the first time that Hylleraas-type functions have been used in a molecular scattering calculation.

CALCULATION

As in the case of the earlier calculation[1,2], the nuclei in the target hydrogen molecule are taken to be fixed in their equilibrium position with internulcear separation, $R = 1.4\ a_0$. The coordinates used are prolate spheroidal (or confocal elliptical) coordinates $(\lambda_i, \mu_i, \phi_i)$ where

$$\lambda_i = \frac{r_{iA} + r_{1B}}{R} \tag{3}$$

$$\mu_i = \frac{r_{iA} - r_{iB}}{R}, \tag{4}$$

where r_{iA}, for example, is the distance of particle i from nucleus A. The third coordinate, ϕ_i, is the usual azimuthal angle. Particles 1 and 2 are the molecular electrons and particle 3 is the positron.

In the earlier calculation, the open-channel function, $\Omega(c, \lambda_3, \mu_3; \tau, a)$, was taken to be of the form

$$\Omega(c, \lambda_3, \mu_3; \tau, a) = \frac{B}{(\lambda_3 - 1)}(\sin[c(\lambda_3 - 1)]\cos \tau$$

$$+ \cos[c(\lambda_3 - 1)](1 - \exp[-\gamma(\lambda_3 - 1)])\sin \tau$$

$$+ a(\cos[c(\lambda_3 - 1)](1 - \exp[-\gamma(\lambda_3 - 1)])\cos \tau$$

$$- \sin[c(\lambda_3 - 1)]\sin \tau)S_{oo}(c, \mu_3), \tag{5}$$

where B and γ are constants and τ and a are variable parameters.

$$c = \tfrac{1}{2}kR, \tag{6}$$

where k is the wavenumber of the positron in atomic units, $S_{oo}(c,\mu_3)$ is the spheroidal μ_3 function associated with the lowest partial wave in the expansion of a plane wave in terms of prolate spheroidal coordinates[17,18]. At low energies, where c is small,

$$S_{oo}(c,\mu_3) \approx P_o(\mu_3) = 1 \tag{7}$$

and it was thus taken to be 1 in the calculation. This form of open-channel function is basically of the form introduced by Massey and Ridley[19]; the parameter τ has been introduced to permit variation of the open-channel function to avoid the anomalous singularities that can arise in applications of the Kohn method[20]. I refer to this use of a more general form of open-channel function as the generalized Kohn method[21].

In this calculation, I have also made use of the more accurate alternative form for the open channel function,

$$
\begin{aligned}
\Omega(c\lambda_3,\mu_3;\tau,a) = D(R_{oo}^{(1)}(c,\lambda_3)\cos\,\tau &- R_{oo}^{(2)}(c,\lambda_3) \\
\times\,(1 - \exp[-\delta(\lambda_3 - 1)])^2\sin\,\tau &+ a[R_{oo}^{(2)}(c,\lambda_3) \\
\times\,(1 - \exp[-\delta(\lambda_3 - 1)])^2\cos\,\tau &+ R_{oo}^{(1)}(c,\lambda_3)\sin\,\tau]) \\
\times\,S_{oo}(c,\mu_3),&
\end{aligned}
\tag{8}
$$

where D and δ are constants and τ and a are variable parameters. $R_{oo}^{(1)}(c,\lambda_3)$ and $R_{oo}^{(2)}(c,\lambda_3)$ are the radial solutions to the free-particle equation in terms of prolate spheroidal coordinates which are regular and irregular, respectively, at $\lambda_3 = 1$ and are associated with $S_{oo}(c,\mu_3)$[17,18]. I used an open-channel function of this type in the calculation of the lowest partial wave of Σ_u^+ symmetry which I presented at Detroit[1].

In the earlier calculation the trial function was taken to be of the form

$$\psi_T = \Omega(c,\lambda_3,\mu_3;\tau,a)\psi_G + \sum_{i=1}^{M} g_i\chi_i\psi_G, \tag{9}$$

where $\{g_i\}$ are variable parameters, ψ_G is an approximate target wavefunction. It was the six term wavefunction first used in my initial e^+H_2 scattering calculation[1,22]. The $\{\chi_i\}$ are short-range correlation functions. They were taken to be of the form

$$\chi_1 = \frac{B}{\lambda_3 - 1}\cos[c(\lambda_3 - 1)](1 - \exp[-\gamma(\lambda_3 - 1)])\exp[-\gamma(\lambda_3 - 1)], \tag{10}$$

$$
\begin{aligned}
\chi_i = (\lambda_1^{a_i}\lambda_2^{b_i}\mu_1^{c_i}\mu_2^{d_i}[M_{13}\cos(\phi_1 - \phi_3)]^{p_i} \\
+ \lambda_1^{b_i}\lambda_2^{a_i}\mu_1^{d_i}\mu_2^{c_i}[M_{23}\cos(\phi_2 - \phi_3)]^{p_i})\exp(-\beta[\lambda_1 + \lambda_2]) \\
\times\,B\lambda_3^{r_i}\mu_3^{s_i}\exp(-\alpha\lambda_3), \quad i > 1,
\end{aligned}
\tag{11}
$$

where

$$M_{13} = [(\lambda_1^2 - 1)(\lambda_3^2 - 1)(1 - \mu_1^2)(1 - \mu_3^2)]^{\tfrac{1}{2}}. \tag{12}$$

a_i, b_i, c_i, d_i, p_i, r_i and s_i are non-negative integers. $c_i + d_i + s_i$ must be even so that χ_i is of overall $\sum_g^+$ symmetry. p_i was taken to have the values 0 and 1. Note that if $p_i = 0$ the electronic part of χ_i is of $\sum$ symmetry whereas if $p_i = 1$, it is of Π symmetry.

The crucial change in the present calculation is that I have added extra short-range correlation functions of the form

$$\chi_i = (\lambda_1^{a_i} \lambda_2^{b_i} \mu_1^{c_i} \mu_2^{d_i} r_{13}^{q_i} + \lambda_1^{b_i} \lambda_2^{a_i} \mu_1^{d_i} \mu_2^{c_i} r_{23}^{q_i})$$

$$\times \exp(-\beta[\lambda_1 + \lambda_2]) B \lambda_3^{r_i} \mu_3^{s_i} \exp(-\alpha\lambda_3), \tag{13}$$

where

$$r_{13} = |r_3 - r_1| \tag{14}$$

and

$$q_i = 1.$$

These are the Hylleraas-type functions containing the positron-electron distance as a linear factor. Once again $c_i + d_i + s_i$ must be even to obtain overall $\sum_g^+$ symmetry.

The inclusion of such functions means that to calculate the matrix elements in the Kohn equations, integrals involving up to three particles have to be evaluated, rather than just two particles as in the case of the separable functions used in the earlier calculation. Furthermore, as James and Coolidge[11] point out, it is not possible to incorporate the inter-particle distance straightforwardly in the variables of integration as was possible in the case of spherical symmetry.

These difficulties are considerable but not unsurmountable. It might be thought that it would be necessary to evaluate integrals involving the very complicated factor, $r_{13}r_{23}/r_{12}$. However, the $1/r_{12}$ factor comes from the target Hamiltonian

$$H_T = -\tfrac{1}{2}\nabla_1^2 - \tfrac{1}{2}\nabla_2^2 - \frac{1}{r_{A1}} - \frac{1}{r_{A2}} - \frac{1}{r_{B1}} - \frac{1}{r_{B2}} + \frac{1}{r_{12}} + \frac{1}{R}. \tag{15}$$

As I make use of the method of models[23] and ψ_G appears as a factor in the trial function, ψ_T, it is not necessary to evaluate matrix elements such as those containing $r_{13}r_{23}/r_{12}$, which involve the electronic potential[23,16]. The price to be paid for using the method of models, however, was that matrix elements had to be evaluated between all the functions $\{\chi_i \eta_j\}$, where

$$\psi_G = \sum_{i=1}^{N} c_i \eta_i \tag{16}$$

and N = 6. This was very time consuming.

It was necessary to evaluate matrix elements involving the factors r_{23}/r_{13} and $r_{13}r_{23}$. This required evaluation of integrals over all configuration space of the form:

$$\iiint \lambda_1^d \lambda_2^e \mu_3^f \mu_1^\ell \mu_2^m \mu_3^n [M_{13} \cos(\phi_i - \phi_3)]^P e^{-A(\lambda_1 + \lambda_2)}$$

$$\times e^{-2\alpha\lambda_3} \frac{r_{23}}{r_{13}} dV_1 dV_2 dV_3 \qquad i = 1 \text{ or } 2, \tag{17}$$

and

$$\iiint \lambda_1^d \lambda_2^e \lambda_3^f \mu_1^\ell \mu_2^m \mu_3^n e^{-C(\lambda_1+\lambda_2)} e^{-\alpha\lambda_3} \times G(c,\lambda_3,\mu_3) \frac{r_{23}}{r_{13}} \, dV_1 dV_2 dV_3 \quad (18)$$

$$\iiint \lambda_1^d \lambda_2^e \lambda_3^f \mu_1^\ell \mu_2^m \mu_2^n e^{-A(\lambda_1+\lambda_2)} e^{-2\alpha\lambda_3} r_{13} r_{23} dV_1 dV_2 dV_3, \quad (19)$$

where e, d, f, ℓ, m, n are non-negative integers, p = 0 or 1, A and C are constants and $G(c,\lambda_3,\mu_3)$ is a constituent function of one of the forms of open-channel function, $\Omega(c,\lambda_3,\mu_3)$, given in equations (5) and (8).

These integrals were evaluated by first expressing r_{23}/r_{13} and $r_{13} r_{23}$ in the form

$$\frac{r_{23}}{r_{13}} = \frac{r_{23}^2}{r_{23} r_{13}} \quad (20)$$

and

$$r_{13} r_{23} = \frac{r_{13}^2 r_{23}^2}{r_{13} r_{23}} . \quad (21)$$

r_{i3}^2 is a separable function and has the form

$$r_{i3}^2 = \tfrac{1}{4} R^2 [\lambda_i^2 + \lambda_3^2 + \mu_i^2 + \mu_3^2 - 2 - 2\lambda_i \lambda_3 \mu_i \mu_3$$
$$- 2M_{i3} \cos(\phi_i - \phi_3)] \quad (22)$$

in terms of prolate spheroidal coordinates. It is well-known (see, for example, James and Coolidge[11] and Rüdenberg[24]) that $1/r_{i3}$ can be expressed in terms of prolate spheroidal coordinates using the Neumann expansion

$$\frac{1}{r_{i3}} = \frac{2}{R} \sum_{\tau=0}^{\infty} \sum_{\nu=0} D_\tau^\nu P_\tau^\nu(\lambda_<) Q_\tau^\nu(\lambda_>) P_\tau^\nu(\mu_i) P_\tau^\nu(\mu_3) \cos \nu(\phi_i - \phi_3), \quad (23)$$

where D_τ^ν is an expansion coefficient and $\lambda_>$ and $\lambda_<$ are the greater and lesser of λ_i and λ_3, respectively. $P_\tau^\nu(\lambda)$ and $Q_\tau^\nu(\lambda)$ are the associated Legendre functions of the first and second kind, respectively.

As the expressions in equations (20) and (21) separate into two factors, one of which involves only electron 1 and the other only electron 2, the double series which result from the substitution of equations (22) and (23) in equations (17)-(19) terminate in every case when integrated term by term. The integrals involving the μ and ϕ variables are straightforward to evaluate exactly, either analytically or by Gaussian quadrature.

The integrals involving the λ variables are of the form

$$\int_1^\infty H_\tau^\nu(\lambda_3;d,M) H_{\bar{\tau}}^{\bar{\nu}}(\lambda_3,e,M) \lambda_3^f [L(c,\lambda_3)]^p e^{-F\lambda_3} d\lambda_3, \quad (24)$$

where

$$H_\tau^\nu(\lambda_3,d,M) = \int_1^{\lambda_3} W^\nu P_\tau^\nu(\lambda) Q_\tau^\nu(\lambda_3) \lambda^d e^{-M\lambda} d\lambda + \int_{\lambda_3}^\infty W^\nu P_\tau^\nu(\lambda_3) Q_\tau^\nu(\lambda) \lambda^d e^{-M\lambda} d\lambda \quad (25)$$

and

$$W = [(\lambda^2 - 1)(\lambda_3^2 - 1)]^{\frac{1}{2}}. \quad (26)$$

d, e and f are non-negative integers and M and F are constants. $L(c,\lambda_3)$ is a constituent function of $G(c,\lambda_3,\mu_3)$ and p = 0 or 1. The evaluation of the functions $\{H_\tau^\nu(\lambda_3;d,M\}$ causes no difficulty as they were required in my earlier calculation[1,2] in the evaluation of the matrix elements of the positron-hydrogen-molecule potential

$$V = \frac{1}{r_{A3}} + \frac{1}{r_{B3}} - \frac{1}{r_{13}} - \frac{1}{r_{23}} \qquad (27)$$

involving open-channel functions. They were evaluated exactly using recurrence relations and a mixture of analytical integration and Gaussian quadrature.

The integral over λ_3 in equation (24) can be accurately evaluated using the 'boundary derivative reduction' method[25,26,14]. The variable of integration is first changed to $x \in [0,1]$ by the transformation[27]

$$\lambda_3 = 1 - P \log(1 - x) \qquad (28)$$

where P is a constant. It is then further changed to $q \in [0,1]$ by the transformation[26,14],

$$x = 30 \int_0^q y^2(1 - y^2)dy \qquad (29)$$

$$= 6q^5 - 15q^4 + 10q^3. \qquad (30)$$

The integral is then evaluated using the trapezoidal rule for numerical integration. The choice of the variable of integration, q, has the effect of making the error in the integral of $O(n^{-4})$, where n is the number of integration points[16,26].

It is easy to adapt the method outlined above to evaluate integrals such as

$$\left(\frac{1}{4\pi}\right)^3 \iiint e^{-\bar{r}_1} e^{-\bar{r}_2} e^{-\bar{r}_3} \, f_{123} dV_1 dV_2 dV_3, \qquad (31)$$

where

$$\bar{r}_i = \frac{r_{Bi}}{R} \qquad i = 1,2,3 \qquad (32)$$

and

$$f_{123} = \frac{r_{23}}{r_{13}} \quad \text{or} \quad r_{13} r_{23} \, . \qquad (33)$$

which have integrands which are spherically symmetric about nucleus B (or A). The results can be compared with the exact values obtained, for example, by the analytical method of Roberts[28]. This shows that the results obtained with 30 integration points are not sensitive to the detailed choice of the constant P in equation (28) and the error in the numerical integration is of the order of $10^{-5}\%$[16].

Integrals involving only one or other r_{13} or r_{23} were evaluated either analytically or using the 'boundary derivative reduction' method. All other integrals were evaluated as in my earlier calculation. Various checks were made to see if representative matrix elements satisfied hermiticity and other required conditions. These were found to be satisfied to five significant figures or better in all cases.

Apart from the inclusion of the extra Hylleraas-type functions, the generalized Kohn calculation was carried out in the same way as in my earlier calculation[1,2].

RESULTS AND DISCUSSION

It is of interest to begin by examining the effect of the inclusion of Hylleraas-type functions in the basis set has on the calculated value of $Z_{eff}(k)$. The results obtained for k in the range 0.1-1.0 a_0^{-1} using various basis sets are given in Table 1. It can be seen that the inclusion of Hylleraas-type functions in the basis set does indeed have a dramatic effect on the calculated value of the contribution to $Z_{eff}(k)$ from the lowest partial wave, particularly at very low energies. The inclusion of just a small number of Hylleraas-type functions, in addition to the functions of $\sum$ and Π electronic symmetry used in my earlier calculation, brings about a large increase in the contribution to $Z_{eff}(k)$ at the lower energy values.

It can be seen that the results are not very sensitive to the choice of the non-linear parameters α and δ, nor to the choice of open-channel function. Investigation[15] shows, however, that the omission from the basis set of the functions of $\sum$ or Π electronic symmetry very much reduces the increase in the contribution to $Z_{eff}(k)$ at the lower energy values[15].

We can see why this should be as follows. The wavefunction from which the contribution to $Z_{eff}(k)$ is calculated is obtained by a generalization of the Kohn variational method. The Kohn method obtains an approximation to the exact wavefunction for the whole of configuration space, and not just for the region where a positron is close to an electron which is important for the calculation of the contribution to $Z_{eff}(k)$. To obtain an approximate wavefunction which is accurate in this region requires a trial function containing not only the Hylleraas-type functions appropriate for representing this region but also the functions of $\sum$ and Π electronic symmetry appropriate for representing the region where the positron and the electrons are more widely separated. If the functions of $\sum$ or Π electronic symmetry are omitted, the accuracy of the Kohn wavefunction in the region where the positron is close to an electron is reduced in an attempt to compensate for the missing functions in the region where the positron and the electrons are more widely separated.

For thermal positrons at 293 K, where kT = 25 meV, the results obtained with α = 0.575 and the first open-channel function (equation (5)) predict a value[15] of 10.2 for $\bar{Z}_{eff}$, the Boltzmann average value of $Z_{eff}(k)$. This is much closer to the experimental value[6] of 14.8 than has been obtained in any previous calculation . Note that the other calculations using Hylleraas-type functions will give values of $\bar{Z}_{eff}$ grouped around this value.

The results obtained from the phase shift for k in the range 0.1-1.0 a_o^{-1} using both types of open-channel function and various basis sets are given in Table 3. The results obtained with only separable functions, i.e. basis sets 1 and 5, are in good overall agreement with the results obtained by Tennyson[4] using the R-matrix method and basis sets made up of separable functions. This has already been discussed by Tennyson.

It can be seen that the inclusion of just a small number of Hylleraas-type functions, in addition to the separable functions, brings about a large increase in the phase shift over the entire k range. This continues the pattern of upward convergence which was a feature of my earlier calculation[1,2]. The results obtained with basis sets 2 and 3 are typical of the way the increase comes about, irrespective of the choice of α value or type of open-channel function. They show that it is the inclusion of the first three Hylleraas-type functions which brings about most of the increase as in the case of $Z_{eff}(k)$. Hylleraas-type basis function 9 is not included in any of the basis sets in Table 3 but its inclusion was found to have only a small effect on the phase shift[16]. The results obtained with α = 0.575 and α = 0.3 are similar except in the case of basis set 4 and k = 1.0 a_o^{-1}, where there is evidence of instability in the Kohn calculation.

Table 1.   Contributions to $Z_{eff}(k)$ from the lowest partial wave in the range k = 0.1 to 1.0 a_o^{-1}.

k (a_o^{-1})	E (eV)	Contribution to $Z_{eff}(k)$							
		1	2	3	4	5	6	7	8
0.1	0.14	3.28	9.42	9.62	9.77	8.46	10.45	8.64	11.50
0.2	0.54	2.78	7.31	7.47	6.85	6.22	7.44	5.38	7.74
0.3	1.2	2.02	5.28	5.40	5.26	4.96	5.57	4.23	6.14
0.4	2.2	1.74	4.43	4.54	4.47	3.78	4.68	3.58	5.05
0.5	3.4	1.62	4.13	4.22	4.29	3.41	3.88	3.08	4.04
0.6	4.9	1.60	3.57	3.64	3.19	2.80	3.12	2.70	3.20
0.7	6.7	1.29	2.74	2.79	2.78	2.37	2.71	2.35	2.73
0.85	9.8	1.15	2.33	2.37	2.62	2.07	2.31	2.06	2.30
1.0	13.6	1.15	2.49	2.49	2.06	3.08	2.00	1.95	1.99

Basis sets

First open-channel function with $\gamma = 0.75$

1. The 64 separable functions used in my earlier calculation with $\alpha = 0.575$.

2. As in 1 plus functions 1-3 in Table 2 with $\alpha = 0.575$.

3. As in 1 plus functions 1-3 and 9 in Table 2 with $\alpha = 0.575$.

4. As in 1 plus functions 1-8 in Table 2 with $\alpha = 0.575$.

5. As in 1 plus functions 1-6 in Table 2 with $\alpha = 0.3$.

Second open-channel function with $\delta = 1.0$

6. As in 4, but without χ_1.

7. As in 5, but without χ_1.

Second open-channel function with $\delta = 0.75$

8. As in 1, but without χ_1, plus functions 1-6 in Table 2 with $\alpha = 0.575$.

In all cases, $\beta = 0.2$.

Table 2. Hylleraas-type basis functions

i	a_i	b_i	c_i	d_i	q_i	r_i	s_i
1	0	0	0	0	1	0	0
2	0	0	0	0	1	1	0
3	1	0	0	0	1	0	0
4	1	0	0	0	1	1	0
5	0	0	1	1	1	0	0
6	0	0	1	1	1	1	0
7	1	1	0	0	1	0	0
8	1	1	0	0	1	1	0
9	0	0	1	0	1	0	1

It can be seen that the increase in the phase shift is much greater in the case of the calculations using the second type of open-channel function. Investigation[16] shows that the two types of open-channel function give similar results when only the first two Hylleraas-type functions are included in the basis set. The addition of the third function, however, brings about a much bigger increase in the phase shift if the second type of open-channel function is used.

The third Hylleraas-type function is of the form

$$(\lambda_1 r_{13} + \lambda_2 r_{23})e^{-\beta(\lambda_1 + \lambda_2)}{}_{Be}^{-\alpha\lambda_3}\psi_G \qquad (34)$$

where ψ_G is the approximate target wavefunction. The presence of the λ_1 and λ_2 factors means that this basis function is well suited to representing an electron at some distance from the nuclei interacting with the positron through the attractive Coulombic potential. The difference in phase shift values is due to a significant difference in the matrix elements between the two types of open-channel function and the Hylleraas-type basis function in (34). The results obtained using basis sets 6, 8 and 9 show that at low k values this difference is sensitive to the value of the shielding exponent, δ.

As pointed out earlier, the second type of open-channel function is more accurate as it contains the regular and irregular solutions to the free particle equation in prolate spheroidal coordinates which are associated with the lowest spheroidal μ_3 function, $S_{00}(c,\mu_3)$, whereas the first type is made up of approximations to these solutions. However, even the second form is an approximation to the exact open-channel function as the Schrödinger equation with the optical potential for the positron-molecule system is not separable in prolate spheroidal or any other coordinates and thus mixing of spheroidal partial waves will occur[1].

I have not taken this into account in this calculation. However, the good overall agreement between the results of Tennyson's calculation[4] using separable basis functions in the R-matrix method, which takes mixing of spherical partial waves into account, and my calculation with separable basis functions, indicates that, in calculations with separable functions, the mixing of spheroidal partial waves is small in the k range

Table 3. Phase shifts in the range k = 0.1 to 1.0 a_o^{-1} obtained using both types of open-channel function

k (a_o^{-1})	E (eV)	1	2	3	4	5
0.1	0.14	0.090	0.157	0.157	0.174	0.088
0.2	0.54	0.104	0.200	0.213	0.219	0.114
0.3	1.2	0.077	0.188	0.199	0.183	0.084
0.4	2.2	0.026	0.142	0.155	0.128	0.028
0.5	3.4	−0.034	0.081	0.066	0.060	−0.038
0.6	4.9	−0.113	−0.002	−0.002	−0.016	−0.104
0.7	6.7	−0.180	−0.066	−0.058	−0.076	−0.168
0.85	9.8	−0.243	−0.138	−0.137	−0.140	−0.249
1.0	13.6	−0.278	−0.210	−0.204	0.055	−0.308

k (a_o^{-1})	E (eV)	6	7	8	9
0.1	0.14	0.229	0.223	0.294	0.209
0.2	0.54	0.293	0.281	0.303	0.285
0.3	1.2	0.277	0.253	0.286	0.267
0.4	2.2	0.215	0.186	0.219	0.208
0.5	3.4	0.140	0.105	0.141	0.134
0.6	4.9	0.064	0.022	0.063	0.061
0.7	6.7	−0.010	−0.055	−0.012	−0.012
0.85	9.8	−0.111	−0.151	−0.112	−0.112
1.0	13.6	−0.193	−0.221	−0.191	−0.194

Basis sets

First type of open-channel function with $\gamma = 0.75$

1. The 64 separable functions used in my earlier calculation with $\alpha = 0.575$[†].
2. As in 1 plus functions 1-3 in Table 2 with $\alpha = 0.575$.
3. As in 1 plus functions 1-8 in Table 2 with $\alpha = 0.575$.
4. As in 1 plus functions 1-6 in Table 2 with $\alpha = 0.3$.

Second type of open-channel function with $\delta = 1.0$

5. As in 1, but without χ_1.
6. As in 3, but without χ_1.
7. As in 4, but without χ_1.

Second type of open-channel function with $\delta = 0.75$

8. As in 1, but without χ_1, plus functions 1-6 in Table 2 with $\alpha = 0.575$.

Second type of open-channel function with $\delta = 1.25$

9. As in 8.

In all cases, $\beta = 0.2$

[†]Corrected results as given in the corrigendum[2].

under consideration. It is possible that the inclusion of Hylleraas-type functions in the basis set may alter this conclusion. This will be explored in future calculations.

The first significant rearrangement threshold is the positronium formation threshold at 8.63 eV[22]. I have made no attempt to take this into account in this calculation. I hope eventually to take it into account, but this will be very difficult, much more difficult than the inclusion of Hylleraas-type functions in the present calculation.

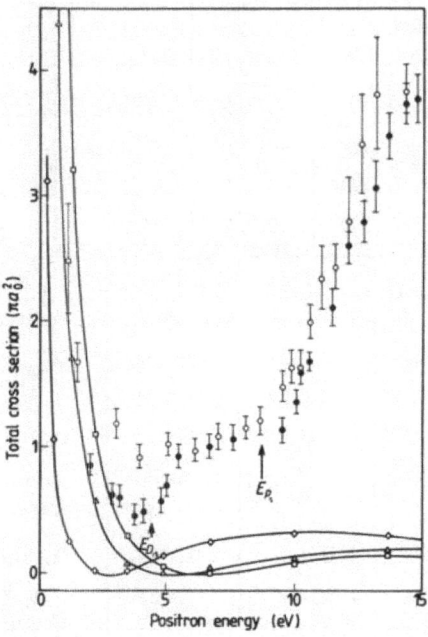

Figure 1. Comparison of experimental and theoretical total cross sections. O, Hoffman et al.[38]: ●, Charlton et al.[39]: Δ, this calculation (1st type of open-channel function and basis set 3 in Table 3.): □, this calculation (2nd type of open-channel function and basis set 6 in Table 3.): ◇, Armour[1,2].

The contribution to the total elastic cross section from the lowest partial wave obtained using both types of open-channel function are compared with experiment in Figure 1. It can be seen that the inclusion of the 8 Hylleraas-type functions very much improves the agreement with experiment at very low energies. The contribution to the total cross section is now sufficient to account for the experimentally observed total cross section up to about 2 eV. This is very much higher than I expected on the basis of my earlier calculation[1,2]. It is also higher than the corresponding result for the s-wave in the case of positron-helium scattering[30]. Note that the minimum in the experimental cross section for positron-helium scattering is at a positron energy of 2 eV, whereas in the case of the hydrogen molecule it is at 5 eV.

The separable functions play an important role in the calculation. Investigation shows that the omission of either the functions of Σ or Π electronic symmetry from the basis set leads to a significant decrease in the phase shift values, even if Hylleraas-type functions are included.

The separable functions are suitable for representing the system when the positron is not close to a molecular electron. In particular, I have shown that they and ψ_G give reasonably accurate values for the permanent quadrupole moment and dipole poarizabilities of the hydrogen molecule[1,2]. Note that they are not suitable for representing the asymptotic form of the exact closed channel function[31,16] but this is not important in the corresponding s-wave positron-helium scattering calculation for the k range 0.1 to 1.0 a_o^{-1}.

As pointed out earlier, the Hylleraas-type functions are suitable for representing the system when the positron is close to a molecular electron. They thus play a role complementary to the separable functions. The positron is attracted by the molecular electrons and this makes accurate representation of the region of configuration space in which the positron is close to an electron more important than in the case of the corresponding region in electron-molecule scattering. It was to be expected that accurate representation of this region would be essential for the calculation of $Z_{eff}(k)$, which measures the value of the wavefunction when the positron and an electron coincide[7,3,15]. My calculation shows that it is also essential if accurate phase shifts and total elastic cross sections are to be obtained.

Though I have been able to include Hylleraas-type functions in this calculation, the difficulties of the calculation have meant that I have only been able to use a rather restricted set of Hylleraas-type functions. I have not examined the effect of including higher members of the set of Hylleraas-type functions, e.g. functions in which r_{13} and r_{23} are replaced by $r_{13}M_{13} \cos(\phi_1 - \phi_2)$ and $r_{23}M_{23} \cos(\phi_2 - \phi_1)$ or $r_{13}r_{12}$ and $r_{23}r_{12}$, respectively, or both are repalced by $r_{13}r_{23}$. Nor have I examined the effect of including separable functions of Δ electronic symmetry. The corresponding functions are all included in Campeanu and Humberston's positron-helium scattering calculation[32,30,13]. I expect that the inclusion of such functions will lead to a further increase in $Z_{eff}(k)$ and the phase shift. However, I do not think that the effect will be as dramatic as when Hylleraas-type functions were first introduced into the basis set. As I mentioned earlier, it is also possible that allowing for mixing of partial waves will have a significant effect on the scattering parameters.

The target wavefunction takes into account 50% of the correlation energy of the hydrogen molecule ground state[22] and is thus considerably better than an SCF target wavefunction. However, it could be improved without complicating the calculation excessively by including products of functions of π symmetry in the molecular basis set used to calculate the target wavefunction. Note that Campeanu and Humberston's helium target wavefunctions H5 and H14[30,13] take into account 77% and 90%, respectively, of the correlation energy of helium.

Finally, though the rotational motion of the molecule has been taken into account within the fixed-nuclei approximation[33], the vibrational motion of the nuclei has not been taken into account. The inclusion of vibrational motion is likely to have a significant effect on the scattering parameters and on $Z_{eff}(k)$[34]. There is scope for a detailed study of the effects of rotational and vibrational motion within the adiabatic nuclei approximation[35]. This could be carried out reasonably straightforwardly by using the present calculation to obtain values of the scattering parameters and $Z_{eff}(k)$ for various values of the internuclear separation R.

I intend to investigate the effect of all these improvements in future calculations. Work is already in progress on the calculation of the lowest partial wave of Σ_u^+ symmetry using basis sets which contain Hylleraas-type functions. An outline of this calculation is given in an abstract elsewhere in these proceedings.

CONCLUSION

I have demonstrated the importance of including Hylleraas-type functions on the basis set if accurate results are to be obtained for the scattering parameters and the annihilation rate determining parameter, $Z_{eff}(k)$, for low energy positron-hydrogen-molecule scattering. This was very much to be expected both on theoretical grounds and in view of the success of the Kohn method with basis sets including such functions in calculations on low energy positron-atom scattering[13]. The value of 10.2 for $\bar{Z}_{eff}$ for thermal positrons at 293 K greatly reduces the discrepancy between the theoretical value and the experimental value[6] of 14.8. This discrepancy was pointed out by Heyland et al.[36] at the first Positron Satellite Conference on positron-gas scattering in Toronto in 1981 and has been the subject of some concern among experimentalists[37].

The calculation of the necessary matrix elements to include the Hylleraas-type functions was complicated but practicable. It was possible to avoid calculation of the most complicated three particle integrals by use of the method of models[23]. The remaining three particle integrals could be evaluated to high accuracy using the 'boundary derivative reduction' method of numerical integration of Handy and Boys[26]. The integral calculations as presently carried out are very time consuming but I can see ways of speeding them up. This should make possible the use of a more extended set of Hylleraas-type functions.

This is the first calculation which I have carried out for positron-hydrogen-molecule scattering which is, in any way, comparable in accuracy with those that have been carried out for positron-atom scattering[13]. Refinement of the method used in this calculation should make it possible to obtain as detailed an understanding of positron-hydrogen-molecule scattering as has already been obtained for positron-helium scattering[30].

ACKNOWLEDGEMENTS

I am grateful to Dr. D. J. Baker for his assistance with the calculation, to Dr. M. Charlton for encouraging me to calculate $\bar{Z}_{eff}$ for positron-hydrogen-molecule scattering and to the SERC(UK) for financial support for this research.

REFERENCES

1. E.A.G. Armour, 'Proc 3rd Int Workshop on Positron(Electron)-Gas Scattering, Detroit, 1985', World Scientific, Singapore, 1986, Invited papers, p.85.

2. E.A.G. Armour, J. Phys. B, 18, 3361 (1985); J. Phys B 20: Corrigendum (1987).

3. E.A.G. Armour and D.J. Baker, J. Phys. B, 18: L845 (1985).

4. J. Tennyson, J. Phys. B, 19, 4255 (1986).

5. P.G. Burke, I. Mackey and I. Shimamura, J. Phys. B, 10: 2497 (1977).

6. J.D. McNutt, S.C. Sharma and R.D. Brisbon, Phys. Rev. A, 20: 347 (1979).

7. D.M. Schrader and R.E. Svetic, Can. J. Phys., 60: 517 (1982).

8.   T. Kato, Commun. Pure Appl. Math., 10: 151 (1957).

9.   R.T. Pack and W. Byers Brown, J. Chem. Phys., 45: 556 (1966).

10. E. A. Hylleraas, Z. für Phys., 54: 347 (1929).

11. H.M. James and A.S. Coolidge, J. Chem. Phys., 1: 825 (1933).

12. C. Schwartz, Phys. Rev., 124: 1468 (1961).

13. J.W. Humberston, Adv. At. Mol. Phys., 15: 101 (1979).

14. D.C. Clary, Mol. Phys., 34: 793 (1977).

15. E.A.G. Armour and D.J. Baker, J. Phys. B, 19: L871 (1986).

16. E.A.G. Armour and D.J. Baker, J. Phys. B, in press.

17. C. Flammer, 'Spheroidal Wavefunctions', Stanford University Press, Stanford CA, 1957.

18. H. Takagi and H. Nakamura, J. Phys. B, 13: 2619 (1980).

19. H.S.W. Massey and R.O. Ridley, Proc. Phys. Soc.(London), A69: 659 (1956).

20. C. Schwartz, Ann. Phys. NY, 16: 36 (1961).

21. T. Kato, Phys. Rev., 80: 475 (1950).

22. E.A.G. Armour, J. Phys. B, 17: L375 (1984).

23. R.J. Drachman, J. Phys. B, 5: L30 (1972).

24. K. Rüdenberg, J. Chem. Phys., 19: 1459 (1951).

25. S.F. Boys and P. Rajagopal, Adv. Quantum Chem., 2: 1 (1965).

26. N.C. Hardy and S.F. Boys, Theo. Chim. Acta, 31: 195 (1973).

27. E.A.G. Armour, Mol. Phys., 26: 1093 (1973).

28. P.J. Roberts, J. Chem. Phys., 43: 3547 (1965).

29. A. Temkin and K.V. Vasavada, Phys. Rev., 160: 109 (1967).

30. R. I. Campeanu and J.W. Humberston, J. Phys. B, 10: L153 (1977).

31. L. Castillejo, I.C. Percival and M.J. Seaton, Proc. Roy. Soc. (London), A254: 259 (1960).

32. J.W. Humberston, J. Phys. B, 6: L305 (1973).

33. E.S. Chang and A. Temkin, Phys. Rev. Letts, 8: 399 (1969).

34. S. Sur and A.S. Ghosh, J. Phys. B, 18: L715 (1985).

35. F.H.M. Faisal and A. Temkin, Phys. Rev. Letters, 28: 203 (1972).

36. G.R. Heyland, M. Charlton, T. C. Griffith and G.L. Wright, Can. J. Phys., 60: 503 (1982).

37. M. Charlton, Private communication (1983).

38. K.R. Hoffman, M.S. Dababneh, Y.-F. Hsieh, W.E. Kauppila, V. Pol, J.H. Smart and T. S. Stein, Phys. Rev., $\underline{A25}$: 1393 (1982).

39. M. Charlton, T.C. Griffith, G.R. Heyland and G.L. Wright, J. Phys. B, $\underline{16}$: 323 (1983).

THE AB INITIO INCLUSION OF POLARISATION EFFECTS IN LOW-ENERGY POSITRON-

MOLECULE COLLISIONS USING THE R-MATRIX METHOD

Jonathan Tennyson and Grahame Danby

Department of Physics and Astronomy
University College London
Gower Street, London WC1E 6BT, U.K.

INTRODUCTION

Positron-molecule collisions at low energy have become a subject of great experimental interest (Charlton, 1985a, 1985b, and other articles in this volume), providing a concomitant theoretical challenge. As pointed out by Massey (1976) and others, collisions involving low-energy positrons provide one of the most stringent tests of our theoretical understanding of scattering. In electron collisions, the static and polarisation contributions to the overall potential are both attractive. In positron scattering, the static potential is generally repulsive leading to a cancellation between interactions which is difficult to represent.

Polarisation effects are particularly difficult to describe in positron scattering because of the attractive nature of the positron-electron interaction. This means that any accurate calculation must allow the motion of the positron and target electrons to be closely correlated. In particular, if a theoretical estimate is required for the annihilation rate of positrons in a gas, the correct representation of the correlated positron-electron motion at zero separation is essential. Satisfactory calculations of this property have only been performed on atomic and molecular systems with very few electrons (Humberston, 1986; Armour and Baker, 1986). These calculations used Hylleraas-type wavefunctions which explicitly included the positron separation from each target electron. Use of such wavefunctions is currently confined to simple systems because of the great difficulty of evaluating the integrals involved.

The calculation of cross-sections for electron or positron collisions with a frozen (static) target is relatively straightforward. However, the static model fails both to give cross-sections with the correct magnitude and the observed structure for many processes, including low-energy elastic scattering of positrons with most systems. To improve upon this it is necessary to consider the way in which the target is polarised by the charged projectile. Polarised calculations for positron-molecule collisions thus need to allow for the correlated motions between the positron and the target electrons. To do this ab initio is difficult; indeed Jain (1986a) recently commented that "calculating a true positron-molecule polarisation interaction by ab initio methods is a prohibitively difficult task".

111

With the exception of low-energy positron scattering from molecular
hydrogen which has been the subject of a series of detailed calculations
by Armour (1984, 1985) and Armour and Baker (1985, 1986), polarisation
effects have generally been included in e^+ - molecule collision calculations in an ad hoc way, see Morrison et al, (1984). For example, phenomenological polarisation potentials have been derived and adjusted to
reproduce experiment (Darewych, 1982; Horbatsch and Darewych, 1983). However,
such methods yield little insight and are of limited predictive value.

More recently Jain (1986a, 1986b, 1986c) has adapted a phenomenological
method of generating polarisation potentials so that the potentials can be
determined a priori. However, the close agreement found between his static
and polarised calculations on a number of systems, a feature not found in
more sophisticated calculations (Humberston, 1986), suggests that his
method neglects most of the polarisation effects.

The R-matrix method is one of a number of methods (see Burke and Noble,
1986) which have been successfully applied to electron-molecule collisions
at low and intermediate scattering energies. In this method short and
intermediate range polarisation effects can be included ab initio in a
number of ways: (1) by building an optical potential using 2-particle -
1-hole excitation of the target electrons. (2) by explicit inclusion of
excited electronic states of the target in the close-coupling expansion.
(3) by the use of so-called pseudo-states in the close-coupling expansion,
generated ab initio to mimic the target polarisability. In principle, a
converged short-range polarisation potential can be obtained by using
either method (1) or (2) although in electron-scattering calculations the
truncated expansions have been found to be complementary. For example,
poor results are obtained for the Feshbach resonances in the e^- - H_2^+ system
unless both are included (Tennyson et al, 1984; Tennyson and Noble, 1985).
Similarly both methods were utilised in recent calculations on elastic
e^- - NO scattering (Tennyson and Noble, 1986).

In this work, the primary method of including polarisation in the
calculation will be via method (1), although other possibilities will be
borne in mind. The inclusion of polarisation effects when the positron and
target are well separated by explicity including the target polarisabilities
in the asymptotic potential will also be investigated.

THEORY

Use of the R-matrix method was originally suggested by Wigner (1946a,b)
and Wigner and Eisenbud (1947) for the study of nuclear reactions. Burke
et al (1971) showed that this theory could be adapted in a natural way to
electron-atom scattering. Schneider (1975a,b) and Burke et al (1977)
pioneered the application of R-matrix theory to electron-molecule collisions.
The molecular R-matrix method has now become one of several tried and tested
methods of performing ab initio electron-molecule scattering calculations.
See for example the comparative studies by Baluja et al (1985), Schneider
and Collins (1985) and Lima et al (1985) on electron-H_2 scattering. The
first application of the R-matrix method to positron-molecule scattering
was by Tennyson (1986) who studied e^+ - H_2 and e^+ - N_2 collisions.

The physics behind the molecular R-matrix method is the division of
space into two regions, see Figure 1. In the internal region there are
complicated, multicentred interactions including the full Coulomb interaction between the charged projectile and the target electrons. It is in
this region that accurate methods of representing polarisation must be
sought. In the external region, it is assumed that all interactions can
be represented using a single-centred expansion given by the asymptotic

charge distribution and polarisability of the target. For this approximation to be valid it is necessary for the charge distribution of the target to be entirely confined to the internal region. This region is bounded by a sphere of radius a centred at the molecular centre of mass, G. Typically, and in all the calculations discussed here, $a = 10a_o$.

In the internal region the wavefunction for a positron-molecule collision can be written

$$\psi(\underline{x}_1, \cdots \underline{x}_N, \underline{r}) = \sum_i \Phi_i(\underline{x}_1, \cdots \underline{x}_N) \, F_i(\underline{r}) + \sum_j \phi_j(\underline{x}_1, \cdots \underline{x}_N) \, \phi'_j(\underline{r}) \, b_j \quad (1)$$

where $\{\underline{x}_n\}$ and $\underline{r}$ are the space-spin coordinates of the N target electrons and the positron respectively. The first summation in (1) runs over products of target electronic states or pseudo-states Φ_i and contin-

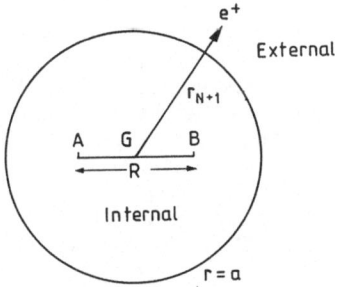

Figure 1. Schematic representation of the R-matrix regions for positron scattering from diatomic AB with centre of mass G.

uum functions, F_i, used to carry the scattered positron. The first term in the sum contains Φ_o which represents the ground state of the target, here given within the Hartree-Fock approximation. Further terms in the sum can be used to include charge polarisation effects by coupling to excited and pseudo states of the target. As will be discussed below, these contribute mainly to the polarisation in the intermediate region.

The final summation in (1) is over L^2 configurations generated to allow for high angular momentum (correlation) and polarisation effects in the region of the nuclei. In this sum ϕ_j and ϕ'_j represent configurations generated from the short-range electronic and positronic orbitals respectively. In practice these orbitals comprise the occupied and virtual target molecular orbitals and the same set is used to carry both the electronic and positronic motions.

The continuum functions are expressed as a partial wave expansion about the molecular centre of mass G in an R-matrix basis

$$F_i(\underline{r}) = \sum_{\ell j} r^{-1} U_{\ell j}(r) Y_{\ell m}(\hat{r}) a_{ij} \tag{2}$$

where $Y_{\ell m}$ is a spherical harmonic. For linear molecules m but not ℓ is a constant of motion. The radial basis functions are obtained by solving numerically the zero-order coupled equations

$$\left(\frac{d^2}{dr^2} - \frac{\ell(\ell+1)}{r^2} + V_o + k_{\ell j}^2 \right) U_{\ell j}(r) = 0 \tag{3}$$

subject to the boundary conditions

$$\left.
\begin{aligned}
U_{\ell j}(0) &= 0 \\
\frac{a}{U_{\ell j}} \frac{dU_{\ell j}}{dr}\bigg|_{r=a} &= 0
\end{aligned}
\right\} \text{ all } \ell \text{ and } j \tag{4}$$

at the origin and R-matrix boundary. These functions are then Schmidt orthogonalised to the short-range positronic molecular orbitals, ϕ_i' in (1). The constraint on the derivative of the function at a necessitates the use of a Buttle (1967) correction.

The functions denoted by eq.(1) are used to solve the $N + 1$ particle Hamiltonian within the R-matrix sphere, ie for $r < a$. This truncation of space leads in general to the introduction of a Bloch term in $\hat{H}_{N+1}$, but with the boundary conditions specified by (4) this is actually a null operator. Diagonalisation of the secular matrix obtained by solving $\hat{H}_{N+1}$ with the specified basis leads to a determination of the variational coefficients a_{ij}, eq.(2), and b_{ij}, eq.(1) and eigenenergies. This portion of the problem is solved using an adapted (Noble, 1982) version of the Quantum Chemistry code ALCHEMY (McClean, 1971).

If the target system is a diatomic molecule, then it is also necessary to consider the internuclear separation R of the nuclei, and remember that these nuclei can both vibrate and rotate. Vibrational motion will not be considered here, although work on this problem (Danby and Tennyson, 1987) is now underway. We will only comment that it is a relatively straightforward matter to adapt a series of fixed nuclei calculations to allow for vibrational motion. This has been done for electron molecule scattering within the R-matrix method using both the adiabatic nuclei approximation (Tennyson, 1987) and a non-adiabatic method (Morgan, 1986). The calculations discussed here all fixed the nuclei at their equilibrium separations. Some mention of the approximate inclusion of rotational motion will be given later.

Once the internal problem has been solved it is possible to construct the R-matrix on the boundary. For a positron scattering with energy E this is defined as (Burke et al, 1977)

$$R_{\ell,\ell'}(a) = \frac{1}{2a} \sum_k \frac{w_{\ell k}^m(a) \, w_{\ell' k}^m(a)}{E_k - E} \tag{5}$$

where $w_{\ell k}^m(a)$ represents the summed amplitude of eigenvector k on the boundary. As only the numerical continuum functions, F_i, have amplitude at a, it is only necessary to consider the amplitude of these functions. Furthermore, as these functions are expressed as a partial wave expansion it is possible to construct the boundary amplitudes, $w_{\ell k}^m(a)$, for each partial wave of the truncated expansion. The resulting R-matrix is a symmetric matrix whose rows and columns can each be associated with a

114

target state and a partial wave, ℓ , of the positron.

Once the R-matrix has been constructed for a given energy, its solutions are found by a mixture of propagation and asymptotic expansion. Standard computer codes are available for this (Baluja et al, 1982; Noble and Nesbet, 1984).

RESULTS AND DISCUSSION

In previous work (Tennyson, 1986) using the R-matrix method for positron-molecule collisions, scattering from both H_2 and N_2 targets was considered. The calculations involving H_2 were performed so the method could be compared with the benchmark Kohn variational calculations of Armour (1985). Very good agreement between the R-matrix and Kohn variational calculation was found for a number of models. This was the more surprising firstly because there were a number of differences between the calculations which one might have expected to be significant and secondly because even the most elaborate of these calculations gave only modest agreement with experiment. For more recent calculations on e^+ - H_2 collisions the reader is referred to Armour and Baker (1986) and Armour's article in this volume.

Table 1 presents elastic cross-sections for low-energy impact of positrons on N_2 in its equilibrium geometry of 2.068 a_0. The results extend those of Tennyson (1986) whose corrected SP4 results are given for comparison. These calculations used 2-particle - 1-hole excitations involving all the N_2 target electrons and virtual orbitals of σ, π and δ symmetry. These polarised results gave cross-sections significantly lower than those calculated within the static approximation. Polarisation of N_2 $1\sigma_g$ and $1\sigma_u$ orbitals and excitation into virtual orbitals of δ symmetry were found to give only minor changes to the cross-section.

Table 1. Elastic cross-sections in a_0^2 for e^+ - N_2 collisions with the four lowest symmetries. Results are given using (a) the "SP4" treatment of polarisation in the internal region; (b) with the asymptotic form of the e^+ - N_2 polarisation interaction included in the external region. These results replace those of Tennyson (1986) which used a quadrupole moment with the wrong sign.

k^2 (Ryd)	Σ_g		Σ_u		Π_u		Π_g	
	a	b	a	b	a	b	a	b
0.01	10.56	1.70	0.82	3.67	0.40	0.40	0.04	0.29
0.04	11.04	9.11	0.51	1.91	0.27	0.26	0.04	0.52
0.09	11.39	11.03	0.24	0.54	0.19	0.22	0.04	0.38
0.16	11.82	11.29	0.07	0.17	0.17	0.24	0.05	0.21
0.25	12.19	11.83	0.02	0.05	0.20	0.20	0.09	0.18
0.36	12.10	11.95	0.18	0.16	0.30	0.27	0.18	0.26
0.49	11.66	11.54	0.55	0.52	0.43	0.41	0.30	0.38
0.64	11.12	11.02	1.04	1.00	0.62	0.59	0.50	0.55

Table 1 gives two extensions of the previous calculations: the inclusion of asymptotic polarisation effects and consideration of the Π_g symmetry. The results concentrate on the region below the positronium formation threshold as this channel was not included in the calculations. Comparison of the previous results (Tennyson, 1986) with experiment (Hoffman et al, 1982; Charlton et al, 1983) showed good agreement for the total $(\Sigma_g + \Sigma_u + \Pi_u)$ cross-sections in this region, except at the lowest energies (below 0.1 Ryd). This was attributed to the neglect of "long-range" polarisation in the calculations.

Within the R-matrix method asymptotic polarisation effects can be accounted for in the external region in a straightforward manner. In this region the target wavefunction is assumed to have zero amplitude and the potential can consequently be expressed as a simple multipole expansion. In Table 1 column a the asymptotic potential consisted only of the N_2 quadrupole moment, - 1.33 a.u., as calculated using the target wave-function (Nesbet, 1964). In Table 1 column b this was augmented by the N_2 isotropic and anisotropic polarisabilities, α_0 = 11.43 a.u. and α_2 = 3.36 a.u. (Morrison and Hay, 1977).

From inspection of Table 1 it is clear that inclusion of the long-range polarisation contributions has little effect on the predicted cross-section at the higher energies considered. At the lowest energies the asymptotic potential is dominant and the asymptotic polarisabilities thus have a large effect. As the polarisation potential is attractive, inclusion of the asymptotic polarisabilities acts to raise the cross-section when the eigenphase sum is positive and lower it otherwise. The general behaviour of Table 1 can then be accounted for by noting that in low-energy positron-molecule collisions the eigenphases generally start positive and decrease, causing the cross-sections to go through a minimum as the eigen-phase sum goes through zero. As noted previously (Tennyson, 1986), the eigenphase sum for Σ_g e^+ - N_2 collisions within the SP4 model is always negative. This is an artefact of the calculation attributable to the lack of both intermediate and long-range polarisation. Only the latter has been added here.

To appreciate this problem fully it is necessary to consider in detail how the R-matrix method treats different regions of space, especially when one is concerned with dynamical polarisation effects. As outlined above, polarisation is most easily represented in the external region. It is this form of the potential, which is insensitive to whether the projectile is a positron or an electron, which has been adopted by Darewych (1982) and Jain (1986a, 1986b, 1986c).

In the internal region, the present calculations utilised 2-particle - 1-hole excitations of the target to represent polarisation. These config-urations are built from the L^2 functions, in our calculations molecular orbitals as expansions of Slater-Type Orbitals centred on the nuclei, which are used to represent the target. The functions are concentrated in the region of the nuclei and are assumed to have zero amplitude at the R-matrix boundary. This means that the functions do not give a good representation of the intermediate region between the target and the R-matrix boundary. In this region the positron flux is carried by the continuum functions, F_i in eq.(1). However, in the calculations presented here no allowance is made for polarisation in this region.

One method that has been commonly used in calculations on scattering from atomic targets is the use of so-called polarised pseudo-states to represent polarisation effects in this intermediate region (eg. Morgan, 1982a, 1982b). This method involves putting extra terms in the close-coupling expansion, as represented by the first sum in eq.(1). These terms are not added to represent actual electronically-excited states of the target, but rather included to allow for the target polarisability. It is through these terms that account can be taken of intermediate-range polarisation effects. Calculations using this method for $e^- - N_2$ scattering are currently being performed (Gillan et al, 1987) and it is our aim to use this procedure for positron-scattering calculations.

Table 1 also shows the cross-section for $e^+ - N_2$ collisions with Π_g symmetry. This symmetry was previously neglected because, as the lowest partial wave is a d wave, it was assumed that its contribution to the total-cross-section would be small at the energies considered. This is indeed true at the lowest energies, but rather surprisingly the cross-section was found to rise quite steeply with scattering energy. Inspection of the eigenphase sums showed that this increase in cross-section was caused by the presence of a broad resonance at 1.54 Ryd (width 0.23 Ryd), well above the positronium formation threshold. No such resonance structure or increase in cross-sections was found in any of the static approximation calculations.

Cross-sections have been computed for positron scattering from a CO target with Σ, Π and Δ symmetry. These calculations are for CO frozen at its equilibrium separation and with no account taken of its rotational motion. This latter approximation leads to an immediate overestimation of the elastic cross-section. This is because CO has a permanent dipole, the effect of which, at long-range, is significantly reduced by rotational motion. A detailed consideration of rotationally-resolved $e^+ - CO$ collisions, with the rotational motion accounted for within the multipole-extracted adiabatic-nuclei (MEAN) approximation of Norcross and Padial (1982), will be published elsewhere (Tennyson and Morgan, 1987). This work will include detailed comparison with the experiments of Kwam et al (1983) and Sueoko and Mori (1984). For the present we will concern ourselves with the effect of including polarisation upon the fixed nuclei cross-sections.

In Table 2 results are presented for $e^+ - CO$ collisions with the static and static plus polarisation (SP) approximation. The polarised calculations correspond closely to the $e^+ - N_2$ SP4 calculations discussed above. Further details of these calculations can be found in Tennyson and Morgan (1987).

Table 2.   Positron-CO collision cross-sections in a_0^2.

k^2 (Ryd)	Σ		Π		
	Static	SP	Static	SP	SP[a]
0.01	38.72	19.31	12.55	12.61	12.57
0.04	28.43	9.32	2.29	2.17	2.14
0.09	27.58	11.03	1.39	0.61	0.74
0.16	25.29	12.36	1.94	0.42	0.64
0.25	23.14	13.45	2.30	0.55	0.48
0.36	21.29	14.06	2.70	0.98	0.50
0.49	19.60	14.24	3.17	1.45	0.62
0.64	18.16	14.25	3.61	1.90	0.97

a
Static plus polarisation (SP) with the $4\sigma \rightarrow 2\pi$ and $5\sigma \rightarrow 2\pi$ target excitations removed.

A feature of Table 2 is that in the SP approximation the cross-section in both Σ and Π symmetries is largest at $k^2 = 0.01$ Ryd; the cross-section drops and then increases. However, the Π cross-section increases very much more rapidly at the higher energies than the Σ cross-sections. Inspection of the corresponding eigenphase sums shows that there are two broad resonances at 1.46 Ryd and 1.81 Ryd. Both these features are well above the positronium formation threshold, but as in the $N_2 \Pi_g$ symmetry, appear to affect the cross-sections below this threshold. Similar behaviour is shown by e^+ - CO calculations with Δ symmetry.

Analysis of the 2-particle - 1-hole configurations which contribute to the polarisation potentials shows that the two Π symmetry resonances are associated with excitations of a target electron from the 4σ and 5σ orbitals into the 2π virtual orbital. Removal of configurations which involve target excitations from $(4\sigma, 5\sigma)$ to 2π from the wavefunction results in the complete disappearance of the resonances and the decrease in cross-sections.

Resonances which occur at intermediate scattering energies, i.e. at energies above which there are open channels which have not explicitly been included in the calculations, are commonly found in scattering calculations. These resonances occur in calculations which allow for polarisation effects and can be thought of as being due to the open channel(s) being included implicitly in certain regions, such as near the nuclei. For example polarised e^- - CO R-matrix calculations by Salvini et al (1984) found a large number of pseudo-resonances which were removed by averaging the T-matrix. Similarly Morgan (1982b) also observed pseudo-resonances in e^- - H calculations. Recently Whelan et al (1987) have shown that the effects of these can be mitigated, without loss of accuracy, by careful treatment of the offending terms in the polarisation potential. Such a procedure has been followed in the final column of Table 2 where results are presented for e^+ - CO SP calculations with Π symmetry but with the 4σ and 5σ to 2π target excitations omitted.

CONCLUSIONS

The inclusion of polarisation effects in low-energy positron-molecule collisions using the R-matrix method has been discussed. It has been shown how, within the R-matrix method, different methods can be used to represent the polarisation potential in three regions which can approximately be labelled short, intermediate and long-range. Sample calculations have shown that polarisation must be represented in all 3 regions if accurate results are desired over the whole energy range.

Intermediate-range polarisation effects can be represented by the use of polarised pseudo-states. In the current calculations dynamical polarisation effects at short-range have been represented by an optical potential constructed from two-particle one-hole excitations of the target. This method has been successfully employed in electron-scattering calculations but has the disadvantage that it is difficult to demonstrate convergence with it.

One effect of the inclusion of polarisation is that pseudo-resonances can occur in the region above the threshold to any channel which has not been explicitly included in the calculation. In positron scattering positronium formation always offers a low-lying channel above which pseudo-resonances may occur. If, as found here, these effects extend below the positronium formation threshold then these effects may be thought of as due to virtual positronium formation. However, experience (Seiler et al, 1971; Drachman, 1975) has shown that such resonance effects are usually an artifact of an incomplete calculation and are best removed. This is the procedure adopted here.

Finally we note that the first resonance structure to be observed in positron-molecule collisions has been recently reported (Katayama et al, 1987). As the static potential for e^+ collisions is generally repulsive, any resonance effect should be intimately related to polarisation effects. It is our intention to study positron scattering from a number of candidate molecule targets in the hope of identifying possible resonances and even bound states.

References

Armour, E.A.G., 1984, Application of a generalisation of the Kohn variational method to the calculation of cross-sections for low-energy positron-hydrogen-molecule scattering, J.Phys.B: At.Mol.Phys., 17:L375.

Armour, E.A.G., 1985, The inclusion of π functions in the treatment of low-energy positron-hydrogen-molecule scattering by a generalisation of the Kohn method, J.Phys.B: At.Mol.Phys., 18:3361.

Armour, E.A.G. and Baker, D.J., 1985, A calculation of Z_{eff} for low-energy positron-hydrogen-molecule scattering, J.Phys.B: At.Mol.Phys., 18:L845.

Armour, E.A.G. and Baker, D.J., 1986, An improved theoretical value for Z_{eff} for low-energy positron-hydrogen-molecule scattering, J.Phys.B: At.Mol. Phys., 19:L871.

Baluja, K.L., Burke, P.G. and Morgan, L.A., 1982, R-matrix propagation program for solving coupled second-order differential equations, Comput. Phys. Comms., 27:299.

Baluja, K.L., Noble, C.J. and Tennyson, J., 1985, Spin-forbidden electronic excitation in e^- - H_2 collisions, J.Phys.B: At.Mol.Phys., 18:L851.

Burke, P.G. and Noble, C.J.,1986, Theory of electron collisions with diatomic molecules, Comments At.Mol.Phys., 18:181.

Burke, P.G., Hibbert, A. and Robb, W.D., 1971, Electron scattering by complex atoms, J.Phys.B: At.Mol.Phys., 4:153.

Burke, P.G., Mackay, I. and Shimamura, I., 1977, R-matrix theory of electron-molecule scattering, J.Phys.B: At.Mol.Phys., 10:2497.

Buttle, P.J.A., 1967, Solution of Coupled equations by R-matrix techniques, Phys. Rev., 160:719.

Charlton, M., 1985a, Atomic Physics with Positrons and Positronium, Comments in At.Mol.Phys., 16:133.

Charlton, M., 1985b, Experimental studies of positron scattering in gases, Rep.Prog.Phys., 48:737.

Charlton, M., Griffith, T.C., Heyland, G.R. and Wright, G.L., 1983, Total scattering cross-sections for low-energy positrons in molecular gases H_2,N_2,CO_2,O_2 and CH_4, J.Phys.B: At.Mol.Phys., 16:323.

Danby, G. and Tennyson, J., 1987, to be published.

Darewych, J.W., 1982, Elastic scattering and annihilation of low-energy positrons by molecular nitrogen, J.Phys.B: At.Mol.Phys., 15:L415.

Drachman, R.J., 1975,Feshbach resonances in positron-hydrogen scattering, Phys.Rev.A., 12:340.

Gillan, C.J., Noble, C.J. and Burke, P.G., 1987, to be published.

Hoffman, K.R., Dababneh, M.S., Hsieh Y.-F., Kauppila, W.E., Pol, V., Smart, J.H. and Stein, T.S., 1982, Total cross-section measurements for positrons and electrons colliding with H_2, N_2 and CO_2, Phys.Rev.A, 25:1393.

Horbatsch, M. and Darewych, J.W., 1983, Model potential description of low-energy e^+ - CO_2 scattering, J.Phys.B: At.Mol.Phys., 16:4059.

Humberston, J.W., 1986, Positronium - its formation and interaction with simple systems, Adv.At.Mol.Phys., 32:1.

Jain, A., 1986a, Positron-CO scattering below the positronium threshold, J.Phys.B: At.Mol.Phys., 19:L105.

Jain, A., 1986b, Vibrational excitation of $v' = 1$ and 2 levels of CO molecules by positron impact below the positronium formation threshold, J.Phys.B: At.Mol.Phys., 19:L379.

Jain, A., 1986c, Positron-monosilhane (SiH_4) collisions at low, intermediate and high energies using a spherical complex optical potential approach, J.Phys.B: At.Mol.Phys., 19:L807.

Katayama, V., Sueoka, O. and Mori, S., 1987, Inelastic cross-section measurements for slow positron-O_2 collisions, J.Phys.B: At.Mol.Phys., 20:1645-57.

Kwam, Ch.K., Hsieh, Y.F., Kauppila, W.E., Smith, S.J., Stein, T.S., Uddin, M.N. and Dababneh, M.S., 1983, $e^{\pm}$ - CO and $e^{\pm}$ - CO_2 total cross-section measurements, Phys.Rev.A., 27:1328.

Lima, M.A.P., Gibson, T.L., Huo, W.M. and McKoy, V., 1985, Cross-sections for electron impact excitation of the $b\ ^3\Sigma_u^+$ state of H_2; an application of the Schwinger multichannel variational method, J.Phys.B: At.Mol.Phys., 18:L865.

McClean, A.D., 1971, in: "Conference on Potential Energy Surfaces in Chemistry", W.A. Lester, Jr., ed., IBM Research Laboratory, San Jose.

Morgan, L.A., 1982a, Positron impact excitation of the $n = 2$ levels of hydrogen, J.Phys.B: At.Mol.Phys., 15:L25.

Morgan, L.A., 1982b, Angular distributions for the electron impact excitation of the $n = 2$ levels of atomic hydrogen, J.Phys.B: At. Mol.Phys., 15:4247.

Morgan, L.A., 1986, Resonant vibrational excitation of N_2 by low-energy electron impact, J.Phys.B: At.Mol.Phys., 19:L439.

Morrison, M.A. and Hay, P.J., 1977, Ab initio static polarisabilities of N_2 and linear symmetric CO_2 in the Hartree-Fock approximation: variation with internuclear separation, J.Phys.B: At.Mol.Phys., 10, L647.

Morrison, M.A., Gibson, T.L. and Austin, D., 1984, Polarisation potentials for positron-molecule collisions: positron-H_2 scattering, J.Phys.B: At.Mol.Phys., 17:2725.

Nesbet, R.K., 1964, Electronic structure of N_2, CO and BF, J.Chem.Phys., 40:3619.

Noble, C.J., 1982, "The ALCHEMY linear molecule integral generator", Daresbury Laboratory Technical Memorandum DL/SCI/TMT33T.

Noble, C.J. and Nesbet, R.K., CFASYM, a program for the calculation of the asymptotic solutions of the coupled equations of electron collision theory, Comput. Phys. Comms., 33:399.

Norcross, D.W. and Padial, N.T., 1982, The multipole-extracted adiabatic-nuclei approximation for electron-molecule collisions, Phys.Rev.A., 25:226.

Salvini, S., Burke, P.G. and Noble, C.J., 1984, Electron scattering by polar molecules using the R-matrix method, J.Phys.B: At.Mol.Phys., 17:2549.

Schneider, B.I., 1975a, R-matrix theory for electron-atom and electron-molecule collisions using analytic basis set expansions, Chem.Phys. Letts., 31:237.

Schneider, B.I., 1975b, R-matrix theory for electron-molecule collisions using analytic basis set expansions II electron-H_2 scattering in the static-exchange model, Phys.Rev.A., 11:1957.

Schneider, B.I. and Collins, L.A., 1985, Electronic excitation of the $b\ ^3\Sigma_u^+$ state of H_2 by electron impact in the linear algebraic approach, J.Phys.B: At.Mol.Phys., 18:L857.

Seiler, G.J., Oberoi, R.S. and Callaway, J., 1971, Algebraic close-coupling calculations of the scattering of electrons and positrons by hydrogen, Phys.Rev.A., 3:2006.

Sueoko, O. and Mori, S., 1984, Total cross-sections for electrons and positrons colliding with N_2, CO and CO_2 molecules, J.Phys.Soc. Japan, 53:2491.

Tennyson, J., 1986, Low-energy, elastic positron-molecule collisions using the R-matrix method: e^+ - H_2 and e^+ - N_2 scattering, J.Phys.B: At.Mol.Phys., 19:4255.

Tennyson, J., 1987, Fully vibrationally resolved photoionisation of H_2 and D_2, J.Phys.B: At.Mol.Phys., 20:L375.

Tennyson, J. and Morgan, L.A., 1987, Rotational and polarisation effects on low-energy positron-CO scattering using the R-matrix method, J.Phys.B: At.Mol.Phys., in press.

Tennyson, J. and Noble, C.J., 1985, Low-energy electron - H_2^+ scattering: variation of resonance parameters with internuclear separation, J.Phys.B: At.Mol.Phys., 18:155.

Tennyson, J., Noble, C.J. and Salvini, S., 1984, Low-energy electron - H_2^+ collisions using the R-matrix method, J.Phys.B: At.Mol.Phys., 17:905.

Whelan, C.T., McDowell, M.R.C. and Edmunds, P.W., 1987, Electron impact excitation of atomic hydrogen, J.Phys.B: At.Mol.Phys., 20:1587.

Wigner, E.P., 1946a, Resonance reactions and anomolous scattering, Phys.Rev., 70:15.

Wigner, E.P., 1946b, Resonance reactions, Phys.Rev., 70:606.

Wigner, E.P. and Eisenbud, L., 1947, Higher angular momenta and long-range interactions in resonance reactions, Phys.Rev., 72:29.

POSITRON–MOLECULE COLLISIONS: ELASTIC, INELASTIC AND DIFFERENTIAL CROSS

SECTION CALCULATIONS

Ashok Jain

Physics Department, Cardwell Hall
Kansas State University
Manhattan, KS 66506 USA

INTRODUCTION

There are a whole variety of simple diatomic (H_2, D_2, N_2, O_2, CO) to polyatomic (CO_2, CH_4, N_2O, NH_3, H_2O, SiH_4) molecules for which positron (e^+) scattering has been investigated in the laboratory (see a recent review by Charlton[1] and references therein). Here we are concerned only with the non-linear systems mainly the e^+ collisions with CH_4, SiH_4, NH_3 and H_2O molecules. For CH_4, there are quite a few experimental studies[2-6] for the total (σ_t) cross sections, while for the other three gases only the measurements of Sueoka and Mori[7-9] are available for comparison with theoretical results.

We divide this article in two parts depending upon the incident e^+ energy E (eV): 1. Low-energy region, i.e., $E < E_{Ps}$, where E_{Ps} is the positronium (Ps) formation threshold of the molecule. 2. Intermediate-energy region, i.e., $E_{Ps} < E <$ few hundred eV. The reason of this division is quite obvious, since above E_{Ps} energy, the σ_t are dominated by several rearrangement channels (Ps formation, ionization, dissociation, electronic excitation, etc.). However, below E_{Ps} energy, elastic channel dominates over inelastic (mainly the rotational and/or vibrational) processes and, therefore, calculated integral (or rotationally summed) cross sections (σ_i) can easily be compared with experimental σ_t. There are only a few theoretical studies on the rotational and vibrational transitions in molecules due to positron impact (see Jain[10-12] and references therein), while the experiments have still to be performed for the nuclear excitation channels.

For low energy e^+ – molecule scattering the absence of exchange-correlation makes calculation simpler but at the same time the dynamic correlations in the target electrons in the presence of incoming e^+ (the socalled polarization force, V_p) is not known exactly except its asymptotic form ($- \alpha_0/2r^4$, α_0 is the polarizability of the target; note that we have anisotropic part also depending upon the molecular point group symmetry). The extrapolation of the asymptotic V_p in the molecular region is still a challenging problem in $e^{\pm}$ – molecule collisions. Since the V_p^+ is still attractive for e^+, it has cancellation effects with the repulsive static Coulomb term (V_{st}^+). At low energies, this cancellation introduces a zero potential point and a weak attractive well in the total optical potential in the outer region of the molecular charge density (for example, in case of SiH_4, this well occurs at about 4.5 au radial distance with an energy of .025 au). Consequently the scattering parameters strongly depend upon the structure of this well specifically at very low energies. The non-adiabatic effects in the polarization interaction also become excessively important near the origin and at high impact energies.

In almost all the e^+ – molecule calculations, published so far, an e^- polarization potential (V_p^-)has been invoked for the e^+ case, with the only exception for the H_2 molecule by Morrision and coworkers[13] who obtained approximate e^+ – H_2 polarization potential from ab initio theory and demonstrated that indeed the cross sections are sensitive to the charge of the incoming particle. The issue of a true e^+ – molecule polarization interaction has recently been discussed thoroughly by Morrision.[14] Owing to obvious pragmatic difficulties in evaluating a charge–dependent parameter–free polarization potential for the present e^+–polyatomic molecules, we however, still apply the same V_p^- for the following results. This limitation in the present work is not serious except near the lower end of the present energy regime.

Here we have chosen two non–polar (CH_4 and SiH_4) and two polar (NH_3 and H_2O) gases for e^+ scattering. It is a well known fact that the e^- scattering with polar gases is much more difficult than the corresponding scattering with non–polar gases both in theory and experiment. The strong long–range dipole interaction presents severe problems in determining the σ_i ; in the fixed–nuclei theory, the differential cross sections (DCS) become infinite at zero angle leading to an undefined σ_i. The same problem is encountered in the present e^+ case too. However, at intermediate energies the elastic channel takes over the dipole scattering which, as a weak effect, can be included via the first Born approximation (FBA) in an incoherent manner. For the e^+ case, the FBA prescription may be questioned for the simple reason that the repulsive dipole positron term should give different cross sections, while the FBA cross sections do not depend upon the charge of the particle. Nevertheless, the error introduced should be small at these higher energies without affecting the qualitative behaviour of the cross sections as compared with measured quantities.

Several interesting points will be emphasized in the following discussion for e^+–molecule scattering with respect to their counterpart e^- scattering. For example, the $e^\pm$ DCS differ greatly with each other for all the four molecules studied here. This stimulates the question of merging the two sets of cross sections at higher energies. We also noticed that $e^\pm$–SiH_4 collision properties are very close in quality with the $e^\pm$–Ar system.

THEORETICAL MODEL

In order to set up close–coupling (CC) scattering equations for the scattered e^+ function F(r), we need several simplifying assumptions: for example, only ground electronic state of the target is considered; rigid–rotator approximation is valid; the effect of virtual or real Ps formation channel is neglected; below electronic excitation threshold, the virtual electronic transitions are taken into account via a local real polarization potential, etc. The infinite set of coupled differential equations for the F(r) for a particular irreducible representation (IR) ($p\mu$) of the molecular point group in the body-fixed coordinate frame is written as,[15]

$$\left[\frac{d^2}{dr^2} - \frac{\ell(\ell+1)}{r^2} + k^2\right] F_{\ell h}^{p\mu}(r) = 2 \sum_{\ell'h'} \left(V_{\ell h,\ell'h'}^{A_1,ST}(r) + V_{\ell h,\ell'h'}^{A_1,POL}(r)\right) F_{\ell'h'}^{p\mu}(r)$$

(1)

where the direct potential matrix $V_{\ell h, \ell' h'}^{A_1, ST}(r)$ defines the coupling
between two channel functions through the operator of Coulomb forces of
all the charge particles and $V_{\ell h, \ell' h'}^{A_1, POL}$ is the model polarization interaction. Note that both the potential functions belong to totally
symmetric $^1A_1$ IR (in the following, we drop the index A_1). In our
approximation for the model potentials, $V_{\ell h, \ell' h'}^{ST}(r)$ is generated from
near-Hartree Fock one-center-expansion (OCE) wave functions with enough
terms in the expansion to ensure convergence, while for the $V_{\ell h, \ell' h'}^{POL}$ (to
be denoted for simplicity by $V_p(r)$), we employ the e^- potentials derived
approximately ab initio by Jain and Thompson[16] (to be denoted by V_p^{JT}
potential) without involving any adjustable parameter.

Another parameter free model for V_p used in recent years for e^-
scattering is the socalled correlation-polarization potential (CPP)[17] (to
be denoted by V_p^{CPP}), which has been proved to be quite successful to
yield reliable cross sections for e^- molecule (atom) collisions at low
energies. We make use of both of these parameter-free e^- polarization
potentials in e^+ scattering events. In using V_p^- for V_p^+, we can only say
that up to second-order perturbation theory, both the e^- and the e^+
distort the target charge distribution in a similar manner. The third-
order effects (giving rise to different positron term) may be important
at very low energy region. In the later model (V_p^{CPP}), one calculates
correlation energy (which depends only on the target density) at each e^-
position and joins it smoothly with the $-\alpha_0/2r^4$ potential at large
distances where they first cross each other. This simple prescription
for the complicated nature of the polarization potential is apparently
successful particularly in those situations where exchange-correlation is
treated either exactly or very close to exact treatment; otherwise the
success of the CPP potential is only satisfactory. The use of the CPP
for the e^+ case raises further questions: since the e^+ spends less time
in the high-electron-density region near the nucleus due to repulsive
potential, the effects of positron correlation with core electrons are
supposed to be smaller.

The JT and CPP potentials both do not depend upon the incident
energy and include non-adiabatic effects very crudely. At higher
energies (E > 100 eV) however, it is essential to introduce non-adiabatic
energy-dependent polarization potential. We invoke a simple form

$$V_p(r) = \frac{-\alpha_0 r^{2n-4}}{2(r_c^2 + r^2)^n}, \qquad n=0,1,2,3,4\ldots\ldots \qquad (2)$$

which at large distances ($r \gg r_c$) reduces to the usual asymptotic
expression. The value of n is arbitrary, however, n=3 (see Ref. 18) or
higher is more suitable. The value of r_c is determined approximately
from the relation[19]

$$r_c = \frac{2}{3} \frac{\Delta}{k} \qquad (3)$$

where Δ is the mean excitation energy of the target.

It is now straightforward to determine the K-matrices for each
symmetry and impact energy from the solutions of Eq.(1). More details
regarding specifically for polyatomic gases can be found elsewhere.[15]
Thus we can evaluate rotationally elastic, inelastic and summed cross
sections from the body-frame K-matrices. The solutions of Eq.(1) are
feasible only at low energies, where one can truncate the infinite sum
into a finite number of coupled channels; whereas, at intermediate

energies it is impractical to employ CC methods. We, therefore, adapt a simple strategy (see the following paragraph) at energies above E_{Ps}.

Assuming that the rotational and/or vibrational cross sections are very small at $E > E_{Ps}$, we can retain only the (first) dominant spherical term in the OCE of the optical potential for the e^+ (or e^-) molecule scattering system (i.e., $V_{01,01}$ term in Eq. 1). In a partial-wave (PW) decomposition, Eq.(1) then reduces into a simple uncoupled second-order differential equation which can be solved easily to yield the ℓth S-matrix ($S_\ell(E)$) at each energy. This type of model has recently been used successfully by us in a wide energy-range (even at low energies) for both the electron and positron collisions with CH_4 and SiH_4 molecules (see Refs. 20-24). This spherical model is in fact more suitable for molecules with zero dipole and quadrupole moments such as the CH_4 and SiH_4 molecules. Furthermore, this simple methodology has been found to be quite successful for e^- (e^+) scattering also with polar NH_3 and H_2O gases above 60 eV impact energy (see Ref. 25).

Since at intermediate energies, inelastic events (ionization, Ps formation, dissociation, etc.) dominate over the elastic channel, it is essential to invoke a complex potential (spherical-complex-optical-potential (SCOP)); the imaginary part of the SCOP (to be denoted by V_{abs}) takes into account the loss of flux due to all energetically accessible inelastic processes and can be modeled to be local and real for computational ease. There are quite a few V_{abs} model potentials for e^- case (to be represented by V^-_{abs}) (see Ref. 26), we recently explored these V^-_{abs} potentials for the $e^- - CH_4$ (and SiH_4) scattering.[23-24] The corresponding V^-_{abs} is derived semi-emperically (without involving any parameter) as a function of target density, Fermi momentum $k_F(r)$, static-exchange-polarization potential and the mean excitation energy (Δ) of the target. We[23-24] found that for V^+_{abs}, a modified form (in terms of V^-_{abs}) can be written as,

$$V^+_{abs}(r) = 2\ V^-_{abs}\ (r,\ \Delta,\ \rho_p(r),\ k_F(r),\ V_{st} + V_{ex} + V_p)/kr\ . \tag{4}$$

Note that $V_{ex} = 0$ and V_{ST} is repulsive for positrons. Therefore V^-_{abs} is different for the projectiles e^- and e^+.

However, above 100 eV, the V^+_{abs} without the extra factor of kr on the right-hand side (Eq. 4) expression is also successful (with a different value of Δ). As we shall see later, this simple method is quite gratifying in reproducing experimental σ_t cross sections in the E_{Ps} - 400 eV region, particularly the sharp rise in σ_t near the Ps threshold and a bell-shaped structure around 20-50 eV due to mainly the ionization channel. We also found that the use of a polarized charge-density ($\rho_p(r)$; its calculation is described in Ref. 27) in Eq.(4) gives better agreement with experimental data. This choice of polarized density in the evaluation of absorption potential is more sensible since in the course of inelastic processes the target charge cloud is no more static.

LOW-ENERGY ($E < E_{Ps}$) CROSS SECTIONS

Low-energy e^- molecule cross sections are characterized with features such as the shape-resonance (above 1 eV) and Ramsauer-Townsend (RT) minimum (below 1 eV). The question is, can the e^+ also see a gas nearly transparent (the RT effect) or nearly opaque (large cross sections as around a shape-resonance). The low-energy e^+ scattering becomes therefore very interesting to investigate both in theory and experiment. So far there is no evidence of any resonance behaviour in e^+ - molecule collisions at any energy. On the other hand, below 1 eV, no experimental data exist for σ_t for e^+ - molecular gas collisions in order to search for the RT minimum (note that in He and Ne gases, a pronounced RT effect is observed; a very shallow minimum for Ar, but no such structure found for heavy Kr and Xe inert gases, see Ref. 28).

126

From theoretical standpoint, it is very difficult to predict any structure in the total cross sections at low energies due to the fact that such behavior is highly sensitive to the choice of a polarization interaction. Only an ab initio theory with a true e^+ molecule polarization force may, in future, be capable to predict accurate e^+ cross sections. Therefore, theoreticians depend totally upon the measured data in order to provide any meaningful results. In a way, this situation is rather disappointing for e^+–molecule collisions. On the other hand, such a despair state of theoretical research in e^+ molecule case does not exist for the corresponding low energy e^- collisions.

In Fig. 1 we have shown total (rotationally summed) cross sections for e^+ –SiH$_4$ collisions by employing both the V_p^{JT} and V_p^{CPP} parameter-free models for polarization potential (same for electrons and positrons). The JT curve in Fig. 1 is due to V_p^{JT}, while the CP curve make use of the V_p^{CPP} potential. Except at higher energies where the two models are almost identical, in the lower energy region the importance of a using a true e^+ polarization is very clear. The pure static term (curve S in Fig. 1) is greatly changed when polarization is switched on. Below 1 eV the CPP and JT models are in total contrast with each other, while in the 1–5 eV regime, the JT model seems to be better than the CPP one. Note that the JT approximation is found to be better for the CH$_4$ case[27,29] also (not shown) as compared to the CPP model. At present, we are not in position to make any comment about these calculations below 1 eV energy. If data from several measurements, agreeing with each other, are available for e^+ – SiH$_4$ (or CH$_4$) collisions at lower energies, it is possible to determine a reasonable polarization potential by the tuning procedure. However, tuning the experimental DCS rather than the σ_t is much more safe and reliable.

Jain and Thompson[29] reported $e^\pm$ CH$_4$ rotational excitation cross sections below 10 eV and discussed their results with respect to corresponding electron data. Their conclusion was that the elastic channel dominates completely in this small energy region. In case of silane also the rotational excitation cross sections are very small to make any significant contribution to the σ_t. For example, for e^+ – SiH$_4$ case,[30] at 2 eV the (03), (04) and (06) (where (JJ') means rotational

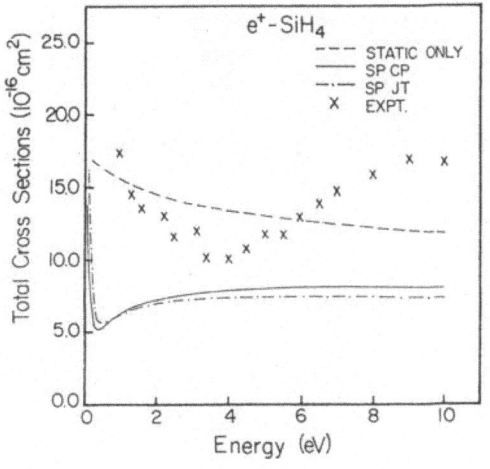

Fig. 1

transition from J to J') channels contribute to the σ_t only .15%, .1% and .008% respectively; at 5 eV, this contribution for the (03), (04) and (06) excitations becomes respectively as 3%, 1% and .04%.

For e^+ scattering with polar molecules (H_2O and NH_3), there is a large discrepancy between theory[31] and experiment.[8] In Fig. 2, we have shown a comparison of experiment[7-8] and theory[31] for $e^+ - NH_3$ (H_2O) total cross sections below 6 eV. The theortical curves are dominated by the dipol FBA term (same for e^- and e^+) which may be a possible source of this large discrepancy. It is in fact a challenging problem to calculate σ_t for polar molecules for either particle (e^- or e^+). If experiments could provide DCS for such tragets, theoretical work may gain some impetus.

Low-energy positron impact cross sections for the rotational ($\sigma_{JJ'}$) and vibrational excitation cross sections ($\sigma_{vv'}$) for CO molecules were recently reported by us.[11-12] The study of vibrational channel due to positron impact is very interesting since, in general, vibrational transition normally proceeds via resonant scattering in the electron case.[32-33] In sharp contrast to the electron case, where $\sigma_{vv'}$ are full of structure, the e^+-molecule $\sigma_{vv'}$ cross sections are expected to be structureless and smaller than the corresponding e^- data. In addition, knowledge of the size of the vibrational excitation cross sections of e^+-molecule scattering is needed, for example, as a possible experimental way of producing low-energy e^+ beams. Earlier attempts to study vibrational process in e^+-molecule are for H_2 molecules.[34-35]

We now discuss briefly the low-energy DCS for e^+-molecule scattering. So far, there are no measurements on the low-energy DCS for either atomic or molecular systems. The corresponding DCS for e^- case have been measured for many molecular gases (see the review by Trajmar et al.[36]). The electron DCS reveal lots of structure in the 1-10 eV region due to dominance of p-, d- and sometimes higher partial waves. It is tempting to know the corresponding DCS for e^+ case. Although all theoretical calculations on the σ_t carry information on the DCS, such

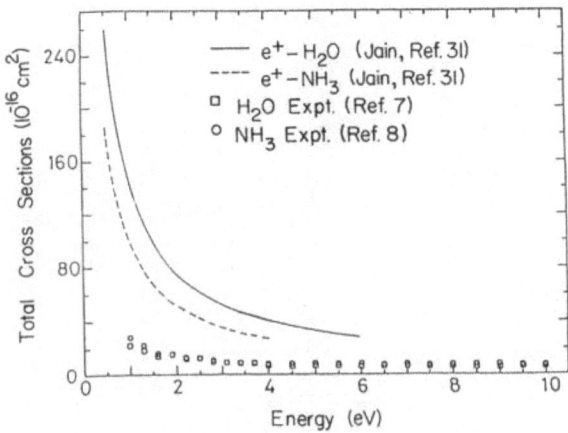

Fig. 2

data must be analyzed very carefully. The reason being that in this low-energy region, the scattering is very sensitive to the correct inclusion of polarization interaction; specifically the DCS are much more sensitive than the integral quantities. The e^+ DCS calculated by employing e^- polarization potential may be misleading. We, therefore, do not think it worth to show any DCS curve for e^+-molecule scattering in this energy region.

INTERMEDIATE ENERGY ($E_{Ps} < E < 400$ eV) CROSS SECTIONS

Total Cross Sections

Almost all e^+ – gas σ_t are characterised by a sharp rise near the Ps threshold following a bell shaped structure around 20–50 eV (depending upon various thresholds; E_{Ps}; ionization, E_{ion}; electronic excitation, E_{exc}; dissociation, E_{dis}, etc.). In general, these thresholds, i.e., E_{Ps}, E_{exc}, E_{ion} and E_{dis} are separated by few eV energy from each other; for example, the ore-gap ($E_{ex} - E_{Ps}$) for CH_4 has about 2 eV energy region where the sharp rise in the σ_t is purely due to the Ps formation channel. This information is valuable in order to estimate σ_{Ps} cross sections near the threshold, thus providing data to determine the positron threshold laws. Another interesting property which can be deduced from near-threshold σ_t values is the behaviour of elastic cross sections (σ_{el}). Since the σ_{el} is lowered with the inclusion of the V_{abs}, as the imaginary part of the SCOP, it is expected that the σ_{el} may show cusp behaviour near the E_{Ps} threshold.[37] In order to search for such a structure in the σ_{el}, one has to carry out a very careful calculation in the ore-gap; any model calculation such as the discussed here may be able to furnish meaningful information in the region of ore-gap provided that the available experimental σ_t are reliable in the ore-gap.

As mentioned earlier, a complex-optical-potential (COP) is essential to study e^+-molecule collisions in this energy region. For molecules such as the CH_4, SiH_4, NH_3 and H_2O, the SCOP model is a good approximation. The anisotropic part of the COP is in fact negligible for CH_4 and SiH_4 targets, while for the polar (NH_3 and H_2O) molecules, we invoke FBA to account for these non-spherical contributions. Certainly some error is introduced in the polar molecule cross sections by FBA quantities; however, the qualitative behavior is not altered and the calculated results, using FBA incoherently, compare fairly with the measurements.

Figure 3 displays e^+ – CH_4 σ_t calculated[24] values at E_{Ps} – 600 eV along with several sets of experimental points.[3-6] As is seen from Figure 3 the SCOP model[24] reproduces both in quality and quantity the sharp rise in the σ_t above E_{Ps} and a bell-shaped structure around 40 eV. Recall that the SCOP model is parameter-free except a little adjustment in the mean excitation energy, Δ, appearing in the original derivation of the V_{abs} potential. For positron case, naturally the Δ is close to E_{Ps} value. Similar tests were repeated for the He atom[24] and for a more polarizable and complicated SiH_4 molecule.[23] In both the later cases, the SCOP model gave a very good agreement with experimental σ_t values in the present energy region.

We therefore assume that the SCOP model may be quite capable to give reliable information on the σ_{Ps} cross section also in the ore-gap and beyond. In this procedure, one has to first tune the experimental σ_t (by adjusting the polarization potential) below E_{Ps} energy and then use the SCOP model with a reasonable V_{abs} potential to yield σ_t in close agreement with measured data. A major difficulty in this method is of course the selection of most accurate experimental data, since there is considerable discrepancy amongst the existing experimental σ_t results (for example in the CH_4 case). The cusp behavior (if there is any) in the elastic cross section can thus be investigated in the region of E_{Ps}.

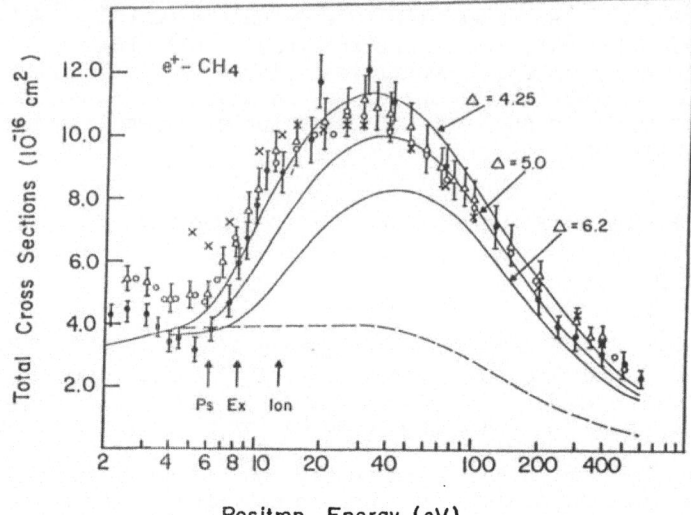

Fig. 3 The total cross sections for e^+–CH_4 scattering at 2–600 eV.
Calculations (Ref. 24): upper solid curve is with Δ=4.25 eV,
middle solid curve with Δ=5.0 eV while the lower solid curve is
obtained with Δ=6.2 eV (E_{Ps} value); the dash curve represents
pure elastic (σ_{el}) cross sections. Experimental data are due
to: Δ, Sueoka; O, Kaupila et al.[4]; X, Floeder et al.[5]; ●,
Charlton et al.[2,3] The positronium (Ps), electronic excitation
(Ex) and ionization (Ion) threshold are shown by arrows.

We now present some new calculations[25] on the e^+ – NH_3 (H_2O)
scattering at 30–400 eV. For comparison, we also provide corresponding
e^- – NH_3 (H_2O) collision cross sections[38] in the same energy region. For
this calculation, again the SCOP model is utilized. As mentioned above
one, should, however, be careful in invoking the SCOP model for polar
gases such as the NH_3 and H_2O with a sizeable dipol moment of .574 and
.728 (in au) respectively. We first note that above 100 eV energy, the
σ_t value due to dipol and higher order moment terms are small (only the
DCS in the forward direction are affected significantly) and therefore
the SCOP model should work fine. This is further confirmed by
calculating the σ_t in the FBA above 100 eV; the FBA gives quite small
values for the σ_t relative to SCOP σ_t values (see Fig. 4). The FBA is
generally a satisfactory approximation for the dipol and quadrupole
transitions from low to high energies. Below 100 eV, the contribution of
these non–spherical terms is considerable and dominating. We do not
recommend the SCOP results below 100 eV to be reliable for polar
molecules NH_3 and H_2O.

Figure 4 displays our $e^\pm$ – NH_3 (H_2O) σ_t cross sections in the SCOP
model (by employing a polarized density in the evaluation of the V_{abs}^+).
The experimental data of Zecca et al.[39] for e^- – H_2O and Sueoka and
Mori[7-8] for $e^\pm$ – NH_3 and $e^\pm$ – H_2O are also shown in Fig. 4. Although for
the polar targets, the σ_t can be defined in a two–potential incoherent
approach as follows,

$$\sigma_t = \sigma_t^{SCOP} + \sigma_t^{FBA}, \tag{5}$$

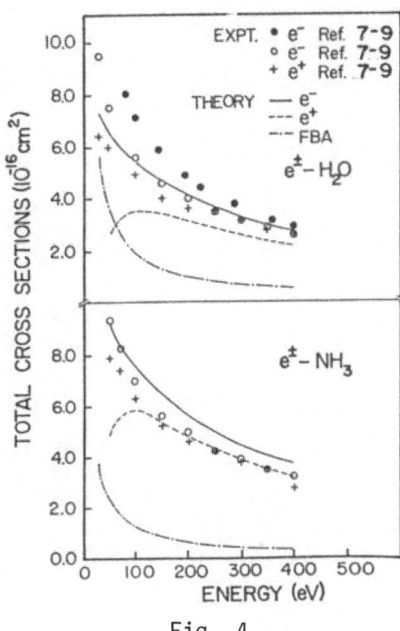

Fig. 4

we have illustrated the σ_t^{FBA} and the σ_t^{SCOP} separately in Fig. 4. The reason of doing this being that the FBA may be overestimating (or under-estimating) the true values of dipole (and quadrupole, etc.) cross sections. We see that the σ_t^{SCOP} values are smaller than the experimental values below 100 eV indicating the importance of higher-order moment terms. We see from Fig. 4 that there is good agreement (particularly above 100 eV) between the present $\sigma_t^{\pm}$ results and the experimental data for both the targets. Keeping in mind that the two sets of observations (Refs. 39 and 7-8) differ considerably with each other, our calculations lie in the middle of these measured points for H_2O case.

Another point of interest in this energy region is the merging of e and e^+ cross sections at high energies. Our results are summarized for all the four molecules in Table 1 for the ratios defined as

$$R_i = \sigma_i^-/\sigma_i^+; \ R_t = \sigma_t^-/\sigma_t^+; \ R_{el} = \sigma_{el}^-/\sigma_{el}^+; \ R_{abs} = \sigma_{abs}^-/\sigma_{abs}^+ \qquad (6)$$

Notice that the σ_{el} and σ_i are elastic integral cross sections with V_{abs} switched on and off respectively. We see that at 1000 eV energy, the σ_{el}^- is about 15-20% larger than the σ_{el}^+ for the CH_4, SiH4, NH_3 and H_2O molecules. The e^- and e^+ σ_t values converge into each other within 12% at 1000 EV energy. The DCS at 1000 eV for e^- and e^+ therefore should not behave very differently at these high energies. It is however interesting to note that the absorption cross sections due to e^- and e^+ are very close to each other at all energies shown in Table 1: for example the ratio R_{abs} ($\sigma_{abs}^-/\sigma_{abs}^+$) for silane at 200, 400 and 1000 eV is respectively 1.08, 1.07 and 1.05. From Table 1, we can conclude that the merging of e^- and e^+ σ_t cross sections occurs (within 10%) around 1000 eV.

Table 1

Ratios (Eq. 6) R_i, R_{el}, R_{abs}, R_t for CH_4, SiH_4, NH_3 and H_2O Molecules

1(a) R_i (pure elastic, absorption off)

Energy (eV)	CH_4	SiH_4	NH_3	H_2O
100	4.25	6.95	2.86	3.29
200	2.00	2.55	1.74	1.88
400	1.42	1.53	1.39	1.46
600	1.28	1.32	1.29	1.44
1000	1.18	1.19	1.21	1.42

1(b) R_{el} (reduced elastic, absorption on)

100	4.7	2.75	1.98	2.60
200	2.34	1.60	1.38	1.59
400	1.51	1.30	1.22	1.29
600	1.29	1.20	1.17	1.22
1000	1.15	1.13	1.12	1.15

1(c) R_{abs}

100	1.13	1.10	1.08	1.08
200	1.10	1.08	1.07	1.08
400	1.09	1.07	1.06	1.09
600	1.08	1.06	1.05	1.09
1000	1.08	1.05	1.05	1.07

1(d) R_t

100	2.10	1.57	1.35(1.28)[*]	1.57(1.37)[*]
200	1.54	1.23	1.24(1.16)	1.23(1.18)
400	1.26	1.14	1.11(1.09)	1.15(1.12)
600	1.18	1.10	1.08(1.08)	1.12(1.10)
1000	1.12	1.08	1.07(1.06)	1.10(1.08)

*Values in the parentheses are the ratios using Eq.(5) for the σ_t.

The issue of $e^{+,-}$ cross sections merging is not much different for the polar and non-polar gases studied here. The only measurements of Sueoka and coworkers for the $e^{\pm}$ – NH_3 (H_2O) collisions indicate that the two sets of σ_t values merge with each other around 400 eV. Our theoretical results show that at 400 eV the ratio R_t is about 1.15 for these polar gases; this means that our calculations predict an earlier merging of $\sigma_t^{\pm}$ cross sections for NH_3 and H_2O in agreement with the observation of Sueoka.[7-8]

Differential Cross Sections

The measurement of DCS provides a more stringent test of any theoretical model than the corresponding observation of the σ_t for that particular scattering mechanism. In recent years, there has been noticeable progress in measuring DCS for e^+ – gas interactions.[40] More recently, Hyder et al.[41] have reported first application of a crossed-beam experiment for e^+ – gas scattering and presented elastic DCS of 100–300 eV e^+ by argon at 30° – 135° angles. These are the first such measurements where elastically scattered e^+ were detected directly at various angles. Although there are no such measurements on the DCS for

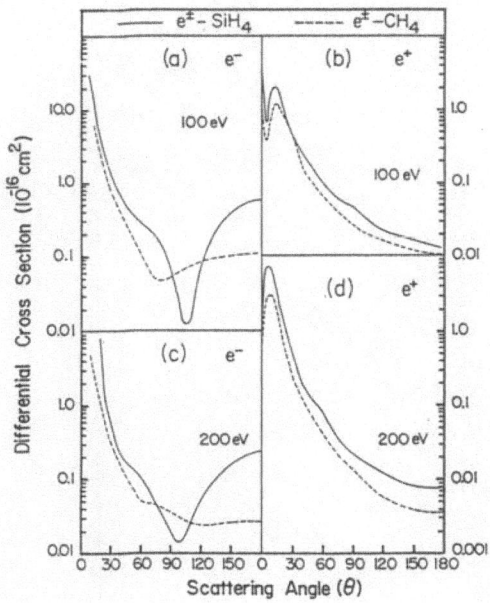

Fig. 5

molecular gases with which we are concerned here, it is interesting to compare the behaviour of e^- and e^+ DCS for Ar, CH_4 and SiH_4 gases. It is well known that the σ_t for both the Ar and CH_4 targets behave quite similarly with each other at low energies.

In order to compute elastic DCS for e^+ and e^- scattering with CH_4 and SiH_4 gases, we switch off the imaginary part of the SCOP and include a large number of partial waves (up to 300 at 300 eV) in order to obtain converged DCS at all angles. In Fig. 5 we have shown our $e^-,^+ - SiH_4$ (and CH_4) DCS at 100 eV and 200 eV energies. There is a striking similarity between our Fig. 5 for silane and Fig. 3 (of Ref. 41) for argon. Note that there is no adjustable parameter in the calculation of our silane DCS. The e^+ DCS are quite different in shape from the corresponding e^- DCS. The CH_4 DCS for e^+ case (Fig. 5) are very similar in shape to silane DCS but smaller in magnitude, which is consistent with the larger polarizability and size of SiH_4 than the CH_4. The DCS for e^+-CH_4 at 300 eV have been measured recently[42] in the angular range 30°–135°; the preliminary data indicate that the $e^{\pm}-CH_4$ DCS are similar in shape in this angular region. As expected there is no structure (not shown) observed in the DCS for e^+-NH_3 (H_2O) collisions in this energy region.

The inclusion of V_{abs} in the optical potential reduces the DCS significantly. The large angle DCS are reduced more as compared to small angle DCS. It is observed for e^- case that the reduced DCS are exposed to more pronounced structure.[22,43]

Thus, in electron case, the integral elastic cross sections are always reduced with V_{abs} on. However, the situation is a little different for e^+ case as depicted in Fig. 6, which displays the e^+-CH_4 DCS at 100, 200 and 400 eV with (to be denoted by RDCS, broken curves) and without (solid curves) the absorption effects. We see that at small angles ($\theta < \theta_c$) the inclusion of V_{abs} increases the DCS, while at higher

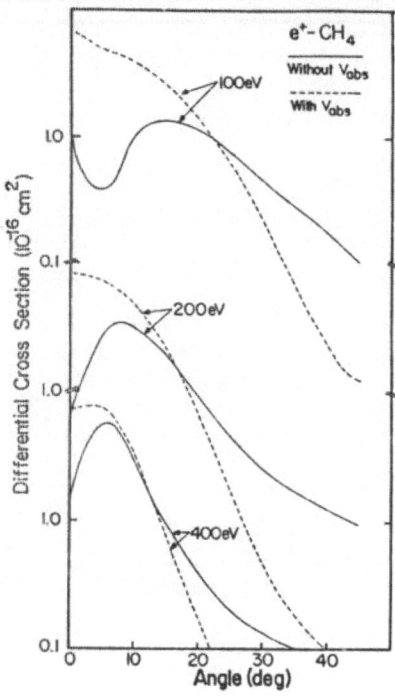

Fig. 6

angles ($\theta > \theta_c$) the DCS are reduced significantly. The value of θ_c depends upon incident energy, for example, $\theta_c = 22°$, 17° and 12° respectively at 100, 200 and 400 eV (see figure 6). The integral cross sections (with absorption), therefore, may not be reduced if θ_c is larger. However, the momentum transfer cross sections are always reduced with the inclusion of absorption due to the fact that forward scattering is eliminated by the weighting factor of $(1-\cos\theta)$. Table 2 summarizes the effect of absorption on the e^+-CH_4 σ_{el} values.

Table 2 e^+-CH_4 Integral Elastic Cross Sections With and Without Absorption Effects

Energy	Without V_{abs}		With V_{abs}	
(eV)	σ_{el}	σ_m	σ_{el}	σ_m
100	1.36	0.39	1.64	0.21
200	1.36	0.21	1.36	0.05
400	0.998	0.09	0.91	0.027

CONCLUSION

From the above discussion, it is clear that in order to understand e^+ interactions in molecules, we need more investigations both in theory and experiment. A major source of uncertainty and error in slow e^+-molecule scattering calculations is the inclusion of polarization effects incorrectly. The work of Morrison and coworkers[13,14] for e^+-H_2

collisions, therefore, should be taken very seriously and tested for other heavy systems in order to comfort the issue of polarization. All previous and present calculations for e^+–molecule collisions with an e^- polarization interaction may work as zeroth–order data for reference and comparison with more realistic models.

We need experimental information on the DCS for e^+–molecules at high as well as at low energies. This will stimulate more theoretical activity in complicated systems such as the discussed here.

ACKNOWLEDGEMENT

This work was supported in part by the U.S. Department of Energy, Office of Energy Research, Division of Chemical Sciences. Helpful conversation with F.A. Gianturco and W.E. Kauppila is thankfully acknowledged. I sincerely thank Ms. Jane Torrey for preparing this manuscript so nicely.

REFERENCES

1. M. Charlton, Rep. Prog. Phys. 48, 737 (1985); and references therein.
2. M. Charlton, T. C. Griffith, G. R. Heyland, K. S. Lines and G. L. Wright, J. Phys. B13, L757–760 (1980).
3. M. Charlton, G. Clark, T. C. Griffith and G. R. Heyland, J. Phys. B16, L465 (1983).
4. Ch. K. Kwan, Y. F. Hsieh, W. E. Kauppila, S. J. Smith, T. S. Stein, M N. Uddin and M. S. Dababneh in "Positron (Electron)–Gas Scattering," ed. by Kauppila et al. (World Scientific, Singapore, 1985), p. 241.
5. K. Floeder, D. Fromme, W. Raith, A. Schwab and G. Sinapius, J. Phys. B18, 3347–59 (1985).
6. O. Sueoka and S. Mori, J. Phys. B19, 4035 (1986).
7. O. Sueoka, S. Mori and Y. Katayama, J. Phys. B19, L373 (1986).
8. O. Sueoka, S. Mori and Y. Katayama, J. Phys. B (in press).
9. O. Sueoka, private communication.
10. A. Jain, in "Positron (Electron)–Gas Scattering," ed. by W. E. Kauppila, et al. (World Scientific, Singapore 1986) p. 110.
11. A. Jain, J. Phys. B19, L105 (1986).
12. A. Jain, J. Phys. B19, L379 (1986).
13. M. A. Morrison, T. L. Gibson and D. Austin, J. Phys. B17, 2725 (1984).
14. M. A. Morrison, in "Positron (Electron)–Gas Scattering," ed. by W. E. Kauppila et al. (World Scientific, Singapore).
15. F. A. Gianturco and A. Jain, Phys. Rep. 143, 347–425 (1986); see also Ref. 32 for general e^-– molecule theory; for the adiabatic–nuclei theory in e^-–molecule scattering, see A. Temkin and K.V. Vasavada, Phys. Rev. 160, 109 (1967).
16. A. Jain and D. G. Thompson, J. Phys. B15, L631 (1982).
17. J. K. O'Connel and N. F. Lane, Phys. Rev. A27, 1893 (1983); N. T. Padial and D. W. Norcross, Phys. Rev. A29, 1742 (1984); for application of such potential to polyatomic gases, see F.A. Gianturco, A. Jain and L.C. Pantano, J. Phys. B20, 517 (1987).
18. B.L. Jhanwar and S.P. Khare, J. Phys. B9, L527 (1986).
19. F.W. Byron and C.J. Joachain, J. Phys. B10, 206 (1977).
20. A. Jain, J. Chem. Phys. 78, 6579 (1983).
21. A. Jain, J. Chem. Phys. 81, 724 (1984).
22. A. Jain, Phys. Rev. A34, 3707 (1986).

23. A. Jain, J. Phys. B19, L807 (1986).
24. A. Jain, Phys. Rev. A (in press, 1987).
25. A. Jain (to be published).
26. See for example G. Staszewska, D. W. Schwenke and D. G. Truhlar, J. Phys. B16, L281 (1984); Phys. Rev. A29, 3078 (1984).
27. A. Jain and D. G. Thompson, J. Phys. B16, I113 (1983).
28. T. S. Stein and W. E. Kauppila, in "Electron and Atomic Collisions," ed. D, C, Lorents, et al. (Elsevier Science Publishers B V 1986).
29. A. Jain and D.G. Thompson, Phys. Rev. A30, 1098 (1984).
30. F. A. Gianturco, A. Jain and L. C. Pantano, Phys. Rev. A (in press).
31. A. Jain, in "Positron (Electron)–Gas Scattering," eds. W. E. Kauppila, et al. (World Scientific, Singapore 1986) p. 238.
32. N.F. Lane, Rev. Mod. Phys. 52, 29 (1980).
33. D.G. Thompson, Adv. At. Mol. Phys. 19, 309 (1983); and reference therein.
34. P. Baille and J.W. Darewych, J. Physique Lett. 32, L243.
35. S. Sur and A.S. Ghosh, J. Phys. B18, L715 (1985).
36. S. Trajmar, D.F. REgister and A. Chutjian, Phys. Rep. 97, 319 (1983).
37. R.I. Campeanu, D. Fromme, G. Kruse, R.P. McEachran, L.A. Parcell, W. Raith, G. Sinapius, A.D. Stauffer (to appear in J. Phys. B, 1987).
38. A. Jain (in preparation).
39. A. Zecca, G. Karwasz, S. Oss, R. Grisenti and R. S. Brusa, J. Phys. B20, L133 (1987).
40. P. G. Coleman and J. D. McNutt, Phys. Rev. Lett. 42, 1130 (1979); see also various articles in "Positron (Electron)–Gas Scattering, ed. by W. E. Kauppila, et al. (World Scientific, Singapore, 1986).
41. G. M. A. Hyder, M. S. Dababneh, Y-F Hsieh, W. E. Kauppila, C. K. Kwan, M. M. Hezaveh, and T. S. Stein, Phys. Rev. Lett. 57, 2252 (1987).
42. S.J. Smith, G.M.A. Hyder, W.E. Kauppila, C.K. Kwan, T.S. Stein, Bull. Am. Phys. Soc. 32, 1272 (1987).
43. A. Jain, J. Chem. Phys. 86, 1289 (1987).
44. A. Jain and D.G. Thompson (unpublished).

FIELD THEORY OF ELECTRON AND POSITRON SCATTERING FROM ATOMIC AND MOLECULAR SYSTEMS

Edgardo Ficocelli Varracchio

Department of Chemistry

University of Bari, Bari, Italy

1. INTRODUCTION

Many processes, in atomic and molecular physics, can be visualized in terms of the excitations of only a 'limited' number of electrons, out of a 'reference' state, usually coinciding with the system ground state. This parallels the situation in high energy physics, where a Field Theoretic (FT) formalism was originally developed, in order to describe processes corresponding to the excitation of particles out of a 'vacuum' state, or to the destruction of such particles into it. The FT approach was the tailored so as to follow the dynamics of such a limited number of particles, rather than getting involved with the full details of the vacuum reference state itself[1].

The previous considerations represent one of the basic motivations that have led, during the last few years, to an increased interest towards the application of FT techniques to the general area of bound state problems in atomic and molecular physics[2].

Much less established is the applicability of the same formalism to scattering processes and, in particular, to a description of collisions of electrons and positrons on atomic or molecular targets[3,4].

In these lectures we shall then try to outline the main features of such a line of development. The emphasis will be, partly, on the formal aspects of the theory, ultimately leading to a set of relationships, among FT amplitudes, containing the full (non perturbative) structure of the dynamics. At the same time, approximation schemes will be suggested and physically motivated, mostly in terms of diagrammatic interpretations for all quantities of interest. Finally, a comparison of the formalism, as developed for both the e^- and e^+ species, will point out that the FT approach is most suited for analyzing, on physical grounds, differences of behaviour for the two systems.

The plan of the lecture is as follows. In Section 2 we shall concentrate on an analysis of e^- - atom, molecule processes. Section 3 will be devoted to a consideration of e^+ - atom, molecule collisions. The numerical behaviour of the theory will be illustrated, at suitable points, in terms of applications to the $e^{\pm}$ - H_2 systems. A concluding summary, in Section 4, will highlight the basic final results of the theory.

2. FT FORMULATION OF e⁻ – ATOM, MOLECULE SCATTERING

2.1 The one-particle Green's function as an alternative to the total system wave function.

The FT approach is most naturally developed in terms of the formalism of second quantization. This deals with the particles involved in the physical process by means of, so called, field operators[5]. For our present consideration of e⁻ – atom, molecule systems, we can correspondingly define $\psi(1)$, $\psi^\dagger(1)$ operators, that destroy, create one electron at (space-spin-time) position $1 = (\vec{r}_1, m_{s1}, t_1)$ (for notational convenience, in all our future equations even and odd numbers will label positron and electron coordinates, respectively). These obey conventional Fermion anticommutation relations ($\{A,B\} = AB + BA$)

$$\{\psi(1), \psi(1')\} = \{\psi^\dagger(1), \psi^\dagger(1')\} = 0$$

$$\{\psi(1), \psi^\dagger(1')\} = \delta(1 - 1') \tag{1}$$

The above operators can be immediately used to build up one of the basic amplitudes of the formalism, the $G(1,1')$ one-particle Green's function (or propagator), defined according to [5]

$$G(1,1') = \frac{1}{i} \; \langle \Psi_o^N | T \, [\psi(1) \, \psi^\dagger(1')] \, | \, \Psi_o^N \rangle =$$

$$= \frac{1}{i} \, \theta(t_1 - t_1') \, \langle \Psi_o^N | \psi(1) \, \psi^\dagger(1') | \, \Psi_o^N \rangle \; - \tag{2}$$

$$- \frac{1}{i} \, \theta(t_1' - t_1) \, \langle \Psi_o^N | \psi^\dagger(1') \, \psi(1) | \, \Psi_o^N \rangle$$

As shown by the first equality, in (2), $G(1,1')$ involves one creation and one destruction field operator acting on the $|\Psi_o^N\rangle$ (exact) ground state of the, N-electron, target. Besides, T is Wick's time ordering operator, that arranges the $[\psi\psi^\dagger]$ product according to decreasing time. The action of T has been made explicit in the second equality of (2), involving the θ step function, which shows that $G(1,1')$ contains two pieces of physical information. The first of these ($t_1 > t_1'$. e.g. $\psi\psi^\dagger$ sequence of field operators) corresponds to creating a particle at $1'$, having it interact with the system, and destroying it at position 1. This is exactly the sequence characterizing a scattering process, so that one should expect that the corresponding term will be proportional to the S-matrix of an e⁻ – atom, molecule (elastic) collision.

The rightmost amplitude, in (2), ($t_1 < t_1'$, e.g. $\psi^\dagger\psi$ ordering of field operators) pictures, instead, the sequence of operations corresponding to destroying an electron, at position 1, and putting one back, into the system at $1'$. This just represents the physical operation of measuring an electronic density, in the target, so that the relevant amplitude will be completely equivalent to the conventional 'first order density matrix' of atomic and molecular physics[6].

The above considerations then clearly show the enormous amount of information concealed within $G(1,1')$. Furthermore, the retrieval of this does not explicitly require, at any stage, having to deal with the total system wave function. The basic quantity of interest is the G amplitude itself, for which Field Theory yields explicit computational tools.

In the next Sections we shall outline a derivation of all FT equations of interest, we shall suggest some physical approximation schemes and we shall analyze the numerical behaviour of the theory, for the e⁻-H_2 system.

2.2 The integral equation defining the $G(1,1')$ propagator.

The derivation of an integral (Dyson) equation for the G one-particle Green's function immediately brings into the picture a consideration of FT amplitudes of a different nature. We shall, correspondingly, formulate the theory in terms of a 'minimal' set of equations, trying to stress the physical interpretation of all quantities involved.

An equation of motion for $G(1,1')$ can be obtained by starting from the equation obeyed by the $\psi(1)$ field in the Heisenberg picture. Using standard results[5], it can be readily shown that

$$\left[i \frac{\partial}{\partial t_1} - H_o (\vec{\tau}_1)\right] G(1,1') = \delta(1-1') +$$

$$+ \frac{i}{2} \int d3 d3' d5 \, \Gamma_o (1\,3,3'\,5) \, G(5\,3', 1'\,3^+) \tag{3}$$

where H_o and Γ_o are just the basic constituents of the system Hamiltonian, defined by (in atomic units)

$$H_o (\vec{\tau}) = - \frac{1}{2} \nabla^2_{\vec{\tau}} + Z/|\vec{\tau}|$$

$$\Gamma_o(1\,3, 1'\,3) = V(\vec{\tau}_1, \vec{\tau}_3)\delta(t_1-t_3)[\delta(1-1')\delta(3-3') - \delta(1-3')\delta(3-1')] \tag{4}$$

with V_e the usual Coulomb repulsion term. Equation (4) is strictly valid for an atomic target, but the derivation is immediately generalizable to molecular systems, with a suitable modification of the H_o term. It should be stressed that we have explicitly antisymmetrized the electron repulsion term through the definition of Γ_o in (4). This step is certainly useful because: a) it leads to a smaller number of terms in diagrammatic representations of the theory; b) it keeps throughout an important symmetry property of the system that might otherwise be lost, when enforcing approximation schemes.

To conclude the analysis of (3), we simply point out that the G amplitude under the integral sign is the two-particle Green's function, defined according to

$$G(1\,3, 1'\,3') = \frac{1}{i^2} \langle\Psi_o^N | T [\psi(1) \, \psi(3) \, \psi^\dagger(3') \, \psi^\dagger(1')] | \Psi_o^N\rangle \tag{5}$$

Equation (3) then couples the one-particle Green's function to the two-particle propagator. It can, formally, be transformed into an integral equation for the $G(1,1')$ amplitude of interest by rearranging the rightmost term in (3) according to

$$\frac{i}{2} \int d3 d3' d5 \, \Gamma_o (1\,3, 3'\,5) \, G(5\,3', 1'\,3^+) = \int d_3 \, \Sigma(1,3) \, G(3,1') \tag{6}$$

In the above expression, Σ can be considered as the FT definition of an optical potential presiding over elastic scattering processes in e^- - atom, molecule collisions. There are two basic approaches in the literature to the derivation of an explicit equation for Σ. The first of these follows from writing $G(1,1')$ in the interaction picture and analyzing its expansion in terms of Feynman diagrams. Such a method is not general enough and it certainly requires that suitable diagrammatic conventions be established for the system of interest.

The second approach[7,8] is based, instead, on the introduction of an arbitrary external potential, U_e, that may act as a source of particles (electrons, in the present situation). Higher order amplitudes will then be related to lower order ones in terms of functional differentiations with respect to such a source. For example, it can be readily shown that

$$G(1\ 3,1'3') = G(1,1')\ G(3,3') - \frac{\delta G(1,1')}{\delta U_e(3',3)} \tag{7}$$

Equation (7) can now be used in (6) in order to obtain an explicit relationship for Σ. It should be stressed that, once all necessary manipulations have been performed, the arbitrary source can be made to disappear by enforcing a $U_e \to 0$ limit.

This second approach, although more formal, is, evidently, much more powerful than the diagrammatic technique. We shall then largely rely on functional differentiation, in the following, as a tool for deriving equations and we shall revert to diagrams for the purpose of physical interpretation whenever necessary.

Using the technique of external sources, it is just a matter of algebraic manipulations to obtain the following closed equation for the optical potential[8]

$$\Sigma(1,1') = i \int d3d3'\ \Gamma_o\ (1\ 3,\ 3'1')\ G(3',3^+) \quad -$$

$$\frac{i}{2} \int d3d3'd5d5'\ \Gamma_o\ (1\ 3,3'5)\ G(5,5') \frac{\delta\ \Sigma(5',1')}{\delta\ U_e(3^+,3')} \tag{8}$$

(in (8) and all future expressions, a limit, $U_e \to 0$, is implicit). In terms of (8), the integral equation obeyed by the one-particle propagator then simply becomes

$$G(1,1') = g_o(1,1') + \int d3d3'\ g_o(1,3)\ \Sigma(3,3')\ G(3',1') \tag{9}$$

where g_o represents the unperturbed system propagator (corresponding to the operator within square brackets on the lhs of (3)). The diagrammatic representation of (9) is given in Figure 1.

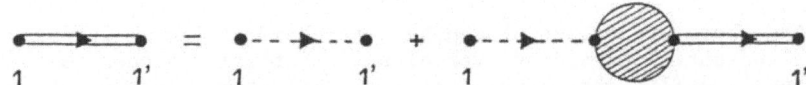

Fig. 1. Graphical representation of Dyson's equation for e^--atom, molecule scattering. The double and broken lines picture the one-particle G and g_o amplitudes, respectively, while the shaded blob corresponds to the Σ optical potential.

Equation (8) for Σ could be solved by means of an iterative procedure-based on the (approximate) knowledge of the first term (static potential) on the rhs as starting approximation. The problem with such an iterative technique of solution is that one has to deal, already at the first iterate, with two-particle (Bethe-Salpeter type of) equations that are rather difficult to handle[8].

We shall consider, instead, a different line of development[9] that relates Σ to a new FT amplitude, whose solution at no stage requires anything more complicated than 'single particle' equations. We define, first of all, a four point function, Γ_e, according to

$$\frac{\delta \, \Sigma(1,1')}{\delta \, U_e(3',3)} = \int d5d5' \, G(3,5) \, \Gamma_e(1\ 5, \ 1'5') \, G(5',3') \tag{10}$$

Substituting this into (8), the following expression for the optical potential is recovered

$$\Sigma(1,1') = i \int d3d3' \, \Gamma_o(1\ 3,3'1') \, G(3',3^+) \ +$$

$$+ \frac{i}{2} \int d3d5d7d3'd5'd7' \, \Gamma_o \, (1\ 3,3'5) \, G(5,5') \ \times \tag{11}$$

$$G(3',7) \, \Gamma_e(5'7,7'1') \, G(7'3)$$

This is pictured in Figure 2, from which the physical interpretation of Γ_e should start becoming evident. The diagram shows, in fact, two particles coming into the interaction region (shaded square) and two getting out of it (the empty dot in 1' will also, finally, be occupied by one propagator), so that Γ_e is expected to contain full information on 'correlation' effects in two-particle motions.

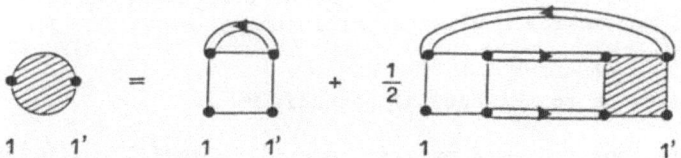

Fig. 2. Graphical representation of the equation defining the optical potential. The empty square pictures the i Γ_o 'bare' interaction while the shaded square represents the Γ_e (four-point) function.

We need now an explicit equation for Γ_e and for this purpose we can write, by extending to functional differentiation the conventional rules of partial differentiation,

$$\frac{\delta \, \Sigma(1,1')}{\delta \, U_e(3',3)} = \int d5d5' \, \frac{\delta \, \Sigma(1,1')}{\delta \, G(5',5)} \, \frac{\delta \, G(5',5)}{\delta \, U_e(3',3)} \tag{12}$$

By defining

$$\Xi_e(1\ 3,1'3') = \frac{\delta\ \Sigma(1,1')}{\delta\ G(3',3)} \tag{13}$$

and using (12) and (13) into (10), it is, then, just a matter of simple algebraic manipulations to show that the following integral equation results for Γ_e

$$\Gamma_e(1\ 3,3'1') = \Xi_e(1\ 3,3'1') - \int d5d7d5'd7'\ \Xi_e(1\ 5,5'1')$$

$$\times\quad G(5',7)\ \Gamma_e(3\ 7,7'3')\ G(7',5) \tag{14}$$

The above equation can be represented in diagrammatic terms, as in Figure 3.

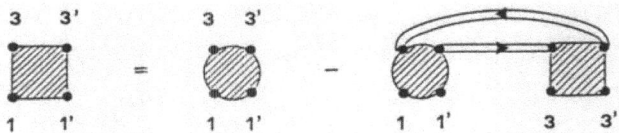

Fig. 3. Integral equation for the Γ_e function. The shaded circle pictures the Ξ_e two-particle interaction. All remaining graphical elements are as in the previous Figures.

Equations (11), (13) and (14) represent a close set of relationships completely determining the optical potential of e^--atom, molecule scattering The difficult (to evaluate) operation of functional differentiation has now been relegated into the definition of the new unknown quantity, Ξ_e, that appears as the inhomogeneous term on the rhs of (14). In such a form functional differentiation can actually be performed 'analytically' to any required degree of accuracy by simply inspecting how the structure of Σ functionally depends upon the G one-particle propagator. This aspect of the theory will become more clear in the next Section, where a convenient computational scheme for Σ will be outlined.

2.3 Approximations to the optical potential

Approximations to the Σ optical potential, in (11), will depend on the level of accuracy within which Γ_e, defined by (14), is known. This requires in turn, the knowledge of some initial approximation for Ξ_e, so that we start by considering such a quantity in more detail. In so doing we will also be able to clarify the physical interpretation of Ξ_e.

From (13) we can write to lowest order

$$\Xi_e(1\ 3,1'3') \simeq \frac{\delta\ \Sigma^{HF}(1,1')}{\delta\ G^{HF}(3',3)} \tag{15}$$

where G^{HF} represents the Hartree-Fock (HF) approximation to the one-particle propagator (this is a conventional starting point in atomic and molecular physics calculations). Σ^{HF} is explicitly given by the first term on the rhs of (11), according to

$$\Sigma^{HF}(1,1') = i \int d3d3' \, \Gamma_o(1\,3,3'1') \, G^{HF}(3',3^+) \qquad (16)$$

Using (16) into (15) and performing the indicated functional differentiation then immediately leads to

$$\Xi_e(1\,3,1'3') \simeq -i \, \Gamma_o(1\,3,1'3') \qquad (17)$$

(the minus sign derives from the antisymmetry of Γ_o). The lowest order approximation, (17), shows, then, that Ξ_e can generally be interpreted as being the 'full' two-particle interaction in the medium. For the present physical situation, due to the 'repulsive' nature of the Coulomb potential, we expect that (17) will be a suitable approximation to Ξ_e. In the next Section we shall obtain a corresponding expression for the full two-body interaction of e^+-atom, molecule systems. In that situation the change of sign of the Coulomb potential will make the counterpart of (17) a very poor approximation and higher order terms will have to be considered.

Substituting (17) into (14), the following integral equation for Γ_e then results

$$\Gamma_e^{RPA}(1\,3,3'1') = -i \, \Gamma_o(1\,3,3'1') +$$

$$i \int d5d7d5'd7' \, \Gamma_o(1\,5,5'1') \, G^{HF}(5',7) \, \Gamma_e^{RPA}(3\,7,7'3') \, G^{HF}(7',5) \quad (18)$$

We have used, in (18), the notation Γ_e^{RPA} to denote that the solution of the corresponding integral equation is completely equivalent to the RPA approximation of the literature[10,11] (this could be easily verified by a diagrammatic representation of the iterative solution of (18)).

Solving for the four-point function (18), would be, anyway, a rather complicated numerical task that, luckily, we do not have to perform. We know, in fact, from the results of the previous Section, that Γ_e^{RPA} contains all two-particle correlation effects to the RPA level of approximation. We might then imagine a procedure whereby such an information is gradually transferred from this amplitude to two of the G^{HF} propagators attached to it. At the end of the process Γ_e^{RPA} would be 'emptied' to the $i \, \Gamma_o$ level (complete loss of information about correlation), whereas the two G^{HF} functions will have transformed into 'renormalized' propagators (G^{RPA}, say) built in terms of 'correlated' orbitals. The whole process is graphically illustrated in Figure 4 and mathematically represented by the equation

Fig. 4.   Diagrammatic representation of the equation defining G^{RPA} (single line carrying a shaded circle). The single lines in the diagram represent Hartree-Fock Green's functions, G^{HF}.

$$G^{HF}(5',7) \; \Gamma_e^{RPA}(3\,7,7'\,3') \; G^{HF}(7',5) =$$

$$- i \; G^{RPA}(5',7) \; \Gamma_o(3\,7,7'3') \; G^{RPA}(7'5) \tag{19}$$

It is not difficult to show from (19) that the following integral equation defining 'correlated' G^{RPA} one-particle propagators follows, in the form

$$G^{RPA}(1,3) \; G^{RPA}(3',1') = G^{HF}(1,3) \; G^{HF}(3',1') - i \int d5d7d5'd7'$$

$$\times \; G^{HF}(1,5) \; G^{RPA}(3',7) \; \Gamma_o(5\,7,7'5') \; G^{RPA}(7',3) \; G^{HF}(5',1') \tag{20}$$

The practical advantage of (20) over (19) is in the fact that we can turn (20) into integral equations for the single 'particle' and 'hole' states, in terms of which G^{RPA} can be expanded. These are very similar to the conventional HF equations of the literature and they can, then, be readily solved by standard numerical procedures. We shall not pursue such aspects any longer, except for pointing out that solving for G^{RPA} has the twofold effect of leading to an explicit representation for Σ (see eq. (21), below), and of giving detailed information on the electronically excited states of the target. We have performed explicit numerical applications[12] for the H_2 molecule and we list, in Table 1, some excitation energies for both singlet and triplet states at the equilibrium internuclear separation. Also shown, for comparison, are results in the Time Dependent Hartree Fock (TDHF) approximation[13] and the exact values of Kolos and Wolniewicz [14,15].

Table 1. Excitation energies (a.u.) for H_2, at equilibrium distance (1,402 a.u.), obtained through a solution for the G^{RPA} one-particle Green's function. Results of the theory are compared to Time Dependent Hartree Fock (TDHF) values (Ref. 13) and to the 'exact' calculations of Kolos and Wolniewicz (Refs. 14 and 15).

EXCITED STATE	KW	G^{RPA}	TDHF
B $^1\Sigma_u^+$	0.4687	0.4696	0.4658
C $^1\pi_u$	0.4860	0.4830	0.4809
E $^1\Sigma_u^+$	0.4826	0.4783	0.4854
B' $^1\Sigma_u^+$	0.5459	0.5369	0.5376
b $^3\Sigma_u^+$	0.3904	0.3804	-------
c $^3\pi_u$	-------	0.4788	-------
a $^3\Sigma_g^+$	0.4609	0.4706	-------

We can now return to the problem of constructing an approximation to Σ, useful for the purpose of scattering calculations. Using $\Gamma_e \simeq \Gamma_e^{RPA}$ on the rhs of (11) and replacing, for consistency, all G s by G^{HF} readily leads to

$$\Sigma^{RPA}(1,1') = \Sigma^{HF}(1,1') + \frac{i}{2} \int d3d5d7d3'd5'd7' \; \Gamma_o(1\;3,3'5)$$

$$\times \; G^{HF}(5,5') \; G^{HF}(3'7) \; \Gamma_e^{RPA}(5'7,7'1') \; G^{HF}(7\,{}_!^!3) \tag{21}$$

where the definition (16) of Σ^{HF} has been used. Enforcing, next, in (21), the identity (19) immediately gives

$$\Sigma^{RPA}(1,1') = \Sigma^{HF}(1,1') + \frac{1}{2} \int d3d5d7d3'd5'd7' \; \Gamma_o(1\;3,3'5)$$

$$\times \; G^{HF}(5,5') \; G^{RPA}(3',7) \; \Gamma_o(5'7,7'1') \; G^{RPA}(7\,{}_!^!3) \tag{22}$$

Equation (22) can be considered as the final working approximation suggested by the FT approach to the optical potential of e^-- atom, molecule elastic collisions. Σ^{RPA} is represented, diagrammatically, in Figure 5, showing that the system excited states will participate in the collision process through the two upper G^{RPA} propagators.

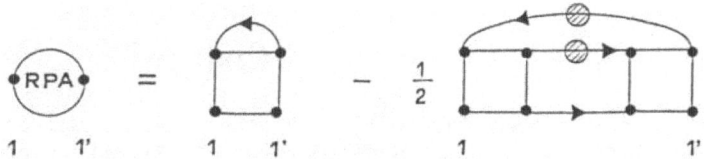

Fig. 5. Graphical representation of the optical potential in the Random Phase Approximation, Σ^{RPA}.

We have considered the numerical evaluation of (22), both in an 'adiabatic' approximation to the G^{RPA} propagators[16] and by evaluating Σ^{RPA} in full[17]. We have also performed elastic scattering calculations for the e^--H_2 system within a 'fixed nuclei' approximation to nuclear motion[18,19]. In Table II we correspondingly list selected values of the Σ_g^+ eigenphase sums so obtained, and we compare these to some recent very accurate R-matrix results of the literature[20].

Table II. $^2\Sigma_g^+$ eigenphase sum for the e^--H_2 system at different collisional energies (Ry.) in the present theory (Σ^{RPA}) and in the R-matrix formalism (Ref. 20).

E(Ry.)	Σ^{RPA}	R-matrix
0.01	3.0316	2.9740
0.04	2.8376	2.7882
0.09	2.6571	2.6025
0.16	2.4628	2.4324
0.25	2.2853	2.2735
0.36	2.1503	2.1228
0.64	2.0729	1.8928

A consideration of the previous numerical values should represent convincing evidence of the accuracy that can be achieved by the FT approach already at the RPA level of approximation in applications to e⁻ –atom, molecule systems. In the next Section we shall investigate collisions involving positrons in order to find out the modifications to the present scheme induced by the new physical situation.

3. FT FORMULATION OF e^+ – ATOM, MOLECULE SCATTERING.

3.1 The one-positron Green's function.

The FT formulation of e^+ – atom, molecule scattering requires that two new field operators be introduced. We can denote these as $\phi(2)$, $\phi^\dagger(2)$, corresponding to the destruction, creation of one positron at (space-spin-time) position $2 = (\vec{r}_2, m_{s2}, t_2)$ (as remarked in Section 2.1, even and odd numbers are reserved to the e^+ and e^- species, respectively). The ϕ $(\phi^\dagger)$ and $\psi(\psi^\dagger)$ fields will anticommute with each other since they refer to different species.

In analogy to the $G(1,1')$ one-electron propagator, we can define[21] the $S(2,2')$ one-positron propagator according to

$$S(2,2') = \frac{1}{i} < \Psi_o^N | T [\phi(2) \phi^\dagger(2')]| \Psi_o^N >$$

$$= \frac{1}{i} \theta(t_2-t_2') < \Psi_o^N| \phi(2) \phi^\dagger(2')| \Psi_o^N > \qquad (23)$$

It is important to stress that (as outlined by the second equality in (23)) only the $t_2 > t_2'$ time ordering can now contribute to the structure of $S(2,2')$. This follows from the fact that the $|\Psi_o^N >$ reference state does not contain any positron to be destroyed by the application of the   destruction operator. At the same time, it should be evident from the discussion of eq. (1) that the rightmost amplitude in (23) is related to the S-matrix of elastic scattering of e^+ off a target in the $|\Psi_o^N >$ state.

In order to develop the theory of these processes we need, then, the equivalent of Dyson's equation for the one-positron propagator, S. The algebraic manipulations to be performed for such a purpose are similar to those considered in Section 2.2, for the G propagator. We shall then omit these completely and simply state the final result[22], represented by the equation

$$S(2,2') = S_o(2,2') + \int d4d4' \ S_o(2,4) \ \Lambda(4,4') \ S(4'2') \qquad (24)$$

In (24) S_o represents the free-positron propagator, while Λ is the optical potential of the elastic scattering process. Once again a functional equation can be obtained for this last amplitude by introducing the same U_e external source of electrons to obtain

$$\Lambda(2,2') = - i \ \delta(2-2') \int d1 \ V_p(1,2) \ G(1,1^+)$$

$$+ i \int d1d4 \ V_p(1,2) \ S(2,4) \ \frac{\delta \ \Lambda(4,2')}{\delta \ U_e(1)} \qquad (25)$$

where $V_p(1,2) = - \delta(t_2 - t_2')/ |\vec{r}_2 - \vec{r}_2'|$ represents the (attractive) $e^+ - e^-$ Coulomb potential. The coupling to the electronic degree of freedom in the structure of Λ appears explicitly in the first (static) term on the rhs of (25). At the same time, it is implicitly contained in the

$\delta \Lambda/\delta U_e$ functional differentiation in the rightmost term of (25). As we shall see, this is the term that shall, finally, be responsible for a description of target 'polarization' effects during the collision process.

Equations (24) and (25) represent the 'formal' solution to the e^+- atom, molecule elastic scattering problem. In the next Section we shall outline a more convenient definition of Λ and we shall explicitly derive a useful numerical scheme of approximation.

3.2 An equation for the Λ optical potential and a numerical scheme of approximation.

Equation (25) for the Λ optical potential can most conveniently be expressed in terms of some new amplitudes more directly related to quantities having a direct physical interpretation. We find convenient, for this purpose, to define a Γ_p (four-point) function according to

$$\frac{\delta \; \Lambda(2,2')}{\delta \; U_e(1)} \; = \int d3d3' \; G(1,3) \; \Gamma_p(2\ 3,3'2') \; G(3',1) \tag{26}$$

Using the above definition in (25) immediately leads to the following alternative expression for Λ

$$\Lambda(2,2') = - \; i \; \delta(2-2') \int d1 \; V_p(1,2) \; G(1,1^+)$$

$$+ \; i \int d1d3d3'd4 \; V_p(1,2) \; S(2,4) \; G(1,3) \; \Gamma_p(4\ 3,3'2') \; G(3',1) \tag{27}$$

Analogously to Γ_e, the Γ_p amplitude contains all correlation effects characterizing the motion of an $e^+ - e^-$ pair and some straightforward algebraic manipulations allow establishing for it the following integral representation

$$\Gamma_p(2\ 1,1'2') = \Xi_p^{(1)}(2\ 1,1'2') - \int d3d5d3'd5' \; \Xi_p^{(1)}(2\ 3,3'2')$$

$$\times \; G(3',5) \; \Gamma_e(1\ 5,5'1') \; G(5',3)$$

$$+ \int d4d6d4'd6' \; \Xi_p^{(2)}(2\ 4,4'2') \; S(4'6) \; \Gamma_p(1\ 6,6'1') \; S(6'4) \tag{28}$$

It could be shown, on the basis of the physical interpretation of the three terms on the rhs of (28), that the third of these is of smaller relevance for the collision process than the first two contributions. We shall then completely neglect it in the following. As for the second term on the rhs of (28), we immediately recognize that this involves a direct coupling to the electronic motion through the Γ_e amplitude, for which we already have a convenient approximation (Γ_e^{RPA}, see (18)). We are then left with a consideration of the $\Xi_p^{(1)}$ amplitude which is explicitly defined by

$$\Xi_p^{(1)}(2\ 3,3'2') \; = \; \frac{\delta \; \Lambda(2,2')}{\delta \; G(3',3)} \tag{29}$$

and that physically represents the full two-body interaction acting between the e^+ and e^- species. The lowest order approximation to this amplitude is correspondingly given by

$$\Xi_p^{(1)}(2\ 3,3'2') = \frac{\delta\ \Lambda(2,2')}{\delta\ G(3',3)} \approx -i\ \delta(2-2')\ \delta(1-1')\ V_p(1,2) \qquad (30)$$

which is the counterpart of (17) of the electron scattering processes. In the present physical situation we have good reasons to believe, anyway, that the lowest order term, (30), will not be a meaningful approximation. This can be seen very simply by considering that there is, now, no Pauli exclusion principle that may prevent the e^+ and e^- species from coming very close together. The attractive V_p interaction will correspondingly become very strong and higher order terms will make non-negligible contributions to $\Xi_p^{(1)}$. By using arguments of a physical nature and relying on functional differentiation techniques, it can be shown that the most convenient definition of $\Xi_p^{(1)}$ is in terms of the following integral equation

$$\Xi_p^{(1)}(2\ 1,1'2') = -i\ \delta(1-1')\ \delta(2-2')\ V_p(1,2)$$

$$+ i \int d3d4\ V_p(1,2)\ S(2,4)\ G(1,3)\ \Xi_p^{(1)}(4\ 3,1'2)$$

$$+ i \int d3d4\ V_p(1',2)\ S(2,4)\ G(3,1')\ \Xi_p^{(1)}(4\ 1,3\ 2') \qquad (31)$$

that corresponds to the infinite summation of the so called 'ladder' diagrams[5]. The physical interpretation of the approximation (31) can be best appreciated from its diagrammatic representation in Figure 6.

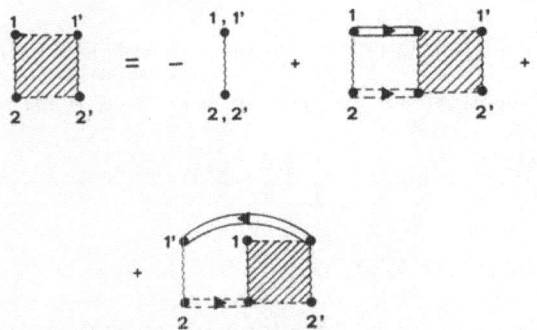

Fig. 6. Diagrammatic representation of the integral equation defining a 'ladder' approximation to $\Xi_p^{(1)}$.

We are now ready to summarize all our previous results in order to obtain a final working expression for the optical potential, (27). By using our previous results and enforcing $S = S_o$ and $G = G^{HF}$, it is not difficult to show that the following equation is recovered

$$\Lambda(2,2') = -i\ \delta(2-2') \int d1\ V_p(1,2)\ G^{HF}(1,1^+)$$

$$+ i \int d3d4d5d5'\ V_p(2,3)\ S_o(2,4)\ G^{RPA}(3,5)$$

$$\times\ \Xi_p^{(1)}(4\ 5,5'2')\ G^{RPA}(5'3) \qquad (32)$$

with $\Xi_p^{(1)}$ given by (31). This final working approximation is represented in graphical terms in Figure 7, from which the physical interpretation of all contributions should be evident. Without going into a detailed discussion of such aspects, here we simply wish to point out that we expect that the coupling to the Positronium formation channel[23] will be represented, somewhat, by the $\Xi_p^{(1)}$ term (shaded area in the Figure).

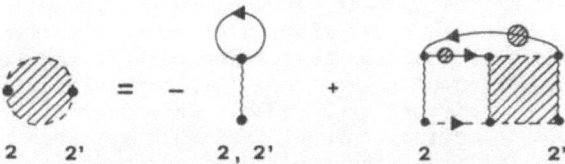

2   2'   2, 2'   2   2'

Fig. 7.   Approximation to the Λ (shaded circle on the left) optical potential of e^+ - atom, molecule scattering. The first term, on the right, represents the static potential in the Hartree-Fock approximation. The second term contains target polarization effects (G^{RPA} propagators) and Positronium formation channel ($\Xi_p^{(1)}$ shaded square).

To summarize, we can say that the theory of e^+ - atom, molecule scattering differs from the corresponding electronic processes for the need to solve the new integral equation (31). Expressions of this type have already been considered in the nuclear physics literature[24], so that we believe that such new equations will also be manageable, numerically, in the present atomic physics situation. For the time being we have not attempted the full solution of (32), but we have simply considered the first term on the rhs, corresponding to the static potential in the HF approximation. Selected results of eigenphase sums for the e^+ - H_2 system are reported in Table III, where we also list numerical values of the literature that include target polarization effects[25,26]. The comparison clearly shows the inadequacy of the static approximation for such processes and the need to introduce the contribution to the Λ optical potential of the second term on the rhs of (32).

Table III. $^2\Sigma_g^+$ eigenphase sum for the e^+-H_2 system in the Hartree-Fock approximation (Λ^{HF}) compared to results of R-matrix (Ref. 25) and Kohn (Ref. 26) methods.

E (Ry)	Λ^{HF}	R-matrix	Kohn
0.01	−0.0748	0.0733	0.0900
0.04	−0.1504	------	------
0.09	−0.2153	------	------
0.16	−0.2789	0.0221	0.0258
0.49	−0.4488	−0.1589	−0.1798
0.64	−0.4973	------	------

4. SUMMARY

In these lectures we have tried to point out how the FT approach can lead to a unified description of $e^{\pm}$ - atom, molecule collision processes. In particular, elastic scattering for the two systems can be described in terms of optical potentials, for which we have suggested working approximations in (21) and (32), respectively. A key quantity of the formalism is represented by 'quasi particle' states directly related to electronic excitations in the RPA approximation (see (20)). Such single particle states appear in both types of optical potentials and, in the language of scattering theory, they are related to a description of target dynamical polarization effects. The theory of positron scattering further requires that a new kind of integral equation be solved (see (31)) describing repeated scatterings of an e^+ - e^- pair off each other. Such a quantity is related to the new (Positronium formation) channel available for such systems and its contribution to the full dynamics still awaits a numerical clarification.

ACKNOWLEDGEMENTS

Many of the calculations referred to in these lectures have been performed in collaboration with Ugo Lamanna (University of Bari). Besides, the author gratefully acknowledges Jurji Darewych, Al Stauffer and Bob McEachran (York University), with whom various aspects of the theory, both of electron and positron scattering, have been discussed. Finally, the partial support of NATO, in the form of an International Grant, is gratefully acknowledged.

REFERENCES

1. J.D. Bjorken and S.D. Drell, "Relativistic Quantum Fields", McGraw-Hill Book Co., New York (1965).
2. I. Lindgren and J. Morrison, "Atomic Many-Body Theory", Springer-Verlag, Heidelberg,(1982).
3. E. Ficocelli Varracchio, J. Phys. B 10:503 (1977).
4. E. Ficocelli Varracchio, J. Phys. B 12:3427 (1979).
5. A.L. Fetter and J.D. Walecka, "Quantum Theory of Many-Particle Systems", McGraw.Hill Book Co., New York (1971).
6. F.L. Pilar, "Elementary Quantum Chemistry", McGraw.Hill Book Co., New York, (1968).
7. L.P. Kadanoff and G. Baym, "Quantum Statistical Mechanics", W.A. Benjamin Inc., Menlo Park (1962).
8. G. Csanak, H.S. Taylor and R. Yaris, Adv. At. Mol. Phys. 7:287 (1971).
9. E. Ficocelli Varracchio (to be published).
10. G. Wendin, in "Photoionization and Other Probes of Many-Electron Interactions", F.J. Wuilleumier ed. NATO Advanced Study Institutes Series, B18, Plenum, New York (1975).
11. E. Ficocelli Varracchio, J. Phys. B 17:L611 (1984).
12. E. Ficocelli Varracchio and U.T. Lamanna (to be published).
13. R.F. Stewart, D.K. Watson and A. Dalgarno, J. Chem. Phys. 65:2104(1976).
14. W. Kolos and L. Wolniewicz, J. Chem. Phys. 48:3672 (1968).
15. W. Kolos and L. Wolniewicz, J. Chem Phys. 43:2429 (1965).
16. E. Ficocelli Varracchio and U.T. Lamanna, J. Phys. B 19:3145 (1986).
17. E. Ficocelli Varracchio, U.T. Lamanne and G. Petrella (to be published).
18. E. Ficocelli Varracchio, J. Phys. B 14:L511 (1981).
19. E. Ficocelli Varracchio and U.T. Lamanne, Chem. Phys. Lett. 101:38(1983).
20. R.K. Nesbet, C.J. Noble and L.A. Morgan,Phys. Rev. A 34:2798 (1986).
21. E. Ficocelli Varracchio, Ann. Phys. (N.Y.) 145:131 (1983).
22. E. Ficocelli Varracchio (to be published).
23. E. Ficocelli Varracchio, J. Phys. B 17:L311 (1984).

24. K.A. Brueckner and J.L. Gammel, Phys. Rev. 109:1023 (1958).
25. J. Tennyson, J. Phys. B (in press).
26. E.A.G. Armour, J. Phys. B 18:3361 (1985).

THE HIGH BRIGHTNESS BEAM AT BRANDEIS

K. F. Canter, G. R. Brandes, T. N. Horsky, and P. H. Lippel

Department of Physics
Brandeis University
Waltham, MA 02254

A. P. Mills, Jr.

AT&T Bell Laboratories
Murray Hills, NJ 07974

At the first workshop in this series, York University - 1981, the "brightness enhancement" proposal[1] which was about a year old at the time was referred to as an "ambitious method" of increasing the brightness-per-volt R_V of a slow positron beam.[2] Today, however, there are several groups actively pursuing or actually using brightness enhancement in their research. This talk will deal mainly with the brightness enhanced beam at Brandeis since it is a particularly high performance beam and uses well documented optics. The latter feature is important since the lack of well documented electrostatic optics for slow positron beams, in particular the immersion lens gun, has often inhibited the use of electrostatic transport in favor of the more intuitively simple, but generally more restrictive, magnetically guided beam. Equally important have been the improvements due to single crystal metal moderators in the planar back-scattering geometry[3] over the large transverse energy spreads of parallel vane moderators.[4] A large transverse energy component E_T (perpendicular to the initial beam direction) can be as injurious to the focussability of the beam as it is to the predictability of the positron trajectories through the system. In addition to briefly reviewing the design and operation of the Brandeis beam, some results will be presented in the context of the current diffraction and microbeam experiments for which the beam is presently being used. Although these experiments fall mainly within the domain of solid state physics, some examples will be given illustrating the extent to which present day brightness enhancement may be of value in typical atomic physics experiments.

A schematic view of the Brandeis beam in its present configuration is shown in Fig. 1. The characteristics of the primary beam up to the first remoderator position in the brightness enhancement chamber were described earlier at the previous workshop in this series.[5] The 10 mm diameter primary moderator is imaged down to 1 mm at the first remoderator by accelerating to 5 keV. The first remoderator, a W(110) single crystal, is located at the "cathode" position of an immersion lens gun which is similar to the primary modified Soa gun.[6] The process is then repeated by imaging the positrons emitted from the first remoderator down to 0.1 mm at the second

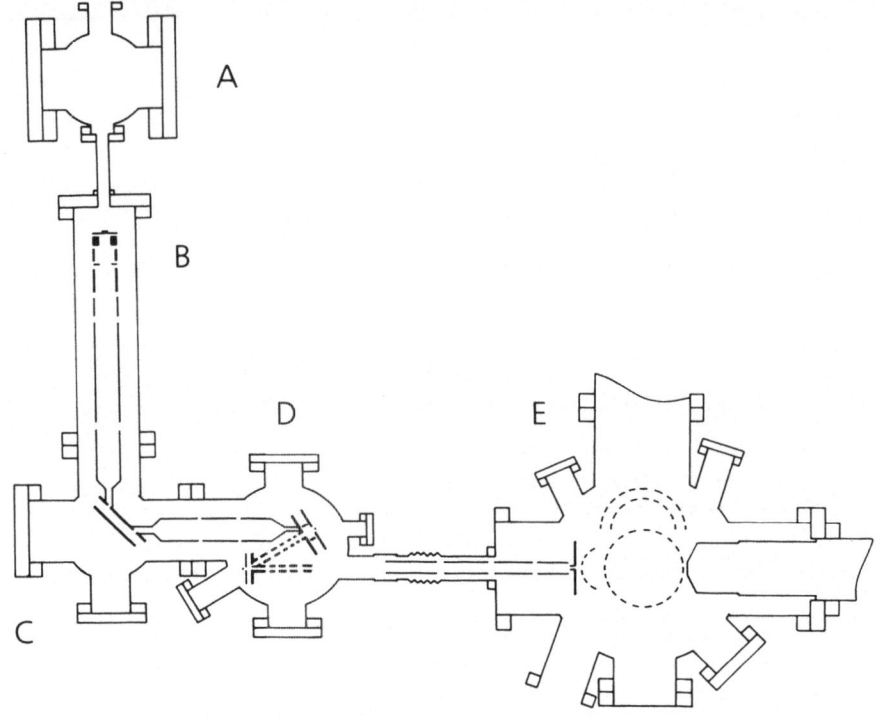

Fig. 1.   The high brightness beam at Brandeis. A) Primary moderator
          annealing chamber.   B) Primary beam gun tube.   C) Cylindrical
          mirror deflector chamber.   D) Brightness enhancement chamber
          with first remoderator gun (upper right in chamber) and second
          remoderator gun (lower left).   E) Target sample chamber.   For
          purposes of scale, the outside diameter of the primary gun
          tube is 10 cm.

W(110) remoderator. Positrons emitted from this 0.1 mm spot constitute our
final beam. Our present overall maximum efficiency for reducing the emitting diameter from 10 mm to 0.1 mm is 5%, which means a corresponding increase in brightness-per-volt by a factor of 500. This is off by only a
factor of two from the theoretical estimate of a factor of 1000 brightness
enhancement using two-stage, weak-focussing, reflection mode brightness
enhancement with room temperature W(110) remoderators.[7] This discrepancy
can be accounted for by the fact that the first remoderator is observed to
have only a 15% remoderation efficiency. A 35% remoderation efficiency is
observed for the second remoderator, which is typical of a well prepared
W(110) crystal.[3] We presently attribute the factor of $\approx$ 2 loss in the first
remoderator to bulk carbon contamination. The remoderation efficiency was
originally 7% and only came up to 15% after several hours of in situ heating
at $\approx$ 800° C in 10^{-4} Pa of O_2. Since this in situ heating tended to contami-
nate some of the insulators in the system, leading to electrical discharges,
we have refrained from further attempts to improve the first remoderator's
efficiency. Aside from this "materials science" aspect of brightness enhancement, there is nothing to indicate that our optics presents any limitation in achieving the factor of 1000 brightness enhancement that the system
was designed for.

     The issue of obtaining the theoretical efficiencies for remoderation
was an important one since we took a gamble by deviating from our original
intention to make the remoderation immersion lens guns exact replicas

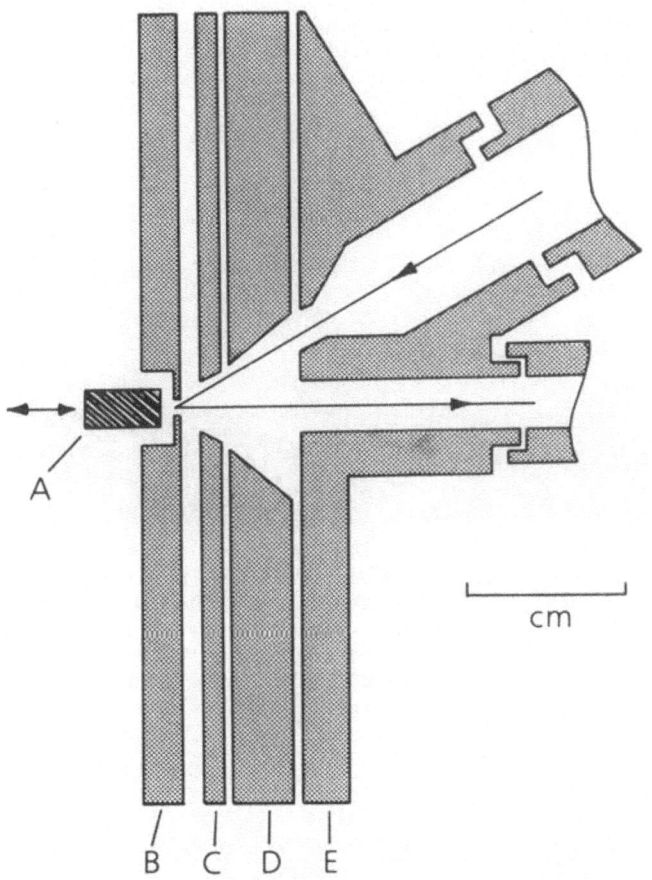

Fig. 2.   Second remoderator gun. Positrons emitted from the
first remoderator are incident on the second remoderator.
A) From the upper right. B) Cathode plane. C) Whenelt
electrode. D) Soa tube. E) Anode electrode. The second
remoderator can be swung away from the cathode for
electron bombardment annealing and otherwise is in close
contact with the cathode aperture.

(except for a scaling factor) of our primary modified Soa gun. Although
the remoderation guns kept the same ratio of the electrode inner diameters
to thickness as for the primary gun, the "Wehnelt" and "Soa Tube" were
tapered to accommodate the beam from the previous stage. This is illustrated by the detail of the second remoderation gun shown in Fig. 2.

Tapering the Wehnelt and Soa Tube electrodes resulted in a gun that
differed enough from the primary gun that new SLAC trajectory calculations
had to be carried out. The output trajectories will be reported in a forth-
coming paper which will present a detailed documentation of the optics. For
the present, it suffices to say that the tapered electrodes result in the
remoderator guns being less forgiving with respect to launch angle from the
cathode than was the case for the primary gun. Launch angles greater than
$\approx 20°$ tend to overfill the remoderation guns. However an angular spread of
15° is usually the maximum range one deals with when using a well prepared
W(110) remoderator with $E_T = 0.2$ eV.[8]

The first demonstration of brightness enhancement made by Frieze et al.

also used two stages of reflection mode remoderation, but with gridded lenses.[9] The fact that they only obtained a factor of 20 in brightness enhancement was due more to the poor quality of their primary beam than their remoderation optics. The possibility of obtaining appreciably more than a

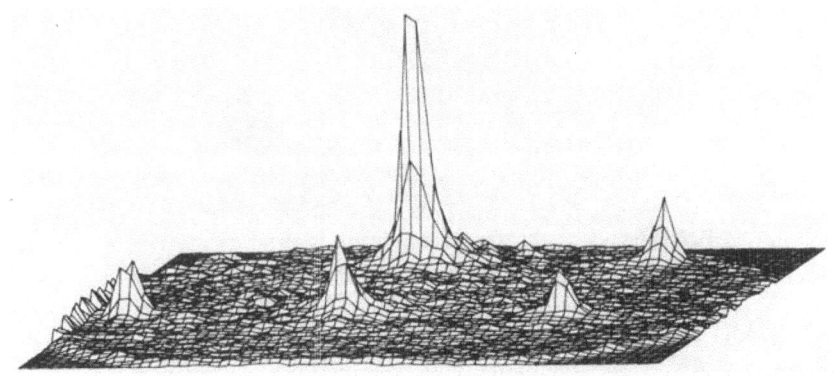

Fig. 3a. Diffraction pattern produced by 185 eV positrons incident at 50° on a Cu(100) crystal located 31 mm above 2D detector. The active area shown here is 40 mm × 40 mm. Data is acquired with 256 × 256 channel resolution, binned-down to 64 × 64 channels and smoothed to produce this 3D perspective of counts (vertical) versus detector position (horizontal plane). The tallest peak is the specular (0,0) peak and the smaller peak in front is the $(\overline{1},\overline{1})$ peak, for example, with $(\overline{2},0)$ and $(0,\overline{2})$ being adjacent.

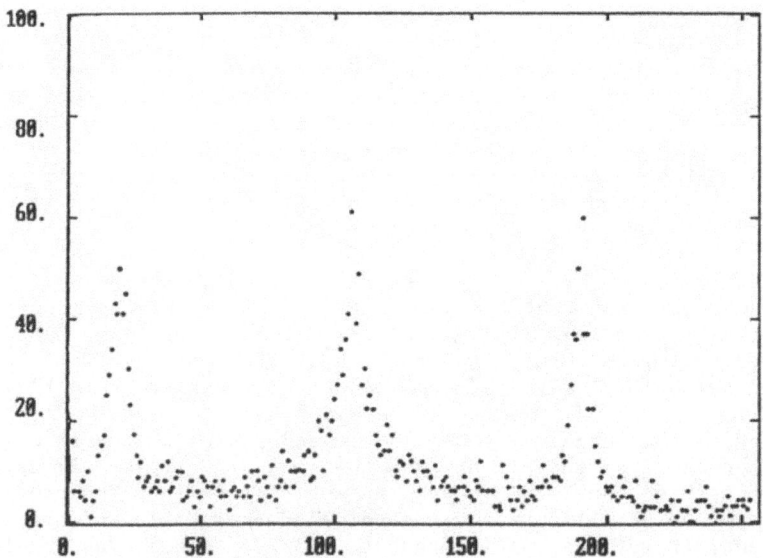

Fig. 3b. A slice through the $(\overline{2},0)$, $(\overline{1},\overline{1})$, and $(0,\overline{2})$ diffraction peaks in the 2D data used for 3b. Counts were integrated over a 9 channel width perpendicular to the strip. Each channel (horizontal axis) represents 0.18 mm. The peaks are much sharper than those in 3b since the data is unsmoothed and is binned only perpendicular to the direction of the slice.

factor of 1000 in brightness enhancement by using stronger focussing into the remoderation guns with a gridded injection lens does not look promising at the moment. Another approach to brightness enhancement, which is easier from the optics point of view but more difficult from the materials science point of view, is transmission mode brightness enhancement. This method was orginally avoided because of the difficulty of obtaining self-supporting, thin (≈ 1000 Å) single crystal metal films. Now that these films are becoming more available, particularly Ni(100)[10] and W(100),[11] the contention that transmission mode is four times less efficient than reflection mode brightness enhancement[12] may be the only remaining minus in using the transmission approach.

The Brandeis beam is presently being used to investigate surface barrier effects near grazing emergence thresholds in low energy positron diffraction (LEPD) and to carry out studies of near surface defect structures of solids on the few micron scale. The application of brightness enhancement in the LEPD experiments which is of particular relevance to this workshop is the ability to produce a beam with a small angular divergence and diameter at low energies. In our LEPD work, a high brightness manifests itself in sharp diffraction peaks. Some of the diffraction peaks resulting from 185 eV positrons incident at 50° on a Cu(100) sample situated 31 mm above a 2D position sensitive detector are shown in Fig. 3a. Although less visually spectacular is the observation of a clear specular diffracted beam at energies as low as 7 eV. The FWHM of such a peak with the same sample and detector arrangement used for the data in Fig. 3b is observed to be 3 mm.

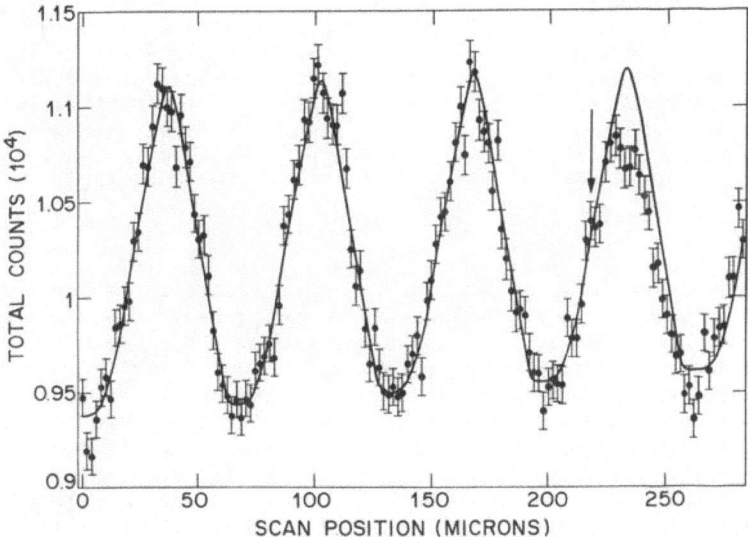

Fig. 4. Detected annihilations produced by rastering a microbeam across a mesh with 40 μm cell openings and 25 μm bar widths. The background is 6.7×10^3 counts in this scan. The solid line is a fit to the data from 0 to 220 μm produced in simulation by rastering a 7 μm × 55 μm beam profile across the mesh. The poor fit past 220 μm, indicated by the arrow, is attributed to a mesh irregularity. Up to the arrow, the fit produced a $\chi^2/\nu = 151/101$.

Another way to demonstrate brightness enhancement is the ability to produce a "micro beam".[13] By accelerating the positrons emitted from a 130±20 μm diameter at the second remoderator of the Brandeis beam to a few keV, the Liouville theorem allows one to obtain, in principle, a few micron diameter focus. The practical difficulites in obtaining a phase space limited focus on this size scale are considerable. Achieving the smallest spot diameter possible requires large angles of convergence. However since the spherical abberation contribution to the spot size scales as the cube of the convergence angle times the condensing lens diameter, one is required to work with very small diameter lens elements in order to minimize abberations. This in turn poses a practical problem due to the mechanical machining tolerances and lens alignment limitations. As a preliminary demonstration of the micro-focussability of the Brandeis beam, and as a precursor to studying defect structures of micron size solid samples, a sectored, two-tube, 1.5 mm diameter, condensing/rastering lens system was introduced into the beam line at the target sample chamber location. The beam was accelerated to 6.7 keV and then rastered across a specimen grid (25 μm bar widths and 40 μm cell openings) viewed by a NaI annihilation γ-ray detector. The resulting oscillations in annihilations as the beam was rastered across the mesh are shown in Fig. 4.

The solid line in Fig. 4 is a simulated scan produced by rastering a 7 μm × 55 μm uniform density beam spot across the mesh. The smaller dimension is in the direction of rastering which is along one of the symmetry axes of the mesh array. The longer dimension of the rectangular spot is oriented 55° with respect to the rastering direction. Although we can fit a range of spot sizes to our one dimensional scans, we are constrained to the rectangular beam shape when trying to account for scans obtained at right angles to the direction used for Fig. 4 as well as accounting for an observed maximum transmission of 75% through the mesh. Measuring the beam profile at different points along the beam axis enables us to estimate a mean angular divergence of 74 mrad, in reasonable agreement with a calculated 86 mrad divergence. The corresponding $\Gamma\sqrt{E}$ product for this beam profile is 0.15 mm-rad $eV^{1/2}$. This is to be compared with $2D\sqrt{E_T} = 0.12$ mm-rad $eV^{1/2}$ for a beam emanating from a D = 130±20 μm diameter at the second remoderator and where $E_T = 0.2$ eV is assumed. The "measured" phase space being larger than the expected value is consistent with our present feeling that the microbeam profile is distorted into a rectangular shape due to aberrations normally seen when the beam is not centered on the condensing lens axis. Once this problem is corrected, a 16-20 μm diameter round spot, with an 86 mrad divergence, should be readily obtained at a beam energy of 6.7 keV.[13]

The most conventional atomic physics experiments for which brightness enhancement is well worth the effort are "crossed-beam" experiments in which a positron beam is incident on a collimated atomic beam.[14] In order to maximize data collection and angular resolution, a small diameter and well collimated positron beam is desired. Figure 5 illustrates the typical design criterion in order to produce a beam having a desired diameter d_2 at the scattering target. The total angular divergence is $(d_1+d_2)/Z_{1,2}$. This is an unambiguous way of describing the angular divergence since no distinction between the "beam" angle and "pencil" angle need be made. The limiting factors in obtaining the final beam are the diameter D of the final emitting moderator and the transverse energy component E_T.[2] If you wish to focus the beam without a loss in flux onto a target of diameter d_2 at a final energy E, then you must observe the condition

$$2D\sqrt{E_T} \leq \frac{d_1 d_2}{Z_{1,2}} \cdot \sqrt{E} \ ,$$

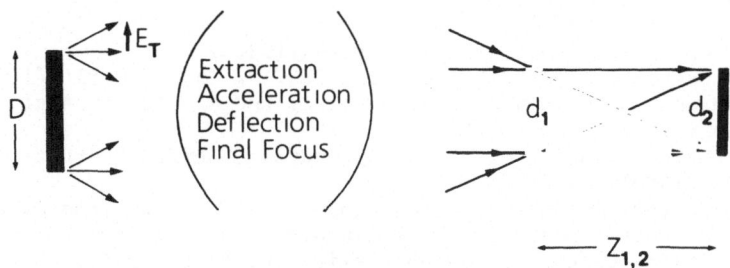

Fig. 5. Schematic representation of the fundamental
quantities D and E_T determining the final
beam parameters at a target of diameter d_2.

where d_1 is the diameter of an aperture (real or imagined) which is upstream
(in a field free region) at a distance $Z_{1,2}$ from the target. Although most
positron experiments do not require an angle-limiting aperture, i.e., the
d_1 aperture, keeping trajectory angles less than 20° and lens filling less
than ≈50%, results in the same constraint as having this aperture present.
When $2D\sqrt{E_T}$ is less than the right hand quantity, one can have a smaller or
more collimated beam focus. Applying the values D = 0.1 mm and E_T = 0.2 eV
at a beam energy of 5 eV, for example, yields a value of $d_1d_2/Z_{1,2}$ equal to
0.04 mm-rad. Thus one could have a final diameter d_2 = 1 mm with a corresponding $d_1/Z_{1,2}$ = 0.04 rad. This would result in a total angular divergence of 48 mrad (2.8°) if one chose d_1 = 5 mm and $Z_{1,2}$ = 125 mm, for example. Except for the extra care necessary to deal with stray fields, we see
that present day brightness enhancement technology makes crossed beam experiments in the few mm-degree range a realistic possibility for energies as
low as 1 eV.

The low energy range is particularly interesting to investigate since
positron-electron total elastic scattering cross-sections in helium, for
example, differ by more than two orders of magnitude at 1-2 eV.[15] How
these differences manifest themselves in the partial differential scattering
cross sections could even be more dramatic at low energies.[14] Decreasing
the positron beam dimensions also has the practical advantage of enabling
one to work with a gas jet, or a gas cell, which is also smaller in the
dimension transverse to the positron beam. This results in a proportional
decrease in the pressure surrounding the gas target for a given differential
pumping capability. The production of monoenergetic Ps beams in the few eV
to ≈ 30 eV range using rare gas targets[16] would be improved in flux and angular definition if the gas target cells could be reduced enough to outweigh
the 90% loss in positron flux due to brightness enhancement.

ACKNOWLEDGEMENT

This work was supported in part by NSF grant DMR-8519524.

References

1. A. P. Mills, Jr., Appl. Phys. 23:189 (1980).
2. K. F. Canter and A. P. Mills, Jr., Can. J. Phys. 60:551 (1982).
3. A. Vehanen, K. G. Lynn, P. J. Schultz, and M. Eldrup, Appl. Phys. A32: 163 (1983).
4. J. M. Dale, L. D. Hulett, and S. Pendyala, Surf. Interface Anal. 2: 199 (1980).
5. K. F. Canter, T. Horsky, P. H. Lippel, W. S. Crane, and A. P. Mills, Jr., "Development of High Brightness Slow Positron Beams," in Positron (Electron)-Gas Scattering, W. E. Kauppila, T. S. Stein, and J. M. Wadehra, eds., World Scientific Press, Singapore, p. 202, 1986.
6. K. F. Canter, P. H. Lippel, W. S. Crane, and A. P. Mills, Jr., "Modified Soa Immersion Lens Positron Gun," in Positrons in Solids, Surfaces and Atoms, A. P. Mills, Jr., W. S. Crane, and K. F. Canter, eds., World Scientific Press, Singapore, p. 199, 1986.
7. K. F. Canter, "Low Energy Positron and Positronium Diffraction," in Positron Annihilation in Gases, J. W. Humberston and M. R. C. McDowell, eds., Plenum Press, NY, p. 219, 1986.
8. D. A. Fischer, K. G. Lynn, and D. W. Gidley, Phys. Rev. B33:4479 (1986).
9. W. E. Frieze, D. W. Gidley, and K. G. Lynn, Phys. Rev. B31:5628 (1985).
10. P. J. Schultz, E. M. Gullikson, and A. P. Mills, Jr., Phys. Rev. B34: 442 (1986).
11. D. M. Chen, K. G. Lynn, R. Pareja, and B. Nielsen, Phys. Rev. B31:4123 (1985).
12. K. F. Canter, "Slow Positron Optics," in Positrons in Solids, Surfaces and Atoms, A. P. Mills, Jr., W. S. Crane, and K. F. Canter, eds., World Scientific Press, Singapore, p. 102, 1986.
13. G. R. Brandes, K. F. Canter, T. N. Horsky, P. H. Lippel, and A. P. Mills, Jr., Bull. Am. Phys. Soc. 54, 1944 (1987). A manuscript on the Microbeam is in preparation for submission to Rev. Sci. Inst.
14. G. M. A. Hyder, M. S. Dababneh, Y.-F. Hsieh, W. E. Kauppila, C. K. Kwan, M. Mahdavi-Hezaveh, and T. S. Stein, Phys. Rev. Lett. 57:2252 (1986).
15. T. S. Stein, W. E. Kauppila, V. Pol, J. H. Smart, and G. Jesion, Phys. Rev. A17:1600 (1978).
16. B. L. Brown, "Creation of Monoenergetic Positronium in a Gas," in Positron Annihilation, P. C. Jain, R. M. Singru, and K. P. Gopinathan, eds., World Scientific Press, Singapore, p. 201, 1985.

A HIGH INTENSITY POSITRON BEAM AT THE BROOKHAVEN REACTOR

K.G. Lynn,[1] M. Weber,[2] L.O. Roellig,[2] A.P. Mills, Jr.[3] and
A.R. Moodenbaugh,[1]

1) Brookhaven National Laboratory, Upton NY 11973 USA
2) City College of the City University of New York, NY 10031 USA
3) A.T. & T. Bell Laboratories, Murray, NJ 07974 USA

ABSTRACT

We describe a high intensity, low energy positron beam utilizing high
specific activity ^{64}Cu sources (870 Ci/g) produced in a reactor with high
thermal neutron flux. Fast-to-slow moderation can be performed in a self
moderation mode or with a transmission moderator. Slow positron rates up
to 1.6×10^8 e$^+$/s with a half life of 12.8 h are calculated. Up to $1.0 \times
10^8$ e$^+$/s have been observed. New developments including a Ne moderator and
an on-line isotope separation process are discussed.

INTRODUCTION

A number of proposed positron experiments require a high intensity slow
e$^+$ beam to attain an acceptable signal to noise ratio.[1] Such a beam could
be used for differential cross section measurements of positrons scattering
in gases.[2] The 2 γ angular correlation measurements of the Fermi surfaces
of metals have begun to be extended to surfaces of solids.[3] The beam has
been used as a source for a monoenergetic positronium beam to study the
possibility of positronium scattering and diffraction from surfaces.[4] If
this proves to be practical, such a new tool would greatly supplement results
on neutral atom beams.[5] A high positron count rate would also make feasible
rare decay studies, such as the measurement of the single photon decay rate
of positrons.[6] The implementation of several stages of remoderation,[7] a
technique that recently has been realized for the first time,[8] could provide
a microscopic beam usable as a positron microprobe or to study matter-anti-
matter systems at high densities.[1]

At present two ways to reach these high count rates exist and are being
further developed. In one approach the bremsstrahlung of a 100 MeV electron
beam of a LINAC hitting a target produces showers of positron-electron pairs.[9]
The positrons are moderated to form the low energy beam. Due to the nature
of the LINAC, a pulsed beam of positrons is available.

The second method, described here, uses thermal neutrons to produce the
positron source ^{64}Cu. A reactor with a high thermal neutron flux such as
the High Flux Beam Reactor (HFBR) at Brookhaven National Laboratory (BNL) is
essential. If available as a single crystal, the copper can be used as the

moderator, a combination, that we call a self moderator. Alternately the positrons can be moderated by a thin single crystal film; a tungsten film has been used successfully as such a transmission moderator.[10,11]

The reactor based beam will be discussed in more detail below. Section 2 contains calculations to predict the performance of the beam, given the available neutron flux. Self moderation and transmission moderation are compared. In section 3, advantages and disadvantages of such a beam are presented. The technical realization of such a beam at BNL is described in section 4. Problems that were encountered during beam development are presented in section 5. The present beam performance, as well as improvements planned or under way, are covered. Section 6 provides an outlook, and an appendix contains the detailed self moderator calculations outlined in section 3. Also the possibility of further increasing the count rate by an on-line enrichment of ^{64}Cu is considered.[12] Finally the ongoing implementation of the recently developed neon moderator is discussed.[13]

CALCULATIONS TO PREDICT THE BEAM PERFORMANCE

Production of the Positron Activity

The source of positrons, ^{64}Cu, is used. It is produced in the High Flux Beam Reactor (HFBR) at Brookhaven by means of the thermal neutron reaction ^{63}Cu (n,γ) ^{64}Cu. A flux of $f = 8.3 \times 10^{14}$ $n_o/cm^2 s$ of thermal neutrons[1] near the core, along with the relative high cross section of $\sigma = 4.5$ b for this reaction, make reasonable amounts of ^{64}Cu possible. The short half life of ^{64}Cu of $t_{1/2} = 12.8$ h necessitates frequent renewal of the source.

With a production rate $\lambda_p = f\sigma$ and decay rate $\lambda_d = \ln2/t_{1/2}$ after an irradiation of time t the positrons activity is

$$A(t) = bp \frac{m}{M} N_a \lambda_p [1 - e^{-\lambda_d t}] \quad . \tag{1}$$

This formula holds for the case $\lambda_p \ll \lambda_d$.

In the equation $b = 19\%$ is the fraction of decays through the positron channel, p the fraction of ^{63}Cu in a sample of mass m, and $M = 63.54$ g/mol the molar weight. N_a is Avogadro's number.

After an irradiation of duration

$$t_m = \frac{1}{\lambda_d} \ln \frac{\lambda_d}{\lambda_p} \tag{2}$$

the maximal activity is reached. Figure 1 shows the activity $A(t)/A(t_m)$ as a function of the irradiation time. For the neutron flux available at the HFBR the 95% of $A(t_m)$ is reached after $\sim$ 55 hours. Figure 2 shows the dependence of the activity on the neutron flux on a log-log scale. Irradiation times are chosen to maximize the activity in each case.

Performance of Copper as a Moderator

A system of Cu(111) on W(110) with a perfect interface is considered. The efficiency ϵ, defined as the ratio of slow positrons to fast positrons, of such a moderator with the source as an integral part of the copper layer is made up of several contributions. These are the efficiencies resulting from a single crystal of copper on a tungsten substrate, the backscattering effect of the tungsten, and the moderation capability of the tungsten crystal.

162

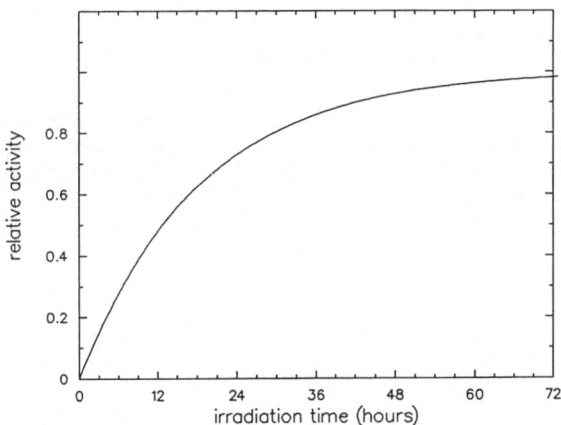

Fig. 1. Fraction of the maximum activity of a copper sample, as a function of the irradiation time in the reactor. The thermal neutron flux is 8.3×10^{14} n_o/cm^2s.

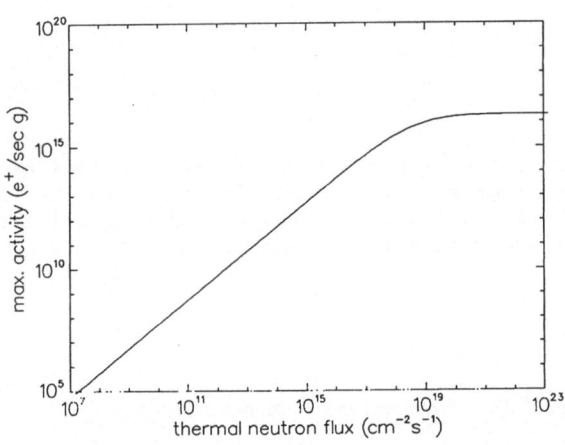

Fig. 2. Maximum positron activity of a sample that can be reached depending on the thermal neutron flux. 99% enriched ^{63}Cu is used.

These contributions are calculated in the appendix and are shown in Figure 3. When the Cu layer is on the order of 10^3 nm thick or more only the contribution of the Cu layer itself remains significant. ε_S can be approximated by

$$\varepsilon_S (d) \approx \frac{1}{2} Y_0 \frac{L_+}{d} \left[L_+\alpha \, \ell n \left(\frac{1 + L_+\alpha}{L - L_+\alpha} \right) + \int_0^1 dt \, \frac{t}{t - L_+\alpha} (1 - e^{-\alpha d/t}) \right] \quad (3)$$

The fraction diffusing back from the W substrate is

$$\epsilon_w(d) = \frac{1}{2} Y_o \frac{L_+}{d} (1-f) \frac{L_w \alpha_w}{L_+ \alpha} \frac{1}{\cosh d/L_+} \int_0^1 dt \frac{t}{t+L_w \alpha_w} (1-e^{-\alpha d/t}) \qquad (4)$$

Here d is the thickness of the copper layer, α and α_w are the products of mass absorption coefficient and density ρ of copper and tungsten respectively,[7,12] f is the backscatter fraction for positrons from the β^+-spectrum off the W–Cu interface, L_+ and L_w are the diffusion lengths for positrons in copper and tungsten, and Y_o is the branching ratio [Y_o = 0.55 for Cu(100) + S].

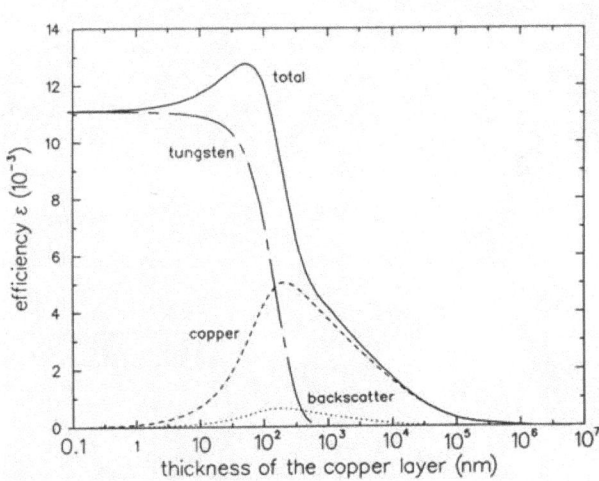

Fig. 3. The various contributions to the efficiency ϵ of the reactor positron source. – – – : the single crystal Cu layer, ... : the fraction due to backscattering off the interface, – — – : the W substrate, and ———— : the total efficiency.

Performance with a W(100) Transmission Moderator

An alternative to self moderation is the recently developed transmission moderator.[10,11] It consists of a thin (< 10^3 nm) single crystal of tungsten. Its moderation efficiency is theoretically estimated to be $\epsilon_o = 4 \times 10^{-3}$ (emitted slow positrons per fast positrons hitting the crystal).[10] Due to self absorption of fast positrons in the source, its activity is reduced by a factor s.

$$s(d) = \frac{1}{\alpha d} \int_0^1 dt \, t \, [1 - e^{-\alpha d/t}] . \qquad (5)$$

Further losses depend on the source–moderator geometry factor G. G is at most 0.5 since only one side of the source faces the moderator. The absolute efficiency with this type of moderator would then be

$$\epsilon_t(d) = Gs(d)\epsilon_o . \qquad (6)$$

Expected Performance of the Reactor Beam

For a discussion of a self moderator, the efficiency as the fraction of the available activity that contributes to the beam is not very useful. With growing thickness of the crystal the efficiency will decrease but this will be more than compensated by the larger amount of available activity.

In the appendix the product

$$E = \epsilon d \qquad (7)$$

is introduced. Figure 4 shows the various components of E again as a function of the thickness of the Cu layer and constant specific activity. Beyond a 10^5 nm thick copper film significant gains in E are no longer possible. At this thickness only the copper contribution is significant. The source-transmission moderator geometry and the value ϵ_o determine which type of moderation is superior. The efficiency of a tungsten single crystal foil was estimated in reference 10 to be $\epsilon_o = 4 \times 10^{-3}$ for a 10^3 nm thick crystal.

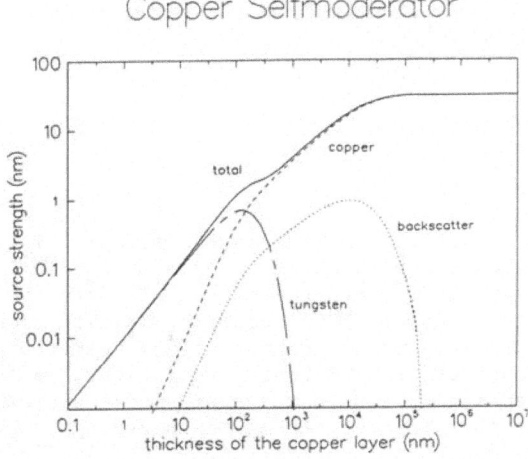

Fig. 4. The source strength on a logarithmic scale as a function of the thickness of the copper layer. The same symbols as in Fig. 3 are used.

In Figure 5 the performances of both moderators are shown, G is assumed to be 0.25. It should be noted that the best reported efficiency for a transmission moderator is 5×10^{-4}.

These values present an upper limit. The thermal neutron flux decreases during a reactor cycle (at BNL about 5%). The capsule absorbs neutrons (< 5%). The beam becomes operational about 2 hours after the new source is removed from the reactor, in which time the copper has decayed to about 90% of its original strength.

A typical source of 100 mg is irradiated for 48 hours. A 10^5 nm thick copper crystal of about 1 cm^2 area is produced. It has a positron-activity of A = 14.9 Ci of positrons. Self moderation is calculated to deliver a beam of A = 4.3 mCie$^+$/s = 1.6 × 10^8 e$^+$/s and a transmission moderator A = 2.5 mCie$^+$=9.3 × 10^7 e$^+$/s with ϵ_t = 10^{-3}.

Copper Selfmoderator

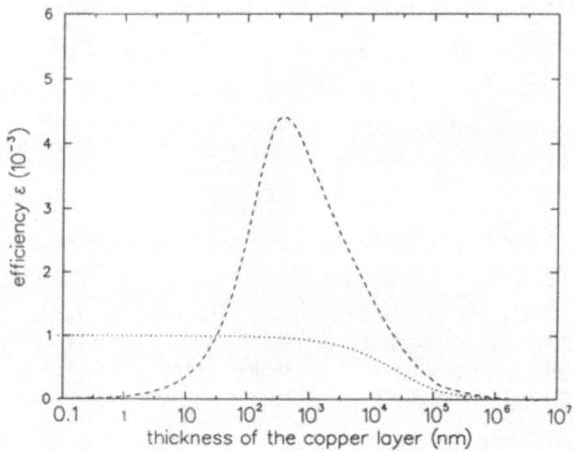

Fig. 5. Comparison of the efficiency ε of a self moderating film of Cu (dashes) and a tranmission moderator when located in front of the Cu film (dots).

DISCUSSION OF THE HFBR BEAM

The high positron activity of the source material is achieved by utilizing the reactor's high thermal neutron flux of 8.3×10^{14} n_o/cm^2 s. A (n,γ) reaction converts ^{63}Cu into the 12.8h half life positron emitter ^{63}Cu. The fraction of 69.1% of usable ^{63}Cu occurring in natural copper is usually increased to about 99% in enriched copper. Thus positron activities of 166 Cie+/g can be achieved with 48 hour irradiations. After irradiation the copper is transferred into the vacuum system, and evaporated onto a W(110) crystal surface. On this surface the epitaxial growth of single crystal copper in the (111) direction is preferred.[16] Cu(111) is the most suitable orientation if used as a moderator.[17] In this case the copper will be the source as well as the moderator.

Periodic irradiation, of sources does not interrupt or disturb work performed by other reactor users. The only interference with a steady supply of new sources are infrequent sample irradiations by other users of this high flux reactor port. In such cases a port with a somewhat lower neutron flux can be used. In contrast most LINACs cannot support the positron beam facility and other experiments simultaneously.

The short half life of the source is advantageous when maintenance work has to be done in the source chamber. In addition to the source material, only a few radioactive isotopes are produced. By using high purity copper, the radioactivity of impurities is kept more than six orders of magnitude below that of the ^{64}Cu isotope. At a LINAC the bremsstrahlung causes a high background radiation and activates material in the vicinity of the positron source. While positrons from pair production are unpolarized, a beam of polarized positrons is possible when they originate in β+-decays as in ^{64}Cu.

The reactor-based beam operates in a pseudo DC mode that avoids the problem of pileup of signals during the short bursts of positrons in a pulsed beam. A pulsed beam is advantageous for time of flight experiments.

THE TECHNICAL REALIZATION

The Reactor Division routinely samples for all kinds of research. They have chosen aluminium capsules as containers that are inserted through thimbles into the reactor core at the end of long aluminium tubes. The apparatus to transfer the copper pellet from the reactor into the UHV system was based on this design. The copper sources produced by the reactor cannot be handled directly. The radiation level is high enough to require a remote control system for the transfer of sources into the vacuum chamber. The chamber must be shielded. A concrete house, referred to as the blockhouse, has been constructed around the source chamber. It is capable of protecting against up to 10 kCi sources.

The capsule is held in a lead pig for about one hour to let the short-lived isotopes decay. It is cut off the Al tube and drops from the lead container on top of the blockhouse, through a drying chamber into a shear mechanism, where the source pellet is removed. In the drying chamber the small amounts of tritium contaminated cooling water from the reactor thimble on the capsule are evaporated and blown through filters in the reactors decontamination facility. Passing an airlock, the copper pellet reaches the crucible, where it is evaporated onto a tungsten crystal and forms the positron source.

In the shear the capsule is broken open to release the source pellet. The shearing process tears flakes off the capsule wall, which can fall into the crucible along with the pellet. These would poison the moderator crystal and greatly reduce the positron intensity of the beam. To prevent this, the part of the capsule that is sheared is made of high purity copper. The empty capsule and the plug are dumped into a lead pig inside the blockhouse.

As it drops out of the airlock, the copper pellet is guided through a tube into an alumina crucible. The copper is evaporated onto a W(110) crystal, which induces the crystalline growth of Cu(111). Annealing and a subsequent H_2S treatment increase the moderator efficiency of the copper. Depending on the initial evaporation, annealing improves the moderator performance by one order of magnitude while the sulfur treatment may gain another 20%. A x-ray study of a copper film verified the Cu(111) structure. By resistive heating of the tungsten crystal, the copper can be evaporated onto a dump after it has decayed and the tungsten crystal is ready for a new source. After several source cycles the dump is replaced to keep the accumulation of radioactive material low.

An overview of the beam line is shown in Figure 6. In the "run" position the copper-coated tungsten faces the beam line. The crystal can be electrically floated to give the beam its transport energy. An accelerator tube in front of this crystal position improves the geometry of the electric field. Here a tungsten single crystal transmission moderator of about 10^3 nm thickness can be rotated into place. A location close to the source is important to improve the geometry on which the efficiency depends to a large extent.

The slow positrons are transported by a magnetic guiding field.[18,19] Two sets of $\vec{E} \times \vec{B}$ plates guide the beam around a γ-ray shield and energy filter the positrons. The positrons can be guided to one of two experimental areas. A set of apertures on one line permits a continuous acceleration up to 16 keV. The other line is equipped with diffusion pumps and a large turbo pump to differentially pump a gas cell where a positronium (Ps) beam is produced.

As the source material, both spectroscopic pure copper metal (69.1% of usable ^{63}Cu) or isotope ^{63}Cu oxide (about 99% of ^{63}Cu) from Oak Ridge

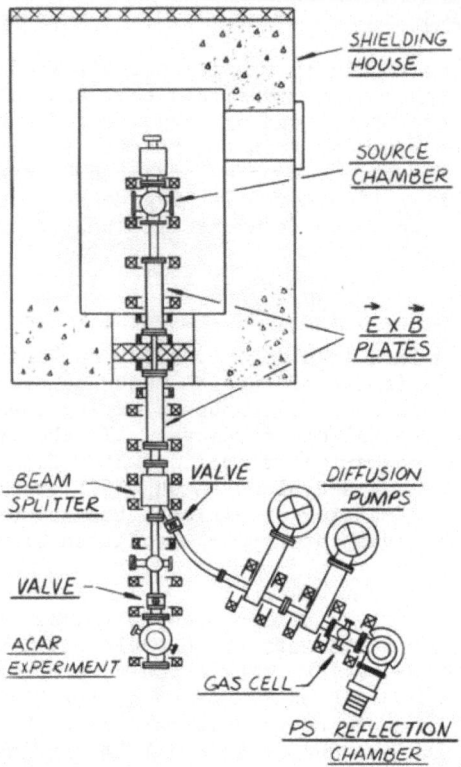

Fig. 6. Top view of the positron beam at the Brookhaven Reactor. Slow positrons are extracted from the source chamber and magnetically guided through two $\vec{E} \times \vec{B}$-regions out of the blockhouse. In the splitter the positrons can be directed to two experimental areas, the ACAR chamber or the Ps-beam line (PS). Shut off valves permit a separation from the main beam line.

National Laboratory (ORNL) can be utilized. The isotopic copper oxide is first reduced to metal. Care must be taken to keep the copper free of impurities that would inhibit its use as a good self moderator.

PROBLEMS, IMPROVEMENTS AND RESULTS

Iron and cobalt in the capsule material caused higher than expected radiation levels, owing to neutron induced activation. Higher purity aluminium is now used as the capsule material. The capsule walls are kept thin to accumulate less radioactive waste.

Impurities and dirt that were transferred with the copper poisoned the moderator. Capsules now have a copper piece where they are sheared. Much of the copper was lost during evaporations from an open wire basket. Now an outgassed alumina insert in the wire basket increases the copper yield on the tungsten. Also, the contamination of the vacuum chamber with copper is greatly reduced.

Of prime concern are impurities that are already contained in the source material. An analysis of the γ-ray spectrum of an irradiated pellet showed lines of several metals. Only high grade spectroscopic copper or carefully enriched and reduced material can be used. ^{64}Cu also decays to ^{64}Zn via a β$^-$-decay. Some of this ^{64}Zn will be converted to ^{65}Zn which emits a 1.115 MeV γ-ray during its decay with a halflife of 245 days. Although the activity of impurities is below 1mCi/g of copper they accumulate to significant amounts due to the longer half lives. For this reason, the copper dump is changed periodically. The evaporation procedure is still being refined to produce copper films with a better crystal structure and fewer defects.

Count rates up to 1×10^8 e$^+$/s have been reached using the self moderation mode. The transmission moderator produced lower rates and it was dismantled as the source evaporation became more reliable. In the gas cell, after the curve in the beam line, the FWHM of the energy spread of the beam was measured to be on the order of 0.5 eV (see Figure 7.)

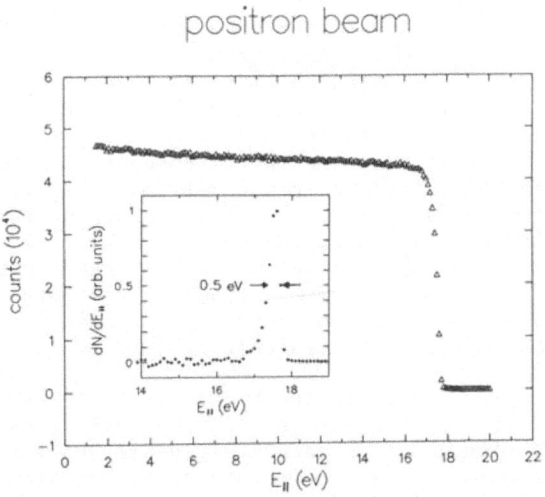

Fig. 7. Retarding potential measurement of the longitudinal energy spread of the positron beam at 18 eV energy. The retarding potential was applied on the gas cell at the end of the Ps-beam line.

OUTLOOK

In order to make measurements and estimates of the conversion efficiency of the moderator possible, a surface barrier detector will be installed to count fast β-particles coming off the source. A neon moderator will be tested and will finally replace the copper self moderator. The implementation is discussed in the appendix.

The possibility of mass-separation of ^{64}Cu from the much more abundant but inactive ^{63}Cu is being investigated. About 2.4×10^{-4} of the copper atoms are ^{64}Cu after a 48 hour irradiation. An improvement of this ratio by a mass separation during the evaporation will increase the specific activity. The source crystal could be made thinner. A higher efficiency ε (see Figure 3) will result in a higher count rate of positrons in the beam. This also will be discussed in the appendix.

A hybrid guiding system is envisioned for a second generation system. The positrons are extracted electrostatically from the source region. For longer distances a magnetic guiding field is planned, which returns to an electrostatic system at the experimental area if required. The source chamber could be floated up to 100 keV. Finally several beam switches are planned to make this beam a multi-user facility.

APPENDIX

The calculation of the efficiency of a self moderator provides the expressions used in section 3. More detailed estimates about the planned neon moderator are presented in the following part. An estimate of possible improvements of the self moderator by incorporating an on-line separation system of ^{64}Cu into the system concludes the appendix.

The self moderator

Positrons originating in β^+-decays of the ^{64}Cu thermalize rapidly and diffuse throughout the crystal. Either they annihilate with electrons or reach the surface. If they do not fall into a surface state or pick up an electron and form Ps they are expelled into the vacuum due to the negative workfunction of the crystal. Modifications to this part result from the fact that the source material is not a free standing foil. The substrate – W(110) in this case – causes a change in the boundary conditions at the interface side. A fraction f of fast positrons will be reflected back into the copper layer. The remaining positrons penetrate the interface and thermalize in the tungsten. They all diffuse throughout the material and have a finite probability of returning to the surface where they can contribute to the positron beam. Since the workfunction of tungsten is more negative than the workfunction for copper this junction acts like a diode which tends to push the thermal positrons out of the tungsten.

The average time for positrons to thermalize (< 10 ps) is very short compared to the half life of the source. Consequently the steady state solution of the diffusion equation adequately describes the situation. In addition the large area of the source moderator in relation to its thickness permits a one dimensional calculation. A perfect interface is assumed in this derivation. Further the possibility of positrons scattering out of the source while they thermalize parallel to and near the surface is neglected. These assumptions tends to overestimate the resulting ε.

From the exponential implantation profile of a source far removed from the moderator the profile for a source moderator can be calculated. Figure 8 illustrates how the reduction to one dimension has been performed. The exponential profile leads to the source moderator profile

$$p_s(x) = \frac{1}{d} \int_0^1 dt \left[1 - e^{-\frac{1}{2}\alpha d/t} \cosh \frac{\alpha(d/2 - x)}{t} \right] ; \; t \equiv \cos\theta. \quad (8a)$$

Alpha can be calculated from the empirical formula[15]

$$\alpha = C \, \rho Z^{0.25} / \bar{E}^{1.58}; \; C = 1.1 \; cm^2/gMeV^{1.58} \quad (9)$$

d is the thickness of the crystal, ρ is the density, Z the atomic number and $\bar{E}$ the mean energy of the β^+-spectrum ($\bar{E}$ = 0.2 MeV for copper). Similarly

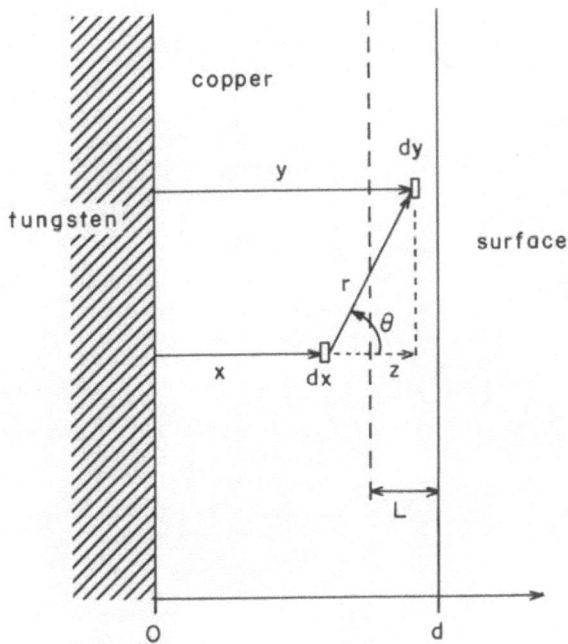

Fig. 8. Schematic to illustrate the calculation of the 1-dimensional implantation profiles. The positrons created at dx stop after a flight path r in the direction θ at dy. Integration across the source yields the implantation profile. The source area is assumed to be large compared to its thickness.

the contributions to the implantation profile due to the back reflection $p_f(x)$, and the implantation into the tungsten $p_w(x)$ are calculated. Figure 9 shows the various implantation profiles.

$$p_f(x) = \frac{1}{2d} f \int_0^1 dt \, (1 - e^{-\alpha d/t}) \, e^{-\alpha x/t} \qquad (8b)$$

$$p_w(x) = \frac{1}{2d} (1 - f) \frac{\alpha_w}{\alpha} \int_0^1 dt \, (1 - e^{-\alpha d/t}) \, e^{+\alpha_w x/t} \qquad (8c)$$

The implantation profiles are used in the diffusion equation for a steady state. The density n(x) of positrons vanishes at the copper surface and deep in the tungsten. The flux j(x) is continuous at the interface.

The fraction Y_0 is emitted as slow positrons. Y_0 and j(d) form the efficiency of the self moderator ε(d). The various contributions amount to

$$\varepsilon_s(d) = \frac{1}{2} Y_0 \frac{L_+}{d} \left[L_+\alpha \int_0^1 dt \, \frac{t}{t^2 - L_+^2\alpha^2} (1 - e^{-\alpha d/t}) \right.$$

$$\left. + \tanh \frac{d}{L_+} \left(L_+\alpha \, \ell n \, \frac{1 + L_+\alpha}{1 - L_+\alpha} + \int_0^1 dt \, \frac{t^2}{t^2 - L_+^2\alpha^2} (1 - e^{-\alpha d/t}) \right) \right] \quad (10a)$$

171

$$+ \frac{1}{\cosh \frac{d}{L_+}} \; L_+\alpha \int_0^1 dt \; \frac{t}{t^2 - L_+^2\alpha^2} \; (1 - e^{-\alpha d/t}) \Bigg]$$

$$\varepsilon_f(d) = \frac{1}{2} Y_0 \; f \Bigg[L_+\alpha \int_0^1 dt \; \frac{t}{t^2 - L_+^2\alpha^2} \; (1 - e^{-\alpha d/t}) \; e^{-\alpha d/t}$$

$$+ \tanh \frac{d}{L_+} \int_0^1 dt \; \frac{t^2}{t^2 - L_+^2\alpha^2} \; (1 - e^{-\alpha d/t}) \; e^{-\alpha d/t} \tag{10b}$$

$$- \frac{1}{\cosh \frac{d}{L_+}} \; L_+\alpha \int_0^1 dt \; \frac{t}{t^2 - L_+^2\alpha^2} \; (1 - e^{-\alpha d/t}) \Bigg]$$

$$\varepsilon_w(d) = \frac{1}{2} Y_0 \; \frac{L_+}{d} \; (1 - f) \; \frac{L_w\alpha_w}{L_+\alpha} \; \frac{1}{\cosh \frac{d}{L_+}} \int_0^1 dt \; \frac{t}{t + L_w\alpha_w} \; (1 - e^{-\alpha d/t}) \tag{10c}$$

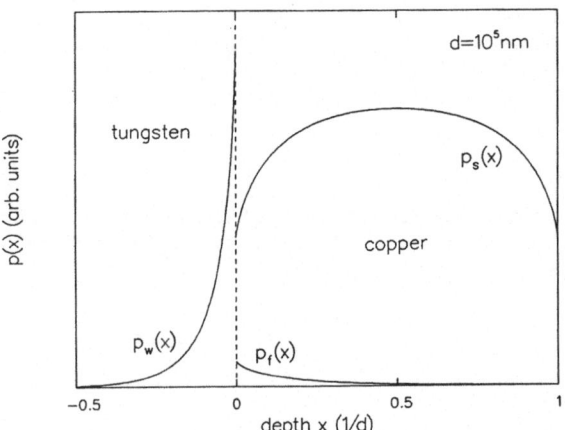

Fig.9. The contributions to the implantation profile of the Cu self
moderator on a single crystal W substrate. Shown are the
profile from the self moderating source $p_s(x)$, the fraction
due to backscattering $p_f(x)$, and the implantation into the W
substrate $p_w(x)$. The Cu layer is 10^5 nm thick.

Figure 3 shows the efficiency of a crystal depending on its thickness d.[20]
For copper $Y_0 = 0.55$ and $f = 0.17$ was estimated for the Cu-W interface.
With growing thickness $\varepsilon(d)$ decreases. On the other hand this is more than
compensated by the larger amount of source activity. In the case of a
source moderator the product $E(d)$ of efficiency and thickness describes the
situation better.

$$E(d) = d\varepsilon(d) \tag{7}$$

The absolute source strength is the product of the specific source activity, the source area, and E(d). Figure 4 shows the various parts of E on a logarithmic scale versus the thickness of the copper layer. At about 10^5 nm E levels off. Additional amounts of copper do not improve the source strength unless the area is increased.

For d of about an order of magnitude larger than $L_+ = 110$ nm

$$\tanh \frac{d}{L_+} \approx 1; \quad \cosh \frac{d}{L_+} < 10^{-4}.$$

which simplifies equations (10) and (7) to

$$E(d) = \frac{1}{2} Y_o L_+ \left[L_+ \alpha \, \ell n \frac{1 + L_+ \alpha}{1 - L_+ \alpha} + \int_0^1 dt \frac{t}{t - L_+ \alpha} (1 - e^{-\alpha d/t}) \right] \quad (11)$$

The Neon Moderator

As mentioned by Mills and Gullikson, rare gas solids and solid neon in particular, can be used as positron moderators.[13] They report a high yield ($Y_o = 0.7$) and a large diffusion length ($L_+ \simeq 10^3$ nm) for solid neon. A moderation efficiency $\epsilon = 7 \times 10^{-3}$ measured by them is higher than any previously known moderator. Neon was condensed onto a cylindrical source geometry to take advantage of the large reflection coefficient of Ne for slow positrons.

The first neon moderated beam[21] performed at the high initial level over a period much larger than the half life of ^{64}Cu. With the cylindrical source geometry self absorption of a ^{64}Cu source can be reduced. The larger surface area of the cylinder permits a thinner copper layer without increasing the beam cross section. The efficiency of a copper source–neon moderator combination was estimated using the same approach as for the efficiency calculation for a self moderator presented in this paper. A 3000 nm thick layer of Ne on a 1.6×10^4 nm thick Cu source result in an efficiency $\epsilon = 0.64\%$. A 100 mg source of enriched ^{63}Cu irradiated for two days and evaporated onto a 7 cm^2 area of a cup (1 cm diameter; 2 cm high) would yield close to 4×10^9/sec slow positrons.

The reported energy spread of 0.58 eV for neon moderators is acceptable. Some advantages of the neon moderator are as follows: 1) no single crystal copper is necessary; 2) no spectroscopically pure material is required; 3) the copper does not have to be removed prior to a new source evaporation. The cup itself can be replaced after several months. The design of a liquid He cooling system inside a LN_2 cooled thermal shield to achieve a temperature of 6K appears to be straightforward. Possible disadvantages are that the base pressure of the vacuum system should be in the 10^{-10} torr range to prevent contamination and the excellent properties of the neon moderator may be affected by the high radiation level of the source. A neon moderator for the reactor beam will be installed in the near future.

On-Line Isotope Separation

The plan is to vaporize the copper and then selectively ionize the ^{64}Cu atoms and only collect these on the tungsten substrate. The ionization could be carried out by means of tunable dye lasers. This method would take advantage of the different hyperfine structure splitting of the atomic energy levels of the various copper isotopes.[12]

At about 200 nm thickness the efficiency of a self moderator peaks, (see Figure 3). A reasonable compromise between source size and positron rate is a layer of this thickness. At present about 1 in 4000 ^{63}Cu atoms will be converted to ^{64}Cu during one irradiation. The proposed separation process enhances the ratio of ^{64}Cu to ^{63}Cu by a factor of 1000. Then up to 24% of the source will be active. The total efficiency at 200 mm thickness is about 5×10^{-3} and the result would be a beam of 1×10^{10} e$^+$/s with a 1 cm^2 source area. To achieve this it would be necessary to irradiate 280 mg of isotopic copper for 3 days.

The separation requires some time. A rate of 10^{-8} g ^{64}Cu/min was estimated. Then evaporation times of 2 - 3 hours are necessary. On the other hand the area of the source could be reduced without losses in brightness and within a shorter irradiation time of the order of an hour. A beam of several mm^2 size and 10^8 e$^+$/s seems to be realistic.

ACKNOWLEDGEMENTS

The authors would like to tha k J. Zahradka for his design efforts and J. Rutherford and P. Schnitzenbaumer for their technical help. J. Hurst provided the capsules with the pellets. Work of D. Rorer and T. Holmquist was essential for the success of the project. D.M. Chen's effort with the transmission moderator is very much appreciated. This work is supported in part by the Division of Materials Sciences, U.S. Dept. of Energy, under Contract No. DE-AC02-76CH00016, and in part by the National Science Foundation through Grant No. DMR-8315691.

REFERENCES

1. For example: A.P. Mills, Jr., in "Positron Scattering in Gases", J.W. Humberston and M.R.C.McDowell eds. Plenum Press, 121 (1984).
2. R.J. Drachman in "Positron Scattering in Gases," ibid, p. 121.
3. R.H. Howell, P. Meyer, I.J. Rosenberg, J.J. Fluss, Phys. Rev. Lett. 54:1698 (1985) and K.G. Lynn, A.P. Mills, Jr., R.N. West, S. Berko, K.F. Canter, L.O. Roellig; Phys. Rev. Lett. 54:1702 (1985).
4. To be presented by L. Roellig in these proceedings.
5. For example: H. Hoinkes, Rev. Mod. Phys. 52:933 (1980) and refs. therein.
6. K.G. Lynn, D.N. Lowy, I.K. McKenzie, J. Phys. C: Solid State Phys. 13:919 (1980).
7. A.P. Mills, Jr., Appl. Phys. Lett. 23:189 (1980).
8. W.E. Frieze, D.W. Gidley, K.G. Lynn, Phys. Rev. B31:5628 (1985).
9. R.H. Howell, R.A. Alvarez in "Positron Scattering Gases", as 1, p.155
10. A. Vehanen, J. Makinen, Appl. Phys. A36:97 (1985).
11. D.M. Chen, K.G. Lynn, R. Pareja, B. Nielsen, Phys. Rev. B31:4133 (1985).
12. R. Engleman, Jr., R.A. Keller, C.M. Miller, N.S. Nogar, F.A. Paisner, Nucl. Inst. and Meth. in Phys. Res. Section B26:448 (1987).
13. E.M. Gullikson, A.P. Mills, Jr., Phys. Rev. Lett. 57:376 (1986).
14. In "HFBR Handbook", S. Shapiro, D.C. Rorer, H. Kuper eds. informal report Brookhaven National Laboratory.
15. M. Mourino, H. Lobl, R. Paulin, Phys. Lett. 71A:106 (1979).
16. A. Vehanen, K.G. Lynn, P.J. Schultz, M. Eldrup, Appl. Phys. A32:163 (1983).
17. A.P. Mills, Jr. in Proc. LXXXIII Intern. School of Physics "Enrico Fermi", W. Brandt, A. Dupasquier eds. (Academic Press, N.Y. 1982).
18. A.P. Mills, Jr., P.M. Platzman, B.L. Brown, Phys.Rev.Lett. 41:1079(1978)
19. A.P. Mills, Jr., Phys. Rev. Lett. 41:1828 (1978).
20. H.E. Hansen, S. Linderoth, K. Petersen, Appl. Phys. A29:99 (1982).
21. A.P. Mills, Jr., E.M. Gullikson, Appl. Phys. Lett. 49:1121 (1986).

FIELD ASSISTED MODERATORS

C.D. Beling, R.I. Simpson and M. Charlton

Department of Physics and Astronomy
Unive sity College London
Gower Street, London WC1E 6BT

INTRODUCTION

Conventional positron moderators rely on diffusion to transport a small
proportion of e^+ implanted from a radioactive source to a surface where they
may be emitted into vacuum with an energy distribution peaked at a value
characteristic of the negative workfunction of that material. The energy
spread of the distribution will depend on the e^+ properties in the bulk and
at the surface of the moderating material. Since the initial conception of
a positron moderator by Madanski and Rasetti[1] and the first practical demonstration by Cherry[2] in 1958 there has been considerable improvement in conversion efficiencies. Of particular significance was the work of Mills who
showed that many clean metal surfaces emit slow positrons and was thus able
to develop higher efficiency moderators[4]. The factors which govern the
suitability of a particular material for use as a positron moderator are
given in the following empirical relationship for the moderator efficiency,
p[1,5],

$$P = y_o \, \alpha \sqrt{D\tau} \tag{1}$$

where y_o is the probability of slow positron emission from the surface (the
branching ratio), α is the β^+ absorption coefficient and $\sqrt{D\tau}$ the e^+ diffusion coefficient with D and τ the positron diffusion constant and lifetime
respectively. To maximise these factors, recently work has largely concentrated on dense single crystal materials especially W(110) for which an efficiency of 3.2×10^{-3} has been reported[6]. The importance of annealing in
removing defects which can lead to e^+ trapping (lowering of D) has also been
realised. The effect of surface treatment which can change y_o has been
studied. Thin layers of S on Cu have been investigated in detail[7] and S
coverage has been related to changes in the surface dipole moment, the S
layer making the workfunction more negative. A thin epitaxial layer of Cu
(111) on W(110) has enabled the high emission probability of Cu(111) to be
combined with the superior stopping power of W[6]. This idea of a positron
rectifier has also been discussed theoretically by Debowska et al[8]. It has
also been suggested that cooling the moderator may lead to higher efficiencies[9] (increases D) as well as reducing the emitted e^+ energy spread[10].

Recently it has been shown that thin layers of solid rare gases can be
used as efficient e^+ moderators, in particular solid neon for which an
efficiency of 0.7% has been reported for a cylindrical geometry[11]. The high
efficiency and energy spread (FWHM = 0.58 eV) are consistent with a 'hot
positron' model[12], in which, due to the slow energy loss rate once the

energy of the e^+ has been moderated below the band gap, there is a high probability of an e^+ diffusing to the surface with sufficient energy to overcome a positive workfunction. It would appear that this is close to the upper limit of efficiency that can be expected from this type of moderator because of the inability of diffusive transport to move more than a small fraction of implanted positrons to the moderator surface.

However in the case of a semiconductor or insulator it may be possible to enhance the number of positrons reaching a surface by the application of an electric field. In this case equation (1) is no longer valid and a new expression incorcorating the e^+ electric field drift must be used. The concept of field assisted (FA) moderation was first suggested by Lynn and McKee[13] who drifted positrons on a conventional Si surface barrier detector with a 200A gold window on the emitting face. The apparent lack of success of this experiment can be attributed to the existence of defects in the non-epitaxially grown gold contact or contamination due to the relatively poor vacuum (5×10^{-7} torr). In a recent publication it was suggested that these problems can be overcome by using thin epitaxially grown metal-silicide layers in UHV[14]. Efficiencies of up to 10% were predicted.

It is the aim of this communication to discuss the requirements of a field assisted moderator with reference to mobility, energetics of the metal-semiconductor/insulator and metal-vacuum interfaces, and the elimination of defects. Experiments currently being carried out on the $NiSi_2$-Si system are also described. It is also hypothesised that, in the case of insulators, problems which may be associated with the thin metal contact might be overcome if the e^+ can be 'heated up' by the application of very large (MV cm^{-1}) electric fields.

REQUIREMENTS FOR FIELD ASSISTED MODERATION

In order to estimate the fraction of e^+ that can be drifted in an insulator or semiconductor to an interface or vacuum surface, Beling et al[14] solved the diffusion equation to obtain expressions for the fractions of positrons $q(0)$ and $q(d)$, which drift to perfectly absorbing boundaries located at $x = 0$ and $x = d$. In the high field case, such that the drift velocity, $v(E)$, at electric field E satisfies $v(E) \gg \sqrt{2D/\tau}$, diffusive terms can be neglected and the e^+ fraction reaching $x = d$ is given by

$$x(d) = \frac{\tau \alpha v(E)(\exp(\alpha d) - \exp(-d/_{(v(E)\tau)}))}{1 - \tau \alpha v(E)} . \qquad (2)$$

This expression relates efficiency to the important positron properties in the semiconducting/insulating bulk. Extra terms describing transmission through the thin contact and the branching ratio at the vacuum surface could also be added. It would also be possible to allow for reflection at the metal-insulator interface. The essential properties of a suitable FA moderator can thus be summarised as being:

1) The bulk material must possess a high positron mobility, $\mu+$, and be capable of sustaining high electric fields.

2) Application of electric field necessitates either a contact or conducting layer through which positrons must pass. Defects in this region will drastically reduce the positron yield. This suggests that the layer should be thin and most probably epitaxial.

3) The positron energy levels in the bulk and contact must be favourable for transmission through the interface and subsequent vacuum emission.

4) The density and lifetime of the bulk material should be large

enough to ensure that a reasonable fraction of e^+ can be stopped and drifted into the contact.

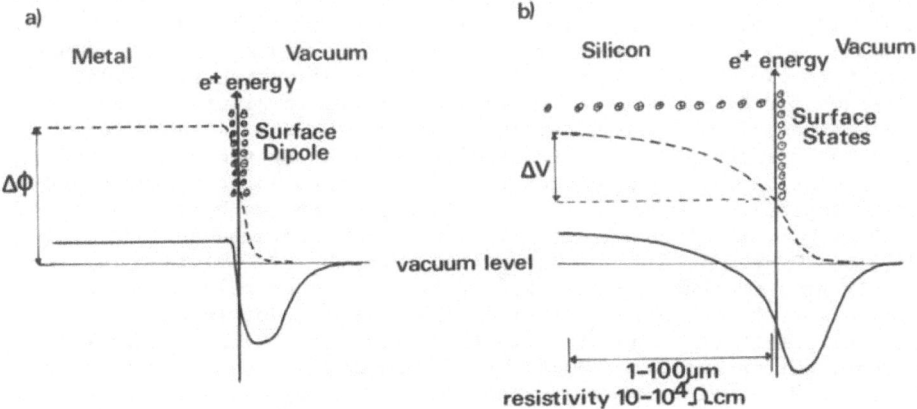

Figure 1. Simplified energy diagram for e^+ at
a) metal surface
b) silicon (semiconductor) surface
Dashed line represents the electrostatic potential. Solid
line represents e^+ energy level.

The main problem in selecting likely options to produce a working FA moderator is the lack of experimental e^+ mobility data for more than a few materials.

THE NiSi$_2$-Si(111) SYSTEM

i) Characteristics

It has been proposed that the NiSi$_2$-Si(111) system might provide the basis for an FA moderator as it appears to satisfy most of the above criteria. Si is one of the few materials for which reliable e^+ mobility measurements have een reported. The non polar nature of the Si lattice gives a weak phonon coupling and thus a high mobility. Using a Doppler technique, Mills and Pfeiffer[14] measured values of $\mu+$ of (460 ± 20) and (173 ± 15) cm^2 V^{-1}s^{-1} at 80 and 184 K respectively. Their results appear to be concordant with an acoustic phonon picture with the possibilty of some impurity scattering. If impurity scattering is significant, silicon has the advantage over many materials in that high purity samples are commercially available.

Referring to point 2 above, silicon interfaces with many other materials have been well studied. Of particular significance are the silicides especially NiSi$_2$ and CoSi$_2$. Sub 100Å layers of NiSi$_2$ with good uniformity and near perfect epitaxy can be grown by evaporation on a Si(111) substrate under UHV conditions[15]. Silicide-silicon structures are thermally stable, so it might be possible to transfer such samples from one system to another, with only a small amount of sputtering and annealing being necessary. It has also been suggested that after evaporation the silicide layer could be given a protective coating of a material with a low evaporation point such as arsenic[16]. Samples could then be transferred to a new chamber or beam line and this layer evaporated by gentle heating to give a clean silicide surface.

Despite the fact that this contact can be fabricated. many aspects of the behaviour of e^+ at semiconductor surfaces are not understood. A negative positron work function for silicon was originally reported by Mills et al[3]. It now appears that these data do not correspond to work function emission from a clean silicon surface and/or the emission observed may have been due to epithermal e^+. In either case it is thought that the e^+ do not have sufficient energy for transmission into vacuum or the silicide contact. In an attempt to understand the behaviour of e^+ at Si surfaces, simplified energy diagrams for Si and a metal surface are shown in Figure 1. In the case of metals, overlap of the electron cloud into vacuum causes a near-surface dipole of magnitude $\Delta\phi$, which assists e^+ leaving the bulk. In silicon, and presumably other semiconductors, e^- fall into surface states leaving an intrinsic depletion region which can extend µm into the bulk. As can be seen in the figure, the band energy levels are bent near the surface creating local E fields which will affect the behaviour of the e^+. If this region is significantly greater than the mean free path for phonon relaxation the e^+ can de-excite losing $\Delta V eV$ before reaching the surface, so, in contrast to metal surfaces positrons do not receive the beneficial 'kick' from a surface dipole moment. It is anticipated that different surface structures and preparations can, by altering the e^- surface states, change both the amount by which the levels bend and thus the local fields. This point has been raised by Mills and Murray[18] and Nielsen et al[19], whose experimental results seem to exhibit effects of this kind. It should also be noted that the presence of a thin silicide layer may also affect the depletion region in a manner that is difficult to predict.

Finally regarding point 3 above, it is known that $CoSi_2$ has a $- 0.46$ eV workfunction[20], so it can be hoped that $NiSi_2$ will likewise prove to be a positron re-emitter.

It can thus be seen that although there are problems associated with using a $NiSi_2$ structure as an FA moderator, it may satisfy most of the requirements outlined above and with present knowledge would appear to be the most likely material for an FA moderator. Thorough experiments on this structure may also throw some light on interface/surface properties of semiconducting materials in general.

(ii) Experimental Details

Two batches of samples have been prepared for these studies. 300Å thick $NiSi_2$ layers were fabricated at Aarhus University in Denmark by electron beam evaporation of Ni on to both sides of a prepared Si(111) substrate. The silicide was then formed by a subsequent rapid thermal anneal at 820° C for 15 s in an inert atmosphere using a xenon lamp[21]. A thick oxide guard ring had previously been grown in an attempt to eliminate leakage currents along the sides of the sample. A second set of samples has been prepared under UHV conditions at the FOM Institute in Amsterdam. 35 - 40Å thick silicide films were fabricated on 1000Ω cm n-type Si(111) by a method similar to that used by Tung [2,23]. Annealing was carried out at 550° C for 5 min. It is intended to form contacts on the rear of these samples by flash evaporation of a gold-antimony mixture. Subsequent sintering at 400° C causes the antimony to diffuse into the silicon forming a heavily doped n^+ layer for ohmic contact. The thinness of the silicide layer necessitates care in making electrical contact so small balls formed on the end of thin gold wires will be used to make pressure contacts. It is intended to carry out further processing on one of the samples to form a mesa structure used in the construction of avalanche photo-diodes[24] (see Figure 2). This will enable electric fields to be established without risk of electrical breakdown at the edges of the contact.

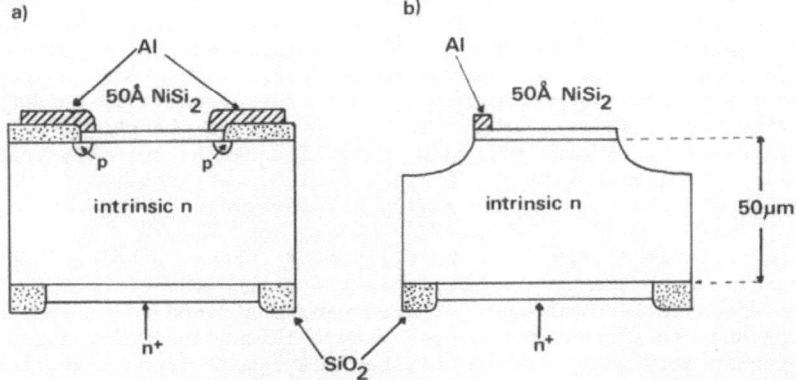

Figure 2: Silicon structures for avalanche photo-diodes.
   a) Guard-ring structure
   b) Mesa structure

Two experimental studies are being carried out. At UCL a NaI-channel-
tron inverted time of flight arrangement has been set up to observe positron
re-emission. A schematic diagram of the experiment is shown in Figure 3.

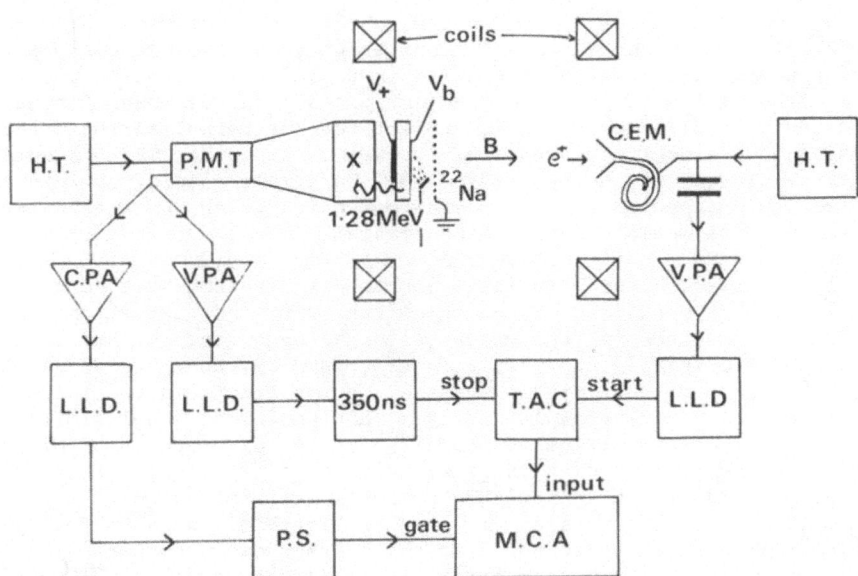

Figure 3: Experimental arrangement at UCL for the measurement of field
emission from Si-NiSi$_2$ structures. H.T: High Tension Supply, P.M.T:
Photomuliplier Tube, X: NaI Crystal, V_b: Beam Voltage, C.E.M: Channel
Electron Multiplier, V.P.A: Voltage Sensitive Preamplifier, C.P.A: Charge
Sensitive Preamplifier, L.L.D: Lower Level Discriminator, T.A.C: Time to
Amplitude Convertor, P.S: Pulse Shaper, M.C.A: Multi-Channel Analyser.

The experiment is carried out in a chamber equipped with diffusion and subli-
mation pumps giving a base pressure of 1×10^{-10} torr. Clean surfaces are
prepared using a 1 - 10 keV ion gun and electron beam annealing. The experiment is carried out in backscattering geometry using a 50 µ Ci ^{22}Na source
deposited on a gold pin. Re-emitted positrons are magnetically guided to
the channeltron, which is located about 10 cm away from the sample. The
system has been set up and calibrated using well annealed poly-crystalline
tungsten foil. Results show that efficiencies 100 times smaller than that
for tungsten ($\sim 2 \times 10^{-4}$) can quite easily be observed.

At UEA a variable energy (0 - 100 eV) positron beam is being used to
study the samples such that e^+ can be implanted beyond the silicon-silicide
interface in a controlled manner. Surfaces can be cleaned using sputtering
and annealing and the system is equipped with LEED and Auger for surface
analysis. Preliminary results from UEA show that there is no significant re-
emission from a dirty $NiSi_2$ surface. Auger spectra indicate that the surface is contaminated with C and O, which suggests the presence of chemisorbed
CO. The thinness of the FOM silicide layers renders surface cleaning difficult, especially since annealing temperatures must be kept below 400° C to
avoid diffusion of Ni from the silicide into the Si bulk.

FIELD ASSISTED MODERATION USING SINGLE CRYSTAL INSULATORS

The behaviour of e^+ in insulators with applied electric fields is at
present little understood. The polar nature of the alkali halides and other
materials such as MgO and SiO_2 is expected to result in low positron mobilities due to the creation of polaron states. This is borne out by the experimental data which does exist. Using a 1-D angular correlation system combined with alternating electric fields, Lang and De Benedetti[25] obtained the
value $\mu+ = (20 \pm 160)$ cm^2 V^{-1} s^{-1} for diamond. Using static fields Sueoka
and Koide[26] have found upper limits on the mobilities of diamond, SiO_2 and
$BaTiO_3$ of 20, 15 and 80 cm^2 V^{-1} s^{-1}. A different technique was employed by
Brandt and Paulin[27] who measured the shift in annihilation profile and
obtained a higher value of $\mu+ = 160 \pm 80$ cm^2 V^{-1} s^{-1} for diamond. It has
been suggested that this technique is subject to pre-thermalization drift
which results in a higher measured mobility value than other methods which
are sensitive uniquely to thermalized e^+ [15]. Positrons slowing down in an
FA moderator would experience this pre-thermal drift which, on the basis of
the discussion above would clearly be beneficial.

Work by Mills and Crane[28] on ionic insulators has demonstrated that
implanted low energy positrons can be re-emitted with energies which, as
shown in Figure 4, extend in some cases up to the band gap. It has been
proposed that the original interpretation of this data, that it was due to
positronium break up at the crystal surface, is incorrect and the large
energy of re-emission can be explained by a 'hot positron' model similar to
that proposed for neon as discussed above[29,12]. If, as proposed by Mills
and Pfeiffer[15], the concept of a large kinetic mobility is correct, then it
may be possible to drift an appreciable proportion of implanted e^+ to a
vacuum interface of such crystals. It is also feasible that at the very
large fields which are close to those at which breakdown occurs (> 5 MV
cm^{-1}) positron motion might become ballistic as is the case for e^- behaviour
in SiO_2 where, at fields > 2 MV cm^{-1}, e^- vacuum emission has been observed
from 100 - 1000Å thick oxide layers[30].

Due to the cubic structure of the alkali halides and some other insulators, epitaxial growth is possible for most materials, for example Ni, Pt and
Pd on an NaCl substrate[31]. MgO[32] and LiF[28], in addition may have negative
workfunctions such that epitaxial systems with favourable energy levels
exist (for example MgO-Al)[33]. It should be noted, however, that if balli-

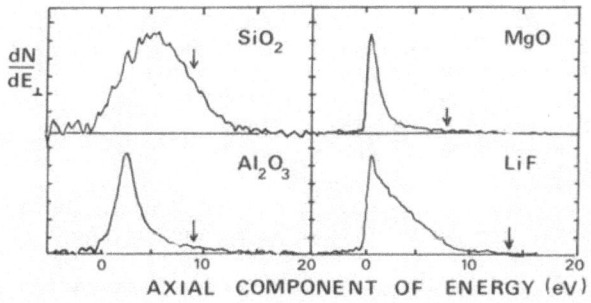

AXIAL COMPONENT OF ENERGY (eV)

Figure 4: Positron emission spectra for SiO_2, Al_2O_3, MgO and LiF single crystals. Positron implantation is at 500 eV. Arrows indicate the position of the conduction band. Taken from the work of Mills and Crane[28].

stic motion does occur in some of these materials, epitaxy may not be so important, although the work by Brorson et al[30] suggests that there is a strong interaction between vacuum emitted electrons and the thin metal exit contact.

CONCLUSION

The possibility of developing a field assisted moderator for low energy beams has been discussed. It has been suggested that a $Si-NiSi_2$ moderator of this type could give efficiencies of up to 10%. The major difficulty appears to concern the energetics of transporting positrons through the silicide and into vacuum. Si and Ge are the only materials for which reliable positron mobility data exists, so an attempt to change the surface and interface energetics by, for example, appropriate doping might be worthwhile. Other possible field assisted moderating systems using insulators have been suggested, but at present, their performance cannot be theoretically evaluated due to the lack of data on the behaviour of e^+ in these materials under the influence of applied electric fields.

ACKNOWLEDGEMENTS

We would like to acknowledge the continuing support of the other collaborators in this project: F.M. Jacobsen and J. Chevallier from Aarhus University, Denmark; J.F. Van der Veen and A. Fischer from FOM, Amsterdam; P.G. Coleman and J. Baker from UEA and J. White and D. Court from Middlesex Polytechnic, London. We would like to thank the SERC for financial support and for the provision of a research studentship for RIS.

REFERENCES

1. L. Madanski and F. Rasetti, An Attempt to Detect Thermal Energy Positrons, Phys. Rev. 79:397 (1950).

2. W. Cherry, Ph.D. dissertation (Princeton University 1958).
3. A.P. Mills, Jr., P.M. Platzman and B.L. Brown, Slow Positron Emission from Metal Surfaces, Phys. Rev. Lett. 41:1076 (1978).
4. A.P. Mills, Jr., Appl. Phys. Lett. 37:1980
5. A.P. Mills, Jr., in "Positron Solid State Physics, Proc. S.I.F. course LXXXIII", W. Brandt and A. Dupasquier, ed. North-Holland, Amsterdam (1983). p440.
6. A. Vehanen, K.G. Lynn, P.J. Schultz and M. Eldrup, Improved Slow Positron Yield Using a Single Crystal W. Moderator, Appl. Phys. A32:163 (1983).
7. C.A. Murray, A.P. Mills, Jr., and J.E. Rowe, Correlation Between Electron and Positron Workfunctions on Copper Surfaces, Surf. Sci. 100: 647 (1980).
8. M. Debowska, R. Ewertoski and W. Swiatkowski, Appl. Phys. A36:47 (1985).
9. As Ref. 4 p.445
10. D.A. Fisher, K.G. Lynn and W.E. Frieze, Re-emitted-Positron Energy-Loss Spectroscopy: a Novel Probe for Absorbate Vibrational Levels, Phys. Rev. Lett. 50:1149 (1983).
11. A.P. Mills, Jr., and E.M. Gullikson, Solid Neon Moderator for Producing Slow Positrons, to be published.
12. E.M. Gullikson and A.P. Mills, Jr., Positron Dynamics in Rare Gas Solids, Phys. Rev. Lett. 57:376 (1986).
13. K.G. Lynn and B.T.A. McKee, Some Investigations of Moderators for Slow Positron Beams, Appl. Phys. 19:247 (1979).
14. C.D. Beling, R.I. Simpson, M. Charlton, F. Jacobsen, T.C. Griffith, P. Moriarty and S. Fung, A Field Assisted Moderator for Low Energy Positron Beams, Appl. Phys. A42:111 (1987).
15. A.P. Mills, Jr. and L. Pfeiffer, Mobility of Positrons in Silicon, Phys. Lett. 63A:118 (1976).
16. E.J. Van Loenen, A.E.M.J. Fischer, J.F. Van der Veen and F. Legoues, High Resolution Studies of $NiSi_2$ Ultrathin Film Formation by Ion Beam Scattering and Cross Section TEM, Surf. Sci. 154:52 (1985).
17. K.G. Lynn, private communication.
18. A.P. Mills, Jr. and C.A. Murray, Diffusion of Positrons to Surfaces Appl. Phys. 21:323 (1980).
19. B. Nielsen, K.G. Lynn, A. Vehanen and P.J. Schultz, Positron Diffusion in Si, Phys. Rev. B32:2296 (1985).
20. E.M. Gullikson, private communication.
21. J. Chevallier and A. Nylandsted Larsen, Epitaxial Nickel and Cobalt Silicide Formation by Rapid Thermal Annealing, Appl. Phys. A39:141 (1986).
22. J.F. Van der Veen, private communication.
23. R.T.Tung,Shottky-Barrier Formation at Single-Crystal Metal Semi-conductor Interface, Phys. Rev. Lett. 52:461 (1984).
24. S.M. Sze, "Physics of Semiconductor Devices", Wiley Intersciences, New York, 2nd Ed. (1981).
25. G. Lang and S. De Benedetti, Angular Correlation of Annihilation Radiation in Various Substances, Phys. Rev. 108:914 (1957).
26. O. Sueoka and S. Koide, Poistron Mobility in Diamond, J. Phys. Soc. Japan, 41:116 (1976).
27. W. Brandt and R. Paulin, Positron Implantation Profile in Solids, Phys. Rev. B15:2511 (1977).
28. A.P. Mills, Jr. and W. Crane, Emission of Band-Gap-Energy Positrons from Surfaces of LiF, NaF and Other Ionic Crystals, Phys. Rev. Lett. 53:2165 (1984).
29. K.G. Lynn and B. Nielsen, comment, in Phys. Rev. Lett. 58:61 (1987).
30. S.D. Brorson, D.J. DiMaria, M.V. Fischetti, F.L. Pesavento, P.M. Solomon and D.W. Dong, Direct measurement of the energy distribution of hot electrons in silicon dioxide, J. Appl. Phys. 53:1302 (1985)
31. J. Chevallier, private communication.

32. J. Van House and P.W. Zitzewitz, Probing the Positron Moderation Process Using High-Intensity, Highly Polarized Slow Positron Beams, Phys. Rev. A29:96 (1984).
33. A.K. Green, J. Dancy and E. Bauer, Insignificance of Lattice Misfit for Epitaxy, J. Vac. Sci. Tech. 7:1 (1969).

PRODUCTION OF SHORT-LIVED POSITRON SOURCES

Günther Sinapius[a] and Helge L. Ravn[b]

[a] Fakultät für Physik, Universität Bielefeld
D-4800 Bielefeld, Fed. Rep. of Germany

[b] CERN, Ch-1211 Geneva, Switzerland

INTRODUCTION

In all fields where low-energy positrons are employed, high intensities are desirable. The low-energy positrons originate from high-energy positrons that are moderated in solids. During the last decade the intensity of moderated positron beams has risen from less than one to 10^9 e$^+$/s. There are two ways to increase the intensity, either by having a higher flux of high-energy positrons hit the moderator or by using more efficient moderators. In this text we will only deal with the first option.

POSITRON SOURCES PRESENTLY IN USE

The β^+ emission rate of commercially available long lived radioactive sources (^{58}Co and ^{22}Na) does not exceed 5 GBq. For the brightness of the positron beam not only the source activity but also its geometrical size and the end-point energy of the β^+ spectrum are crucial. The lower the end-point energy of a source, the more positrons thermalize in a given unit thickness of a moderator and can emerge as low-energy positrons.

High-intensity positron beams are available at electron accelerators[1,2,3] and reactors.[4] Because of their pulse structure present beams at electron accelerators are not useful for applications where coincidences between individual primary positrons and outgoing particles are measured. So far these beams are magnetically guided and for most atomic physics experiments they would have to be extracted out of the guiding field. With a tungsten moderator 200 pA of moderated positrons have been reported.[1] It might be difficult to employ solid neon moderators[5] in close proximity to a converter heated by the primary beam. At the high-flux reactor at the Brookhaven National Laboratory (BNL) ^{64}Cu sources are activated in 48 hour cycles (12.6 h half-life). From $10 \times 10 \times 0.1$ mm^3 sources with 550 GBq β^+ emission rates positron beams of about 25 pA are obtained. About the same current is expected with a solid neon moderator, electrostatic beam transport and one remoderation stage.

The activity of long-lived radioactive sources is rather limited. All these sources are far from pointlike, the reason being the isotope composition and source self absorption. The specific β^+ emission rates reach about 1 GBq/mm^2. Recently, two approaches to producing strong or pointlike radioactive sources have been studied. Before describing the results let us make some general remarks on the production of radioisotopes in nuclear processes.

PRODUCTION OF RADIOISOTOPES

The activity of a source, A(t), that can be collected during an activation time t depends on its decay constant λ. For a production rate R one obtains:

$$A(t) = R \left(1 - \exp(- \lambda t) \right).$$

In order to collect high activities the activation time t should be of the order of the half life. If t is sufficiently long, the achievable activity equals the production rate R. Except for parasitic operation, when beam time is no limitation, one is restricted to short-lived isotopes.

In a nuclear rearrangement process like a (p,n) reaction the production rate depends on the cross section for the process, σ, intensity of the projectiles, I, the density of the target atoms, n, and the penetration depth, ℓ:

$$R \approx \sigma \; n \; \ell \; I.$$

At a proton energy of about 10 MeV the cross sections have the highest value and come close to 10^{-24} cm^2; they are higher for thermal neutrons. The beam intensity is mainly limited by cooling requirements, let us assume it to be 20 μA. If the source is to be used without further processing, self absorption limits the useful thickness to 100 mg/cm^2 which corresponds to   100 μm Cu. With these values the upper limit for the production rate in (p,n) reactions is less then 100 GBq. The choice of the isotope then will be guided by the magnitude of the cross section σ, the β^+ decay fraction, the β^+ end-point energy and the half life.

The minimum diameter of the active area of this kind of source is limited either by cooling considerations for charged projectiles or the lack of focussing for neutrals. The penetration depths of the projectiles reduces the intensity or the brightness. This can be overcome by isotope seperation and selection of the isotope of interest or more easily by brightness enhancement schemes for the moderated psoitron beam. Now let us look at an exmaple for a (p,n) reaction.

PRODUCTION OF ^{48}V

At Karlsruhe the process ^{48}Ti(p,n)) ^{48}V was investigated.[6] ^{48}V is a positron emitter with a half life of 16 d, a β^+ fraction of 56% and an end-point energy of 700 keV. A 13 $\times$ 15 $\times$ 1 mm titanium sample was irradiated with 20 μA of 35 MeV protons for 15 minutes. From the γ-spectrum analysed some days later, the produced activity was determined to A(15 min) $\approx$ 11 MBq. This agrees well with the calculated production rate of R < 25 GBq. With 12 MeV protons a value of R $\approx$ 380 GBq (210 GBq β^+ emission rate) is expected. Due to the self absorption in the sample, however, not all the positrons are available for moderation. If the sample is cut into slices of 50 μm each to arrange them in a matrix, the specific β^+ emission rate will be less than 50 MBq/mm^2.

ON-LINE MASS SEPARATORS

Another approach is to employ on-line isotope separators. In these machines various isotopes are produced simultaneously by spallation, fragmentation and fission through the impact of ions, neutrons or high-energy particles on various targets. The experimental problem is the separation of the produced atoms from the target and the formation of ion beams. Whereas selectivity in the atomic mass can be achieved electromagnetically, the atomic number Z has to be selected chemically. Worldwide 83 elements are available at on-line mass separators.[7] At the isotope separator on-line ISOLDE at CERN radioactive beams of 66 elements can be produced with proton beams.[8] Because of their long range, high-energy protons are favourite projectiles for high yields. When a 1 μA beam of 600 MeV protons passes a 100 g/cm^2 thick target, up to 10^{11} atoms/s of a given species can be produced.

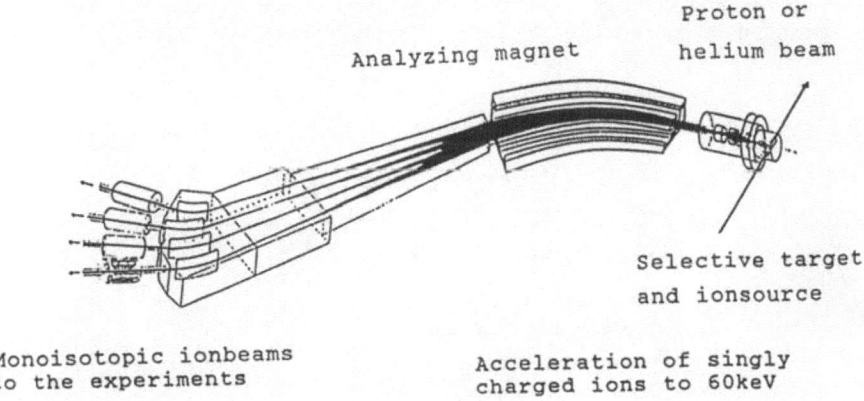

Fig. 1. Lay-out of the ISOLDE on-line mass separator.[7]

The intensity at the experiment is reduced due to the efficiencies for release from the target, ionisation and transport from target to ioniser and finally to the experiment. Overall efficiencies may be as low as 0.01% but reach 80% for elements with high vapour pressure and low ionisation potential like Rb and Cs.[7] The ideally pure isoptope beam can be deposited on diameters well below 1 mm. With a transport energy of 60 keV the penetration depth is several Angstroms only.

COLLECTIONS AT ISOLDE

The first experiment to employ this method for production of strong positron sources was performed at ISOLDE.[9] In this experiment ^{130}Cs and ^{81}Rb were studied. These nuclei were chosen because a 1 μA 600 MeV proton beam and a 50 g/cm^2 thick target yield isotope beams with intensities of above 10^{10} particles/s.[10] Their half-lives of 29 min and 4.7 h and β$^+$ decay branches of 50% and 27%, respectively, allow sufficiently strong sources to be produced. Their endpoint energies are 2 and 1.1 MeV.

Thirteen samples were collected on 6 and 25 μm thick annealed tungsten foils. Their activity was calculated from the beam intensity and the collection time. The activity obtained in a five hours collection later was measured with a Ge-Li γ-detector. It turned out to be 15% less than calculated. This may be due to instabilities in the proton current or sputtering of the sample. In this run 14 GBq of ^{81}Rb were collected on a 1.5 × 3 mm^2 spot, corresponding to a β$^+$ production rate of R ≈ 7 GBq.

After collection the samples were moved out of the beam-line into a separate apparatus (Fig. 2) to study the low-energy positron yield. In different configurations either a separate 6 μm tungsten foil was used as moderator, or the foil with implanted source itself served as moderator. The latter configuration gave maximum yields. Probably due to its lower end-point energy ^{81}Rb showed superior results than ^{130}Cs. The normalised yield, 1.3 × 10^{-5} e$^+$(s Bq$_{\beta^+}$)$^{-1}$, was comparable to the value obtained with a ^{22}Na source and a 6 μm W moderator in the same apparatus.

With improved beam optics the spot size can be reduced to 1 mm diameter. A conservative estimate shows that with a proton current of 2.8 μA and a 300 g/cm^2 target a β$^+$ production rate of 70 GBq can be achieved.

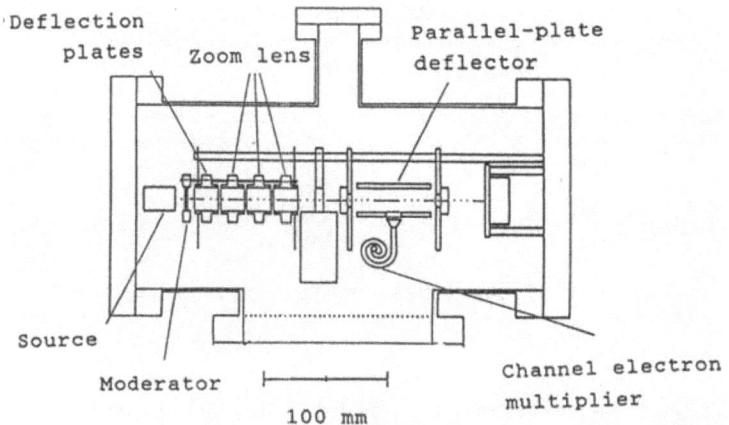

Fig. 2. Sketch of the apparatus for the test of positron sources.[9]

FUTURE DEVELOPMENTS

The synchro-cyclotron that provides the proton beam for ISOLDE can deliver a maximum current of 5 μA. So far the collection of very high activities has not been the major idea and lower currents are employed. With redesigned target containers and modified targets most of them could withstand higher currents. Energy dissipation calculations show that up to 100 μA are feasible without additional cooling.[11] There is only one on-line mass separator at an accelerator that can deliver such proton currents: at TRIUMF in Vancouver.[12] At present this device utilises only about 1 μA proton current, it needs further investment to operate with the maximum current. With a proton current of 100 μA from ^{81}Rb a β$^+$ production rate of about 3.5 TBq should be feasible. For such activities the source diameter increases, because about 10^{15} atoms will be collected during a collection time of one half-life. Sputtering losses will be a severe problem.

The advantage of producing positron source at on-line mass separators lies in the fact that only one specific isotope is collected. Thus intensity losses due to self absorption in the source or large source diameters can be avoided. If operated outside the collection site these sources can be easily used in conjunction with high-yield moderators like solid neon. The main disadvantage is the rather limited beam-time at on-line mass-separators. Because of the frequency switching between different isotope beams the collection of a specific isotope can not be done continuously in a parasitic operation over longer periods.

COMPARISON OF DIFFERENT METHODS TO PRODUCE HIGH β^+ ACTIVITIES

The choice between different ways to produce positrons so far has been mainly guided by the availability of an accelerator or a reactor. No device has been built for this purpose. Thus a comparison can only outline the (dis)advantages of different approaches, the choice will mainly be determined by the technical infrastructure. There is a break-even point between a poor source that works all year and an optimal source with very limited access ("A bird in the hand is worth two in the bush").

Positrons produced via bremsstrahlung and pair-production can be used at the electron accelerator only. The strong radiation background requires a long (magnetic) beam transport. The beam structure is determined by the duty cycle of the accelerator. Suitable LINACS exist at various sites.

At devices where "carry-away" sources can be made, the selection of the specific isotope is a compromise between the achievable activity and parameters relevant for the low energy positron beam. Probably there still are various isotopes worth trying. Worldwide there are several suitable proton accelerators, few on-line mass separators and fewer high-flux reactors. The highest activities are obtained at the BNL high flux reactor. If isotope separation is employed the usable activity can still be drastically increased. From proton accelerators operated at about 10 MeV strong but rather extended sources can be obtained. On-line mass separators are most suitable to produce pointlike sources of virtually every isotope. Due to the limitation on beam time they will be employed mainly where pointlike and short-lived sources are required.

ACKNOWLEDGEMENTS

The authors gratefully acknowledge valuable discussions with G. Spicher and Prof. W. Raith at Bielefeld. The experimental work at CERN has been supported by the ISOLDE Collaboration and the Deutsche Forschungsgemeinschaft.

REFERENCES

1. R.H. Howell, M.J. Fluss, I.J. Rosenberg and P. Meyer, Low Energy, High-Intensity Positron Beam Experiments with a LINAC, Nucl. Instr. Meth. B10/11:373 (1985).

2. G. Graff, R. Ley, A. Osipowicz, G. Werth and J. Ahrens, Intense Source of Slow positrons from Pulsed Electron Accelerators, Appl. Phys. A33:59 (1984).

3. F. Ebel, W. Faust, C. Hahn, S. Langer and H. Schneider, Production of Slow Positrons with the Giessen 65 MeV LINAC, EPOS Symposium Giessen (1986), to be published in Appl. Phys.

4. M. Weber, K.G. Lynn, L.O. Roellig, A.P. Mills Jr., W.E. Frieze and A.R. Moodenbaugh, A High Intensity Positron Beam at the High Flux Beam Reactor, Slow Positron Workshop, UEA, Norwich (1986).

5. A.P. Mills, Jr. and E.M. Gullikson, Solid Neon Moderator for Producing Slow Positrons, Appl. Phys. Lett. 49:1121 (1986).

6. B. Seligmann, Feasibility Study for the Production of an Intense β^+- Source at the Karlsruhe Compact Cyclotron, unpublished.

7. H.L. Ravn, Radioactive Ion-Beams Available at On-Line Mass Separators, to be publ. in: Proc. 11th Int. Conf. on Electromagnetic Separators and Techniques Related to their Applications, Los Alamos 1986, to be publ. in Nucl. Instrum and Meth., preprint CERN-EP/86-128 (1986).

8. T. Bjornstad, E. Hagebo, P. Hogg, O.C. Jonsson, E. Kugler, H.L. Ravn, S. Sundell, B. Vosicki and the ISOLDE Collaboration, Methods for Production of Intense Beams of Unstable Nuclei: New Developments at ISOLDE, Phys. Scr. 34:578 (1986).

9. G. Sinapius, G. Spicher and H.L. Ravn, Positron Sources for Atomic Physics Experiments, J. Phys. E: Sci. Instr. 19:987 (1986).

10. H.J. Kluge (ed), ISOLDE User's Guider, Cern Yellow Report 86:05 (1986).

11. T.W. Eaton, H.L. Ravn and the ISOLDE Collaboration, Beam Heating of Thick Targets for On-Line-Masse-Separators, as ref. 7, preprint CERN-EP/86-135 (1986).

12. K. Oxhorn, L. Buchmann, J. Crawford, J.M. D'Auria, H. Dautet, R. Kokke, J.K.P. Lee, R.B. Morre, V.L. Nikkinen, A. Otter, H. Sprenger and J. Vincent, A Radioactive Beam Facility at TRIUMF, as ref. 7.

ELECTROSTATIC LENSES: HOW TO ROLL YOUR OWN

D.W.O. Heddle

Royal Holloway and Bedford New College
Egham, Surrey TW20 0EX

INTRODUCTION

Electrostatic lenses are used in many fields of physics and it is an
unfortunate fact that the literature of the subject is so diverse that
workers in different fields are often unaware of developments or even basic
notions common in other areas. One sees, for example, papers in Optik, a
popular journal for electron microscopists and lithographers, which report
"new" results which are well known to electron spectroscopists and I have
no doubt that the reverse is also true. Positron physicists are of course
very well read and I would not presume to teach them the basic properties
of electrostatic lenses, but there are a few useful dodges which can help
in the design of lens systems which are perhaps not so well known even
though (and perhaps because) they have been around for some time. There are
two types of approach to the subject: one is from the properties of the
electrostatic field and the other is from the optics. A knowledge of both
aspects is necessary, but one sometimes gets the impression that the optics
is the poor relation. Any really useful lens will consist of more than two
elements and it can be very helpful to look at the system as a combination
of lenses. Even though all electrostatic lenses are thick you can go quite
far with simple, thin lens, ideas though ultimately you will need accurate
data to produce or analyse the final design.

TWO LENSES ARE BETTER THAN ONE

A two electrode lens will form an image of a given object in a
position determined entirely by the voltage ratio. This means that if we
wish to operate between fixed conjugate points - and this is almost
invariably the case - we are stuck with two possible voltage ratios once we
have positioned the lens. The two ratios, one accelerating and the other
decelerating, will give different magnifications and the decelerating lens
will have greater spherical aberration. The focussing behaviour of two
element lenses is very conveniently summarized in the form of P-Q curves
such as are shown in Figure 1 for the lens formed by two coaxial cylinders
spaced by one tenth of their diameter. In passing let me note that cylinder
lenses are generally to be preferred to aperture lenses because they
enclose the beam more fully and shield it from stray electric fields: the
walls of the vacuum chamber are not going to be at the same potential as
all the lens elements! Small gaps are to be preferred for the same reason
and also because the focal properties become somewhat insensitive to the

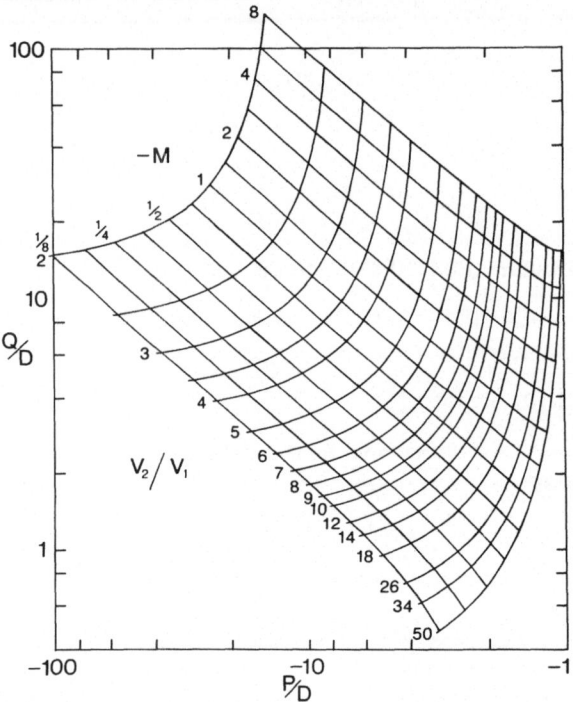

Fig.1. Values of object distance P, and image distance, Q, in terms of ot the lens diameter, D. for two-cylinder lends of spacing D/10.

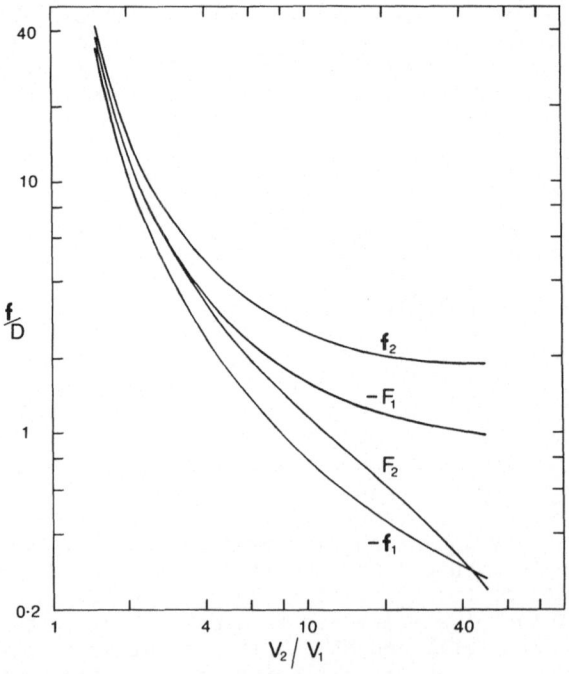

Fig.2. The focal lengths, f, and focal distance, F, in terms of the lens diameter, D, for the two-cylinder lens of spacing D/10.

precise length of the gap. From these curves one can read off the object
(P) and image (Q) distances required to give a specified magnification at a
given voltage ratio. Note that P always refers to the distance on the low
potential side and for a retarding lens becomes the image distance. The
magnification is also shown from the low to the high potential side and one
must take the reciprocal for a retarding lens.

If we wish to preserve an object-image relationship over a range of
voltage ratios then additional lens elements are necessary. A third element
is often sufficient and there is a very simple technique for the design of
three-element lenses which seems to work just as well as more rigorous
methods (Heddle, 1969). We consider the system of three cylinders shown in
Figure 3 and first suppose that the dimensions have been specified
beforehand. We measure all lengths in terms of the cylinder diameter and,
for the moment, assume equal diameters for all three cylinders. L_1 and L_3
are not strictly the lengths of the first and third cylinders, but the
distances from the object and image to the centre of the nearer gap.

We can determine the value of the focussing voltage ratio, V_2/V_1, for
eight values of the overall voltage ratio, V_3/V_1. Suppose we connect the
second and third cylinders together. We now have a two-element lens and can
use the P-Q curves to find the voltage ratios and magnifications
corresponding to $P=L_1$, $Q=L_2+L_3$ (the accelerating case) and $P=L_2+L_3$, $Q=L_1$
(the decelerating case). In each case $V_2=V_3$. Now we connect the first and
second cylinders and find the voltage ratios and magnifications for
$P=L_1+L_2$, $Q=L_3$ and for $P=L_3$, $Q=L_1+L_2$ for which $V_1=V_2$. These ratios and
magnifications are shown in Figure 4 as the points labelled B,F,D and H
respectively. The other four points correspond to situations where the
object and image are each at one of the focal points of the two-element
lens formed by the first (or third) and second cylinders. We can use the
focal distances shown in Figure 2 to find the value of V_2/V_1 for which
$F_1=L_1$ and of V_2/V_3 for which $F_1' = L_3$ and so determine the coordinates of the
point A of Figure 4a. Figure 2 also shows the corresponding values of f_1
and f_1', the focal lengths, from which we can immediately calculate the
angular magnification as f_1/f_1' and, using the law of Helmholtz and
Lagrange, the (transverse) magnification as $(f_1' /f_1)(V_3/V_1)^{1/2}$. This gives us
the point A of Figure 4b. Points C, E and G correspond to the similar
situations with L_1,L_3 equal to F_1,F_2', F_2,F_2', and F_2,F_1' respectively.

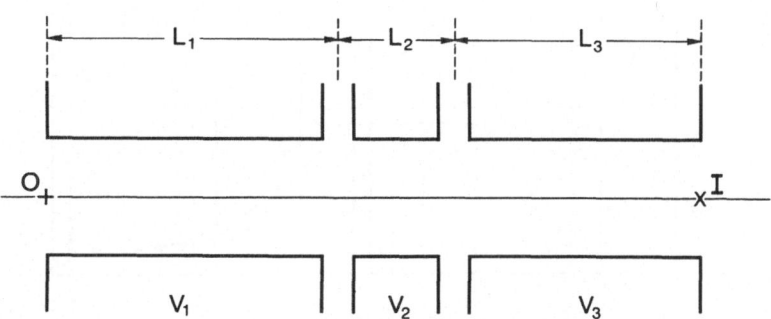

Fig.3. A three-element lens consisting of coaxial cylinders. The lengths
of the cylinders in units of their diameter, D, are indicated. The
lens produces an image at I of the object at O.

Figure 4 also shows measured values (Heddle et al, 1982) of the voltage ratios and magnification for the lens and indicates that for values of V_2 which are less than both V_1 and V_3 a smaller value is needed than the calculations indicate, though at the "two-element" points the agreement remains good. This is because it is not safe to assume that the L_1-L_2 and L_2-L_3 lenses are independent if L_2 is so short. We have found that for L_2 >1.3 the behaviour is satisfactory throughout. It is always preferable to operate in the H-A-B region of the characteristics because the aberrations are smaller and it is safe for practical purposes to use this method for values of L_2 of one or more.

It is almost as simple to design a three-element lens to a specification. This will usually take the form of a magnification to be maintained within practical limits over a range of voltage ratios and the design approach is to consider the maximum ratio to correspond to point B and enter Figure 1 with the specified voltage ratio and magnification and read off values of P ($=L_1$) and Q ($=L_2+L_3$). If one tries to do the same thing with a second, retarding, ratio the sum of the values of P and Q will not usually be the same. There are then two courses of action. As P is really a shorthand for P/D the total length can be kept constant by making the first and third elements (and each half of the centre element) have different diameters. This is discussed by Heddle (1969) but is rarely worth the bother. It is easier, and often quite adequate, to relax the second voltage ratio specification and to enter Figure 1 along the line of chosen magnification (remember that it is the reciprocal value) seeking values of

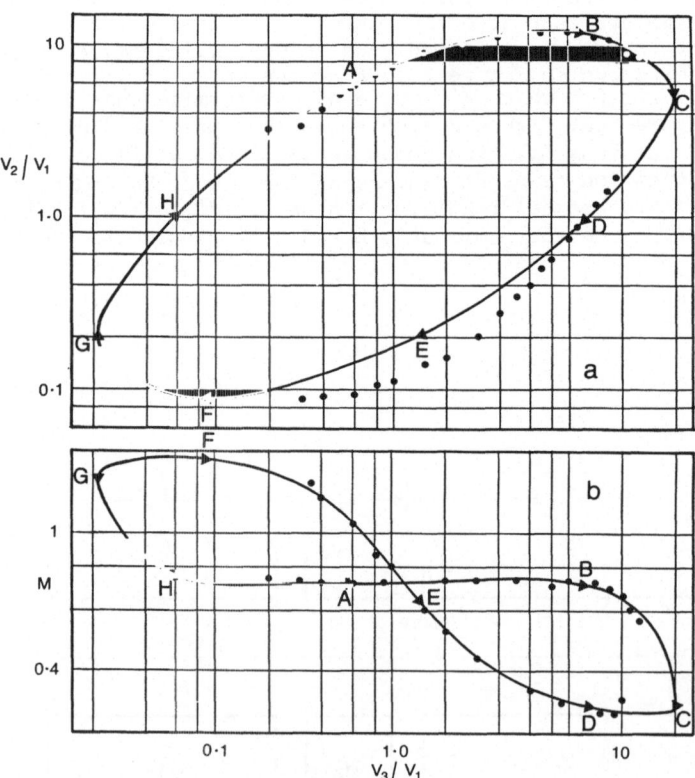

Fig.4. The focal (a) and magnification (b) loci of a three-element lens with L_1 = 2.50, L_2 = 1.00 and L_3 = 1.68 in units of the lens diameter.

P and Q which do have the correct sum and accepting the corresponding voltage ratio.

THE MORE THE MERRIER?

If the constancy of magnification offered by a three-element lens is inadequate or if independent control of magnification is required then additional elements must be added to the lens. Much greater constancy of magnification can be obtained from a four-element lens and it is possible, though rather tedious, to extend the graphical approach used for the three-element lens to this situation. However, a fifth element can provide a much wider range of magnifications, and is amenable to quite simple analysis. Figure 5 shows a simple system of two thin lenses separated by a distance d. We have shown the system as imaging an object d/2 to the left of the first lens at a point d/2 to the right of the second because it simplifies some of the equations and also because it corresponds to a rather special situation to which we shall return later. Each of these lenses is of variable power and it is clear that if all the power is concentrated in the left hand lens the system will have a magnification of 3 in the thin lens approximation. In a similar way the magnification will be 0.33 if all the power is concentrated in the right hand lens. For intermediate situations a detailed analysis is most conveniently done using matrix methods. We consider a ray at the object plane specified by the column vector $\begin{pmatrix} r_1 \\ r_1' \end{pmatrix}$ and act on this with five matrices in succession,

$$\begin{pmatrix} r_2 \\ r_2' \end{pmatrix} = \begin{pmatrix} 1 & d/2 \\ 0 & 1 \end{pmatrix} \begin{pmatrix} 1 & 0 \\ -1/f_2 & 1 \end{pmatrix} \begin{pmatrix} 1 & d \\ 0 & 1 \end{pmatrix} \begin{pmatrix} 1 & 0 \\ -1/f_1 & 1 \end{pmatrix} \begin{pmatrix} 1 & d/2 \\ 0 & 1 \end{pmatrix} \begin{pmatrix} r_1 \\ r_1' \end{pmatrix}$$

$$\left(= \begin{pmatrix} M & 0 \\ -1/f & M_\alpha \end{pmatrix} \begin{pmatrix} r_1 \\ r_1' \end{pmatrix} \right) \qquad \text{for conjugate points)}$$

first a translation matrix then a bending matrix representing the action of the thin lens and further translation and bending matrices to give the coordinate and slope of the ray in the image plane. The condition for conjugate points (no dependence of r_2 on r_1') gives a relationship between the focal lengths of the two lenses which is shown in Figure 6, and either of them can be expressed in terms of the magnification.

$$f_1 = d/(3 + M) \qquad\qquad f_2 = d/(3 + 1/M)$$

In terms of particle optics each of the lenses of variable power requires three elements, but one element of each of these will be the same

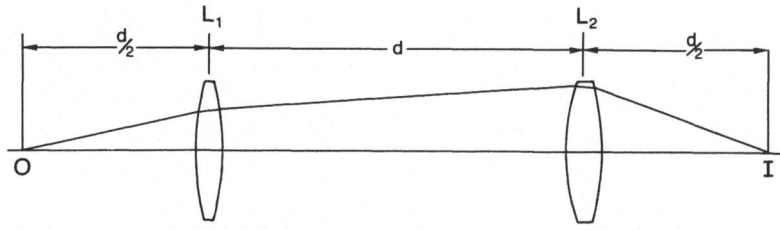

Fig.5. The formation of an image by two lenses of adjustable power.

(otherwise there would be further lens action) and we have a five-element lens. This is rather a lot for graphical analysis, but very simple using matrix methods. It is a little more complicated than the thin photon lens as each gap between elements has to be treated as a lens and the imaging equation involves the product of five translation matrices and four bending ones. The bending matrices have to take account of the thick nature of particle lenses and of the different potentials (refractive indices) on each side of

the lens and become: $\begin{pmatrix} F_2/f_2 & (F_1F_2 - f_1f_2)/f_2 \\ -1/f_2 & -F_1/f_2 \end{pmatrix}$

If you are going to use matrix methods then it is much better to work in terms of the matrix elements themselves rather than the focal lengths and distances, because for weak lens gaps the latter go off towards infinity while the matrix elements are continuous through a voltage ratio of unity as shown in Figure 7. There is a useful paper by DiChio et al (1974a) which tabulates values of the matrix elements over an enormous range of voltage ratios. For voltage ratios close to one the matrix elements can be adequately approximated in terms of the fourth root of the reversed voltage ratio, $\gamma = (V_1/V_2)^{1\!/\!4}$, by

$$a_{11} = a_{22} = \gamma \qquad a_{12} = 0.123 \, (1-\gamma)^2 \qquad a_{21} = -2.595 \, (1-\gamma)^2.$$

Be careful; a second paper by DiChio et al (1974b) defines γ as the reciprocal of their other definition!

For a five-element lens of given proportions there are four voltage ratios and a particular value of magnification may be obtained by many different sets of these. The most common design problem will be to find those ratios which give the required magnification at a specified value of the overall voltage ratio. It is quite difficult to represent all the possible combinations on a simple graph, but if we apply one constraint we can obtain a diagram which is very easy to use. The constraint I have adopted is to make the ratios V_3/V_1 and V_5/V_3 equal leaving only the V_2/V_1 and V_4/V_1 ratios as adjustable parameters of the lens. Figure 8 shows the behaviour of a lens with the length of its centre element 3D and of its other four elements 1.5D. The data is based on measurements on an electron lens and is presented in the form most useful for design purposes. On axes

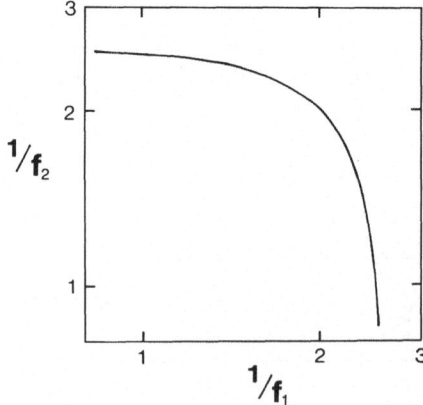

Fig.6.   The inter-relation of the focal lengths of the lenses of
         Figure 5.

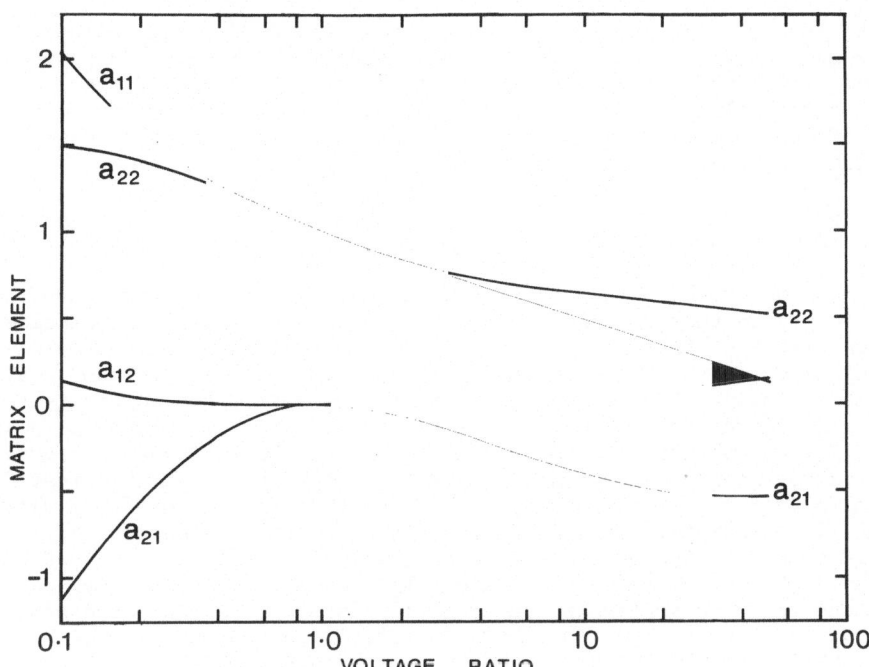

Fig.7. Matrix elements for the two-cylinder lens.

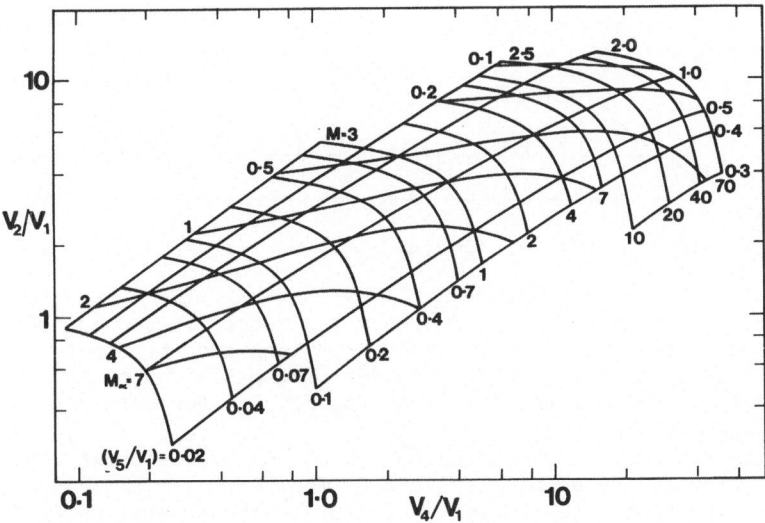

Fig.8. Lines of constant magnification, angular magnification and overall
voltage ratio for the five-element lens discussed in the text.

of V_2/V_1 versus V_4/V_1 are drawn lines of constant V_5/V_1, magnification and angular magnification and it is a simple matter to select the appropriate focussing voltages given the overall voltage ratio and one of the magnification values. Exercise caution in using this diagram for extreme values of the magnification as some data has been extrapolated. Note the resemblance of the lines of constant V_5/V_1 to Figure 6.

PERHAPS TWO ARE ENOUGH

Let us now consider a rather special situation in which we have two lenses so placed that the second focal point of the first lens coincides in position with the first focal point of the second lens. A ray incident parallel to the axis will pass through this common focal point and leave the lens system still parallel to the axis. The system has no finite foci and we call it an afocal lens. This situation is shown in Figure 9a with a sketch of the practical realisation using a five-element lens as Figure 9b. These figures are taken from Heddle (1971). If we place an object at 0, a distance p from the first focal point of the first lens, this lens will form an (intermediate) image at I, a distance q from its second focal point given by $q = f_1 f_2/p$ where f_1 and f_2 are the first and second focal lengths of this first lens. This image lies beyond the first focal point of the second lens and is itself imaged as I' a distance q' from the second focal point of the second lens where $q' = f'_1 f'_2 /q$ and the primed quantities all relate to the second lens. Combining these two relationships gives us $q' = p \times (f'_1 f'_2 /f_1 f_2)$ and if the two lenses do not just share a common focal point but are identical then $q' = p$ which means

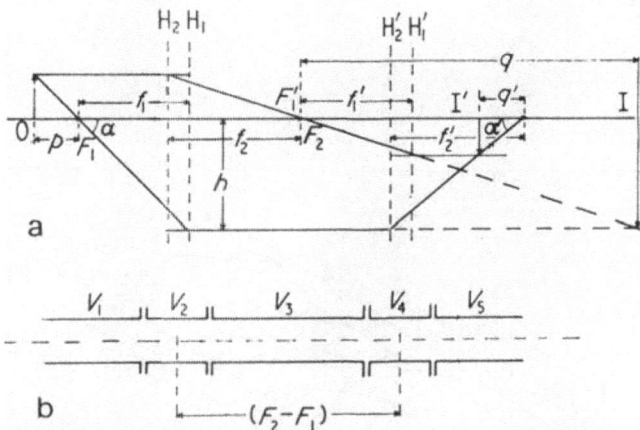

Fig.9. (a) The formation of an image by two lenses spaced so that the second focal point, F_2, of the first lens and the first focal point, F'_1, of the second coincide. The positions of the principal planes, H_1, H_2, and focal points, F_1, F_2, for both lenses are marked and the focal lengths, f_1, f_2, and the object and image distances from the focal points, p and q, are shown. Values relating to the second lens are primed.
(b) The afocal lens of five elements. The separation of the components is equal to the distance between the focal points of either.

that the distance between the object and the final image is just equal to twice the separation of the focal points of either lens and the actual position of the object does not matter. A window and pupil in the object space will be imaged with an identical separation and the hazard of too close approach which can cause severe aberrations or loss of beam is avoided. This lens system has a further property which has perhaps a special application in brightness enhancement applications. The magnification can be calculated without recourse to any knowledge of the potential distribution in the lens. From Figure 9a it is easy to see that the magnification is given by $M = f_1' /f_2 = f_1/f_2$ if the lenses are identical. The angular magnification can also be seen to be given by $M_\alpha = f_1/f_2' = f/f_2 = M$. The law of Helmholtz and Lagrange relates the magnification, the angular magnification and the overall voltage ratio as $M \times M_\alpha \times (V_5/V_1)^{1/2} = 1$ and so for this afocal lens we have $M = M_\alpha = (V_5/V_1)^{-1/4}$. This really works; Figure 10 shows the measured magnification of a five-element lens operated in the afocal mode and it is clear that the $-1/4$ power relation is very well obeyed. The identical nature of the two three-element lenses means that $V_5/V_3 = V_3/V_1$ and this is the origin of the choice of constraint in an earlier section. If a range of overall voltage ratios is not needed then two two-element lenses can be used. For example, two 16:1 lenses can form an afocal lens of overall ratio 256:1 with a magnification of -0.25 and a total length of 4.1D. There are certain hazards with greater voltage ratios, because the end elements become so short that any real source object or image target will "short circuit" the potential variation across the cylinder diameter and affect the focussing properties. In any case it is as well to divide the available length of the end elements so that the low potential end is longer than the other. This is because the bending action of a two-element lens occurs mainly on the low potential side. It should be possible to operate with extremely large voltage ratios by setting four lenses in sequence with the correct spacing when the magnification will be positive, but still be simply related to the overall voltage ratio. I have never operated such a strong lens but ray-tracing makes it look reasonable.

BACK TO BASICS

So far I have discussed ways of using available data on two-element lenses to design more useful lens systems. The more complex the system, the less convenient is the representation of all the parameters involved. Nowhere will you find precisely the lens situation that you actually need and I am about to suggest that you should design your lens systems from scratch. This is very much easier than you perhaps suppose. It is easy to calculate the axial potential and field in a system of closely spaced cylinder lenses by superposing values for each contiguous pair and it is

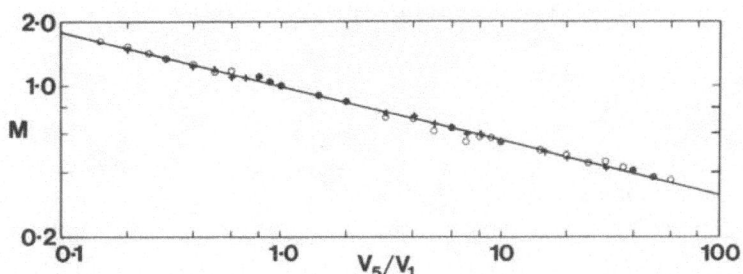

Fig.10. The magnification of a five-element lens as a function of overall voltage ratio.

then just a matter of integrating the Picht equation and adjusting the potential of one or more elements to give the desired focus condition.

Let me start at the end of this sequence. The Picht equation, sometimes called the reduced equation of motion, describes the relationship of the radial and axial positions of a charged particle moving in an axisymmetric potential, $V(z)$. It uses a reduced radius, $R = rV^{\frac{1}{4}}$ and has the form $\dfrac{d^2R}{dz^2} + TR = 0$, where $T = 0.1875\left(\dfrac{V'}{V}\right)^2$ and can be integrated very simply by Numerov's algorithm. This expresses the value of R at the nth point along the axis in terms of its values at the two preceding points.

$$R(n+2) = 2R(n+1) - R(n) - (h^2/12)(T(n)R(n) + 10T(n+1)R(n+1))/(1 + (h^2/12)T(n+2))$$

where h is the step size at which the potential is known.
Provided that you know the potential with sufficient accuracy it is not necessary for h to be unduly small. You can calculate perfectly acceptable focal lengths and distances with $h = D/10$ though the aberration coefficients would be 5-10% low: $D/100$ will give those to 1%. You need to provide the first two values of R: if you wish to calculate the focal lengths then $R(0) = 1$, $R(1) = 0$ are suitable, but if you are looking for an imaging condition then take $R(0) = 0$ and $R(1) = h$. In the latter case you will need to vary some potential until the absolute value of R is less than some suitably small amount at the image plane.

If you know the axial potential and field for a two-cylinder lens with potentials 0 and 1 it is a straightforward matter to find the values for any number of coaxial cylinders at potentials V_1, V_2, V_3 etc so the only remaining problem lies in obtaining the two-cylinder values. Table 1 gives these values for $0 > z/D > -1$ together with expressions for other axial positions. For values of z less than $-D$ it is perfectly safe to use a simple exponential: the exponent is the same for both potential and field. Values for positive z are easy to calculate because the field is symmetric and the potential antisymmetric.

The traditional technique for calculating potentials is the relaxation method. This will always work - eventually - but it requires much computer time and memory and the precision is determined by the choice of mesh size. For cylinder lenses there is a much simpler analytic method based on the expansion of the potential and its derivatives in terms of Bessel functions. Cook and Heddle (1976) applied a variational principle to derive

TABLE 1

Axial potential in a two cylinder lens

z/D	V(z)	V'(z)	
0.00	0.500000	1.316583	For z < -1
-0.10	0.371468	1.224679	
-0.20	0.259628	0.997572	$V(z) = V(-1)\exp(4.808(z+1))$
-0.30	0.173176	0.732612	$V'(z) = V'(-1)\exp(4.808(z+1))$
-0.40	0.111929	0.501239	
-0.50	0.070953	0.328177	
-0.60	0.044475	0.209420	For z > 0
-0.70	0.027702	0.131722	
-0.80	0.017195	0.082195	$V(z) = 1 - V(-z)$
-0.90	0.010653	0.051069	$V'(z) = V'(-z)$
-1.00	0.006593	0.031656	

the approximate expressions, but were constrained by the need to approximate the potential in the gap of the lens by a linear variation. Bonjour (1979) showed that a cubic gave slightly better results. I have since found that it is easy to allow for the non-linear variation of potential in the gap by increasing the value used in the calculation by a small amount. The Appendix gives the listing of a PASCAL program to calculate the axial potential, $V(z)$, and the field $E(z) = -V'(z)$, at intervals of D/100 from -3D to + 3D with a precision of 1 part in 10^7 and an accuracy which I estimate from relaxation calculations with a smaller step size to be ± 1 in the sixth decimal place. The input data required are the zeros, $k(c)$, of the zero order Bessel function and the corresponding values, $j(c)$, of the first order Bessel function. These are tabulated in the British Association Mathematical Tables (1958) to 10 decimal places. A simple extension involving Bessel functions evaluated at points off axis allows the potential and its derivatives to be found anywhere within the lens and very accurate ray tracing can be done.

REFERENCES

Bonjour, P 1979, A simple accurate expression of the potential in electrostatic lenses. Part 1: Two cylinder lenses Rev.Phys.Appl. 14: 533.

British Association Mathematical Tables Vol.6. 1958. University Press, Cambridge.

Cook, R.D. and Heddle, D.W.O., 1976, The simple accurate calculation of cylinder lens potentials and focal properties. J.Phys. E.Sci.Instrum. 9:279.

DiChio, D., Natali, S.V. Kuyatt, C.E. and Galejs, A., 1974a, Use of matrices to represent electron lenses. Matrices for the two-tube electrostatic lens. Rev.Sci.Instrum. 45:566.

DiChio, D., Natali, S.V. and Kuyatt, C.E., 1974b, Focal properties of the two-tube electrostatic lens for large and near-unity voltage ratios. Rev.Sci.Instrum. 45:559.

Heddle, D.W.O., 1969, The design of three-element electrostatic lenses. J.Phys. E:Sci.Instrum. 2:1046.

Heddle, D.W.O., 1971, An afocal electrostatic lens J.Phys.E:Sci.Instrum. 4:981.

Heddle, D.W.O. and Papadovassilakis, N., 1984, The magnification behaviour of a five-element electrostatic lens. J.Phys.E:Sci.Instrum. 17:599.

Heddle, D.W.O.; Papadovassilakis, N. and Yateem, A.M., 1982, Measurements of the magnification behaviour of some three-element electrostatic lenses. J.Phys.E:Sci.Instrum. 15:1210.

APPENDIX

PASCAL program to calculate the axial potential in a cylinder lens

```pascal
PROGRAM CylinderLensPotential;
CONST   radius=50;                          { In units of step size }
        length=300;                         { Two cylinders each 3D long }
        gap=0.1;                            { With a gap of D/10 }
VAR     j,k,q,r:ARRAY [1..150] OF real;     { Bessel functions: see text }
        V,E:ARRAY [-length..length] OF real; { Potential and Field }
        sV,sE:ARRAY [0..156] OF real;       { Sums to 'c' terms }

PROCEDURE ReadData;                         { Modify to read from a file }
VAR  c:integer;
BEGIN       FOR c:=1 TO 150 DO BEGIN
   read(k[c],j[c]);   r[c]:=1/(j[c]*k[c]);  q[c]:=r[c]/(2*k[c]);  END;  END;

PROCEDURE Smooth(c:integer);                { Used to force convergence }
BEGIN sV[c+3]:=((sV[c]*sV[c-2]-sqr(sV[c-1]))/(sV[c]-2*sV[c-1]+sV[c-2]));
   sE[c+3]:=((sE[c]*sE[c-2]-sqr(sE[c-1]))/(sE[c]-2*sE[c-1]+sE[c-2]));   END;

PROCEDURE CalculatePotential;
VAR     l,c:integer;
        g,z,z1,z2,V0,E0,a1,a2,limit:real;   { Cylinder potentials,V0=0,1 }
        dV,dE:ARRAY [0..150] OF real;        { Increments in V and E }
BEGIN
   limit:=0.0000001;   g:=1.082*gap; { Allows for non-uniform field in gap }
   FOR l:=-length TO length DO
   BEGIN
      z:=l/(2*radius);     z1:=2*z+g;      z2:=2*z-g;
      IF z > g/2 THEN BEGIN     V0:=1;     E0:=0;     END
                 ELSE IF z < -g/2 THEN BEGIN     V0:=0;   E0:=0;          END
                       ELSE BEGIN V0:=0.5 + z/g; E0:=-1/g;        END;
      c:=1;    sv[0]:=0;   se[0]:=0;          dE[0]:=1; { Or it will not start }
      WHILE abs(dE[c-1]) > limit DO           { Sum the series }
      BEGIN
         IF z > 0 THEN BEGIN  a1:=exp(-k[c]*z1); a2:=exp(-k[c]*z2);      END
                  ELSE BEGIN  a1:=exp(k[c]*z1);  a2:=exp(k[c]*z2);     END;
         IF abs(z) > g/2 THEN BEGIN    dV[c]:=q[c]*(a1-a2)/g;
                                dE[c]:=dV[c]*2*k[c]*(z/abs(z));       END
                   ELSE IF z > 0 THEN BEGIN  dV[c]:=q[c]*(a1-1/a2)/g;
                                      dE[c]:=r[c]*(a1+1/a2)/g;    END
                         ELSE BEGIN   dV[c]:=q[c]*(1/a1 - a2)/g;
                                dE[c]:=r[c]*(1/a1 + a2)/g;    END;
         sV[c]:=sV[c-1]+dV[c];      V[l]:=V0+sV[c];
         sE[c]:=sE[c-1]+dE[c];      E[l]:=E0+sE[c];
         IF c < 150 THEN c:=c+1  ELSE BEGIN  Smooth(148);  { Not converged }
                                      Smooth(149);
                                      Smooth(150);
                                      Smooth(153);
                                      V[l]:=V0+sV[156];
                                      E[l]:=E0+sE[156];
                                      dV[149]:=0;
                                      dE[149]:=0;     END;
      END;
   END;
END;

BEGIN   ReadData; CalculatePotential; END.    { Modify to write to file }
```

THEORETICAL ASPECTS OF POSITRONIUM COLLISIONS

Richard J. Drachman

Laboratory for Astronomy and Solar Physics
Goddard Space Flight Center
Greenbelt, MD 20771, USA

INTRODUCTION

From its former position as a mathematical curiosity, study of the positronium atom (Ps) has gradually evolved into a subject of quite active experimental and theoretical interest. Ps, especially in the ortho (triplet spin) state, is relatively long-lived and can be considered just like any ordinary atom in most of its interactions with other systems. The unique characteristic of Ps, its annihilation into gamma rays, can usually be treated as a small perturbation superimposed on the dominant Coulomb interaction which determines its static properties and the cross-sections for collisions with electrons, various atoms, molecules, and ions. Although Ps is an atom, in some sense an isotope of hydrogen, its mass is so much less than that of any other that the standard techniques of atomic scattering theory are not applicable, and those of electronic scattering theory may be more appropriate. Nevertheless, the internal structure of Ps, its extension, and the Pauli principle combine to complicate the theory. In this report I will try to summarize the calculations already performed on the scattering of Ps from the simplest atomic systems, describing some of the methods used to analyze the collisions. I will also discuss the long-range effective potential acting between Ps and these other systems; here the low mass of Ps brings about some interesting modifications of the conventional van der Waals force. Next, the interesting "poly-electrons" Ps^- and Ps_2 will be described. Finally, some astrophysical and exotic applications will be mentioned.

THE POSITRONIUM-HELIUM SYSTEM

Although this system contains three electrons and is therefore rather difficult to handle theoretically, it is easy to describe and may be a good introduction to the general subject. At collision energies below 5.10 eV (the first excitation energy of Ps) only elastic scattering is energetically permitted, so only a single channel formalism is needed. Since the helium atom has a spin-singlet ground state no spin-exchange is possible, and our non-relativistic approximation omits spin-orbit coupling. In addition, there is no bound state of the He^- ion or of PsHe.

Fraser[1] carried out the first ab initio calculations of the low-energy scattering of Ps from helium using the static-exchange approximation. In this approximation a variational trial function is constructed from an antisymmetrized product of the ground-state wave functions of the two atoms:

$$\Psi = v \{\tfrac{1}{2} (\vec{r}_1 + \vec{r}_p)\} \psi(|\vec{r}_1 - \vec{r}_p|) \psi(r_2, r_3) \chi_1^1(s_p, s_1) \chi_0^0(s_2, s_3) + \ldots \tag{1}$$

where electron 1 is bound to the positron (p), electrons 2 and 3 are in the Helium atom, and the spin functions represent the triplet state for Ps and the singlet state for helium. The function v represents the motion of the center of mass of the Ps atom relative to the fixed nucleus of the other atom. It is from the necessity of describing the Ps this way when the potentials do not depend on the center of mass coordinate that much of the difficulty and the challenge of Ps calculations arises. Fraser and Kraidy[2] used the simplest one-term Hylleraas closed-shell function to describe the helium ground-state and derived variationally the integro-differential equation satisfied by v, retaining three partial waves. Since the helium function is not exact, there is some ambiguity concerning the formation of the integral equation. This leads to a spread in the calculated cross-sections; at zero energy the variation is between 14.2 and 11.8, in units of πa_0^2; the latter is perhaps the better value although the very existence of the spread is a measure of the inaccuracy of the helium wave function.

This calculation is incomplete since it does not include the effect of the distortion or polarization of the two colliding atoms. Two quite different attempts to rectify this omission have been reported. The first[3] concentrated on the long-range distortion due to the van der Waals potential acting between the two systems. This was found to have the approximate asymptotic form $- 19.3/R^6$ and to approach zero following a certain variational form for small R. In fact, the original motivation for performing this calculation was the measurement of pickoff annihilation in Ps-He collisions, that is, the observation[4] of a density dependence of the lifetime of ortho Ps in helium gas due to the bound positron annihilating with one of the helium electrons during the collision. It was the disagreement between the pickoff rate calculated from the static-exchange approximation and the observed value that clearly showed the inadequancy of that approximation. By adding this potential term to Fraser's integro-differential equation it was possible to increase the pickoff rate but not enough to give good agreement with experiment.

The same motivation led Houston and me along another path.[5] We concentrated on the short-range distortion induced in the Ps by the exchange force acting between the helium closed shell and the Ps electron. This force is expected to re-orient and stretch the Ps, and thus to reduce the scattering length; it is easy to see qualitatively that the sort of distortion described above also increases the overlap between the positron in Ps and the atomic electrons and leads to an increase in the annihilation rate. Both effects should be included in a correct calculation, and I will return to the van der Waals problem later.

Instead of adding correlation terms to the trial function (1), the rigorous way to proceed, we modeled the exchange interaction as a local potential, $V(r) = V_0 \exp(-\alpha r)$, where r represents the electron in the Ps atom, the helium atom is frozen, and antisymmetry between the incoming and target electrons is not included. The two parameters were adjusted to give the no-correlation values of the scattering length and the pickoff rate (at zero energy) obtained previously by Fraser and Kraidy.[2] Then correlation was introduced and the parameters of the resulting wave function were determined by the Kohn variational method. That is, we chose a trial function of the following form:

$$\times(R,\rho,r) = R^{-1} [a_T \{1-\exp(-\delta R)\}-R]$$

$$+ \exp(-\delta R)\exp(-\beta\rho)\exp(-\gamma r) \sum_{i=1}^{N} a_i R^{\ell_i} \rho^{m_i} r^{n_i}. \quad (2)$$

Here $\vec{R} = (\vec{r} + \vec{r}_p)/2$ and $\vec{\rho} = \vec{r} - \vec{r}_p$ are the center of mass and relative coordinates, respectively. As expected, the variational solution to this model problem gives a more satisfactory answer: the cross-section is reduced to 7.72 πa_0^2 (corresponding to a scattering length a = 1.389 a_0) and the pickoff annihilation rate is increased by a factor of 2.3, bringing it close to the experimental value. It seems that this model contained much of the relevant physics and indicated the right direction for future more rigorous work, which has unfortunately not yet appeared.

According to the recent discussions[6] of the possibility of producing high brightness, low-energy beams of Ps it should not be long before scattering cross-sections are available, and another check on the basic correctness of our pictures of correlation in Ps He collisions can be made. In the meantime, there exists only a semiquantitative measurement of the scattering length; it involves interesting physics and is worth mentioning here, although the simple analysis I will describe is almost certainly incomplete. Since the Ps-He scattering length is positive due to the exchange interaction discussed above, it may be energetically favorable for the Ps atom in sufficiently dense Helium to be localized in a cavity in the medium. This "bubble" is thought of as maintained against collapse by the (zero-point) pressure of the Ps atom whose translational motion corresponds to the lowest eigenstate of a particle in a spherical square-well potential. Two pieces of experimental data can fix the radius and depth of this well. First, the angular correlation[7] of gamma-rays from para-Ps in the bubble measures the momentum distribution of the translational motion. It is mainly sensitive to the radius of the well. (The radius can also be obtained by an analysis of the equilibrium of the bubble). Second, the pickoff rate[8] for ortho-Ps in the bubble gives a measure of how much the wave function extends outside the well; this can give information on the depth of the potential when combined with the previous information on its radius. In turn, the depth V_0 depends on the scattering length, but there is controversy over the correct dependence. The simplest proposed relationship[10] is the following:

$$V_0 = (\hbar^2/2m_{Ps}) 4\pi\rho a,$$

where ρ is the He density and a is the scattering length, and this should be adequate for moderately low densities. I like this analysis because it leads to a value[11] of the scattering length a = 1.42 ± .10 a_0 for both He^3 and He^4, in excellent agreement with the result mentioned above! In Ref. 11 the theoretical range of validity of the simple linear relation between well depth and density is considered carefully, and only data in the appropriate density range are used in extracting a. This good agreement is encouraging; even more desirable would be a direct, conventional scattering measurement.[12]

THE POSITRONIUM-HYDROGEN SYSTEM

The Ps-H system has several simplifications when compared to the Ps-He system: both atoms have exactly-known wave-functions and only two electrons are involved. Nevertheless, this system has several features which make it quite complex and interesting. The electronic triplet spin state of the system is much like the Ps-He system, but the singlet state has, in addition, one true bound state (Positronium Hydride or PsH) and an infinite set of scattering resonances. Needless to say, this system is more difficult experimentally than Ps–He.

Again the static–exchange calculation for this system was carried out by Fraser,[13] using a trial function analogous to that in Eq. (1), with the obvious difference that the target atom has only one electron. In this approximation the singlet and triplet scattering lengths are a_s = 7.275 a_o and a_t = 2.476 a_o respectively. From these two quantities the total scattering cross-section at zero energy is found to be σ = 71.32 πa_o^2 while the cross-section for ortho–para conversion is σ_c = 5.76 πa_o^2. Several improvements have been made involving different ways of including correlation; these have been sufficiently discussed previously.[14-16] Their main result is a significant reduction in both the above cross-sections. (I have been concentrating here on the very low energy results mainly because they are likely to be relevant to certain astrophysical processes in neutral hydrogen; for the low temperatures needed to maintain the neutral medium the collision energies are quite low). It is remarkable that the first experiment[17] ever performed on a system containing positrons and atomic hydrogen involved the exchange quenching of ortho Ps in collisions with H; only an upper bound compatible with the theory was obtained.

Although it was not explicitly noted, the static–exchange calculation if combined with effective–range theory would be sufficient to deduce the existence of a bound singlet state of PsH with a modest degree of binding. Since then quite a few elaborate calculations of the binding have been carried out; the latest and best of these is the recent work of Ho[18] which gives a binding energy B = 1.0597 eV. PsH is clearly a strongly–bound system, completely stable against everything except annihilation. However, it has not been seen experimentally, and no reliable method of producing and detecting it has yet been proposed.

Some years ago the complex–rotation method and the stabilization method were applied to the singlet Ps–H system,[14,19] and one s–wave resonance was detected at a bombarding energy of about 4 eV. The structure of this resonance is rather interesting; it represents the "2s" state of the $e^+ - H^-$ hydrogenic system. Imagine the potential acting between a positron and the ground state of H^-. At large distances it falls off like an ordinary Coulomb potential, and hence should possess a complete set of hydrogenic levels, whose energies are approximately

$$E_n = E(H^-) - 1/(n-\mu)^2,$$

where μ is the quantum defect, a slowly varying function of n that measures the effect of the short–range deviation of the potential from pure Coulomb form. These levels would be stable bound states except for the fact that they are energetically capable of decaying into Ps+H; thus they are actually Feshbach resonances[20] in the Ps–H continuum and have complex quantum defects.[21] In this picture PsH itself is the 1s state of this Coulomb system and is the only state whose quantum defect is real. Because the short–range deviation from the Coulomb potential is repulsive the real parts of the quantum defects are negative and p–states lie below corresponding s–states; the lowest resonance should be the 2p. As one increases n some of the excited states of Ps become accessible; this should increase the resonance widths, but no direct computations have yet been done in this region. The full characterization of this series is an interesting project for the future, and eventual observation of these resonances would be desirable.

An interesting by-product is the fact that these complex quantum defects can be extrapolated to the e^+-H^- threshold, and the cross-sections for both elastic [$e^+ + H^- \rightarrow e^+ + H^-$] and inelastic [$e^+ + H^- \rightarrow Ps + H$] scattering slightly above threshold can be estimated. The inelastic reaction is of some interest since it is exothermic, and even zero-energy positrons colliding with H^- ions can produce Ps with non-zero energy. It is not hard to imagine applications for this process, both in astrophysical context and as a possible source of Ps atoms of controlled energy. From the best[19] calculated parameters of the 2s resonance, $E = -1.205$ Ry and $\Gamma = (5.5 \pm 2) \times 10^{-3}$ Ry, one obtains a quantum defect $\mu = -0.5863 + (0.0237 \pm .0087)i$ from which (using the relation $\delta = \pi\mu$ for the phase shift) the low-energy s-wave cross-section is estimated to be

$$\sigma_{Ps}^0 = \frac{0.26 \pm .08}{k^2} \ \pi a_o^2 .$$

Similar results could be obtained for other partial waves to be compared with the more direct results expected from distorted-wave,[22] coupled-channel, and Fock-Tani calculations.[23]

LONG-RANGE FORCES

Most of the foregoing material is quite old, well known, and previously reviewed.[24] It is, however, the basis for a program of Ps-atom interaction experiments and as such needs to be emphasized. In the present section I will discuss in greater detail a relatively recent development in Ps-atom theory: the calculation of van der Waals forces acting between one Ps atom and a second atom (which may also be Ps).

I mentioned that one attempt[3] to improve agreement between theory and experiment in the Ps-He system involved the inclusion of long-range van der Waals (vdW) distortion. To do this, a variational principle was used along with a certain reasonable trial function to describe the distortion as the two atoms approach each other. But this was done in the Born-Oppenheimer or adiabatic approximation appropriate to the description of interactions between ordinary atoms. Later a perturbation calculation of the vdW interactions involving Ps and both H and He was carried out using the pseudostate method.[25] It gave an essentially exact value for the coefficient C_6 of the leading asymptotic term in the vdW potential but was not designed to examine the shorter-range parts of the potential.

Manson and Ritchie[26] must be credited with the important insight that, for neutral systems containing Ps, "recoil" and "non-adiabatic" effects must be included in the derivation of the effective potential. This observation opened the door to a whole set of new calculations[27] analogous to some done years ago for the electron-atom system.[28] The point is that, although Ps can be thought of as an atom, its mass is only twice that of an electron, and hence much of the kinematics of electron scattering is appropriate to Ps scattering.

To introduce the discussion I will review the method used by Barker and Bransden for Ps-He scattering which was mentioned above, but will apply it instead to the fairly general problem of Ps interaction with another hydrogenic atom consisting of an electron and some singly charged positive particle of mass M. In Rydberg units the Hamiltonian is

$$H = -\frac{1}{M} \nabla_{+1}^2 - \nabla_{+2}^2 - \nabla_{-1}^2 - \nabla_{-2}^2 + \frac{2}{|\vec{r}_{+1} - \vec{r}_{+2}|} + \frac{2}{|\vec{r}_{-1} - \vec{r}_{-2}|} - \sum_{i,j=1}^{2} \frac{2}{|\vec{r}_{-i} - \vec{r}_{+j}|}. \quad (3)$$

Since we are interested in long-range forces, we can assume that the otherwise identical particles are distinguishable, and since all the particles have finite masses it is important to measure the positions of the two atoms from their centers of mass. This is accomplished by transforming to the following Jacobi coordinates:

$$\vec{R} = \frac{M\vec{r}_{+1}+\vec{r}_{+2}+\vec{r}_{-1}+\vec{r}_{-2}}{M+3}, \quad \vec{x} = \frac{\vec{r}_{+2}+\vec{r}_{-2}}{2} - \frac{M\vec{r}_{+1}+\vec{r}_{-1}}{M+1}, \quad \vec{\rho}_i = \vec{r}_{-i} - \vec{r}_{+i}. \tag{4}$$

Here $\vec{R}$ is the center of mass of the whole system, and $\vec{x}$ is the vector joining the centers of mass of the two atoms. The idea is to derive an effective Schroedinger equation in the coordinate $\vec{x}$ to describe the scattering.

In these transformed coordinates $H = H_o + T + V$, where

$$H_o = - (1+\tfrac{1}{M})\nabla_1^2 - \frac{2}{\rho_1} - 2\nabla_2^2 - \frac{2}{\rho_2}, \quad T = - \frac{1}{\mu} \nabla_x^2, \quad \text{and } \mu = 2 \frac{1+M}{3+M}. \tag{5}$$

(I have omitted the kinetic energy of the total center of mass which can be set equal to zero). The interatomic potential energy is

$$V = 2 \sum_{i=1}^{2} (-1)^i \left\{ \frac{1}{\left|x - \frac{M}{1+M}\vec{\rho}_1+(-1)^i\vec{\rho}_2/2\right|} - \frac{1}{\left|x + \frac{1}{1+M}\vec{\rho}_1+(-1)^i\vec{\rho}_2/2\right|} \right\}. \tag{6}$$

In the adiabatic picture, where the centers of mass of the two atoms are initially held fixed, one can determine the interatomic potential by a variational calculation involving only $\vec{\rho}_1$ and $\vec{\rho}_2$ as dynamical variables while $\vec{x}$ is treated as a parameter. Expanding the potential V for large values of x we obtain the leading (dipole-dipole) term

$$V_{dd} \sim \frac{2}{x^3} [\vec{\rho}_1 \cdot \vec{\rho}_2 - 3(\vec{\rho}_1 \cdot x)(\vec{\rho}_2 \cdot x)] \tag{7}$$

which is independent of mass, and the next order (quadrupole-dipole) term

$$V_{qd} \sim \frac{3}{x^4} \left(\frac{1-M}{1+M}\right) [5(\vec{\rho}_1 \cdot \hat{x})^2(\vec{\rho}_2 \cdot \hat{x}) - \rho_1^2(\vec{\rho}_2 \cdot \hat{x}) - 2(\vec{\rho}_1 \cdot \hat{x})(\vec{\rho}_1 \cdot \vec{\rho}_2)] \tag{8}$$

which is seen to vanish when M = 1, the Ps-Ps case. A simple, physically reasonable trial function can be written in the following form:

$$\phi_{pol} = \psi_M(\rho_1) \phi_{Ps}(\rho_2) \{1 + A(x) V_{dd}\}. \tag{9}$$

Use of this trial function in a variational calculation with A(x) as a parameter yields the leading term $- C_6/x^6$ in the asymptotic interatomic potential, and the numerical value of C_6 obtained this way is only 9% below the exact one. The scattering problem can now be treated by constructing the wave function

$$\Psi_{scat} = F(\vec{x}) \phi_{pol} \chi, \tag{10}$$

where χ is the spin function, and F describes the relative motion of the two atoms. Barker and Bransden[3] discussed the use of this function in a Kohn variational principle, but for simplicity they actually used a method akin to the polarized-orbital method:[29]

$$\iint d\vec{\rho}_1 d\vec{\rho}_2 \; \psi_M(\rho_1) \; \phi_{Ps}(\rho_2) \; (H-E) \; \Psi_{scat} = 0. \tag{11}$$

The resulting equation for F is essentially the same as the static-exchange equation of Fraser except that the vdW potential discussed above is now added. Since this potential is attractive there is a decrease in the scattering length, but since (11) is not variational it is difficult to assess the reliability of the result. On the other hand, if the Kohn principle had been used additional terms would have appeared in the scattering equation. The most interesting of these are explicitly dependent upon the masses of the two atoms and are generally called non-adiabatic terms.

The usual way to derive the vdW potential is by perturbation theory, and it is probably sufficient to do the calculation up to second order. C. K. Au and I have described[27] a detailed optical potential formalism, based on the Feshbach operator technique,[20] which enables one to keep track of the bookkeeping involved in the perturbation series, and in particular to take açcount of the non-adiabatic corrections. The two leading terms in the adiabatic potential, due to the dipole-dipole (7) and quadrupole-dipole (8) terms in the multipole expansion, are the following:

$$U_{Ad}(x) = -\frac{C_6}{x^6} - \frac{C_8}{x^8}, \quad C_6 = 24 \frac{|<1s_1|P_1(1)\rho_1|n_1><1s_2|P_1(2)\rho_2|n_2>|^2}{\Delta(n_1n_2)},$$

$$\tag{12}$$

$$C_8 = 60 \left(\frac{1-M}{1+M}\right)^2 \frac{|<1s_1|P_2(1)\rho_1^2|n_1><1s_2|P_1(2)\rho_2|n_2>|^2}{\Delta(n_1n_2)}.$$

Here $\Delta(n_1n_2) = E(n_1) + E(n_2) - E_0$, and summation over the complete sets of hydrogenic states for both atoms is understood. The leading non-adiabatic term has the form

$$U_{Nad}(x) = \frac{f^*(n_1n_2)\, h\, f(n_1n_2)}{[\Delta(n_1n_2)]^2} \qquad f(n_1n_2) \equiv <n_1n_2|V_{dd}|1s_11s_2>,$$

$$h \equiv -\frac{1}{\mu}(\nabla_x^2 + k^2).$$

$$\tag{13}$$

(This can be understood as resulting from the expansion of the energy denominators). Since the optical potential is to be used only in the scattering equation, we can commute h to the right in Eq. (13) and allow it to operate on $F(\vec{x})$. Since the leading term in $U_{Ad}(x)$ goes like x^{-6}, we may set $hF(\vec{x}) = 0$ whenever it appears. In addition, $\nabla_x^2 V = 0$ since Laplace's equation holds in the asymptotic region. In this way we obtain the expression

$$U_{Nad}(x) = -\frac{1}{\mu}\frac{f^*(n_1n_2)\, 2\nabla_x f(n_1n_2)\cdot\nabla_x}{[\Delta(n_1n_2)]^2} = -\frac{1}{\mu}\frac{\nabla_x[f(n_1n_2)]^2\cdot\nabla_x}{[\Delta(n_1n_2)]^2}. \tag{14}$$

The second form of this expression requires only that $f(n_1n_2)$ be real, and once the summation has been carried out no angular parts remain. Finally,

$$U_{Nad}(x) = -\frac{D}{\mu}\nabla_x\left(\frac{1}{x^6}\right)\cdot\nabla_x = \frac{6D}{\mu x^7}\frac{\partial}{\partial x}, \tag{15}$$

where D is obtained from C_6 by squaring the denominators in Eq. (12). In a similar way an energy-dependent term in the effective potential can be derived and the scattering equation with the complete "vdW" potential up to order x^{-8} is

$$-\frac{1}{\mu}\left(\nabla^2 + k^2\right)F(\vec{x}) + \left(-\frac{C_6}{x^6} + \frac{6D}{\mu x^7}\frac{\partial}{\partial x} - \frac{C_8}{x^8} + \frac{48Gk^2}{\mu^2 x^8}\right)F(\vec{x}) = 0. \tag{16}$$

The derivative term in Eq. (16) can be eliminated by making the transformation to a new scattering function: $F(\vec{x}) = \exp(-D/2x^6)\, G(\vec{x})$, where the phase shift is unchanged since $G = F$ for large values of x. To order x^{-8} the only effect of this transformation is to modify the derivative term in Eq. (16) which becomes:

$$-\frac{1}{\mu}\left(\nabla^2 + k^2\right)G(\vec{x}) - \frac{C_6}{x^6}G(\vec{x}) + \frac{\left(-C_8 + \frac{15D}{\mu} + \frac{48Gk^2}{\mu^2}\right)}{x^8}G(\vec{x}) = 0. \quad (17)$$

Thus it turns out that there are additional terms in the long-range potential which depend on the reduced-mass of the pair of atoms interacting, and if at least one of them is Ps the new terms will not be negligible. To evaluate the coefficients appearing in (17) a good method is the pseudo-state expansion technique,[25,30] which converges surprisingly fast. Using only ten terms in the expansion, it is possible to get convergence to about ten significant figures. The final numerical form of the vdW potential is the following:

$$U_{Ps-H}(x) = -\frac{69.6702}{x^6} + \frac{(503.626k^2 - 237.384)}{x^8},$$

$$U_{Ps-Ps}(x) = -\frac{415.938}{x^6} + \frac{(7102.68 + 26766.4k^2)}{x^8}$$

It may be interesting to find the range in x over which these asymptotic expansions are valid and useful, where the second term is much smaller than the first. For the Ps-H case $x \gg 1.84$ and for Ps-Ps $x \gg 4.13$ are the ranges of validity at $k^2 = 0$. It is clear then that omitting the vdW force gives an incomplete description of the scattering, while including only the leading term is also not consistent. Other new developments include an analysis of the scattering of excited-state Ps where the degeneracy of the levels introduce novel terms in the scattering amplitude.[31]

THE POSITRONIUM ION AND MOLECULE

Recently the Ps$^-$ ion, consisting of two electrons in the singlet spin state and one positron, has become quite popular, largely due to the elegant experiments of Mills. He succeeded in producing the ion[32] and subsequently measured its annihilation lifetime.[33] This encouraged theorists to apply modern computing techniques to evaluate a range of properties of this formerly exotic system. Let me briefly review the theoretical situation.

The electron affinity of this ion has been calculated repeatedly with gradually increasing accuracy. The best result[34] uses a Hylleraas variational trial function with 220 terms, giving an affinity of 0.024010113 Ry (0.3266769 eV) along with an annihilation lifetime of 0.47936 ns. This lifetime includes corrections of order α and agrees well with the experimental value $0.478 \pm .02$ ns, although the experimental precision does not yet permit a critical test.[33] Mills[35] noticed, by analogy with the H$^-$ ion,[36] that there might be a "non-relativistically bound" $^3P^e$ Ps$^-$ state, which would be in effect an excited bound state. This state is particularly interesting, since it would be quite long-lived against annihilation because all pairs of particles are in relative p-states. He found, however, that no such state was bound in a moderately extensive variational calculation; that conclusion was supported by a more complete calculation[34] which found moreover that systems of two negative and one positive particle possess a $^3P^e$ state if the mass of the positive particle is greater than 16.1 m$_-$ or less than 0.4047 m$_-$. This range includes such interesting systems as H$^-$, Mu$^-$, H$_2^+$ and the muonic hydrogen molecular ions (pμp, dμd, tμt) but excludes Ps$^-$. Recently[37] these states were re-examined using the popular hyperspherical coordinates; this analysis gives some insight into the qualitative behavior of the wave functions and agrees with the previous results. More recently the effect of the mass-polarization term has been reported[38] for a suite of three-body Coulomb systems, again giving some additional feeling for the effects of correlation but no new quantitative results. Ho has investigated the existence of doubly-excited auto-detaching states of Ps$^-$, and found some in analogy with the well-known resonances in e$^-$ - H scattering.[39]

Mills also suggested[32] that Ps$^-$ could be used to generate Ps beams of controlled energy; this would involve acceleration of Ps$^-$ ions and photo-detachment of one electron. This suggestion resulted in several calculations of the photodetachment cross-section. Bhatia and I[40] made several convenient approximations to simplify things considerably; in particular the function describing the final state (e$^-$ + Ps) was assumed to be a plane wave. Ward et al[41] undertook to calculate the final-state wave function more exactly, and used the Kohn variational principle with a well-correlated trial function. Although the photodetachment cross-sections obtained by the two methods are quite similar, the scattering parameters (including those of the triplet spin state) should be useful in future direct experiments, and their astrophysical application will be discussed below. For this application I note the zero-energy results: $a(1) = 12.38 \pm .07\ a_0$ and $a(3) = 5.0 \pm .2\ a_0$, which leads to the cross-sections $\sigma(\text{tot}) = 228 \pm 6\ \pi a_0^2$ and $\sigma(3 \rightarrow 1) = 13.6 \pm .8\ \pi a_0^2$. A certain consistency check can be made: From the electron affinity of Ps$^-$ and the asymptotic normalization of the bound-state wave function it is possible to estimate the singlet scattering length. In this way we[40] found the value $a(1) = 12.233 \pm .006\ a_0$, to be compared with the direct variational result.

The most difficult "simple" system to deal with experimentally is the Ps$_2$ molecule since it requires two positrons to appear simultaneously within a very small distance. The bound state has been investigated theoretically for many years, with somewhat contradictory results. After the initial proof[42] that the molecule is actually bound (by at least 0.1 eV) a series of calculations of increased sophistication were performed. Unexpectedly, these did not lead to gradually increasing lower bounds on the dissociation energy as might have been expected. Instead, high results[43] (0.948 eV and 0.846 eV) have been interspersed with lower one[44] (0.186 eV, 0.197 eV, and 0.221 eV), and it is hard to tell whether there are actual errors in some of the calculations or whether the apparently better results are due to better coordinate systems or trial functions. The most recent apparently reliable and consistent results[45] are one by Lee, using a Monta Carlo technique and giving $E(\text{diss}) = 0.412 \pm .007$ eV, and one by Ho, using a variational method and giving $E(\text{diss}) = 0.411$ eV. The first of these methods is often very efficient but has an inevitable statistical error. In the second calculation, Ho adapted his PsH wave function to the Ps$_2$ problem by simply changing the mass of the proton without specifically enforcing the symmetry of the (now) identical positive particles. The levels obtained in this way include both symmetric and antisymmetric types; the lowest-lying level is symmetric. This technique still gives a lower bound on the dissociation energy, but it requires about twice as many terms as would a properly symmetrized function. So the problem of Ps$_2$ binding is nearly solved but is not quite definitive. Note that no estimate of the effect of the long-range forces has been attempted. Using the same type of wave function in a complex-rotation calculation Ho[46] obtained the first evidence for a resonance in Ps-Ps scattering; the collision energy is 4.65 eV (total energy is 0.6583 Ry) and the width is 0.11 eV. If this resonance is similar to those found in PsH then it represents the 2s member in a series of hydrogenic bound states of e$^+$ + Ps$^-$, whose limit lies at $-$ 0.524 Ry, the binding energy of Ps$^-$. In this case the real part of the quantum defect would be $-$ 0.366, some 60% of the value in PsH; one must be careful with the reduced mass. In the PsH case the quantum defect comes from the potential due to the charge density of the H^- ion, which is repulsive at short distances near the nucleus. In the Ps_2 case, which has no heavy central core, most of the effective repulsion must come from the exchange interaction between the two positrons.

ASTROPHYSICAL APPLICATIONS, EXOTICA AND CONCLUSIONS

In the analysis of positron slowing down and eventual annihilation in astrophysical context, it is necessary to include as a leading source of gamma rays the decay of Ps. Especially in the Sun, where annihilation radiation has been observed during flares, it would be a useful diagnostic of the properties of the annihilation region if one could measure the ratio R of three-photon to two-photon annihilation.[47] This quantity would determine the amount of Ps produced provided that annihilation of the Ps occurs before anything can disrupt the normal ortho-para ratio. In a region composed of neutral H, electrons, and protons several of the processes I discussed above are relevant. The most important reactions that can quench the longer-lived ortho-Ps are

(a) $^3Ps + H \rightarrow {}^1Ps + H,$

(b) $^3Ps + e^- \rightarrow {}^1Ps + e^-,$ and $\qquad\qquad$ (18)

(c) $^3Ps + p \rightarrow e^+ + H.$

The first two of these are spin-exchange reactions due to the symmetry of the scattering wave functions under interchange of two electrons, leading to a decrease of the lifetime of the Ps by a factor of about 1000, and increasing the two-photon component of the observed annihilation radiation. Reaction (c) simply destroys the Ps entirely, and since it is an exothermic reaction (since the binding energy of H is twice that of Ps) it can occur at all energies down to zero. (I am intentionally concentrating on the low energy region, below the inelastic threshold in Ps).

The idea is the following: If the rate of destruction of ^3Ps is comparable to the decay rate, then R will be reduced below its statistical value of 3. That is,

$$N_o = 1/v\sigma\tau, \text{ where } \tau = 1.4 \times 10^{-7} \text{ sec.} \qquad (19)$$

Clearly, for reactions (a) and (b) the critical density N_o increases without limit as the energy decreases, and so for low enough temperature these quenching reactions are negligible. For higher energies the ortho-para cross-sections discussed above are important; it appears that the cross-section for reaction (b) is significantly larger than (a), but in a weakly ionized region there may not be many electrons available.

More is known about reaction (c), since it is simply related to the Ps-formation reaction which has been quite thoroughly investigated, especially by Humberston.[48] In this case the cross-section goes like 1/v for low energy, and the critical density remains finite. I carried out a thermal average over an analytically-fitted form for the cross-section. The critical density N_o reaches 8×10^{15} cm^{-3} at zero temperature, falls rapidly to 10^{15} cm^{-3} at about 5000 K and very slowly decreases to 3.6×10^{14} cm^{-3} at 5×10^4 K. At this temperature other inelastic processes are not quite negligible.

Recently, reaction (c) has been used in a very clever way to make a quantitative proposal of the most exotic sort.[49] With the advent of relatively intense beams of both Ps and anti-protons ($\bar{p}$) it may be possible to produce anti-hydrogen atoms. These could be used to test the TCP theorem and to carry out experiments involving collisions between $\bar{H}$ and H, leading to rearrangement into protonium and Ps atoms,[50] followed by annihilation of both atoms. This reaction is of interest in the baryon-symmetric cosmology,[51] where it controls the annihilation at the boundary between huge clouds of matter and anti-matter. Even more speculative is the often-suggested possibility of using annihilation to power and propel interstellar spacecraft. In any case, it was noticed that reaction (c) above can be converted by charge conjugation to the following form:

$$(\bar{c}) \quad {}^3Ps + \bar{p} \rightarrow e^- + \bar{H} , \tag{20}$$

where every particle is replaced by its antiparticle; Ps is its own antiparticle. In this way, the cross-section for $(\bar{c})$ was obtained without additional work, and it is asserted that the cross-section is large enough to make the experiment seem feasible.

It seems to me most appropriate to conclude on that note of optimism, since the whole field of positron physics has lived for years on optimism. Each outrageous theoretical suggestion has been matched by marvelously clever and persistent experimental implementation, and there is no reason why that exciting sequence should be interrupted now.

REFERENCES

1. P.A. Fraser, Proc. Phys. Soc. 79:721 (1961); J. Phys. B 1:1006 (1968).
2. P.A. Fraser and M. Kraidy, Proc. Phys. Soc. 89:533 (1966).
3. M.I. Barker and B.H. Bransden, J. Phys. B. 1:1109 (1968); 2:730 (1969).
4. B.G. Duff and F.F. Heymann, Proc. Roy. Soc. A 270:517 (1962).
5. R.J. Drachman and S.K. Houston, J. Phys. B 3:1657 (1970).
6. K.F. Canter, T. Horsky, P.H. Lippel, W.S. Crane and A.P. Mills, Jr., in "Positron (Electron)-Gas Scattering," W.E. Kauppila, T.S. Stein and J.M. Wadehra, eds. World Scientific, Singapore (1986).
7. C.V. Briscoe, S.I. Choi and A.T. Stewart, Phys. Rev. Lett. 20:493 (1968).
8. K.F. Canter, J.D. McNutt and L.O. Roellig, Phys. Rev. A 12:375 (1975).
9. J.F. Hernandez, Phys. Rev. A 14:1579 (1976); R.M. Nieminen, I. Valimaa, M. Manninen and P. Hautojarvi, Phys. Rev. A 21:1677 (1980); C.D. Beling and F.A. Smith, Chem. Phys. 49:417 (1980).
10. M.H. Coopersmith, Phys. Rev. 139:A1359 (1965).
11. K. Rytsola, J. Vettenranta, and P. Hautojarvi, J. Phys. B. 17:3359 (1984).
12. D.M. Spektor and D.A.L. Paul, Can. J. Phys. 53:13 (1975).
13. P.A. Fraser, Proc. Phys. Soc. 78:329 (1961), S. Hara and P.A. Fraser J. Phys. B 8:L472 (1975).
14. R.J. Drachman and S.K. Houston, Phys. Rev. A 12:885 (1975).
15. R.J. Drachman and S.K. Houston, Phys. Rev. A 14:894 (1976).
16. B.A. P. Page, J. Phys. B 9:1111 (1976).
17. P.J. Karol and R.L. Klobuchar, unpublished (1975).
18. Y.K. Ho, Phys. Rev. A 34:609 (1986).
19. Y.K. Ho, Phys. Rev. A 17:1675 (1978).
20. H. Feshbach, Ann. Phys. (N.Y.) 19:287 (1962).
21. M.J. Seaton, in: "Atomic Scattering Theory", J. Nuttall, ed., Univ. of Western Ontario, London (1978).
22. K.B. Choudhury, A. Mukherjee and D.P. Sural, Phys. Rev. A 33:2358 (1986).
23. M.D. Girardeau, Phys. Rev. A 26:217 (1982); J.C. Straton, Unpublished (1987).
24. R.J. Drachman, Can. J. Phys. 60:494 (1982); J.W. Humberston, Adv. At. Mol. Phys. 22:1 (1986).
25. D.W. Marton and P.A. Fraser, J. Phys. B 13:3383 (1980).
26. J.R. Manson and R.H. Ritchie, Phys. Rev. Lett. 54:785 (1985).
27. C.K. Au and R.J. Drachman, Phys. Rev. Lett. 56:324 (1986).

28. R.J. Drachman, J. Phys. B 12:L699 (1979).

29. A. Temkin, Phys. Rev. 107:1004 (1957).

30. R.E. Johnson, S.T. Epstein and W.J. Meath, J. Chem Phys. 45:1271 (1967); A. Dalgarno and S.T. Epstein, J. Chem. Phys. 50:2837 (1969); J.F. Bukta and W.J. Meath, Molec. Phys. 27:1235 (1974).

31. G. Feinberg and J. Sucher, Phys. Rev. A 36:40 (1987).

32. A.P. Mills, Jr., Phys. Rev. Lett. 46:717 (1981).

33. A.P. Mills, Jr., Phys. Rev. Lett. 50:671 (1983).

34. A.K. Bhatia and R.J. Drachman, Phys. Rev. A 28:2523 (1983).

35. A.P. Mills, Jr., Phys. Rev. A 24:3242 (1981).

36. A.K. Bhatia, Phys. Rev. A 2:1667 (1970); G.W.F. Drake, Phys. Rev. Lett. 24:126 (1970).

37. J. Botero and C.H. Greene, Phys. Rev. Lett. 56:1366 (1986).

38. A.K. Bhatia and R.J. Drachman, Phys. Rev. A 35:4051 (1987).

39. Y.K. Ho, Phys. Rev. A 19:2347 (1979); Phys. Lett. 102A:348 (1984).

40. A.K. Bhatia and R.J. Drachman, Phys. Rev. A 32:3745 (1985).

41. S.J. Ward, J.W. Humberston and M.R.C. McDowell, J. Phys. B 18:L525 (1985); 20:127 (1987).

42. E.A. Hylleraas and A. Ore, Phys. Rev. 71:493 (1947).

43. R.R. Sharma, Phys. Rev. 171:36 (1968); W.-T. Huang, Phys. Stat. Sol. (b)60:309 (1973).

44. O. Akimoto and E. Hanamura, J. Phys. Soc. (Japan) 33:1537 (1972); W.F. Brinkman, T.M. Rice and B. Bell, Phys. Rev. B 8:1570 (1973); F.R. Vukajlovic and S.I. Vinitsky, Phys. Lett. 118A:185 (1986).

45. Y.K. Ho, Phys. Rev. A 33:3584 (1986) and references therein.

46. Y.K. Ho, Phys. Rev. A 34:1768 (1986).

47. C.J. Crannell, G. Joyce, R. Ramaty and C. Werntz, Astrophys. J. 210:582 (1976).

48. R.J. Drachman, K. Omidvar and J.H. McGuire, Phys. Rev. A 14:100 (1976); J.R. Winick and W.P. Reinhardt, Phys. Rev. A 18:925 (1978); C.J. Brown and J.W. Humberston, J. Phys. B 17:L423 (1984); 18:L401 (1985).

49. J.W. Humberston, M. Charlton, F.M. Jacobsen and B.I. Deutch, J. Phys. B 20:L25 (1987).

50. D.L. Morgan, Jr., Phys. Rev. A 7:1811 (1973); W. Kolos, D.L. Morgan, Jr., D.M. Schrader and L. Wolniewicz, Phys. Rev. A 11:1792 (1975).

51. F.W. Stecker, Nature 273:493 (1978).

ELECTRON CAPTURE FROM SOLIDS BY POSITRONS

R. H. Howell

Lawrence Livermore National Laboratory
Livermore, CA 94550

Since the atomic system of positronium is intrinsically interesting and the electron pickup process is analogous to electron capture by other charged species, the study of positronium formation in gasses was one of the first areas of positron research . It was recognized later that positronium was formed in the interaction of positrons with solids. Electron capture by positrons from solids results in the formation of positronium in both the ground and excited states. Positronium has been observed in insulators in the solid matrix and in the vacuum emitted from metal and semiconductor surfaces.

The capture of electrons in solids is modified from that in gasses by several factors. The most important is the collective interaction of the electrons which results in a density of electron states in the solid in wide bands. Also the high density of electrons in many solids gives a high frequency of interaction as compared to gasses, and quickly destroys any electron-positron states in the metal matrix. Consequently, most positrons implanted in a metal will rapidly thermalize, and unless they reach the surface will annihilate with an electron in an uncorrelated state. Positronium formation from positrons scattered at a metal surface is analogous to ion neutralization,[1] however, most of the positronium comes from positrons passing through the surface from the bulk. The dominant motivation for studying positronium formation has been the hope that the distribution of the electrons at the surface would be obtained through the annihilation properties of positrons trapped at the surface or through analysis of the energy and angular distributions of the positronium emitted into the vacuum.

There are three mechanisms known to emit positronium into the vacuum from metal surfaces, and they are differentiated by the state of the positron at the surface. A thermalized positron may elastically pass through the surface or trap in a surface state. Positrons may also traverse the surface with epithermal energies. Those positrons that pass directly through the surface can form positronium by capturing an electron from the filled electron bands. The energy of the positronium will then be determined by the electron and positron work functions and the energy of the electron relative to the Fermi level. If the positrons have epithermal energies, then that is added to the positronium energy as well.

Like the ejection of photoelectrons, positronium formation by positrons passing through the surface conserves energy and total momentum in the direction parallel to the surface. This results in distinctive energy and angular distributions for the positronium emitted into thee vacuum. There are two measurements that can be used to measure these properties of positronium. They are two-dimensional angular correlation of annihilation radiation, ACAR, and positronium velocity measured by time-of-flight methods, PS-TOF.

Angular correlation of annihilation radiation measurements, ACAR, provide a direct measurement of the momentum of the annihilating electron-positron pair, and can be used to describe electron densities and defect properties in bulk materials. ACAR measurements for positrons trapped at the surface and for positronium emitted into the vacuum have been performed[2-3] with position sensitive detectors that allow the angular deviation to be measured in two dimensions. These axes for the momenta in these measurements have been chosen with one parallel to the surface of the sample and one perpendicular to that surface. From the first experiments we learned that the surface trap potential well results in positron-electron angular correlations that are very similar to those measured for trapped positrons in the bulk. The distributions were nearly isotropic. This was in contradiction to the predictions of models of the surface trap based on perfect surfaces that predicted larger angular deviations for the direction perpendicular to the sample surface. At present, the resolution of this contradiction is incomplete, but trapping at surface defects offers a plausible alternative.[4]

Angular correlation measurements were the first demonstration that the positronium formed at a metal surface obeyed the simple kinetic rules described above. If we assume that the pickup process removes an electron from a free electron gas, then we can make a qualitative description of the momentum distribution of the positronium. Good qualitative agreement was obtained for both Cu and Al. Later data on different surfaces of carefully cleaned Al showed deviations from free electron behavior that can be related to the band filling at the surface of the samples.[5] A complete theoretical description of these measurements is now under development.[6-7]

More recently, the linac beam at Livermore[8] has been used to perform measurements of positronium velocity distributions by measuring the time-of-flight of positronium in a vacuum. The positronium time-of-flight was determined by measuring the time difference between the positron beam arrival and radiation from the spontaneous annihilation of long lived, triplet positronium in the active volume of a highly collimated detector placed a known distance from the sample. The TOF detector collimator excludes most annihilation radiation from outside a planar volume 10.5 +/ 0.35 cm from the sample collimator. Details of the apparatus can be seen below in Figure 1.

Experiments were performed in an ultrahigh vacuum chamber having 1×10^{-10} torr base pressure on samples that were sputter cleaned, annealed and Auger analyzed. The temperature of a sample could be set during the experiment from room temperature to 900°C. Positronium time-of-flight and lifetime spectra were obtained for incident positron energies from 0 to 3 keV set by the bias of the sample to the transport energy. The running time and total linac electron beam charge were recorded for each run.

Many of the established facts have been determined using positronium time-of-flight techniques. The positronium formed with positrons desorbed from the surface trap was shown to be in a thermal distribution using Ps-TOF and a bunched radioactive beam.[9-10] The demonstration

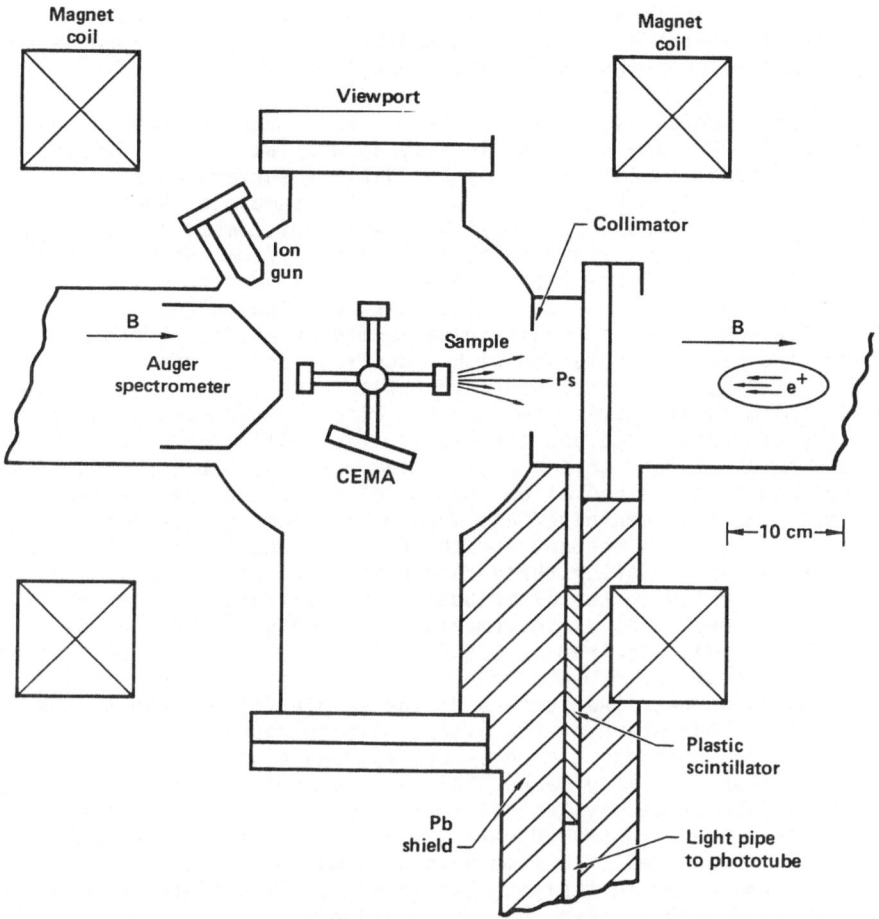

Fig.1 Schematic of the experimental chamber and time-of-flight detection
system.

that energy was conserved so that there is a work function in positronium
production, and the subsequent supposition that positronium is formed by
a positron that picks up an electron and leaves behind a single electron
hole, also resulted from measurements with that beam.[11]

With Ps-FOF measurements, using the linac beam, we can obtain high
statistics spectra very quickly so that we can take good quality spectra
for many sample or beam conditions. Using this tool, we have studied
single-crystalline samples of several metals, and have confirmed that
positronium emission by a work function process is a general property of
metals.[12] This allows the determination of positive values for
positron work functions that can not be measured from observation of
emitted positrons.[13] We have seen that the work function value is
insensitive to surface conditions, but depends sensitively on the
temperature of the sample.[14]

We have measured the energy distribution of positronium for energies
between zero and the work function to test the relationship between the
positronium and electron energy distributions. With Ps-TOF measurements,
we have observed energetic positronium resulting from electron pickup by

non-thermalized positrons that have scattered out of the metal samples.[16] Energetic positronium formed by transmission through a very thin carbon foil was reported in reference.[17]

The experimental evidence for these effects can be seen in the spectra in Fig. 2. The spectrum in Fig. 2a was taken with 2000 eV positrons incident on a clean, annealed Ni(100) surface at room temperature, and Fig. 2b was taken on the same sample at 900°C. The hot sample produced more positronium with slow flight times, i.e., low energies due to the contribution to the spectrum from thermal desorption, and the flight time of the fastest positronium is noticeably longer due to the thermally-induced shift in the work function value. In fig. 2c, the contribution from fast positronium formed by backscattered positrons dominates the spectrum obtained with a positron energy of 50 eV. The data in Fig. 2 have been corrected to reflect the velocity distribution of positronium as it leaves the sample. First, a time-independent background was subtracted and the result was multiplied by $e^{(t/142ns)}$ for the decay in flight of the positronium. A second background, due to scattered annihilation radiation from triplet positronium, was then subtracted, and the remaining data were multiplied by a factor proportional to t^{-1} to account for the velocity dependent part of the detector efficiency. After these corrections, the spectrum in Fig. 2 corresponds directly to the distribution of positronium as it leaves the sample when binned in equal increments of velocity.$^{-1}$ The narrow peak at the left of this spectrum arises from the detection of scattered gamma rays produced from the prompt annihilation of positrons and singlet positronium at or near the surface of the sample. The shape of the prompt peak is indicative of the overall timing resolution of the production, transport and timing circuitry, and also serves as a useful zero time marker. In many cases, this is the best representation to use to view the data as the system resolution is nearly constant as a function of time-of-flight. However, when the positronium energy distribution is desired, the time axis data may be converted into bins per unit of energy by multiplying the intensities by t^3 and plotting on an energy scale such as seen in Figs. 3 and 4.

One potential application of positronium measurements is to determine the electron density of states at the metal surface. This will be possible if a single electron hole is left behind in the metal. We have measured positronium energy distributions for several metals that are thought to have different electron density of states near the surface. The data and a model fit based on a free electron density are shown in Fig. 3. The strength and work function values were the only parameters adjusted in the model. This simple model describes the positronium energy distribution near the work function energy well, but fails to describe the spectra at lower positronium energies where more positronium is measured than predicted. The electron band structure would lead us to expect larger differences in the spectra than were observed. Near the surface, copper and aluminum are free electron like while nickel has sharp structure. The structure in the nickel bands would be seen in the data if it were transferred directly to the positronium.[12]

The energy of the most energetic positronium is a linear function of the sample temperature for all of the samples observed to date.[14] The slope of the work function-temperature line is similar for all of the samples ranging from 5 to 9 meV/°K. This is remarkable consistency, and suggests that there is a simple relationship between the the temperature-induced changes in the positronium work function values and the basic structure of the metal.

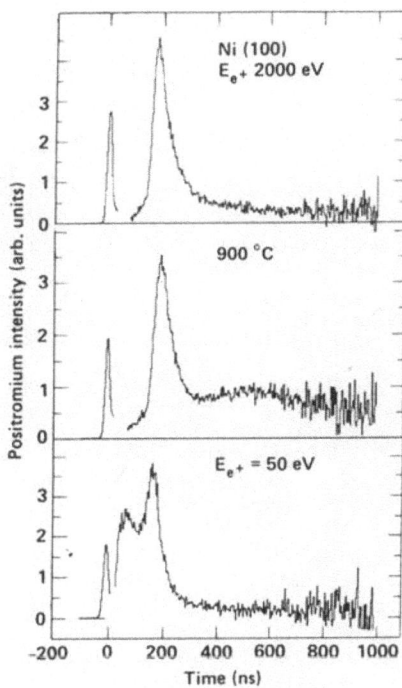

Fig. 2 Time-of-flight
spectra taken on a
clean, well annealed
Ni sample for initial
positron energies of
2000 and 50 eV. In
the spectrum at
900°C, the
contribution from
thermal positronium
is seen at long
times. At lower
initial positron
energies, the
contribution to
positronium formation
from backscattered
positrons becomes
large.

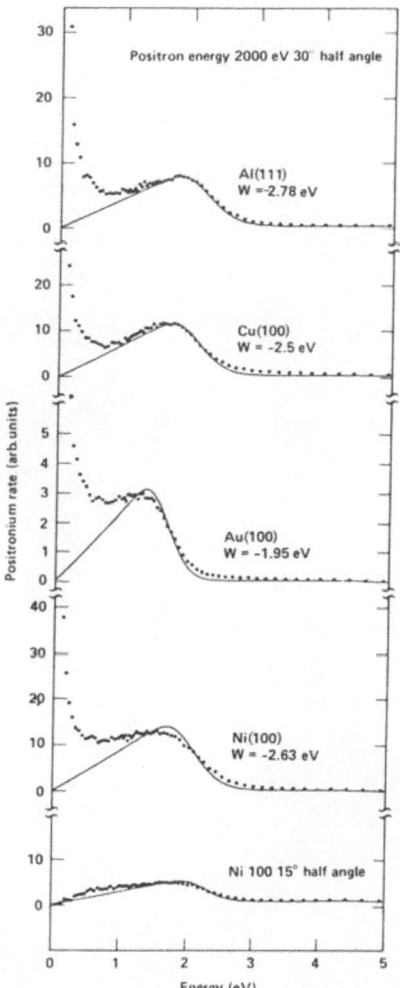

Fig. 3 Energy distributions
of positronium
emitted from clean
metal samples. The
solid curves are
modeled on a free
electron gas density
of states.

At low initial positron energies there is a large flux of
non-thermalized positrons back scattering out of the sample.[15] Many of
these pick up an electron and form positronium with energies higher than
the work function. Positronium formed in this manner will account for
over half of the total amount at low initial positron energies, and is
non-negligible, even at incident energies of three keV. The energy
spectra of positronium formed from bombardment of lead is seen for
several incident positron energies in Fig. 4. The positronium formed by
work function emission is seen as a peak below one eV, and the
positronium formed by pickup of non-thermal positronium is seen at higher
energies. There is more higher energy positronium at low incident
positron energies since there are more positrons backscattered from the
sample at energies that efficiently form positronium. The positronium

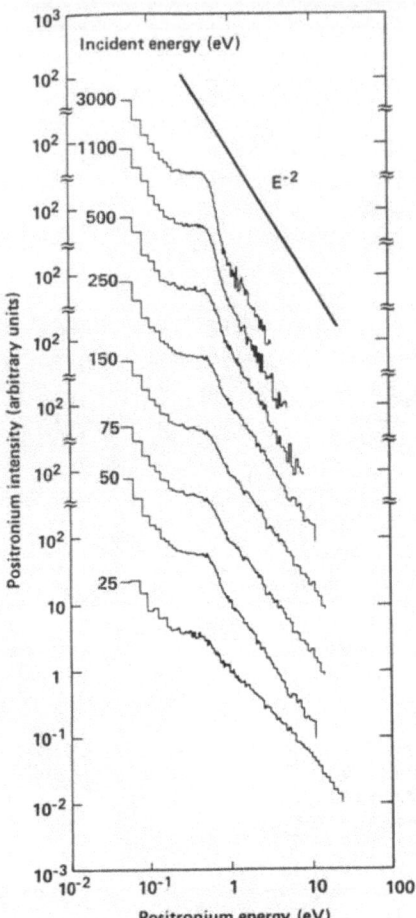

Fig. 4   Positronium energy
distributions taken
on a clean, well
annealed Pb sample
for initial positron
energies of 3000 to
25 eV showing the
E^{-2} variation in
high energy
positronium formation.

intensity appears to depend on the positronium energy by a power law
relationship near. E^{-2} The exact power of energy differs for different
positron beam energies, but an exponent of –2 is a close value for all
cases. We could obtain the value of the pickup matrix element if these
data were combined with careful calculations and measurements of the
epithermal positron emission with matching momentum limits.

The energy and sample dependence of the total intensity of
positronium formed by backscattered positrons can be seen in Fig. 5. All
samples have less high energy positronium formed at high incident
positron energies, but there are sample-to-sample variations. Epithermal
positrons produce more positronium from aluminum at low incident energies
and gold produces more at high incident energies.

Due to the general similarities in the scattering functions, the number of positrons that backscatter at some incident positron energy will be closely related to the number of electrons that backscatter at the same incident energy. Since there are extensive data available for the total backscattering fraction, B, for electrons we have developed a simple model for scaling the positronium formation due to epithermal, backscattered positrons to the total electron backscattering coefficient. Positronium is only efficiently formed by positrons that have relatively low energies compared to the incident positron energy and the energy spectrum of backscattered positrons or electrons spans the whole range from the incident energy down to thermalization. This suggests that the simplest scaling of positronium production to the backscatter coefficient is obtained by taking the fraction of positrons backscattering with low energy to the total. For simple energy distributions of backscattered positrons, this fraction can be approximated by dE/E for a fixed, small value of dE. This results in an E^{-1} dependence of the positronium intensity with incident positron energy if the backscatter coefficient is slowly varying. The positronium produced from backscattered positrons would then be proportional to BdE/E measured for electrons. Positronium intensities calculated using this model normalized at one point are shown in the inset for Fig. 5.

There is a remarkable correspondence between the results of the model scaled to electron data and our measurements. The good performance of this model indicates that the positron pickup matrix element does not vary much from sample-to-sample, and that the variations in the positronium production are the result of differences in the flux of epithermal positrons. Also the ratio of positrons emitted bare to those forming positronium is nearly constant for all positrons leaving the sample. This suggests that the pickup matrix element is slowly varying with positron energy, and that the energy distribution of the epithermal positrons sets the positronium energy distribution.

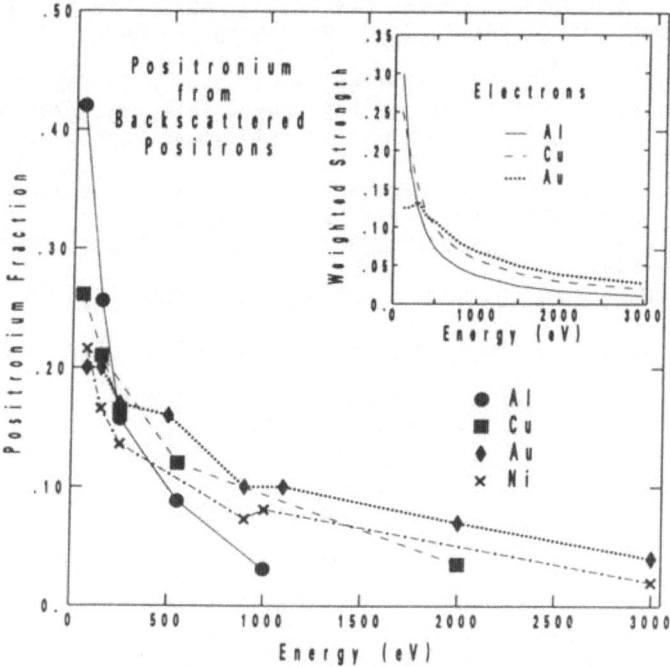

Fig. 5 Total strength of the high energy
positronium distribution for several
samples.

The model will be an assistance to those researchers that wish to use positronium fraction measurements to indicate positron diffusion to the sample surface in diffusion or defect profiling studies. Failure to account for the contribution from backscattered positrons can lead to large differences in the parameters derived from the data. Using available electron backscattering coefficients positronium fraction values can now be corrected to remove the contribution from non-thermal positrons at the surface of many samples.

The help of M. Connor and L. Bernardez in collecting the data, A. Coombs in developing the instrumentation and J. Kimbrough in data acquisition software and hardware development, is gratefully acknowledged as is the contributions of my several collaborators who are coauthors of the descriptions of the work described above. This work was supported by the U.S. DOE, under contract number W-7405-ENG-48, LLNL.

REFERENCES

1. D. W. Gidley, R. Mayer, W. E. Frieze, K. G. Lynn, Phys. Rev. Lett. 58, 595 (1987).
2. K. G. Lynn, A. P. Mills, Jr., R. N. West, S. Berko, K. F. Canter, L. O. Roellig, Phys. Rev. Lett. 54, 1702 (1985).
3. R. H. Howell, P. Meyer, I. J. Rosenberg, M. J. Fluss, Phys. Rev. Lett. 54, 1698 (1985).
4. A. R. Koymen, D. W. Gidley, T. W. Capehart, Phys. Rev. B 35, 1034 (1987).
5. D. M. Chen, S. Berko, K. F. Canter, K. G. Lynn, A. P. Mills, Jr., L. O.. Roellig, P. Sferlazzo, M. Weinert, R. N. West, Phys. Rev. Lett. 58, 921 (1987).
6. A. B. Walker and R. M. Nieminen, J. Phys. F 16, L295 (1986).
7. A. Ishii and S. Shindo, Phys. Rev. B 35, 6521 (1987).
8. R. H. Howell, M. J. Fluss, I. J. Rosenberg, P. Meyer, Nucl. Inst. Meth. B10, 373 (1985).
9. A. P. mills, Jr., and L. Pfeiffer, Phys. Rev. B 32, 53 (1985).
10. A. P. Mills, Jr., L. Pfeiffer, Phys. Rev. Lett. 26, 1961 (1979).
11. A. P. Mills, Jr., L. Pfeiffer, P. M. Platzman, Phys. Rev. Lett. 51, 1085 (1983).
12. R. H. Howell, I. J. Rosenberg, M. J. Fluss, R. E. Goldberg and R. B. Laughlin, Phys. Rev. B 35, 5303 (1987).
13. R. H. Howell, I. J. Rosenberg, M. J. Fluss, P. Meyer, Phys. Rev. B 35, 4555 (1987).
14. I. J. Rosenberg, R. H. Howell, M. J. Fluss, Phys. Rev. B 35, 2083 (1987).
15. R. H. Howell, I. J. Rosenberg, M. J. Fluss, Phys. Rev. B 34, 3069 (1986).
16. A. P. Mills, Jr., and W. S. Crane, Phys. Rev. A 31, 593 (1985).
17. H. J. Fitting, Phys. Stat. Sol. a26, 525 (1974).

ON THE PRODUCTION OF A TIMED POSITRONIUM BEAM BY POSITRON-GAS SCATTERING

G. Laricchia, S.A. Davies, M. Charlton and T.C. Griffith

Department of Physics and Astronomy

University College London, London WC1E 6BT, U.K.

ABSTRACT

In this communication we describe the technique we are currently developing for the purpose of measuring the energy distribution of positronium formed by positron-gas scattering. Preliminary results are presented which demonstrate the tunability of our Ps beam.

INTRODUCTION

The positronium atom (Ps) is of interest as an object of, and as a probe for, investigations. The recent progress in the development of intense positron (e^+) beams have rendered realistic a number of experiments which employ Ps for studies spanning from fundamental atomic processes to surface structures. Ps is believed to possess definite advantages over other surface structure probes such as e^-, e^+ and He atoms[1] and the first low energy Ps diffraction (LEPSD) experiments will probably be accomplished in the near future[2].

Regarding the interaction of Ps with neutral or charged particles, a number of theoretical predictions await experimental tests: positronic complexes with atoms and negative ions[3], resonances[4], total and partial cross sections for Ps-atom scattering (Ps-H[5], Ps-He[6]), Ps-$e^\pm$ scattering[7]. Ps collision studies would facilitate the understanding of e^+ slowing down phenomena[8] and, perhaps even, of the discrepancy existing between the different measurements of the Ps formation cross sections[9].

For investigations of atomic processes, Ps may be used either as a projectile or as a target. In the latter case, Ps could be produced by e^+-solid interactions where high e^+-Ps conversion efficiencies have been obtained[10,11] with Ps energies extending down to thermal energies[11] and depending on the incident e^+ energy[12] and on the converter material, temperature and surface condition[11]. Ps densities of $O(20 \text{ cm}^{-2})$ are believed to be achievable[13,14] by the use of a 0.5 Ci ^{22}Na source in conjunction with a W(100) moderator and an Al (e^+-Ps) converter. Although demanding, this type of experimental technique might be the only viable one for the investigation of Ps scattering at thermal energies. At higher energies ($O(1 \text{ eV})$), it is feasible to use Ps as a projectile. For this purpose, the production of characterised Ps beams has become a new experimental challenge. The techniques which have been developed in the last couple of years employ either a solid or a gaseous target for the conversion of e^+ into Ps. A summary of the experimental progress in this field is given in Table 1, where it should be noticed that until now evidence for tunable Ps beam production has been confined to e^+-gas scattering.

Table 1. Experimental Progess in Field of Ps Beams

Method	Observation	Reference
Bombardment of 50Å thick C film by keV e^+	10-500 eV Ps Efficiency $\sim 5\times10^{-3}$ x detector solid angle	15
e^+ – He scattering	evidence for mono-energetic forward going p-Ps	16
e^+ – Ar scattering	evidence for forward going o-Ps. Efficiency $\sim$26% of all Ps within 7° cone.	17
e^+- Al(111),Cu(100), Ni(100), Au(100), TOF Technique	Energetic backscatter-formed Ps	12
e^+- Ar, He scattering	4% of scattered e^+ detected as o-Ps collimated in a 6° cone	19
e^+ – Ar scattering	10^{-4} of incident e^+ converted to o-Ps	25

The efficiency of the charge-exchange process

$$e^+ + A \rightarrow Ps + A^+$$

(A being the target atom/molecule) for producing collimated tunable Ps beams relies on the behaviour of the differential Ps formation cross section, $d\sigma_{Ps}/d\Omega$, which has been shown, theoretically, to become substantially peaked in the incident e^+ direction from a few eV above the Ps formation threshold[20,21]. An approximate comparison of expected fluxes with experimental results has been performed[19] and is shown in Figure 1. Here the variation of the fraction of positrons

$$F_{Ps}(\theta') = 2 \quad (\int_0^{\theta'} d\sigma_{Ps}/d\Omega \sin\theta \quad d\theta)/\sigma_T \qquad (1)$$

which are emitted as Ps within an angle $\theta' = 5^O$, 10° and 20° about the incident e^+ direction in e^+-He scattering, is plotted against the Ps energy. The latter was deduced assuming a single threshold picture. These results show that approximately 3 - 4% of the scattered positrons will be converted to Ps atoms confined within a cone of 5° about the incident e^+ direction.

If the gas density, ρ, is such that

$$e^{-\rho\sigma_T l} \simeq 1 - \rho\sigma_T l \qquad (2)$$

where σ_T is the e^+ total scattering cross section and l is the gas cell length, positron multiple scattering should be negligible. Under these conditions, the energy spread of the Ps beam will be determined, primarily by:- (i) the intrinsic e^+ beam energy width, (ii) the probability for Ps scattering, (iii) the possibility of Ps formation in an excited state and (iv) Ps formation simultaneous to another e⁺ energy loss process. Considering these factors in turn:-

(i) Typical e⁺ beam energy spreads are of O(1 - 3 eV).
     It has, however, been demonstrated[22] that they can be reduced by a factor of ∿ 50 by using Ni(111) or W(110) moderators cooled down to ∿ 20 K.

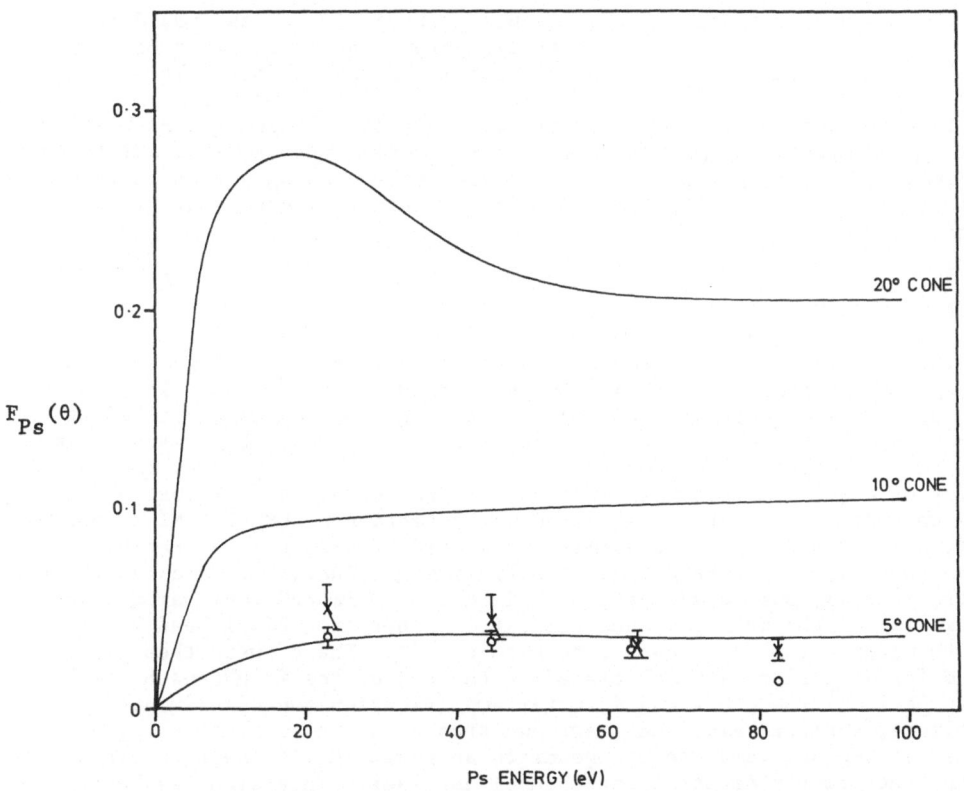

Fig. 1. Comparison between theory (solid curves) and experiment (points) for He gas of the variation of the fraction of positrons, $F_{Ps}(\theta)$, which are emitted as Ps in 5, 10 and 20° angular ranges about the incident e⁺ direction, for various Ps energies (Ref. 19).

(ii) Ps collisions with the target gas will result in beam degradation, i.e. wider angular and/or energy distribution(s). For this reason checks of the linearity between the Ps beam intensity and gas pressure should be performed.

(iii) Theoretically, the formation of Ps at high incident e⁺ energies in a state of principal quantum number, n, is expected to follow approximately the n^{-3} law, such that ≃ 12, 4 and 1.6% of Ps in the state n = 1 are expected to the n = 2, 3 and 4 states respectively. Experimentally, Ps formed in the 2P state by e⁺-gas scattering has been detected[23] with yields up to $\simeq$ 10% in Ar and H_2. The 2P state decays to the ground state with a 3.2 ns life-time and will be present in the Ps beam at $\simeq$ 6% level and with kinetic energies 5.1 eV below the Ps formed in the ground state if the dependence of the differential cross section with e^+ energy for the 2P state is similar to that for the 1S.

(iv) Recent results[24] on $e^+ - CO_2$ scattering have suggested that reactions such as

$$e^+ + CO_2 \rightarrow Ps + (CO_2^+)^*$$

might occur with an appreciable probability. These would result in Ps with kinetic energies E_{ex} below that of the main Ps beam (E_{ex} being the excitation threshold of the ion/molecule).

In view of the above considerations, characterization of a Ps beam is necessary if meaningful Ps scattering cross section measurements are to be performed. For this purpose, we are currently developing a time of flight (TOF) technique at UCL with the aim of determining the energy distribution of our Ps beam.

EXPERIMENTAL DETAILS

The experimental arrangement is schematically shown in Figure 2. A slow e^+ beam of approximately 4×10^4 s^{-1} produced by β^+ from a 16 m Ci ^{22}Na source was magnetically transported through an 8 mm hole drilled at the centre of a Channel Electron Multiplier Array (CEMA1), then made to impinge on a secondary moderator (M2). Both this, and the primary moderator, consist of a set of overlapping W meshes. The secondary e^- liberated at M2 by the incident e^+ are accelerated for detection by CEMA1. M2 is positively biassed and the 90% transmission W grid, G_1, is held at earth. The remoderated and transmitted beam ($\simeq$ 26%) is magnetically transported through a differentially pumped gas cell of length $1 = 20$ mm and containing 8 mm ϕ apertures at the beam entrance and exit. A baratron 220-1 head has been used to measure the gas pressure to within $\simeq$ 5%. The e^+ were then transported for detection by CEMA2 placed at the end of the flight path. The pulses derived from CEMA2 and 1, after amplification and discrimination, constitute, respectively, the start and stop of a conventional delayed coincidence system and were used to generate an inverted TOF spectrum via a time-to-amplitude converter and multi-channel analyser. Inversion has been employed because the count rates of the two detectors are, when detecting Ps, 1:2000 respectively. A timing resolution of 3 ns FWHM was measured by placing a ^{22}Na γ-ray source between the two CEMAs (their efficiency to γ-ray detection is known to be $O(10^{-2}-10^{-3})$). The FWHM of the peak for the remoderated e^+ was found to vary from 5 to 30 ns in the energy range 60 – 20 eV respectively. At energies above 60 eV the timing FWHM remained approximately constant. The combined remoderation and transport efficiency did not vary appreciably with incident e^+ energies between 100 – 500 eV, being only marginally better at 8% at $\simeq$ 425 eV. An additional peak in the TOF spectra has been found to correspond to, primarily, the transmitted beam via one of the secondary e^- released at CEMA2 by the e^+. The overall timing efficiency of the system for both primary and secondary e^+ was found to be $\simeq$ 13%.

To time o-Ps produced in the gas cell, both positively and negatively charged particles were repelled by means of two electrostatic plates (one at earth   and the other held at + 500 V) and a 90% transmission W grid, G_2, held at + 500 V placed in front of CEMA2. The acceptance angle subtended by CEMA2 to the centre of the cell was $\simeq$ 12°.

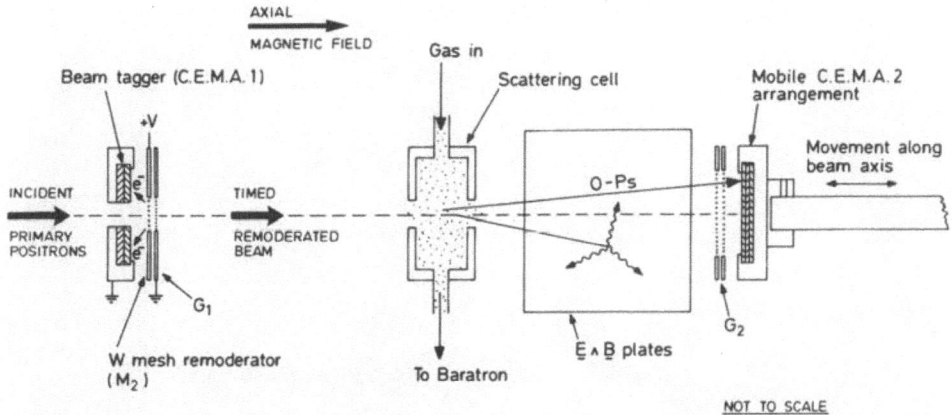

Fig. 2. Schematic illustration of the experimental apparatus.

RESULTS

At each energy investigated, alternate vacuum and gas spectra were collected. After appropriate background subtraction and normalization, the vacuum spectrum was subtracted from its corresponding gas spectrum, yielding a 'raw' o-Ps TOF spectrum. Examples for each type of spectrum are given in Figure 3. The spectra refer to an applied potential of 40 V to M2, the gas is Ar and the run time was $\simeq 10^5$ s. In the vacuum spectrum 4 peaks are distinguishable and are, at present, interpreted as follows: P_1 corresponds to positrons of sufficient energy to overcome the retarding field and are timed via an electron liberated from CEMA2 and detected by CEMA1; P_2 corresponds to t = 0 and is thought to comprise $e^--\gamma$, $\gamma-\gamma$ and e^--(fast) $e^\pm$ events; P_3 and P_4 correspond respectively to the TOF of the primary and remoderated positrons whose detection is provided either by a γ-ray or Ps produced when they strike a surface close to CEMA2. The position of t = 0 was checked by replacing M2 by a shutter and detecting in coincidence the secondary e^- at CEMA1 and the γ-rays, striking CEMA2, produced by the e^+ beam annihilating at the shutter. Figure 3b shows the gas spectrum where, superimposed on the four vacuum peaks, 2 additional ones are found. They both correspond to o-Ps formed in the gas cell by the remoderated positrons. The time delay between the two arises because P_6 is caused by e^+ which, after reflection at CEMA2 and M2, form Ps on the 3rd traversal of the gas region. P_6 is $\simeq$ 30% of P_5 reflecting the attenuation that the reflected beam suffers both at the grids (overall attenuation coefficient $\simeq$ 40%) and in the gas cell. Both peaks are approximately centred at ($<E_+>-E_{Ps}$) where $<E_+>$ is the mean e^+ energy and $E_{Ps} \simeq$ 9 eV is the Ps formation threshold in Ar. Further examples of o-Ps TOF spectra, illustrating the tunability of the beam are shown in Figure 4.

The Ps energy distribution can and will be extracted from the TOF spectra after allowance is made for the intrinsic timing resolution of the system. In addition, an apparant timing spread (of the order of few %) is introduced by the energy width of the remoderated positron beam. This has been found to vary with primary positron incident energy and to comprise a high energy tail corresponding to those positrons ejected from the remoderator prior to thermalisation.

The fraction of epithermal positrons has been found to decrease with increasing energy in agreement with other studies[25] being $\simeq$ 10% at 800 eV incident e^+ energy. Higher e^+ incident energies, possibly coupled to a thin Ni foil remoderator, will be used in future to reduce the energy width of the remoderated beam.

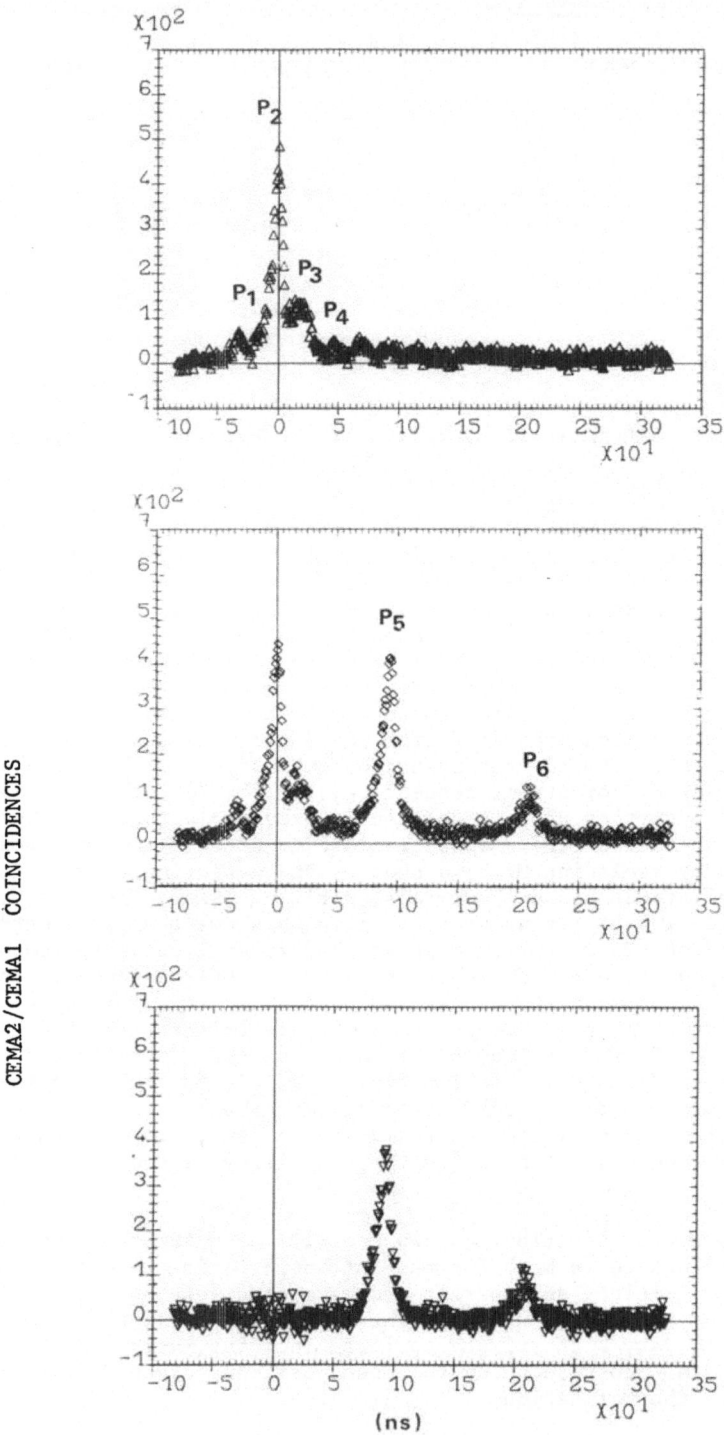

Fig. 3. Examples of (a) vacuum, (b) gas and (c) gas-vacuum spectra obtained in Ar with 40 V applied to M2. P_5 and P_6 are the o-Ps peaks.

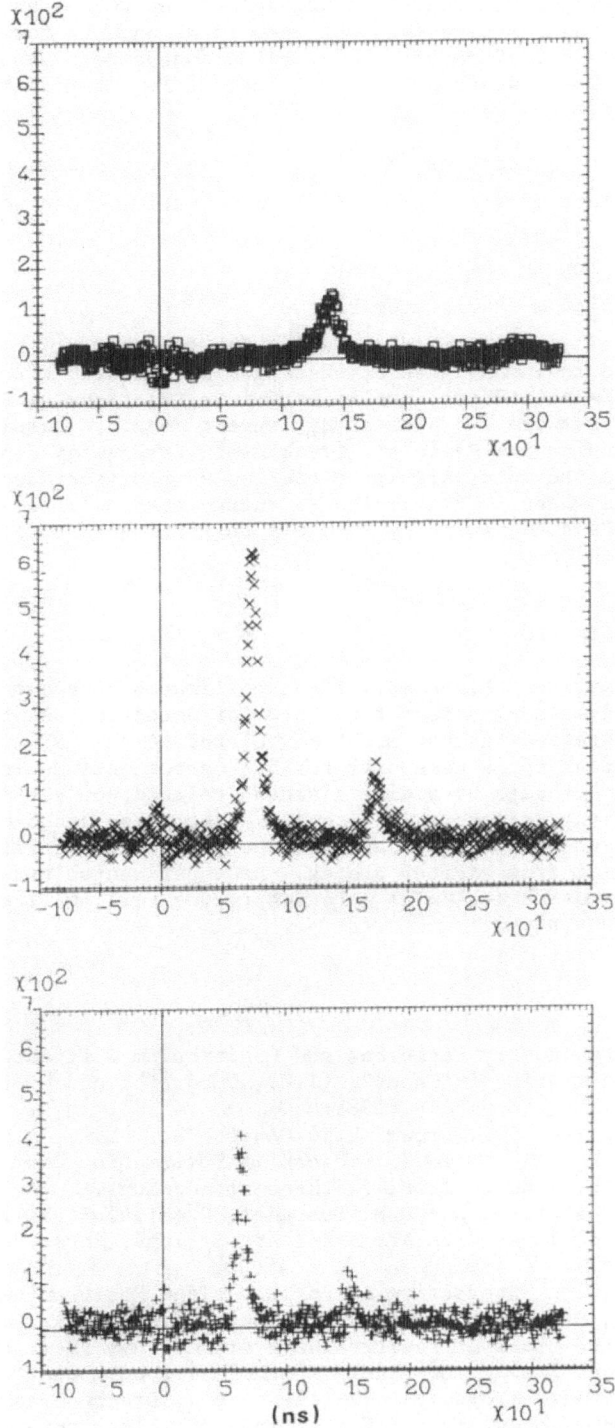

Fig. 4. Further o-Ps TOF spectra illustrating the tunability of the Ps beam. They correspond to: (a) 20 V, (b) 60 V and (c) 80 V applied to M2.

Notably, a Ps signal threshold occurred in Ar between 6 and 8 eV Ps energies. Similar results have been obtained in Ne and CH_4 for which E_{Ps} is equal to 14.8 and 6.2 eV respectively. Also independent results at Brookhaven[2], where a channeltron is used to detect Ps, confirm the existence of a detection threshold.

The origin of this threshold might be related to the following:-

i) The behaviour of the differential Ps formation cross section as a function of E_+

ii) Ps scattering in the gas cell.

iii) the detection mechanism for Ps.

With reference to (i), no calculations exist for the case of Ar, Ne or CH_4, we therefore intend to investigate He where comparison with theory should be possible. As for Ps scattering, preliminary investigations at $E_+ \sim 22.5$ eV in Ar at pressures between $1 - 4$ μm of Hg suggest negative results. Finally the detection mechanism for Ps is still unknown. Its energy dependence could be related to the intrinsic efficiency of Ps for secondary e^- emission from the detector surface. However the Ps energy threshold for detection is very close to the Ps binding energy (6.8 eV) suggesting that Ps might have to break up for detection.

CONCLUSION AND FUTURE PROSPECTS

We have demonstrated the tunability of collimated Ps beams produced by e^+-gas scattering. A Ps energy threshold for detection has been observed and it has been suggested that it might be related to the Ps binding energy implying that Ps break up is necessary for its detection. We intend to investigate this hypothesis by adding a second coincidence system between the tagger and a γ-ray detector. Refinement of both the technique and method of analysis are in progress and we hope these will reveal the contribution to our Ps beam from excited states. An experimental arrangement, containing a Ps formation cell plus a Ps scattering cell, will shortly be incorporated in our system.

REFERENCES

1. K.F. Canter, Low Energy Positrons and Positronium Diffraction in "Positron Scattering in Gases", (J.W. Humberston and M.R.C.McDowell, eds. Planum, New York) 219 (1984).
2. M. Weber, S. Berko, B.L. Brown, K.F. Canter, K.G. Lynn, A.P. Mills, Jr., L.O. Roellig, S. Tang and A. Viescas, A Positronium Beam and Positronium Reflection from LiF, in these proceedings.
3. M.W. Karl, H. Nakanishi and D.M. Schrader, Chemical Stability of Positronic Complexes with Atoms and Atomic Ions, Phys. Rev. A30:1624 (1984).
4. R.J. Drachman, The Interaction of Positrons and Positronium with Small Atoms, Can. J. Phys. 60:494 (1982); Y.K. Ho, A Resonant State and the Ground State of Positronium Hydride, Phys. Rev. A17:1675 (1978)
5. S. Hara and P.A. Fraser, Low Energy Orth-Positronium Scattering by Hydrogen Atoms, J. Phys. 8:L472 (1975); R.J. Drachman and S. Houston, Positronium-Hydrogen Elastic Scattering: The Electronic S = 1 State, Phys. Rev. A14:894 (1976); B.A.P. Page, Positronium-Hydrogen Scattering Lengths, J. Phys. B9:1111 (1976).
6. R.J. Drachman and S. Houston, Simplified Model for Positronium-Helium Scattering, J. Phys. B3:1657 (1970); G. Peach, Elastic Scattering of

Positronium by Helium, presented at NATO Advanced Workshop of e^+-gas Collisions (1984).

7. S. Ward, J.W. Humberston and M.R.C.McDowell, Scattering of Electrons (or Positrons) from Positronium and the Photodetachment of the Positronium Negative Ion, J. Phys. B20:127 (1987).

8. F.M. Jacobsen, On Positronium Formation in Low Density Gases: Slowing Down of Positrons, Chem. Phys. 101:259 (1986).

9. M. Charlton, G. Clark, T.C. Griffith and G.R. Heyland, Positronium Formation Cross Sections in the Inert Gases, J. Phys. B16:465 (1983); L.S. Fornari, L.M. Diana and P.G. Coleman, Positronium Formation in Colisions of Positrons with He, Ar, H_2, Phys. Rev. Lett. 51:2276 (1983); G. Sinapius, D. Fromme and W. Raith, Positron Impact Ionisation Cross Sections of Helium, as in (16), 61.

10. K.G. Lynn and H. Lutz, Slow Positrons in Single-Crystal Samples of Al and Al-AlxOy, Phys. Rev. B22:4143 (1980).

11. A.P. Mills, Jr. and L. Pfeiffer, Velocity Spectrum of Positronium Thermally Desorbed from An Al(111) Surface, Phys. Rev. B32:53 (1985).

12. R.H. Howell, I.J. Rosenberg and M.J. Fluss, Production of Energetic Positronium at Metal Surfaces, Phys. Rev. B34:3069 (1986).

13. B.I. Deutch, F.M. Jacobsen, P. Hvelplund, H. Knudsen, L.H. Anderson, M. Holzscheiter, M. Charlton and G. Laricchia, Antihydrogen by Positronium-Antiproton Collisions in an Ion Trap, to be published (1987).

14. F.M. Jacobsen, L.H. Anderson, B.I. Deutch, P. Hvelplund, H. Knudsen, M. Charlton, G. Laricchia, M. Holzscheiter, On Antihydrogen Production, in these proceedings.

15. A.P. Mills, Jr. and W.S. Crane, Beam Foil Production of Fast Positronium, Phys. Rev. A31:593 (1985).

16. B.L. Brown, Creation of a Monoenergetic Positronium Beam in a Gas and Measurement of Positron Survival Fraction in a Gas, "Positron (Electron) - Gas Scattering", W.E. Kauppila, T.S. Stein and J.M. Wadehra, eds: World Scientific, Singapore, 212 (1986).

17. G. Laricchia, M. Charlton, T.C. Griffith and F.M. Jacobsen, Preliminary Results on the Angular Dependence of Ps Emission in e^+ - Gas Collisions, as in (16) 313 (1986).

18. D.W. Gidley, R. Mayer, W.E. Frieze and K.G. Lynn, Glancing-Angle Scattering and Neutralization of a Positron Beam at Metal Surfaces, Phys. Rev. Lett. 58:595 (1987).

19. G. Laricchia, M. Charlton, S.A. Davies, C.D. Beling and T.C. Griffith, The Production of Collimated Beams of o-Ps Using Charge Exchange in Positron - Gas Collisions, J. Phys. B20:L99 (1987).

20. P. Mandal, S. Guha and N. Sil, Positronium Formation in Positron Scattering from Hydrogen and Helium Atoms: The Distorted-Wave Approximation, J. Phys. B12:2913 (1979).

21. C.J. Brown and J.W. Humberston, Positronium Formation in Positron-Hydrogen Scattering, J. Phys. B18:L401 (1985).

22. e.g. (16) or D.A. Fischer, K.G. Lynn, D.W. Gidley, High-Resolution Angle;Resolved Positron Reemission Spectra from Metal Surfaces, Phys. Rev. B33:4479 (1986).

23. G. Laricchia, M. Charlton, G. Clark and T.C. Griffith, Excited State Positronium Formation in Low Density Gases, Phys. Lett. A109:97 (1985).

24. M. Charlton, G. Laricchia, N. Zafar and F.M. Jacobsen, Inelastic Positron Collisions in Gases, in these proceedings.

25. e.g. B. Nielsen, K.G. Lynn and Yen-C. Chen, Study of Solids by Use of Non Thermalised Positrons, Phys. Rev. Lett. 57:1789 (1986).

A POSITRONIUM BEAM AND POSITRONIUM REFLECTION FROM LiF

L.O. Roellig,[1] M. Weber,[1] S. Berko,[2] B.L. Brown,[3] K.F. Canter,[2]
K.G. Lynn,[4] A.P. Mills, Jr.,[3] S. Tang,[1] and A. Viescas[1]

[1]City College of City University of New York, NY, NY 10031
[2]Brandeis University, Waltham, MA 02254
[3]A.T.&T. Bell Laboratories, Murray Hill, NJ 07974
[4]Brookhaven National Laboratory, Upton, NY 11973

INTRODUCTION

At the Brookhaven National Laboratory we have constructed a
positronium (Ps) beam by transmitting monoenergetic, low energy positrons
through a gas cell containing either Ar or He which provide an electron to
form positronium. A description of the positron beam and of the Ps
formation mechanisms are found in these Proceedings (see M. Weber, et al.
and B. L. Brown). The positrons were obtained by magnetically deflecting
positrons in the straight section of the positron beamline (see Fig. 1)
into a beamline which contained the gas cell and a Ps detection chamber.
By having two beamlines we are able to switch from an experiment which
uses positrons (a study of the angular correlation of annihilating
radiation--ACAR) to one which uses Ps atoms without breaking vacuum, nor
moving equipment. This, however, put a constraint on the placement of the
Ps beamline because it could not interrupt the annihilation gamma ray in
its long flight from the target chamber to a gamma ray position imaging
detector (Anger camera). At present this constraint has resulted in a
degradation of the positron beam intensity and energy resolution in the Ps
beamline. Efforts are presently underway to eliminate this problem.

Very preliminary information has been obtained on the characteristics
of the Ps beamline and on the reflection of Ps from a LiF crystal.

Characteristics of the positron beam

(a) The intensity of the positron beam is approximately 8×10^6 e^+/sec.
    This is a reduction of more than an order of magnitude from its
    intensity in the straight section.

(b) The energy resolution (energy width between 90% and 10% of the
    intensity) is 20% at a positron energy of 200 eV, 4% at a
    positron energy of 120 eV 2.8% at a positron energy of 18 eV (see
    Fig. 2 and Fig. 3).

It is felt that both (a) and (b) can be improved by increasing the radius of curvature of the positron trajectory at the beam splitter shown in Fig. 1.

Characteristics of the Ps beam

A channeltron was placed 3 cm from the gas cell and a 3"x3" NaI detector was placed 6 cm above the channeltron and perpendicular to the positronium beam which exits from the gas cell. The potential on the gas cell was varied from 170 V to 200 V, and the positrons had an energy of 200 eV before entering the cell. Voltages were placed on grids near the exit of the gas cell and on the channeltron to prevent positrons and electrons from reaching the channeltron. In Fig. 4 the coincidence rate between the channeltron and NaI detector is shown as a function of the potential on the gas cell with Ar gas in the cell at a pressure of 5×10^{-4} torr (gas on) and with no gas in the cell (gas off). In Fig. 5 the coincidence rate is shown as a function of the potential on the gas cell when Ar gas at a pressure of 5×10^{-4} torr was in the cell. The diamond shaped points along zero coincidence counts are the rates when the positron beam was turned off in the block house, and the crosses are when 200 eV positrons entered the gas cell. It should be noted that the data

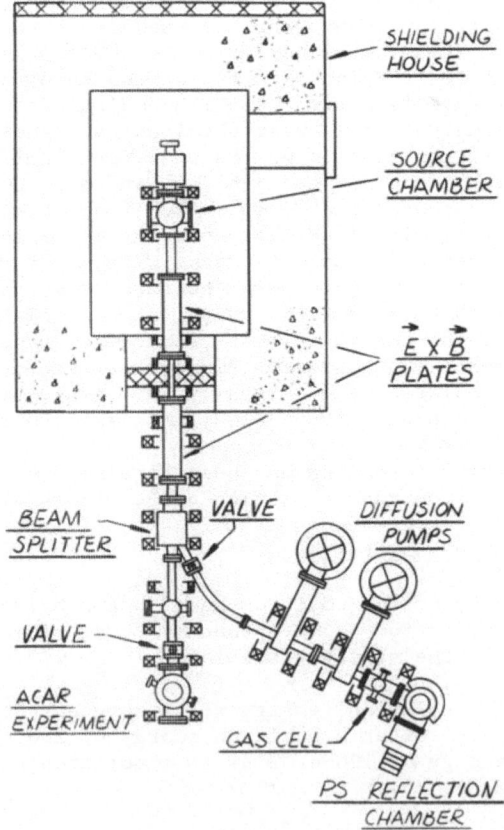

Fig. 1 Positron and positronium beamline.

shown in Fig. 5 is in agreement with the experimental results of L. S. Fornari, et al.[1] Their ratio of the Ps formation cross section, Q_{Ps}, in argon at 20 eV and 10 eV is

$$\frac{Q_{Ps}(20 \text{ eV})}{Q_{Ps}(10 \text{ eV})} = 1.9 \pm 0.3$$

This ratio obtained from the data in Fig. 5 is

$$\frac{\text{coincidence counts at 20 eV}}{\text{coincidence counts at 10 eV}} = 1.6 \pm 0.2$$

It is observed in both figures that Ps is not detected until the positron has an energy 200 eV - 184 eV ≈ 16 eV. The positron must have an energy greater than 8.9 eV to form Ps in Ar, and when it is formed it has a binding energy of 6.8 eV. It appears that the channeltron is not

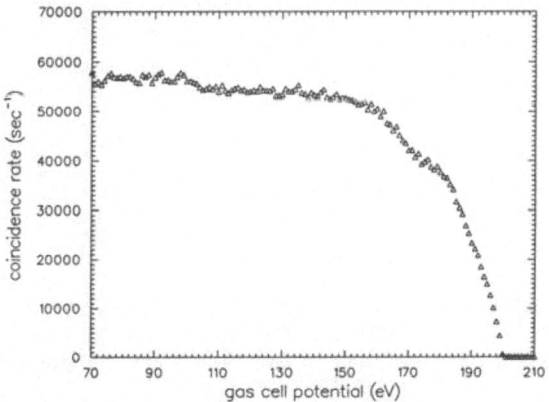

Fig. 2 Coincidence rate between a channeltron placed 3 cm from the gas cell and a 3"x3" NaI detector placed 6 cm from the channeltron and perpendicular to the positron beam exiting the gas cell. The gas cell is devoid of gas and its potential is varied from 70 to 210 volts. The positrons had an energy of 200 eV upon entering the empty gas cell.

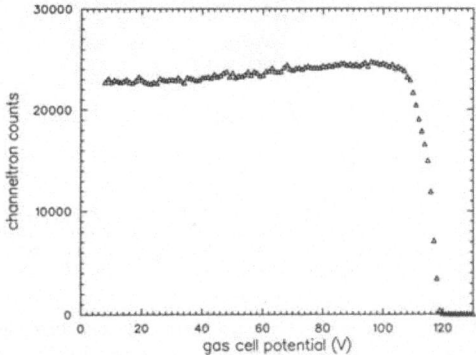

Fig. 3 A measurement similar to Fig. 2 except the positrons had an energy of 120 eV upon entering the empty gas cell.

sensitive to Ps unless the Ps energy is great enough to dissociate it. Thus Ps is not detected if the energy of the positron entering the gas cell is below 9 eV + 8.9 eV ≈ 15.7 eV. It should be noted that in Figs. 2 and 3 the energy resolution shown is the transverse energy spread mainly due to magnetically deflecting positrons into the positronium beamline. The total energy spread will be narrower. This is indicated by the sharp increase in positronium formation as a function of energy shown in Figs. 4 and 5. Of course this results in a reduction of Ps emanating from the gas cell.

The rate of Ps atoms in the Ps beam emerging from the gas cell is estimated to be several thousand per second.

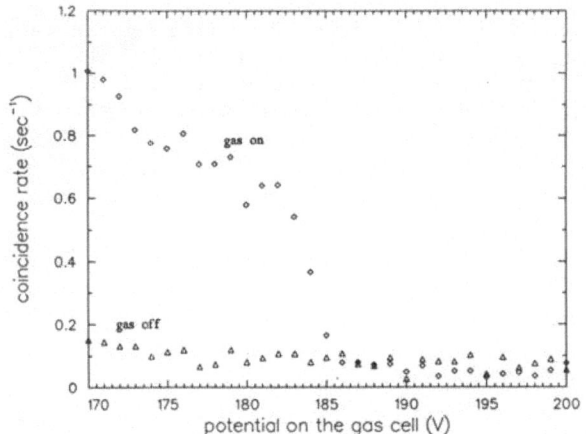

Fig. 4   The coincident rate between the channeltron and NaI detector vs. the potential on the gas cell with Ar gas in the cell at a pressure of 5×10^{-4} torr (gas on) and with no gas in the cell (gas off).

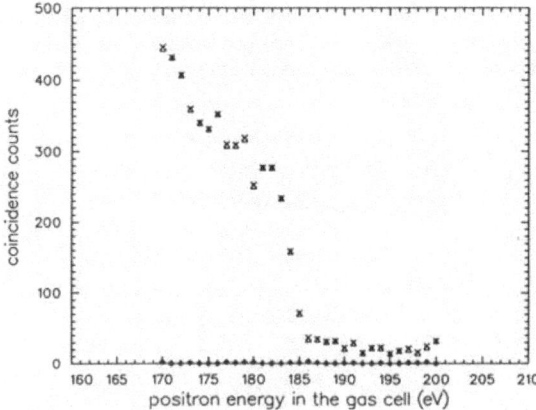

Fig. 5   The coincident rate vs. the potential on the gas cell when Ar gas at a pressure of 5×10^{-4} torr is in the cell. The diamond shaped points along the zero counts are the rate when the positron beam is turned off in the block house, and the crosses are when 200 eV positrons enter the gas cell.

Reflection of Positronium

(a) A schematic diagram of the apparatus for measuring the reflection
of Ps from LiF (100) is shown in Fig. 6. The crystal was cleaved
in air by the supplier, and not treated in any way. Specular
reflection of Ps by an angle $\theta = 45°$ from the crystal face to the
Ps beam direction was measured by observing the annihilation of
Ps on a stainless steel plate by two BGO ($Bi_4Ge_3O_{12}$) detectors in
coincidence. The background was obtained by rotating the crystal
to an angle $\theta = -20°$. The first very preliminary data of this
experiment is shown in Fig. 7. We are presently pursuing
systematic checks on the Ps beamline and our data acquisition
system. It is apparent, however, that we are observing the
reflection of Ps from LiF.

(b) We have constructed apparatus to measure the reflection
coefficient of positronium from various crystalline surfaces.
A diagram of the gas cell and the reflection chamber is shown in
Fig. 8. The sample face can be rotated with respect to the
direction of the Ps beam and a channeltron or channel plate
located in the vacuum chamber can also be rotated about the same
axis. Outside of the vacuum chamber are two BGO detectors, one

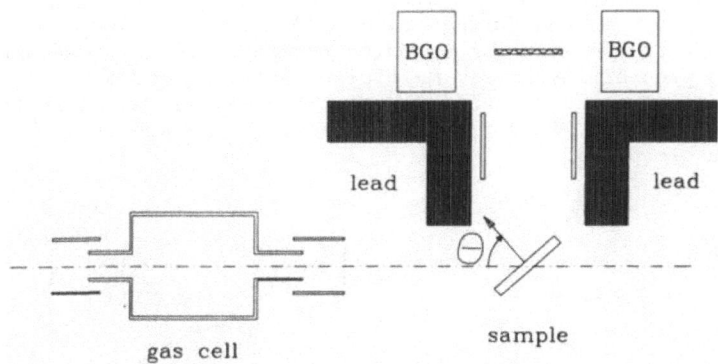

Fig. 6 Ps reflection apparatus utilized to obtain data given in Fig. 7.

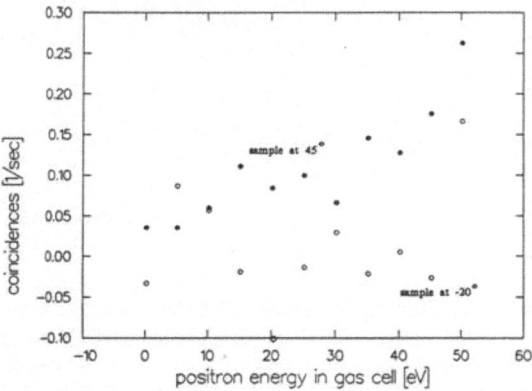

Fig. 7 Positronium reflection from LiF(100) at $\theta = 45°$ (solid circles)
and coincidence rate at $\theta = -20°$ (open cirlces).

directly above and one directly below the channeltron or channel plate to detect the annihilation gamma rays. With this arrangement we have the option to require a threefold coincidence between the detectors to measure the rate of reflected Ps atoms.

Simultaneously with this effort to measure the reflection coefficient modifications of the apparatus are being made to increase the intensity of the Ps beam.

Purpose

The purpose of these measurements is to attempt to measure Ps diffraction (LEPSD) from crystalline surfaces. Ps diffraction is somewhat similar to He atom diffraction,[2] which is a powerful tool in surface structure determination because it is only sensitive to the outer surface layer. However, the savings in complexity in He atom diffraction by not having to treat multiple scattering from subsurface layers (as in the case of LEED) are somewhat mitigated by having to deal with long range forces that dominate in He diffraction. The 0.02 eV energy necessary for He atoms to have ≈ 1Å de Broglie wavelength results in the He atoms having classical turning radii far enough from the individual ion cores that the scattering is mainly due to the average potential presented by the surface. This requires an accurate treatment of the average potential for intensity analysis.[2] In order for Ps to have a ≈ 1Å wavelength, its energy must be on the order of ≈ 75 eV. At this energy, Ps atoms would be oblivious to the mean surface potential and only undergo elastic reflection in close encounters with the ion cores. Because of the large break-up probability of Ps, multiple scattering and other subsurface contributions to the elastically scattered Ps are expected to be negligible. Thus, Ps diffraction offers the possibility of being a novel and valuable probe.

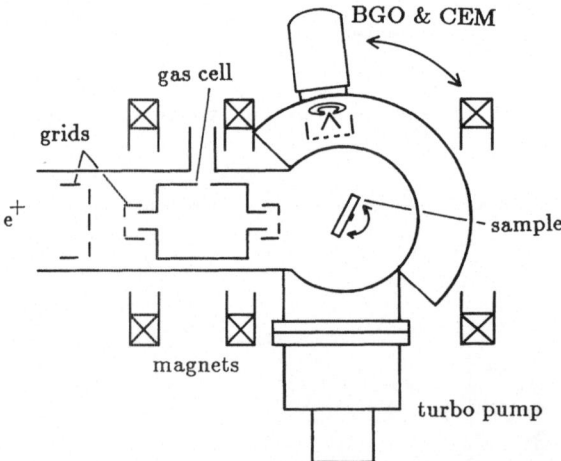

Fig. 8 Apparatus to measure reflected-diffracted Ps from various crystalline surfaces. The channeltron (CEM) is located inside the vacuum chamber and two BGO detectors, one directly above and one directly below the channeltron, to detect annihilation gamma rays are located outside the vacuum chamber. All three detectors can be rotated independently about an axis which is perpendicular to the Ps beam axis.

The degree to which Ps scatters only from the outer surface layer is determined mainly by the interstitial density of valence or conduction electrons of the material. Because of the low mass of weakly-bound electrons, and hence large recoil, elastic Ps-e$^-$ collisions destroy the coherence of the scattered Ps and thus must be regarded as a source of attenuation of the diffracted Ps beam. Typical elastic cross sections in the 10 eV region for Ps-free e$^-$ collisions are on the order of 10^{-15} cm^2.[3] Thus for solids having typical interstitial electron densities of $\approx 10^{23}$ cm^{-3},[3] a mean free path of ≈ 1Å for the Ps can be expected. Consequently, LEPSD from a solid surface would yield diffracted Ps intensities versus incident energy (i.e., "I(V)" curves) which would be dominated by the elastic scattering from only the outer layer atomic distribution. In the case of an ordered adsorbate overlayer, chemisorbed to a surface, however, the incident Ps could easily penetrate the relatively open spaces between the adsorbate atoms. This would lead to the interesting case of interference between Ps scattering from the adsorbate and from the outer surface with a high sensitivity to the structure of the adsorbate layer and outer surface.

This work is supported in part by the National Science Foundation, Grant No. DMR-8315691 and the U.S. Department of Energy, Division of Materials Science, Contract No. DE-AC02-76CH00016.

References

1. L. S. Fornasi, L. M. Diana, and P. G. Coleman, Phys. Rev. Lett. 51, 2276 (1983).
2. K. H. Rieder and N. Garcia, Phys. Rev. Lett. 49, 43 (1982).
3. S. J. Ward, J. W. Humberston, and M. R. C. McDowell, J. Phys. B 18, L525 (1985).
4. P. J. E. Alfred and M. Hart, Proc. R. Soc. Lond. A 332, 239 (1973). J. A. Golovchenko, D. E. Cox, and A. N. Goland, Phys. Rev. B 26, 2335 (1982). B. Dawson, Proc. R. Soc. Lond. A 298, 395 (1967). A. J. Freeman in Electron and Magnetization Densities in Molecules and Crystals, edited by P. Becker (Plenum Press, New York and London, 1980) p. 83.

MONOENERGETIC Ps CREATED IN A GAS AND Ps-He COLLISIONS CROSS SECTION

MEASUREMENT

Benjamin L. Brown[*]

AT & T Bell Laboratories

Murray Hill, NJ 07974, U.S.A.

A source of energetic Ps atoms in the laboratory may make it possible to perform several new experiments. Some of these are Ps gas scattering cross section measurements[1], a search for Ps gas collision resonances[2], Ps molecular bound states[3] and Ps surface diffraction[4]. Recent methods of energetic laboratory Ps production include Ps beam-foil production demonstrated by Mills and Crane[5] and Ps formation on a clean crystal surface demonstrated by Gidley et al[6]. One additional method has been proposed where Ps- is formed in a foil and then ionized to produce variable energy Ps[7].

The method of Ps beam production in a gas was first detailed in late 1984 and early 1985[8]. It was argued that a monoenergetic, well collimated beam of Ps is produced when positrons are incident on gas atoms at sufficiently low densities. The efficiency was estimated in He using available theoretical cross sections for Ps formation and an observation of the p-Ps annihilation spectrum was made showing the general features expected. The energy resolution of the Ps beam was predicted to be limited only by the energy width of the slow positron beam (typically 80 meV). The beam is practical and is relatively efficient compared to other methods. The o-Ps beam has been demonstrated at Bell Labs.[9] and at the University College London[10]. It will be interesting to study Ps-surface interactions with this highly monoenergetic beam[12]. Recent results at Brookhaven show an encouraging Ps reflection coefficient for LiF. (See Roellig et al elsewhere in this volume).

The two experimental arrangements that have been used for o-Ps production, at Bell Laboratories and with the consortium at Brookhaven Laboratory, have been described elsewhere[13]. In this paper a measurement of the Ps-He total collision cross section is presented as part of a detailed discussion of Ps beam efficiency. A careful comparison of the measured efficiency to the theory shows good agreement when the collision cross section is taken into account. A discussion of the possible contamination of the ground state o-Ps beam by n = 2 Ps[*] excited state is also given in some detail.

---

* Present address: Department of Physics, Harvard University, Cambridge, MA 02138, U.S.A.

Ps Beam Characteristics

The efficiency of the Ps beam has been estimated in a regime where much less than one positron-gas collision is anticipated in a gas cell or jet[8-10]. An expression for the efficiency is

$$D \; \rho \; 2\pi \int_{o}^{\theta} (d\sigma_{ps}/d\omega) \sin\theta d\theta$$

where D is the positron path length in the gas, $d\sigma_{ps}/d\omega$ is the differential Ps formation cross section and σ is the total positron collision cross section. The angle θ in the integral is the beam half angle. The differential Ps formation cross section $d\sigma_{ps}/d\omega$ is peaked in the forward direction for H_2 and He according to theory[8,11]. Note however that the intensity per unit angle θ, or $\sin\theta$ ($d\theta/d\omega$), peaks at an angle other than zero degrees. The beam intensity per sterradian is greatest in the forward direction. The beam efficiency is calculated for a 5 degree cone (2.5 degree half angle) as shown in Figure 1. from the calculations of Mandal, Guha and Sil[11].

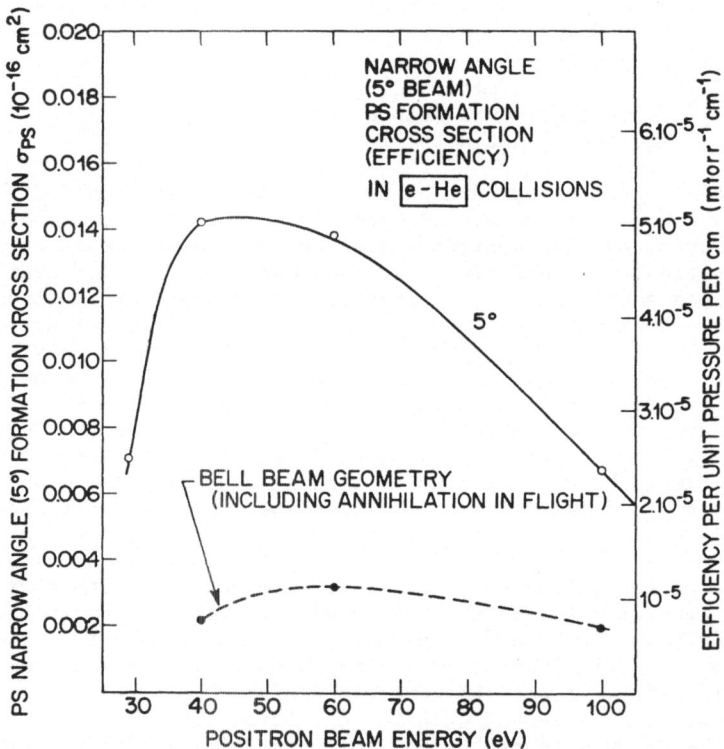

Fig. 1. The beam efficiency calculated for a low density regime for a 5 degree Ps cone (upper curve) and for the Bell Beam geometry (lower curve). The cross sections from Ref. 11 were used to create an analytic model for the differential cross section which was then numerically integrated. The right hand axis is given in ($torr^{-1}$ cm^{-1}) and multiplication by the gas cell pressure and length yields an efficiency for Ps production. The triplet Ps to total Ps ratio, a factor of 3/4, is included in the lower curve only.

One additional consideration in designing a beam is of course the finite lifetime of the o-Ps: $\tau = 1.4 \times 10^{-7}$ s. This reduces the beam efficiency because some of the Ps triplet annihilates in flight. Attenuation due to in-flight annihilation can be written:

$$\exp\left[\frac{-x}{\tau}\left(\frac{m}{E_{ps}}\right)^{1/2}\right]$$

where x is the distance from the Ps formation region to a detector.

Since the gas cell length is not short compared to the Ps flight path, a correction must be made both for solid angle and decay in flight for the efficiency estimate. For simplicity this expression is given as an attenuation coefficient for the already calculated 5 degree Ps beam (2.5 degree half angle = θ) efficiency curve discussed above. Thus α, an attenuation coefficient, can be expressed as

$$\alpha = \frac{h^2}{\sin^2\theta\, D}\int_{L}^{D+L}\frac{e^{-x/\ell}}{x^2 + h^2}\, dx$$

$$\approx \frac{-h^2}{\sin^2\theta\, D}\left[\left(\frac{e^{-x/\ell}}{x}\right) + \frac{1}{\ell}\left(\log x - \frac{x}{\ell} + \frac{x^2}{2!2\,\ell^2} - \frac{x^3}{3!3\,\ell^3} + \frac{x^4}{4!4\,\ell^4} - ..\right)\right]_{L}^{D+L}$$

where the gas cell length is D (20 cm), the distance from the end of the gas cell to the annihilation plate is L (25 cm as shown in Figure 2), the radius of the annihilation plate is h, and the annihilation length at a given energy for the triplet Ps is ℓ.

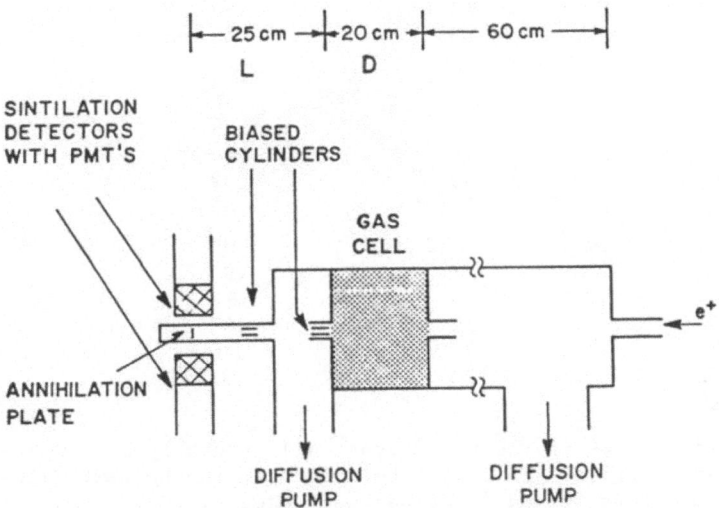

Fig. 2. The Bell Laboratories Ps beam geometry. The finite gas cell dimensions are given and the positron rejection baffles are shown.

Still another additional consideration on the beam efficiency is the orientation of the positrons when they make Ps in a magnetically guided beam. A spread in the integral positron longitudinal energy spectrum of more than ~ 1 eV is generally due to the transverse energy gained by the positrons in the transport of the positrons from the source to the target region. An expression for the reduction in efficiency due to this effect will be given, with certain limiting assumptions. If α' is taken to be the attenuation factor then

$$\alpha' = \frac{\int_0^{90} e^{-\theta^2 \ln 2/H^2}\, e^{-E_o \sin^2\theta \ln 10/\Delta E}\, d\theta}{\int_0^{90} e^{-E_o \sin^2\theta \ln 10/\Delta E}\, d\theta}$$

where H is the HWHM defined by the Ps distribution and ΔE is the longitudinal energy spread of the beam defined as the energy width from the energy at 90% on the integral spectrum to the maximum energy E_o. This assumes an integral spectrum of the form

$$1 - e^{-(E_o - E)\ln 10/\Delta E}$$

with $E < E_o$, which is a reasonable approximation for a typical spectrum. In Table 1 the various values for α' are given for typical beam energies. Below 40 eV the differential cross section is not well modeled by a Gaussian, and so the expresssion will not be accurate. The coefficient α' is generally > 80% with $E_o/\Delta E > 20$.

Table 1. The attenuation coefficient α' due to a spread in the longitudinal energy ΔE of the incoming positrons.

E	ΔE	α'
40	0.2	.99
	2	.89
	5	.77
	10	.64
	20	.49
60	0.2	.99
	2	.90
	5	.78
	10	.66
	20	.52
	30	.43
100	0.2	.99
	2	.91
	5	.81
	10	.69
	20	.56
	50	.37

The formation of Ps in the n = 2 state will probably be somewhat significant at higher energies. This will be discussed in the next section. An observation of n = 2 Ps formed in He has been made[14].

Taking into account the detailed model above, including α and α', the efficiency with the Bell Beam geometry is plotted on the lower curve in Figure 1. An additional attenuation factor of 3/4 is included since the

5 degree efficiency calculations do not include the triplet Ps to total Ps ratio. We assume little reduction in beam intensity over the length of the gas cell. At the highest pressure used with He, 1.6 mtorr, the incident positron beam reduction due to collisions is estimated to be ∿ 10% over the length D (20 cm). The gas pressure differential on each side of the gas baffles is approximately 50 to 1.

The experimental results are shown in Figure 3 for two different pressures. The low pressure run seems to coincide fairly well with theory. This correspondence is encouraging in that the process of Ps formation would seem to be well understood in the low pressure regime. The higher pressure run however deviates significantly. This is thought to be due to the Ps making collisions with the He gas in the cell itself. It does not seem that the gas in the region preceding the gas cell could have produced significant collisions with the incoming positrons.

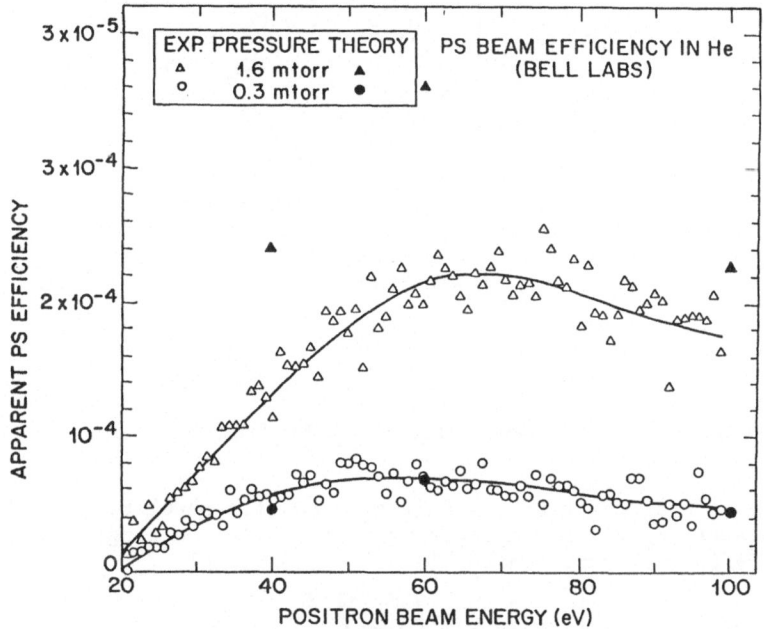

Fig. 3. Ps beam produced by positrons incident on He and detected with coincident NaI detectors. The upper curve shows a reduction in efficiency due to Ps-He scattering. A careful interpretation of the results yields Ps-He total cross sectional values.

Ps Beam Contamination

It is interesting to consider the details of the n = 2 state contamination of the Ps beam. We will consider the effect of this contamination on both the 511 keV singlet annihilation spectrum and on the triplet Ps beam. The annihilation spectrum and the triplet beam have been measured in separate experimental arrangements[8,13]. The short lived states can be significant for the gamma-ray line spectrum of a Ps beam and the long lived states can be significant in the Ps beam itself. Since the binding energy for n = 2 Ps is B' = 1.7 eV, the kinetic energy of the n = 2 components are lower than the n = 1 components by 5.1 eV. This can be seen easily from the conservation of energy for an endothermic Ps forming collisions[8]. Therefore the n = 2 Ps can be regarded as an unwanted contamination in the beam, for present purposes. The energy levels of the various n = 2 states are shown in Figure 4 (Ref. 15). Many of the n = 2 states radiatively decay via a Lyman-alpha photon to the n = 1 state.

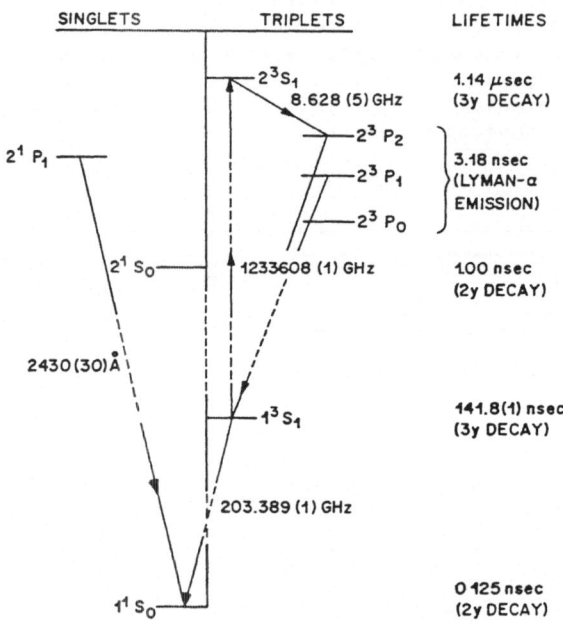

Fig. 4. Energy levels for n = 2 Ps from Ref. 16.

Of the two 2S states that are initially equally populated, the 2^1S_0 state annihilates rapidly (1 ns) giving an annihilation length of .4 cm at 100 eV. We define the annihilation length, ℓ, as simply

$$\ell \;=\; \tau v \;=\; \tau \left(\frac{E_{ps}}{m}\right)^{1/2} \;,$$

where E_{ps} is chosen to be 100 eV as a convenient example. One quarter of the 2S Ps is formed in the 2^1S_0 and three quarters is formed in the 2^3S_1 state, since the one singlet and three triplet states are formed with equal probability. The 2^3S_1 annihilates slowly however (1.14 µs) with an annihilation length of 479 cm. However this state in practice could be quenched in a laboratory Ps beam to the triplet ground state by the electric fields in the biased cylinder region. This biased cylinder or grid is designed to keep positrons from getting into the detection region. An electric field of 1 KeV per cm is sufficient for quenching. We will assume for the present estimate that the 2^3P_1 annihilates at the triplet rate after formation. (This is equivalent to the assumption that the gas cell is short compared to the beam flight path).

The 2P state has a short radiative decay time (3.18 ns) yielding a decay length (analogous to the annihilation length) of 1.32 cm at 100 eV. The 2^1P_1 state is not significant in the beam for target distances greater than 10 cm, for example, since it radiatively decays (3.18 ns) to the singlet ground state, which has a short annihilation lifetime (0.125 ns). It would contribute to the line spectrum. The 2^3P_2, 2^3P_1, and 2^3P_0 states radiatively decay (3.18 ns) to the longer lived triplet ground state (142 ns) and could produce significant beam contamination. The cross sections have been calculated for the 2S and 2P states[16].

We can summarize the discussion above by considering the spectrum and beam contaminations, again assuming a short gas cell. If the number of positrons formed per unit time in a state is I, then in obvious notation we have for the spectrum a contamination ratio R_s for $(n = 2)/(n = 1)$ of

$$ R_s = \frac{\frac{1}{4} 2P}{\frac{1}{4} 1S} + \frac{\frac{1}{4} 2S}{\frac{1}{4} 1S} = \frac{2P + 2S}{1S} $$

and for the Ps beam we have

$$ R_b = \frac{\frac{3}{4} 2P}{\frac{3}{4} 1S} + \frac{\frac{3}{4} 2S}{\frac{3}{4} 1S} = \frac{2P + 2S}{1S} \quad . $$

Thus the ratio of the formation cross sections for the 1S, 2S and 2P states will determine the number of Ps atoms formed in these states. The theoretical cross sections that determine the contamination are taken from Ref.16.

The angular distribution of each state plays a role in the actual n = 2 contamination of the beam. Above ∿ 100 eV the angular distributions are probably similar for the 1S, 2S and 2P states. At lower energies the n = 2 state probably has a wider distribution. The HWHM of the calculated differential cross section of the 1S state is ∿ 11.25 deg. and for the 2P state it is ∿ 18 deg. at an incident positron energy of 60 eV[16]. Thus the n = 2 beam is not as narrow as the n = 1 beam and this would reduce the expected contamination in Figure 5 at energies < 100 eV.

The possibility of Ps collisions in the gas formation region can significantly reduce the contamination of n = 2 Ps. This would be expected because the n = 2 Ps undoubtedly has a larger collision cross section with the gas than the n = 1 Ps. A simple estimate of the radii of the n = 2 states, < r >, yields ∿ 4 a_0 where a_0 is the radius (center of mass) of the ground state of Ps (the Bohr radius). Thus the n = 2 collision cross section would be ∿ 16 times the ground state cross section. This is a filtering effect that will be significant in practical applications. The pressure can be set to a point where the beam efficiency is good and where the n = 2 Ps is also greatly reduced.

Ps-He Total Collision Cross Section

From the experimental data given in the preceding section, we can infer that the Ps lost at high pressures is due to some Ps collisions in the gas cell. Since the pressure was measured accurately (an ion gauge calibrated with a Mercury manometer) and the total positron collision cross sections are known accurately (see Stein et al. elsewhere in this volume and references therein), the beam reduction over what is expected is assumed to be

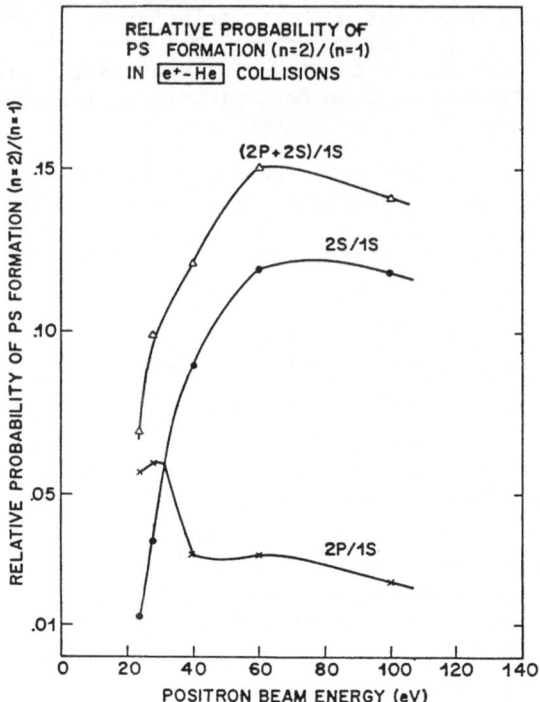

Fig. 5. The relative contamination of the Ps beam and spectrum by
n = 2 Ps is given by the ratio of the calculated cross sections from
Ref. 16. In practice much of the n = 2 component in the beam may be
greatly reduced by collisions in the gas.

due to Ps-He collisions. Using this assumption, with a simple classical
model the Ps-He cross sections can be inferred.

If the detector (annihilation plate in Figure 2) radius is h then an
attenuation coefficient α can be obtained by including Ps beam extinction
due to Ps-He collisions with a cross section σ_c. The expression for α is
thus:

$$\alpha = \frac{h^2}{D\sin^2\theta} \int_L^{D+L} \frac{e^{-x/\ell}}{x^2 + h^2} e^{(L-x)\rho\sigma_c} e^{-(D+L-x)\rho\sigma_t} dx \ .$$

The total positron collision cross section, σ_t, is included in an exponential term under the integral to account for the positrons which make a
collision in the gas. These are assumed to be lost in the present model.
The integral can be performed analytically and a solution for σ_c found
numerically by taking the experimental attenuation coefficient from the
upper higher pressure curve in Figure 3. The determination of σ_c with this
numerical solution is rather specific in that a small artifical change in
the measured pressure or in the experimental attenuation produces a noticeable change in σ_c. Therefore, even though the model is simple we have
confidence in the results.

The results are shown in Figure 6. The cross sections have been calculated at lower energies than those shown here, but at present we are unaware of any prediction in the range of energies presented[1,17,18].

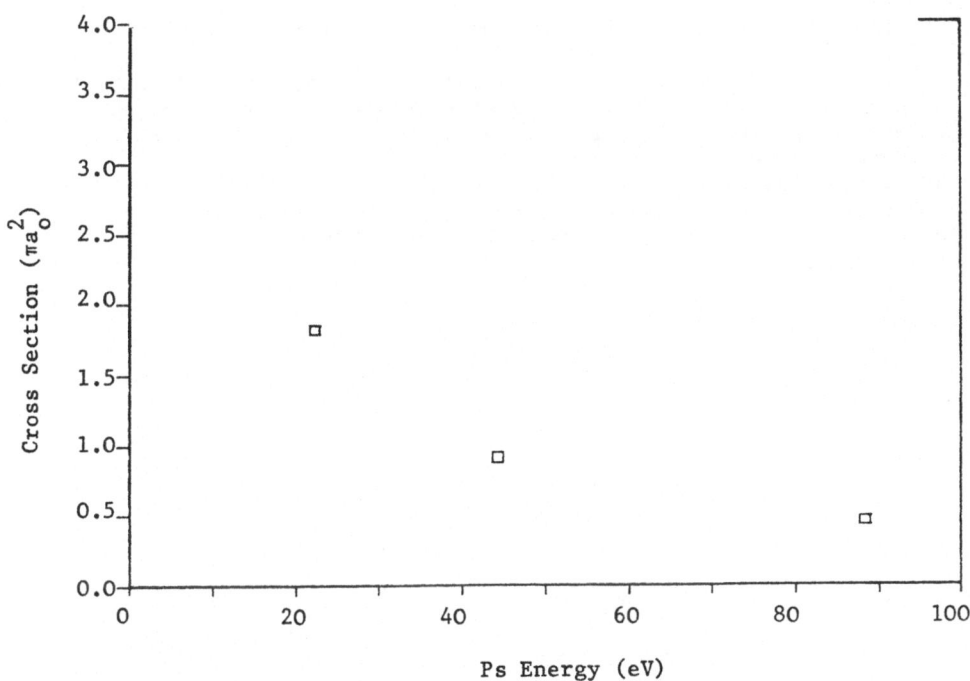

Fig. 6. Ps-He total collision cross section measured from the beam extinction at high pressure (see Fig. 3). Theoretical predictions have been made at lower energies than those measured here.

REFERENCES

1. P.A. Fraser, Proc. Phys. Sco. 79:721 (1961).
2. R.J. Drachman, Can. J. Phys. 60:494 (1982).
3. P.M. Schrader and R.E. Svetic, Can. J. Phys. 60:517 (1982).
4. K.F. Canter in Positron Scattering in Gases, Ed. J.W. Humberston and M.R.C. McDowell (Plenum, New York, 1983).
5. A.P. Mills, Jr. and W.S. Crane, Phys. Rev. A 31:493 (1985).
6. D.W. Gidley, W. Frieze, R. Mayer and K.G. Lynn, Third International Workshop on Positron (Electron)-Gas Scattering, ed. by W.E. Kauppila, T.S. Stein and J.M. Wadehra (World Press, Singapore, 1985).
7. A.P. Mills, Jr. Phys. Rev. Lett. 46:717 (1981).
8. B.L. Brown in Positron Studies of Solids, Surfaces and Atoms, a Symposium to celebrate Stephan Berko's 60th Birthday, December 12, 1984, ed. by A.P. Mills, Jr, W.S. Crane and K.F. Canter, (World Press, Singapore, 1986), and also in Seventh International Conference on Positron Annihilation, ed. by R.M. Singru and P.C. Jain (World Press, Singapore, 1985).
9. B.L. Brown presented at 2nd International Conference on e+ at Surfaces, U.E.A., Norwich, England, July, 1986.
10. G. Laricchia, M. Charlton, S.A. Davies, C.D. Beling and T.C. Griffith, ibid, and also in J. Phys. B B20:L99 (1987).
11. P. Mandal, S. Guha and N.C. Sil, J. Phys. B 12:2913 (1979). P. Khan and A.S. Ghosh, Phys. Rev. A 28:2181 (1983).

12. S. Berko, K.F. Canter, A.P. Mills, Jr., K.G. Lynn, L.O. Roellig and M. Weber in Seventh International Conference on Positron Annihilation, ed. by R.M. Singru and P.C. Jain, (World Press, Singapore, 1985).
13. B.L. Brown in Ref. 6.
14. M. Charlton, G. Clark, T.C. Griffith and G.R. Heyland, Phys. Lett. 16:L465 (1983).
15. A.P. Mills, Jr., S. Berko and K.F. Canter, Phys. Rev. Lett. 34:1541 (1975), and A.P. Mills, Jr., and S. Chu, Proceedings of the 8th International Conference on Atomic Physics, Goteberg, Sweden, Aug. 1982 and references therein.
16. P. Khan, P.S. Masumdar and A.S. Ghosh, J. Phys. B. 17:4785 (1984).
17. M.J. Barker and B.H. Bransden, J. Phys. B 1:1006 (1986), and M.J. Barker and B.H. Bransden, J. Phys. B 2:1109 (1969).
18. G. Peach quoted in Ref. 10.

POSITRON- (AND ELECTRON-) ALKALI ATOM TOTAL SCATTERING MEASUREMENTS

T.S. Stein, M.S. Dababneh,* W.E. Kauppila,
C.K. Kwan, and Y.J. Wan**

Department of Physics and Astronomy
Wayne State University
Detroit, Michigan 48202, U.S.A.

INTRODUCTION

From the outset of scattering experiments with low energy positron beams, there has been a natural tendency to make comparisons between the scattering of positrons and electrons by the same target atoms and molecules. Since positrons, being the antiparticles of electrons, have the same magnitudes for the mass, charge, and spin as the electron, but have the opposite sign of charge, comparison measurements of the scattering of positrons and electrons by atoms and molecules can reveal interesting differences and similarities that arise from the basic interactions which contribute to scattering. The exchange interaction contributes to electron scattering (due to the indistinguishability of the projectile and electrons in the target atoms) but does not play a role in positron scattering. The static interaction (associated with the interaction of the projectile with the Coulomb field of the undistorted atom) is attractive for the electron and repulsive for the positron, while the polarization interaction (resulting from the distortion of the atom by the charged projectile) is attractive for both projectiles. The net effect of the static and polarization interactions is that they add to each other in electron scattering whereas they tend to cancel each other in positron scattering. In general, this results in smaller total scattering cross sections (Q_T's) for positrons than for electrons at low energies. As the projectile energy is increased, the polarization and exchange interactions eventually become negligible compared with the static interaction, and the expected result is a merging of the corresponding positron and electron scattering cross sections at sufficiently high projectile energies. Two scattering channels that are open only to positrons are annihilation which is negligible[1] for the positron energies (>0.2 eV) that have been used in positron-beam scattering experiments, and positronium formation, which has a threshold energy 6.8 eV below the ionization threshold of the target atoms.

---

*Permanent address: Department of Physics, Yarmouk University, Irbid, Jordan.
**Permanent address: Nanjing Institute of Technology, Nanjing, Jiangsu, The People's Republic of China.

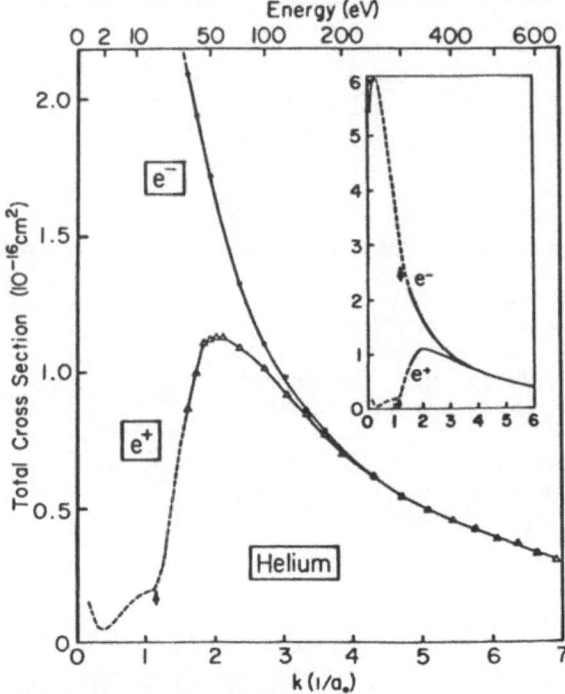

Fig. 1. Comparison of positron-He and electron-He total cross
sections up to intermediate energies. The lowest inelastic
thresholds for each projectile are indicated by arrows.
(From Kauppila et al., Ref. 2)

The general trends observed in comparisons of the total scattering of
positrons and electrons by the room-temperature gases that have been
investigated appear to be consistent with a fairly simple description in
terms of the interactions described above. To illustrate some of these
general trends, Q_T values for positrons and electrons colliding with
helium are shown in Fig. 1 where the measurements for both projectiles
have been performed using the same apparatus and experimental technique.[2]
As the discussion above would suggest, at low energies, the electron-He
Q_T's are significantly larger (more than 100 times as large near 2 eV)
than the positron-He Q_T's. There is also a clear indication in the
positron-He Q_T curve of the onset of positronium (Ps) formation near the
predicted Ps formation threshold (17.7 eV). As the projectile energy is
increased, the positron and electron Q_T's approach each other, and they
are observed to merge (to within 2%) near 200 eV. Similar general
features to those illustrated by the case of helium have been observed for
many other room-temperature gases[3] with respect to (1) a tendency for the
positron Q_T's to be significantly lower than the electron Q_T's at low
energies; (2) clear indications in the positron Q_T curves of the onset of
Ps formation near the predicted Ps formation thresholds; (3) a tendency
for the positron and electron Q_T's to approach each other as the
projectile energy is increased to sufficiently high energies, although,
with the exception of molecular hydrogen[4] and water[5], an actual merging of
the cross sections to the degree that has been observed near 200 eV in
helium has not been observed for other room-temperature gases up to the
highest energies studied (several hundred eV).[3]

Although the general Q_T trends for positrons and electrons colliding with helium and the other room-temperature gases that have been investigated appear to be consistent with expectations based on the interaction arguments discussed above, there are some interesting puzzles that have been pointed out in the case of helium.[2,6] The merging of the positron- and electron-He Q_T's was not expected to occur at such low energies.[2] The positron- and electron-He distorted wave second Born approximation (DW) calculations of Dewangen and Walters[7] do not merge (to within 2%) until 2000 eV. At 200 eV, where the measurements of Kauppila et al.[2] indicate a merging to within 2%, the DW electron calculations are 21% higher than the corresponding positron calculations. Another curious aspect of the observed merging of the positron and electron Q_T's above 200 eV is that the theoretical calculations of Dewangen and Walters[7] suggest that at 200 eV, the electron total elastic cross section is about 2.4 times as large as the positron total elastic cross section. This raises the question of how the Q_T's for positron- and electron-He collisions can merge near 200 eV when these projectiles appear to be behaving much differently with respect to elastic scattering near that energy. Is there some overriding consideration that governs the merging of positron and electron Q_T's at an energy where the separate processes (elastic and inelastic) that contribute to Q_T may still be exhibiting very dissimilar behavior? It is interesting to keep this question in mind in light of the recent measurements that have been performed on a non-room temperature gas, potassium, which will be discussed in this paper.

Prior to the measurements by our group[8] of positron-potassium Q_T's in 1985, the only target atoms for which Q_T's had been measured for positrons were the inert gas atoms. From a theoretical point of view, it would be of considerable interest to measure Q_T's for positron-atomic hydrogen collisions because of the relative simplicity of this system. However, the combined experimental difficulties of obtaining a sufficient number density of atomic hydrogen and a sufficiently intense positron beam have been deterrents to performing this experiment. The alkali metal atoms also have a relatively simple structure, since the ground state of an alkali atom is $^2S_{1/2}$, with a relatively weakly bound valence (n_0s) electron moving outside of a core of closed shells. Although there is some similarity between the single valence alkali atoms and the hydrogen atom, it has been pointed out[9] that it is not necessarily the case that approximation schemes developed for hydrogenic targets will be equally successful for the alkali atoms since the ground states of these atoms have different characteristics. The cross sections for single electron excitation of the alkali atoms are generally larger than those of hydrogen because of the greater size of the alkali atoms. In addition, there is a substantial difference in the relative energies of the ground state and the first excited p level. The 2p level in hydrogen is 10.2 eV above the ground state, whereas the first excited p levels (n_0p levels) of the alkali metals are much closer to the ground state in energy (1.85 eV for Li, 2.10 eV for Na, 1.61 eV for K, 1.56 eV for Rb, and 1.39 eV for Cs). The large coupling between the n_0s and n_0p levels influences significantly the behavior of both elastic and inelastic scattering.[9]

A unique feature of positron-alkali metal collisions is that since the alkali metals all have ionization thresholds less than the binding energy (6.8 eV) of positronium (Ps) in its ground state, a positron with arbitrarily small kinetic energy can form Ps, whereas all of the other atoms and molecules for which Q_T's have been measured have Ps formation thresholds of at least several eV. Another distinguishing feature of the alkali metals is that their polarizabilities are much larger than those of any of the room-temperature gases that have been used as targets for positrons. As an example, the polarizability of K is about 26 times as large as that of Ar, the inert gas which just precedes K in the periodic table of elements.

The main purposes of this paper are to (1) point out what has been accomplished so far in studies of positron collisions with alkali metal atoms, (2) discuss what we consider to be some of the most interesting observations and puzzles in these studies, and (3) suggest some additional investigations in the area of positron-alkali metal atom scattering research which we believe are needed in order to better understand positron- (and electron-) atom collisions.

EXPERIMENTAL TECHNIQUE

The first Q_T measurements for positrons colliding with a non-room-temperature gas (potassium) were reported by Stein et al.[8] who used a beam transmission technique. For these experiments, the positron (and electron) beam apparatus shown in Fig. 2, which had been previously used for our room-temperature Q_T measurements,[2] was also used to generate the projectile beams for the non-room temperature Q_T measurements, but rather than using the curved axial magnetic field region shown in Fig. 2 as the scattering cell (as was done in our room-temperature measurements), the curved magnetic field region is used to transport the projectile beam to the alkali-metal scattering cell shown in Fig. 3 (located in the detector chamber of the system shown in Fig. 2) where the actual beam transmission measurements are made. The weak, guiding axial magnetic

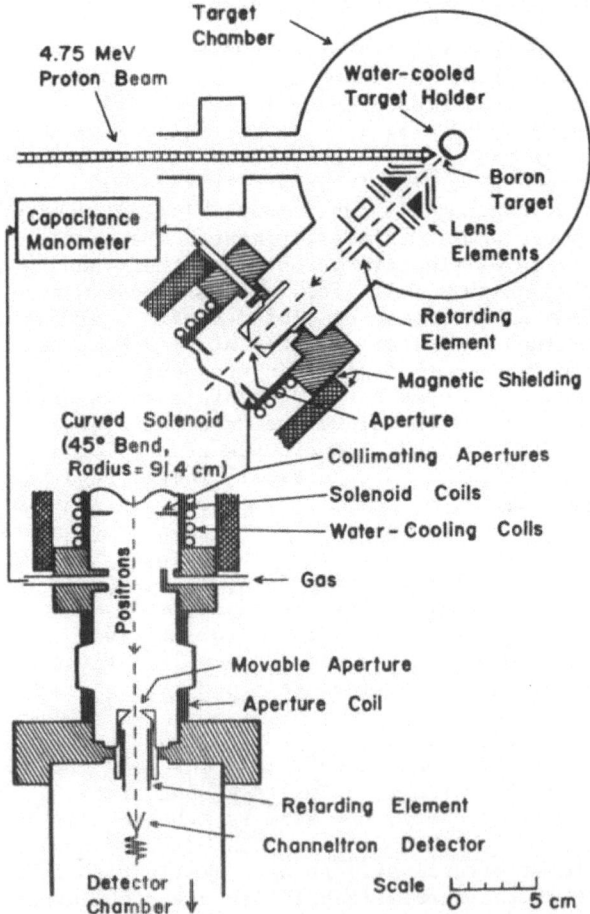

Fig. 2. Experimental setup for measuring total cross sections for room-temperature gases. (From Kauppila et al., Ref. 2)

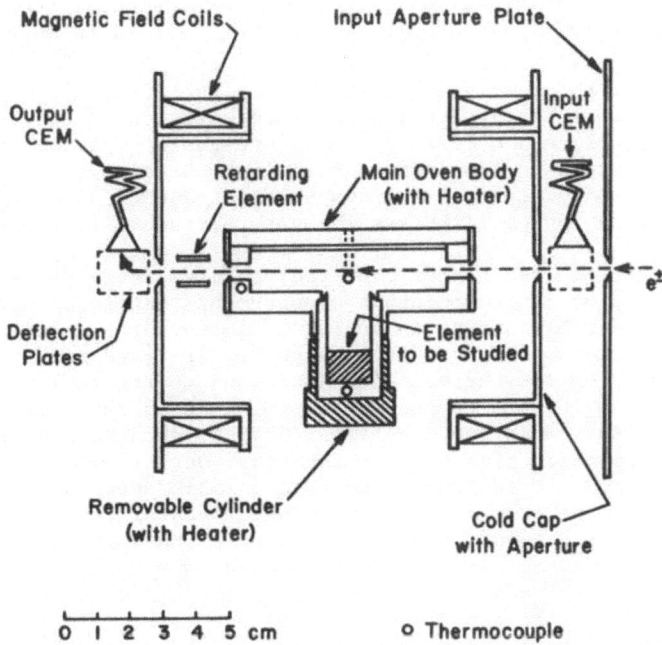

Fig. 3. Experimental setup for measuring total scattering cross
sections for alkali atoms. (From Stein et al., Ref. 8)

field produced by the curved solenoid is extended into the detector
chamber by means of two coils located concentrically with the entrance and
exit apertures of the scattering cell which is a thermally isolated oven.
A Channeltron electron multiplier (CEM) on the input side of the oven
serves (when its front end is biased appropriately) as a detector for
positrons or electrons about to enter the oven. When the cone (front end)
of that detector is placed at ground potential, the projectile beam is
permitted to pass through the oven and the transmitted beam is detected by
the second CEM at the output end of the oven. A retarding element (which
becomes coated with the alkali metal effusing from the oven) located
between the oven and the output CEM is used to measure the projectile
energy as well as to provide additional discrimination[2] (beyond
geometrical considerations) against projectiles scattered through small
angles in forward directions.

In order to determine Q_T's, measurements are made of (1) the ratio,
R_{cold}, of the output CEM to the input CEM counts with negligible vapor in
the oven (oven at room temperature), and (2) the ratio, R_{hot}, of the
output CEM to the input CEM counts with sufficient vapor (oven at an
elevated temperature) to attenuate appreciably the projectile beam. The
purpose of using the ratio of the output CEM to the input CEM counts is to
normalize the transmitted beam intensity with respect to the incident beam
intensity. Determinations of (1) the beam transmission ratio, R_{hot}/R_{cold},
(2) the number density, n, of the alkali atoms, which is determined by
measuring the temperature of the oven at three different locations and by
using published vapor pressure data,[10,11] and (3) the beam path length, L,
of the projectiles through the oven, can be used with the relationship

$$R_{hot} = R_{cold} e^{-nLQ_T}$$

to obtain absolute Q_T values for positrons and electrons colliding with the alkali atoms. Additional details of the experimental technique and a discussion of potential systematic errors are provided in Refs. 8 and 12.

Stein et al. had mentioned in their 1985 publication[8] on positron (electron)-K Q_T measurements that a major potential uncertainty in their experimental technique was related to the uncertainty in the determination of the vapor pressure in the oven which is very sensitive to the oven temperature. As an example, a temperature increase of 10 C^0 in potassium (say from 140^{0}C to 150^{0}C) increases the vapor pressure by roughly a factor of two. In order to prepare for Q_T measurements on the lighter alkali metals where the required higher oven temperatures would place more stringent demands on our temperature measurement techniques, we have invested a considerable effort in trying to improve the accuracy of the oven temperature measurements. The changes are related mainly to reducing heat conduction from the thermocouple junctions to the region outside of the oven, and securing the thermocouple junctions in the oven walls in such a way that the junctions are less likely to change their positions after the thermocouples have been calibrated. Our latest measured positron- (electron)-K results[12] described in this paper were obtained using essentially the same apparatus and experimental technique described in our earlier publication[8], but are based upon these improved oven temperature measurements, and as such supersede our earlier results.[8] In addition, we have extended the energy range of our positron measurements up to 98.5 eV (our earlier positron-K Q_T values[8] were only measured up to 48.6 eV), and we have measured electron-Na Q_T's in preparation for the corresponding positron-Na measurements.

TOTAL CROSS SECTION RESULTS AND DISCUSSION

Our present electron-K and -Na Q_T measurements are shown in Figs. 4 and 5 respectively with prior measurements[13-19] and theoretical results.[20] The present electron results were obtained using the same apparatus and technique as was used for our positron measurements. Neither the prior electron-K measurements nor the prior electron-Na measurements are in particularly good agreement with each other except for the electron-K values of Kasdan et al.[14] and Visconti et al.[15] at low energies (less than 10 eV), so it appears that more Q_T measurements are needed even for electrons colliding with alkali atoms. Of the prior electron-K Q_T values, the indirect determinations of Vuskovic and Srivastava[16] (who used their own crossed-beam measurements of differential cross sections for elastic scattering and for a number of different transitions from the ground state, and ionization cross sections measured by other groups) are in the closest agreement with the present electron-K Q_T measurements. The electron-Na Q_T determinations of the same group (Srivastava and Vuskovic[17]) are also in quite good agreement with the present electron-Na results.

Walters[20] has obtained Q_T's for electron-K and Na collisions by adding the partial cross sections that he selected from existing theoretical and experimental results for the elastic (Q_E), resonance excitation (Q_R), the sum of all the other discrete excitations (Q_D), and the ionization (Q_I) cross sections. Since Walters reported these Q_T values, Q_R and cross sections for many of the other discrete excitations have been measured by Phelps et al.[18] for K and by Phelps and Lin[19] for Na, and we have added these more recent excitation cross section results (rather than the Q_R and Q_D values used by Walters[20]) to the values of Q_E and Q_I selected by Walters, to obtain the Q_T curves shown in Figs. 4 and 5 for K and Na. The resulting Q_T curves (which we will refer to as "Walters-Phelps curves") are close in their shapes and absolute values to our measured absolute Q_T values for these atoms from 20 eV up to the

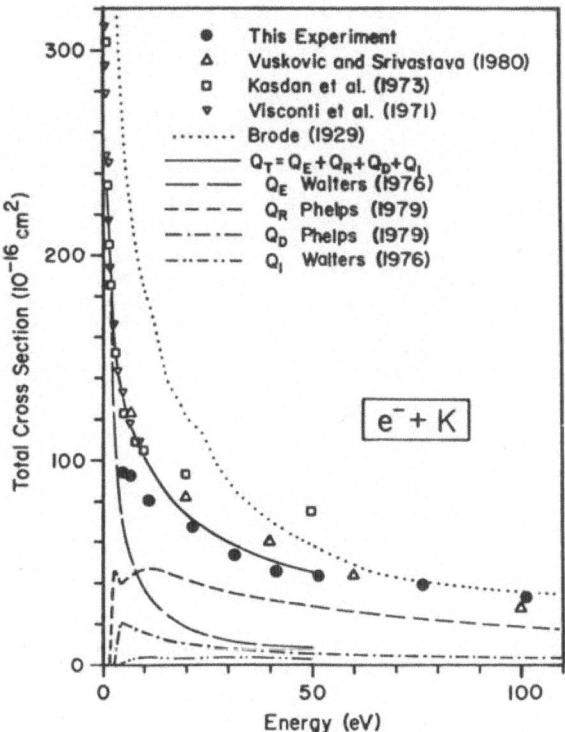

Fig. 4. Electron–K Q_T values.

highest energy of overlap (near 50 eV). Below 20 eV there is a tendency
for our measured electron–K and –Na Q_T's to fall somewhat below the
Walters–Phelps curves as the electron energy is lowered. We feel that the
explanation for this trend in K and Na is as follows. The bias on the
retarding element shown in Fig. 3 is always set within 1.25 V of the
"cut–off" retarding voltage for the projectiles, and since the K and Na
excitation thresholds are 1.61 eV and 2.10 eV respectively, there should
be 100% discrimination against all inelastically scattered projectiles.
In the vicinity of 20 eV for K and for Na, Figs. 4 and 5 show that the
elastic scattering cross section (Q_E) is about 25% of Q_T for K and about
20% of Q_T for Na, and becomes an even smaller fraction of Q_T as the
projectile energy increases toward 50 eV. Below 20 eV on the other hand,
Q_E rapidly becomes a progressively larger fraction of Q_T as the projectile
energy is lowered, and at 5 eV, Q_E accounts for more than 50% of Q_T for
both K and Na. In addition, the angular discrimination of our
apparatus[8,12] against elastically scattered projectiles becomes poorer as
the projectile energy decreases. For instance, the angular discrimination
for electrons is estimated to be about 13° near 5 eV, 9° near 10 eV, 7°
near 20 eV, and is about 5° or less from 30 eV to 100 eV. (The angular
discrimination for elastically scattered positrons is somewhat poorer than
that for electrons, but behaves in a similar way, being about 13° near 10
eV, 11° near 20 eV, 9° near 30 eV, and continuing to improve with
increasing energy, reaching about 5° from 75 to 100 eV.) Our estimates of
errors[12] introduced into the electron–K Q_T's due to an inability to

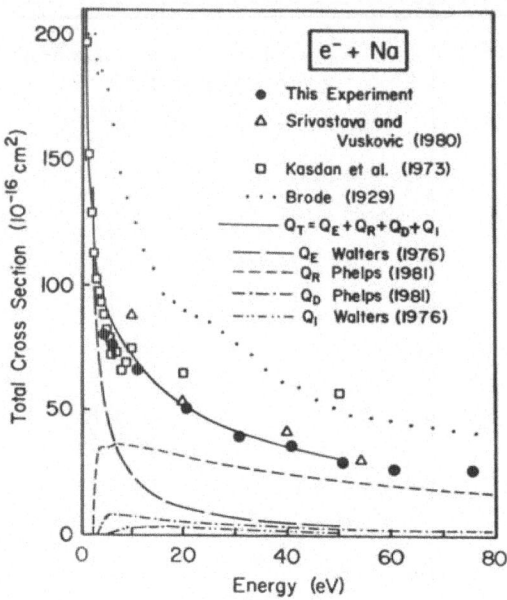

Fig. 5. Electron-Na Q_T values. The present results are
preliminary. Final values will be presented in Ref. 12.

discriminate against projectiles elastically scattered through small
angles in the forward direction suggest that as the projectile energy is
lowered, the increasing ratio of Q_E to Q_T, and the poorer angular
discrimination account for our measured Q_T values falling somewhat below
the Walters-Phelps Q_T curve below 20 eV. At 20 eV and above on the other
hand, we estimate that the error in our measured Q_T values should be of
the order of 10% or less for electron-K collisions and this is also
consistent with our results being slightly lower than the Walters-Phelps
curve near 20 eV and above. We have not yet had an opportunity to
determine the extent of the small-angle forward scattering problem for
electron-Na collisions but the results shown in Fig. 5 appear to be
consistent with the explanation given above, in that the agreement of our
measured Q_T values with the Walters-Phelps curve is very good from 20 eV
to the highest energy of overlap (near 50 eV), with a tendency for our
values to fall slightly below the Walters-Phelps curve below 20 eV,
although it appears that there is even less of a problem in Na than there
is in K. This may indicate that the differential scattering cross section
for low energy (<20 eV) electrons elastically scattered from Na is less
strongly peaked in the forward direction than that for K. It should be
noted that we have not been able to estimate the extent of the small angle
forward scattering problem for positron-K collisions due to a lack of
theoretical calculations of positron-K elastic differential scattering
cross sections.

The closeness of the Walters-Phelps Q_T curves to our own measured electron Q_T values for K and Na gives us some confidence that our experimental technique and apparatus for measuring electron-alkali Q_T's is basically sound, and since the same apparatus and technique is used for the positron measurements, we feel that the positron measurements should not be greatly in error.

The present measured positron-K Q_T's are shown in Fig. 6 along with the corresponding present measured electron results for comparison, and theoretical calculations of Q_T for positrons and electrons,[21] elastic cross sections[22,23] (Q_E), and Ps formation cross sections[23] (Q_{Ps}). The second Born (SB) approximation and modified Glauber (MG) approximation calculations of Q_T by Gien[21] are based upon a "full model potential" (FMP). Neither the positron nor electron theoretical Q_T results of Gien are close to our measured values, although the disagreement may not be too surprising in view of the tendency of these types of calculations to not work as well at the lower energies where comparisons are being made with our results. The electron-K Q_T results of unpublished modified Glauber calculations reported by Khare[24] (not shown in Fig. 6) based on an "inert core" are somewhat lower (within 30%) than the present positron- and electron-K Q_T's, and the "single particle scattering model" results[24] (which do not take multiple scattering terms into account) are appreciably higher than the present values. Our measured positron Q_T's are more than seven times as large as the theoretical estimates of Q_E by Guha and Mandal[23] (who used a pseudopotential formalism) at the energies of overlap. The first Born approximation and distorted wave approximation calculations of Q_{Ps} by Guha and Mandal[23] suggest that Ps formation accounts for a relatively small fraction of the present Q_T's above 10 eV. The discussion above suggests the possibility that for positrons between 10 and 100 eV,

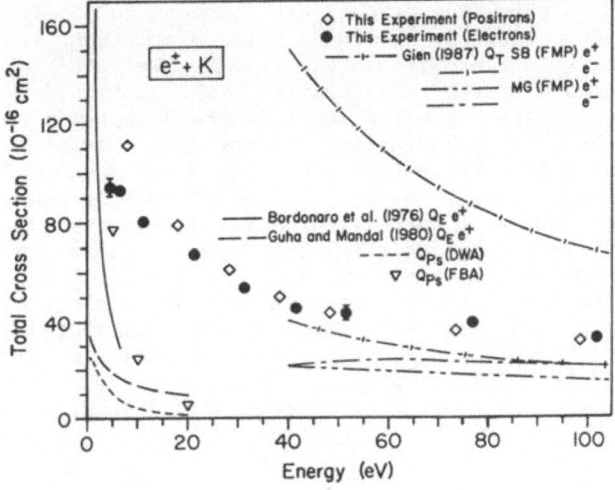

Fig. 6. Comparisons of positron-K and electron-K Q_T values.

excitation or perhaps ionization (for both of which no measurements or calculations of cross sections yet exist for K) may make the major contribution to Q_T (as is the case for electron-K collisions as shown in Fig. 4).

Perhaps the most striking aspect of the results shown in Fig. 6 is the very close proximity of the positron and electron-K Q_T's over the entire energy range that has been investigated. Stein et al.[8] had observed in 1985 that their positron-K Q_T results were closer to the corresponding electron values over the energy range from 5 to 50 eV than had been the case for any other gases for which such comparisons had been made.[2,6,25] The present positron- and electron-K Q_T results[12] which have been obtained with improved temperature measurements suggest that the corresponding Q_T values for positrons and electrons are essentially merged to within the statistical uncertainties of the measurements from near 30 eV to the highest energy which we have investigated (about 100 eV). It is interesting that the only clear difference between our positron and electron results appears as the projectile energy is lowered below 30 eV, where the positron Q_T's become progressively higher than the corresponding electron values as the projectile energy is decreased. The shapes of each of the sets of theoretical Q_{Ps} results[23] in Fig. 6, which show monotonically increasing Q_{Ps} values as the positron energy is lowered, suggest that a significant part of the increasing difference between the positron and electron Q_T values at low energies could be due to the increasing role of Ps formation as the positron energy is lowered. However, differences in the extent of small-angle forward scattering problems for each projectile may also be playing a role in the separation of the respective Q_T curves at low energies.

In a discussion of their original positron- and electron-K measurements, Stein et al.[8] had suggested that one possible interpretation of the proximity of the positron and electron results at low energies is that the polarizability of K is so large that it could be overwhelming the static interaction at the low energies used in those experiments. As a result, the tendency of the static and polarization interactions to cancel each other in the case of positron scattering and to add in the case of electron scattering may not differentiate between these projectiles to the same degree for K as it does in scattering from targets of much lower polarizability. However, theoretical investigations by Walters[20,26] of electron-alkali atom scattering indicate that with increasing energy beyond the first excitation threshold, there is a change-over from a situation where polarization effects are dominant to one in which flux loss[26] becomes dominant. He has shown that calculations which ignore this change run into drastic problems. For example, Walters[20,26] has shown that a calculation[27] of the total elastic cross section for electron-lithium scattering which rests on the assumption that this cross section is dominated by polarization effects up to at least 50 eV, and does not take account of dynamical effects such as the opening of inelastic channels, leads to a result at 50 eV that is nearly an order of magnitude larger than that calculated in the close-coupling and Glauber approximations. In view of Walters' findings, it would seem that (1) the close proximity of our observed positron- and electron-K Q_T's at relatively low energies, but at energies well above the excitation threshold of potassium (1.61 eV), should not be attributed to a dominance of polarization effects, and (2) our earlier speculation[8] is inappropriate that there could be a diverging of the positron and electron Q_T values at higher energies with an eventual remerging at even higher energies.

FUTURE DIRECTIONS

We feel that our observations of the essentially merged values of the positron- and electron-K Q_T's near and above 30 eV, and the higher Q_T's for positrons (than for electrons) below 30 eV raise a number of interesting questions and suggest a need for several additional investigations (experimental and theoretical), a few of which are listed below. (1) Is there a simple theoretical explanation for the observed merging of the positron- and electron-K Q_T's at the low energies investigated in the present measurements? In relation to this question, it is of interest that a theoretical analysis by Dewangen[28] related to higher order Born amplitudes calculated in the closure approximation has been shown to imply[26,29] that if electron exchange can be ignored in the electron-scattering case, and if the closure approximation is valid, then a merging (or near-merging) of positron- and electron-atom Q_T's can occur at energies considerably lower than the asymptotic energies at which the first Born approximation is valid. (2) Is the low energy merging of the positron and electron Q_T's for potassium unique or is it also a characteristic of other alkali metal atoms? It would be of definite interest to measure positron and electron Q_T's for the other alkali atoms to investigate this matter further. (3) To what extent is the increasing difference between the positron- and electron-K Q_T's as the projectile energy is lowered, due to the increasing role of Ps formation, as we have suggested as a possible explanation for this low energy difference? It would be helpful to have direct measurements of Q_{Ps} and additional, reliable Q_{Ps} calculations for potassium (and for the other alkali atoms) to help answer this question. (4) How does the Ps formation cross section for positron-alkali atom collisions behave as the projectile energy approaches zero? As was mentioned above, it is possible to form Ps in collisions with alkali metal atoms at arbitrarily small positron energies. Does Q_{Ps} increase without limit as the positron energy approaches zero? Measurements of Q_T down to lower energies for positron-alkali atom collisions may help to shed some light on the behavior of Q_{Ps} as the positron energy approaches zero.

An additional point to be considered in regard to future directions is that, as was mentioned above, in the case of He, where the positron and electron Q_T's have been observed[2] to merge near 200 eV, the partial contributions (such as Q_E) to Q_T apparently are behaving much differently for positrons than for electrons. What is the case for potassium where the Q_T's appear to have merged near 30 eV? Is Q_E for electrons also significantly larger than Q_E for positrons in this case? The two sets of theoretical positron-K Q_E values shown in Fig. 6 differ from each other too much to draw any definite conclusions regarding how positron- and electron- Q_E values compare with each other. It would be helpful to have additional, reliable theoretical calculations of Q_E values for positron-K collisions (and for other alkali atoms) in order to answer questions such as this. If it is found that there are several examples of mergings of positron- and electron-atom Q_T values at relatively low energies, where the partial contributions to Q_T from Q_E and inelastic cross sections are much different for positrons than they are for electrons, this again raises a question concerning whether there is something more fundamental about Q_T than has been appreciated in the past.

Retarding potential curves, such as that shown in Fig. 7 may shed some additional light on the relative roles of elastic and inelastic scattering for positron- and electron-alkali atom collisions. Fig. 7 shows plots of (1) the electron beam current transmitted through a "cold oven" (no appreciable K vapor in the oven) versus the retarding potential applied to the retarding element shown in Fig. 3, and (2) the corresponding curve for a "hot oven" (appreciable K vapor in the oven).

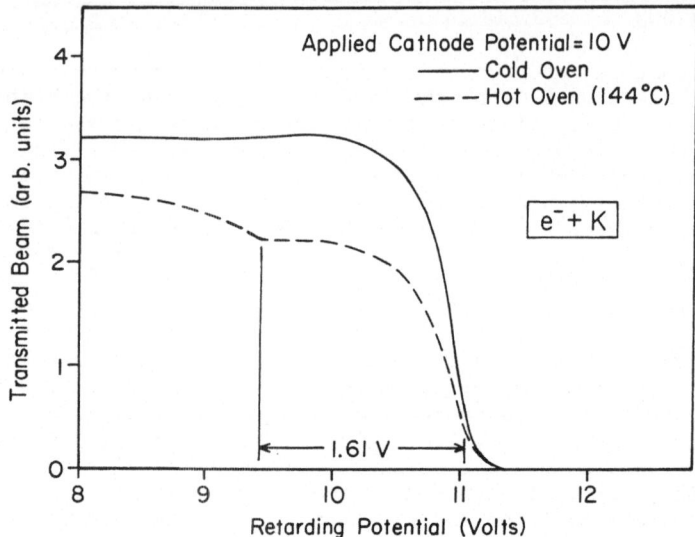

Fig. 7. Comparison of electron retarding potential curves taken with
no K vapor in scattering cell (cold oven) and with K vapor in
scattering cell (hot oven).

(The cold oven curve is actually an average of retarding potential curves
taken before and after the corresponding hot-oven curve.) The hot oven
curve shows a noticeable reduction in transmitted beam current with a
cut-off about 1.6 V prior to the cut-off which was present in the cold
oven curve, which is consistent with the evidence shown in Fig. 4 that an
appreciable fraction of the electrons have scattered inelastically,
producing the "resonance excitation" in potassium which corresponds to an
excitation of the valence electron to an energy level 1.61 eV above its
ground state energy. Although there is angular distribution information
that must be folded in for a proper interpretation of such a set of
retarding potential curves, we feel that retarding potential curves such
as this will yield useful information on the relative roles of the
resonance excitation, versus elastic scattering for positrons compared
with electrons, and we plan to obtain the corresponding positron retarding
potential curves in the near future for this purpose. It would also be
very helpful to have calculations of excitation and ionization cross
sections for positron-alkali atom collisions for comparison with the
corresponding electron cross sections.

ACKNOWLEDGMENTS

    We would like to acknowledge the contributions of R. Lukaszew and S.
Parikh to taking data in these experiments. The Wayne State University
positron (electron) scattering group is supported by the National Science
Foundation (Grant #PHY83-11705).

REFERENCES

1. H.S.W. Massey, Phys. Today 29 (3):42 (1976).
2. W.E. Kauppila, T.S. Stein, J.H. Smart, M.S. Dababneh, Y.K. Ho,
      J.P. Downing, and V. Pol, Phys. Rev. A 24:725 (1981).
3. W.E. Kauppila and T.S. Stein, Can. J. Phys. 60:471 (1982).

4. K.R. Hoffman, M.S. Dababneh, Y.-F. Hsieh, W.E. Kauppila, V. Pol, J.H. Smart, and T.S. Stein, Phys. Rev. A 25:1393 (1982).

5. O. Sueoka, S. Mori, and Y. Katayama, J. Phys. B 19:L373 (1986).

6. T.S. Stein and W.E. Kauppila, in: Electronic and Atomic Collisions, D.C. Lorents, W.E. Meyerhof, and J.R. Peterson, eds. (North-Holland, Amsterdam, 1986), pp. 105–123.

7. D.P. Dewangen and H.R.J. Walters, J. Phys. B 10:637 (1977).

8. T.S. Stein, R.D. Gomez, Y.-F. Hsieh, W.E. Kauppila, C.K. Kwan, and Y.J. Wan, Phys. Rev. Lett. 55:488 (1985).

9. B.H. Bransden and M.R.C. McDowell, Phys. Rep. 46:249 (1978).

10. B. Shirinzadeh and C.C. Wang, Appl. Opt. 22:3265 (1983).

11. R.E. Honig, RCA Rev. 18:195 (1957).

12. C.K. Kwan, M.S. Dababneh, W.E. Kauppila, T.S. Stein, and Y.J. Wan, to be submitted for publication.

13. R.B. Brode, Phys. Rev. 34:673 (1929).

14. A. Kasdan, T.M. Miller, and B. Bederson, Phys. Rev. A 8:1562 (1973).

15. P.J. Visconti, J.A. Slevin, and K. Rubin, Phys. Rev. A 3:1310 (1971).

16. L. Vuskovic and S.K. Srivastava, J. Phys. B 13:4849 (1980).

17. S.K. Srivastava and L. Vuskovic, J. Phys. B 13:2633 (1980).

18. J.O. Phelps, J.E. Solomon, D.F. Korff, C.C. Lin, and E.T.P. Lee, Phys. Rev. A 20:1418 (1979).

19. J.O. Phelps and C.C. Lin, Phys. Rev. A 24:1299 (1981).

20. H.R.J. Walters, J. Phys. B 9:227 (1976).

21. T.T. Gien, Phys. Rev. A 35:2026 (1987).

22. G. Bordonaro, G. Ferrante, M. Zarcone, and P. Cavaliere, Nuovo Cimento Soc. Ital. Fis. 35 B:349 (1976).

23. S. Guha and P. Mandal, J. Phys. B 13:1919 (1980).

24. S.P. Khare, in: Positron (Electron)-Gas Scattering, W.E. Kauppila, T.S. Stein, and J.M. Wadehra, eds., (World Scientific, Singapore, 1986), pp. 131–139.

25. Ch.K. Kwan, Y.-F. Hsieh, W.E. Kauppila, Steven J. Smith, T.S. Stein, M.N. Uddin, and M.S. Dababneh, Phys. Rev. Lett. 52:1417 (1984).

26. H.R.J. Walters, Physics Reports 116:1 (1984).

27. M. Inokuti and M.R.C. McDowell, J. Phys. B 7:2382 (1974).

28. D.P. Dewangen, J. Phys. B 13:L595 (1980).

29. F.W. Byron, Jr, C.J. Joachain, and R.M. Potvliege, J. Phys. B 15:3915 (1982).

THEORETICAL STUDIES OF LOW-ENERGY POSITRON-ALKALI ATOM SCATTERING

S.J. Ward, M. Horbatsch, R.P. McEachran and A.D. Stauffer

Physics Department

York University, Toronto, M3J 1P3, Canada

INTRODUCTION

The recent experimental total cross-section measurements for e^+-K scattering (Stein et al. 1985) have provided a stimulus to obtain reliable theoretical results for low-energy positron scattering by the alkali atoms. The fact that for positron scattering from these atoms, the rearrangement channel, namely positronium formation, is open right down to zero energy provides an additional reason to consider this class of atoms. With previously studied systems, for instance positron scattering by hydrogen or the inert gases, this threshold becomes open only at a finite non-zero energy. It should be interesting to examine the effect of this anomalous theshold on low-energy scattering.

THEORETICAL CONSIDERATIONS

Positron scattering from the alkali atoms is one of the few systems which have been studied both theoretically and experimentally. These atoms are basically "one-electron" atoms and very good wave functions exist to describe such systems, for instance the frozen-core Hartree-Fock wave functions (Cohen and McEachran, 1980). In any theoretical treatment, the polarizabilities of the target should be well represented since these atoms have extremely large polarizabilities. For example, the static dipole polarizability (αd) of Li is approximately 164 a_0^3 (Miller and Bederson, 1977), which is about 15 times that of Ar. It is noted, however, that the first excited $^2P^0$ state accounts for the majority of this polarization in alkali atoms. For instance, the $2p^2P^0$ state of Li accounts for 98% of αd.

Since the ionization potentials of the alkali atoms are low, ranging from 5.4 to 3.9 eV as we go from Li to Cs, absorption into the excited states will be important even for low-energy scattering. It is thus necessary that at least a few of the lowest states, especially the lowest $^2P^0$ state, be explicitly included in the wave function of the whole system.

Positronium formation into its ground-state should also be included, especially for very low-energy scattering, but the effect of positronium formation into its excited states needs to be considered as well. Positronium has a fairly large polarizability, $36a_0^3$, and thus allowance should ideally be made for the effect of polarization, perhaps by explicitly

including the 2p state of positronium by using a polarized orbital in the
Ps-formation channel.

BRIEF REVIEW OF THEORETICAL CALCULATIONS

Over the last twenty years there has been much effort to determine cross-
sections for e^+-alkali atom scattering. However, a large proportion of these
calculations have been concerned with very high energies, elastic scattering
and mostly with the target atom Li.

Bordonaro et al. (1976) and Ferrante et al. (1978) have used the semiclassical JWKB approximation to determine elastic cross-sections for scattering positrons from Li, Na, K, Rb and Cs for energies less then 7 eV. Sarkar
et al. (1973) have used the first Born approximation (FBA), the polarized
FBA and the modified eikonal method, for the energy range 0.8 to 500 eV, to
determine the cross-section for Li. Both calculations predict a deep minimum
in the differential cross-section at low energies near a scattering angle of
90°. Wadehra (1982) has also predicted the elastic cross-section for Li,
Na, K and Rb by using the FBA with a polarization plus static interaction for
the energy range 500 to 1000 eV. All the above methods are high energy
approximations and are thus not likely to be accurate in the low-energy
region.

Bui and Stauffer (1971) have used the polarized orbital method described
by Temkin (Temkin 1957, 1959; Temkin and Lamkin, 1961) with the P-wave
equation correctly derived by Sloan (1964). They determined the elastic and
diffusion cross-sections as well as Z_{eff} for e^+-Li scattering up to an energy
of 7 eV. This procedure was later used by Bui (1975) for Na.

In all these calculations, no allowance has been made for the rearrangement channel, namely Ps-formation. However, there has been a series of calculations to determine the Ps formation cross-section for e^+-alkali atom scattering. Guha and Mandal (1980) have used a pseudo potential formulation in
which the valence electron is assumed to move in the field of a model potential. Two channels were considered, namely Ps formation and elastic scattering. The equations were solved for Na, K, Rb and Cs in the energy range 0.5
to 20 eV using the distorted wave approximation (Mandal et al. 1979) and the
FBA. These authors extended the calculation to consider Ps-formation in
excited ns-states, for the incident positron energies 5 to 50 eV, but only
using the FBA (Mandal and Guha, 1980). A similar calculation was performed
for Li for the energy range 10 to 500 eV by Guha and Saha (1980). From their
results, they show that the asymptotic Ps-formation cross-section obeys an
inverse n^3 law irrespective of energy. It should be noted however, that the
FBA (or any first order approximation) is not suitable to describe a rearrangement process at high energies since it is the second Born approximation to
which the cross-section converges.

A two-state approximation has been used by Guha and Ghosh (1981) to
consider the problem of e^+-Li scattering for the energy range 0.5 to 10 eV.
The two channels they considered were elastic scattering and Ps-formation.
Polarization potentials derived by the method of Temkin and Lamkin (1961)
were included in both channels. The resulting equations were solved using
an integral approach in momentum space. Their findings show that at and
below 5 eV the differential cross-sections for the two channels are comparable in magnitude. Above the ionization threshold (5.39 eV), the effect of
Ps-formation on the scattering is less significant. However, this calculation suffers from the disadvantage that no allowance has been made for the
excitation of the target atom. The effect of excitation into the $2p^2p^0$ state,
whose threshold lies at 1.85 eV above the ground-state, will be of importance.
Indeed, excitation into the other excited states should ideally be repre-

sented because of the low ionization potentials of the alkalis. Also, since the alkalis have extremely high polarizabilities, the form of the polarization potential should be carefully considered (Bhatia et al. 1978). Mazumdar and Ghosh (1986) felt it necessary to use the polarized orbital equations of Bhatia et al. (1978) since this gives a good representation of the polarization in the incident channel. Their calculation was performed for Li and for the incident positron energies 2.0 to 100 eV. They used the distorted wave model of Khan and Ghosh (1983 a,b) since this proved to be satisfactory for both e^+-H and e^+-He scattering.

Intermediate energy (50 - 300 eV) calculations have been performed by Nahar and Wadehra (1987) for Li and Na in the FBA and DWBA to determine the Ps-formation cross-section. Gien (1987 a) has performed a calculation for the intermediate energy range (40 - 1000 eV) to evaluate the total cross-section for Li, Na and K using the modified Glauber and the second Born approximation. The former calculation has been extended using the core-corrected modified Glauber approximation (Gien, 1987 b).

Only recently have excitation cross-sections been determined for the e^+-Li system. Khan et al. (1987) have evaluated elastic and excitation cross-sections for low-energy e^+-Li scattering using a five state close-coupling approximation. For the ground and excited states of Li they used analytical Hartree-Fock wavefunctions (Weiss, 1963). They neglected Ps-formation since it was felt (Guha and Ghosh, 1981) to only have a marginal effect on the total cross-section above 5 eV. Results were obtained for incident positron energies 2.0 to 10.0 eV. In order to test convergence of this cross-section with respect to the number of eigenstates included in the expansion of the wave function, they performed both a two-state and a four-state calculation. It was concluded that convergence with respect to the number of eigenstates was slow for this energy-range. This is consistent with the results of McEachran and Fraser (1965) for e^+-H scattering where they steadily increased the number of hydrogen states from one to six. Khan et al. (1987) also showed that convergence was slow with respect to the total angular momentum L. The (2s-2p) partial contribution to the total cross-section is not negligible even at L = 30 for an incident energy of 10 eV. Their results indicate that inelastic scattering gives rise to the dominant contribution to the total cross-section. They suggested from their large excitation cross-sections and from the variations of the elastic and Ps formation cross-sections below 2 eV obtained by Guha and Ghosh (1981), that there is likely to be a Ramsauer minimum in the total e^+-Li cross-section. Comparison has been made of the (2s-2p) excitation cross-section for positron impact using a two-state approximation with that for electron impact (Burke and Taylor 1969). The e^+ results are considerably higher.

PRESENT CALCULATIONS

Our principle reason for theoretically treating this problem of e^+-alkali scattering was to provide a comparison with the experimental results, although we are also interested in examining the theoretical implications of the Ps-formation threshold being open at zero energy.

Since the alkalis are highly polarizable, it is essential to treat properly the polarization of the target atom. We have thus performed a polarized orbital calculation as described further in the text which allowed us to determine the elastic scattering cross-section for positron scattering from Li, Na and K.

A close-coupling calculation has also been performed in which the scattering wavefunction has been expanded in terms of the target states. This enables both elastic and excitation cross-sections to be evaluated.

However, a full treatment of the problem should include both the excited states of the target atom as well as the positronium states. In view of the results of Guha and Ghosh (1981) which showed that ground-state Ps-formation has only a marginal effect above 5 eV, we felt that initially it was reasonable to neglect the Ps-formation channel. We have applied the close-coupling method to positron scattering from Li, Na and K. With higher members of this series, relativistic effects become increasingly important.

Polarized Orbital Method

It has been stated in a number of papers, for instance by Walters (1976), Bhatia et al. (1978), that for the alkali atoms, which are highly polarizable systems and in which the various inelastic thresholds occur at low energies, the orthodox formalism of Temkin and Lamkin (1961) is inadequate. Thus, we have employed the method of Stone (1966) to determine our polarized orbital. This method is not based on first order perturbation theory, but rather minimizes the adiabatic energy of the perturbed system in which the wave function is written as a linear combination of the ground state and first excited $^2p^0$ state of the atom. The scattering equation was determined by two methods. In the first approach the Schrödinger equation for the total wave function was projected onto the unperturbed ground state of the atom as originally formulated by Temkin (1957). This gives rise to both a static and a dipole polarization potential. In the second approach we have used the so-called extended polarized orbital method (Callaway et al. 1968) in which the Schrödinger equation is projected onto the perturbed ground-state of the atom; this gives rise to an additional repulsive distortion potential which is finite at the origin and behaves asymptotically as r^{-6}. In addition we have separately determined a quadrupole polarization potential via Stone's method and performed calculations where this latter potential is also included.

Using these procedures we have obtained elastic cross-sections for e^+ scattering from Li, Na and K for the energy range 0.05 to 60.0 eV. Separate calculations have been made in which only the dipole potential (P01), both the dipole and distortion potentials (P02), and finally the dipole, distortion and quadrupole potentials (P03) are included.

Close-Coupling Method

The experimental values (Stein et al. 1987) of the total cross-section for K are remarkably similar for positron and electron scattering in the energy range considered (7.89 - 98.50 eV). This feature has not been observed experimentally for previously studied atoms or molecules. It would be of interest to investigate whether the individual elastic and inelastic cross-sections also exhibit this behaviour. The close-coupling method enables the various elastic and excitation cross-sections to be evaluated, as well as taking into account the majority of the static dipole polarizability through inclusion of the first excited $^2p^0$ state.

We have begun the calculation by neglecting Ps-formation both in its ground and excited states. Also, we have not considered the ionization of the target atom. This is a reasonable approximation since for e^--Li scattering, the experimental ionization cross-section (Jain et al. 1973 and Brink 1962) is very much smaller than the corresponding elastic and resonant excitation cross-sections.

To a good approximation, the alkali atoms can be treated as "one-electron" systems consisting of a valence electron moving outside a frozen core. Numerical Hartree-Fock functions are used to describe both the core and the valence electrons.

The equations for the radial scattering wave function take the same form as that for hydrogen (Percival and Seaton, 1957) provided the core potential (Salmona and Seaton, 1961, Burke and Seaton, 1971) is included in the Hamiltonian. Thus, for e^+-alkali atom scattering, we solve the set of equations

$$h_\nu F_{\nu L}(r) = -2 \sum_{\nu'} V(\nu,\nu')_L F_{\nu'L}(r) \tag{1}$$

where $\nu = n_1 l_1 l_2$,

$$V(\nu,\nu')_L = \sum_\lambda f_\lambda(l_1 l_2, l_1' l_2'; L) \; y_\lambda(n_1 l_1, n_1' l_1'; r) \tag{2}$$

$$h_\nu = d^2/dr^2 - l_2(l_2+1)/r^2 + k_\nu^2 - 2v_c \tag{3}$$

and

$$v_c = Z/r - \sum_{nl} 2(2l+1) \; y_0(nl, nl; r) \tag{4}$$

It should be carefully noted that the frozen core approximation means that we are using the same core orbitals for all states of the valence electron. Thus the summations encountered in equations (1) and (2) are only over the valence electron and the summation for the core potential in equation (4) is only over the core electrons.

These coupled equations were solved by using the matching algorithm described by Smith (1971). This method consists of integrating the equations both outwards from the origin to a radial positron r_0, small enough so that the increasing exponential function has not dominated the solution, and inwards from a value r_B in the asymptotic region, to the value r_0. At the radial position r_0, the inward and outward solutions are matched to determine the K-matrix. Since the polarization potential, which is of long-range, is large, we used the asymptotic expansion given by Burke and Schey (1962) and Burke and Seaton (1971), in order to use as small a value for r_B, the extraction point, as possible.

This program was checked by solving the equivalent equations for positron-hydrogen scattering. Agreement was obtained with McEachran and Fraser (1965) below the n = 2 threshold for the close-coupling approximations 1s, 1s-2s, 1s-2s-2p, ..., 1s-2s-2p-3s-3p-3d and for the total angular momentum L = 0, 1 and 2. Also, above threshold, agreement was obtained with Burke and Smith (1962) for 1s-2s-2p, L = 0, 1 and 2. We then extended the program to the alkali atoms by the inclusion of the core potentials derived by the frozen-core Hartree-Fock treatment.

We have obtained elastic and excitation cross-sections for e^+-scattering from Li, Na and K, for the energy range 1 – 30 eV, using a varying number of states in the expansion of the scattering wavefunction. The states were added in order of their excitation energy. In the Li calculation we have so far included 1, 2 and 4-states from the set (2s, 2p, 3s, 3p). Both for Na and K, two states have been included, namely 3s, 3p and 4s, 4p respectively. Preliminary results are also reported in this paper for a 4-state (4s, 4p, 5s, 3d) calculation for K. A more detailed presentation of our results will be found elsewhere (Ward et al. 1988). This work represents the early stages of our investigation of the e^+-alkali system using the close-coupling (CC) expansion in which the number of states is systematically increased in an attempt to obtain convergence. Where possible, comparison is made with the theoretical results of Khan et al. (1987) and the experimental results of Stein et al. (1987).

RESULTS AND DISCUSSION

In the polarized orbital calculation (PO), the elastic cross-section
for Li, Na and K was very sensitive to the form of the interaction potential.
This can be clearly seen in Figure 1. where Q_{el} was determined for e^+-Li
by the PO1, PO2 and PO3 methods. The inclusion of the repulsive distortion
potential reduces this cross-section at very low energies by more than a
factor of two. It is also to be noted that this cross-section is enormous
at very low-energies. A comparison with experimental results for elastic
scattering, if such could be obtained, would provide a very sensitive test
of the details of the interaction potential.

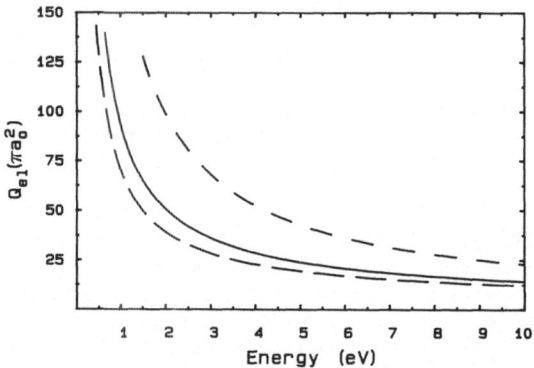

Figure 1. Elastic cross-section (Q_{el}) for e^+-Li scattering determined by
the polarized orbital method with different interaction potentials:
PO1 ($-$ $-$ $-$); PO2 ($-$ $-$) and PO3 ($\underline{\quad\quad}$) .

In our CC calculation, the $3d\,^2D$ state of Li has not been included. The
coupling of this state with the ground-state gives rise to a quadrupole
potential. Therefore, the appropriate PO results with which to compare Q_{el}
evaluated by the 2-state (2s-2p) or 4-state (2s-2p-3s-3p) CC calculations
are those determined by the PO1 method. This is because in this PO calculation neither the quadrupole potential nor the distortion potential has been
included. In Figure 2, Q_{el} determined by the PO1 method is compared with
the 4-state CC results. The agreement between these two sets of results is
reasonable, which is to be expected since it is the coupling of the lowest
P-state alone with the ground-state that accounts for majority of the static
dipole polarizability. This can be seen in Figure 3 since Q_{el} only changes
very slightly as the number of states in the CC expansion is increased from
two to four. The total cross-section, evaluated by summing over the elastic

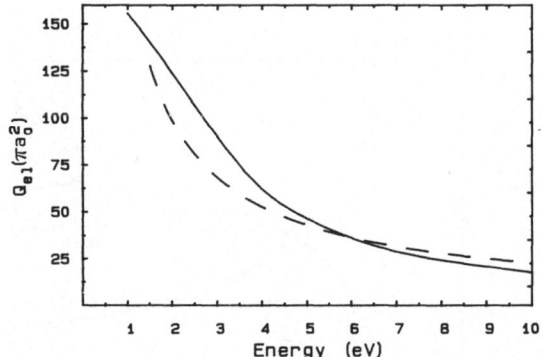

Figure 2. A comparison of the elastic cross-section (Q_{el}) for e^{+}-Li
scattering evaluated by the PO1 (— — —) and the 4-state (2s-2p-3s-3p)
close-coupling (————) calculations.

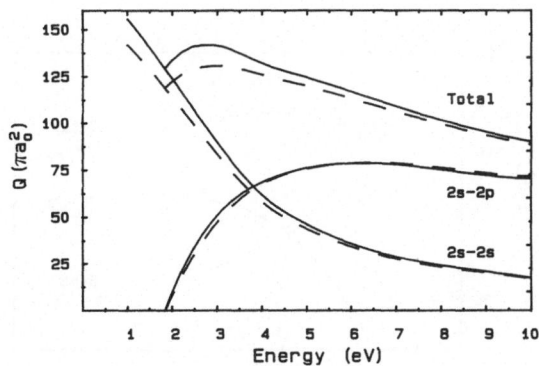

Figure 3. A comparison of the elastic, 2s-2p excitation and total cross-
section for e^{+}-Li scattering evaluated by the 2-state (— — —) and 4-state
(————) close-coupling calculations.

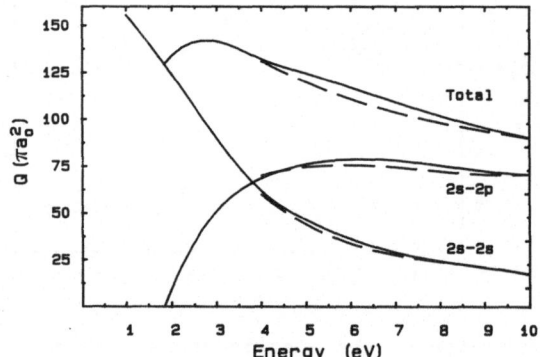

Figure 4. A comparison of the elastic, 2s-2p excitation and total cross-sections for e$^+$-Li scattering determined by Khan et al (1987) using the 4-state calculation (— —) with the present 4-state results (———).

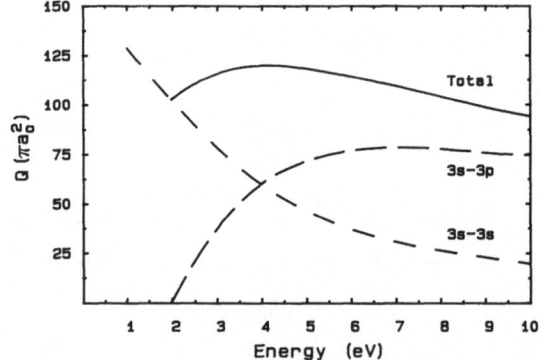

Figure 5. Elastic, 3s-3p excitation and total cross-sections for e$^+$-Na scattering evaluated by using the 2-state (3s-3p) close-coupling calculations.

and various inelastic cross-sections, is slightly larger for the 4-state than the 2-state case for the energy range 1 - 10 eV. By 10 eV, the total cross-section determined by the two calculations merge to within 2%. However, as shown by Khan et al. (1987) the inclusion of the $3d^2D$ state will increase the total cross-section somewhat. In Figure 4, the elastic 2s-2p excitation and total cross-sections evaluated by our 4-state CC calculations are compared with the 4-state results of Khan et al. (1987). The slight difference between the two calculations could be attributed to the fact that we used numerical frozen-core Hartree-Fock wavefunctions for the ground and excited states of the target atom, whereas Khan et al. (1987) used analytical Hartree-Fock wavefunctions given by Weiss (1963).

As shown in Figure 5, the 2-state (3s-3p) CC elastic, the resonant excitation and total cross-sections for e^+-Na scattering have a similar behaviour to that for e^+-Li. However, Sarkar et al. (1987) have some preliminary results in which they showed that the 4-state (3s-3p-4s-3d) CC elastic cross-sections are considerably larger than the 2-state cross-sections, for energies up to 40 eV. The 3s-3p excitation cross-section is somewhat larger in the 4-state calculation. The increase in the total cross-section in going from a 2-state to a 4-state calculation is much more significant in Na than in Li. This is because in the 4-state calculation of Na the $3d^2D$ state has been included which contributes to a long-range quadrupole potential. Sarkar et al. (1987) have also performed a 5-state (3s, 3p, 4s, 3d, 4p) calculation for e^+-Na scattering.

We have performed a 2-state CC calculation for e^+-K scattering for the energy-range 1 to 30 eV to determine the elastic, 4s-4p excitation and total cross-sections. The 4s-4p excitation cross-section converged very slowly with total angular momentum L. This is because K is highly polarizable, the static dipole polarizability being $293a_o^3$. In Figure 6, the total cross-section is compared with experimental data. For energies greater than 10 eV, the theoretical cross-section is shown as a dashed line to indicate that the cross-section, which was calculated with twenty partial waves, had not fully converged with L. The 2-state cross-section falls below the experimental results with increasing positron energy. However, it is possible that the fully converged cross-section will be slightly greater than the experimental results. Also, if the trend in the total cross-section as the number of target states in the CC expansion is increased is similar to that obtained for e^+-Li and e^+-Na, we would expect that the total cross-section for e^+-K scattering to increase as more states are included. We have obtained some preliminary results using a 4-state (4s-4p-5s-3d) expansion for the energy-range 1 - 10 eV and this was the case. For electron scattering, the total cross-section derived from theoretical calculations is slightly larger than the experimental results for electron energies up to 50 eV (see Stein et al. 1987 and references cited therein).

In Figure 7, the elastic and 4s-4p cross-sections evaluated by the 4-state calculation are compared with the 2-state results for the energy range 1 - 10 eV. The inclusion of the $3d^2D$ state affects quite considerably the 4s-4p excitation cross-section. As well, preliminary results indicate that the 4s-3d cross-section is between 29% and 50% of the 4s-4p cross-section. The net effect is an increase in the total cross-section at all energies calculated. We shall report the 4-state partial and total cross-section for e^+-K scattering in more detail elsewhere (Ward et al. 1988).

FURTHER WORK

We are in the process of extending the polarized orbital work to the calculation of excitation cross-sections within the framework of a distorted wave model. With regards to the close-coupling calculation we have so far

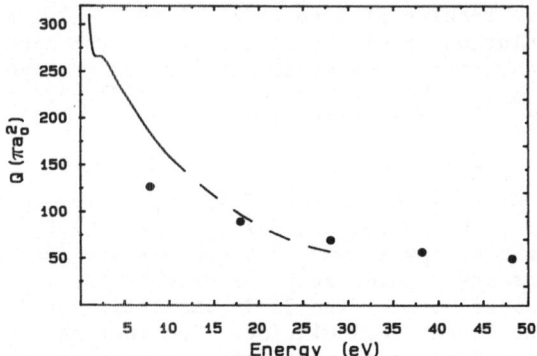

Figure 6. Total cross-section for e^+-K scattering using the 2-state (4s-4p)
close-coupling calculation compared with the experimental results ($\bullet$)
(Stein et al. 1987). The dashed line (— — —) indicates that the cross-
section had not fully converged with angular momentum L.

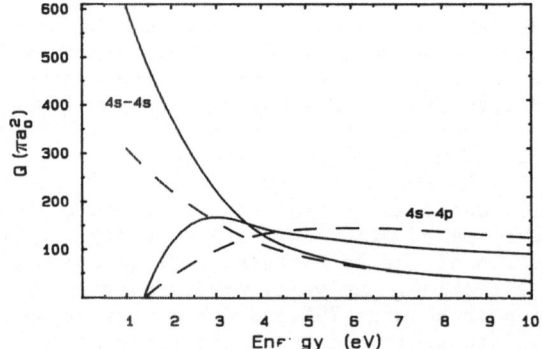

Figure 7. A comparison of the elastic and 4s-4p excitation cross-sections
for e^+-K scattering evaluated by the 2-state (— — —) and 4-state (————)
close-coupling calculations.

reported our results in which a few target states have been included in the expansion of the scattering wavefunction. The number of excited states needs to be systematically increased and pseudostates added until convergence in the total cross-section is obtained.

However, we feel from a physical point of view, the most important step would be to perform an expansion about both the target and positronium atoms. This work is in hand.

REFERENCES

A.K. Bhatia, A. Temkin, A. Silver and E.C. Sullivan, Phys. Rev. A 18:1935 (1978).
G. Bordonaro, G. Ferrante, M.Zarcone and P. Cavaliere, Nuovo Cimento 35B:349 (1976).
G.O. Brink, Phys. Rev. 127:1204 (1962).
T.D. Bui, J. Phys. B 8:L153 (1975).
T.D. Bui and A.D. Stauffer, Can. J. Phys. 49:2527 (1971).
P.G. Burke and H.M. Schey, Phys. Rev. 126:147 (1962).
P.G. Burke and M.J. Seaton, Methods in Computational Physics, Vol. 10, eds. B. Alder, S. Fernbach and M. Rotenberg (Academic Press, New York) p. 1. (1971).
P.G. Burke and K. Smith, Rev. Mid. Phys. 34:458 (1962).
P.G. Burke and A.J. Taylor, J. Phys. B 2:869 (1969).
J. Callaway, R.W. LaBahn, R.T. Pu and W.M. Duxler, Phys. Rev. 168:12 (1968).
M. Cohen and R.P. McEachran, Advances in Atomic and Molecular Physics, vol.16 eds. D.R. Bates and B. Bederson (Academic Press, New York) p.1 (1980).
G. Ferrante, L.L.O. Cascio and M. Zarcone, Nuovo Cimento 44B:99 (1978).
T.T. Gien, Phys. Rev. A 35:2026 (1987a).
T.T. Gien, Chem. Phys. Lett. (1987b) in press.
S. Guha and A.S. Ghosh, Phys. Rev. A 23:743 (1981).
S. Guha and P.J. Mandal, J. Phys. B 13:1919 (1980).
S. Guha and B.C. Saha, Phys. Rev. A 21:564 (1980).
R. Jain, R. Hageman and R. Botter, J.Chem. Phys. 59:952 (1973).
P. Khan and A.S. Ghosh, Phys. Rev. A 27:1904 (1983a)
P. Khan and A.S. Ghosh, Phys. Rev. A 28:2181 (1983b).
P. Khan, S. Dutta and A.S. Ghosh, J. Phys. B 20:2927 (1987).
P. Mandal and S. Guha, J. Phys. B 13:1337 (1980).
P. Mandal, S. Guha and N.C. Sil, J. Phys. B 12:2913 (1979).
P.S. Mazumdar and A.S. Ghosh, Phys. Rev. A 34:4433 (1986).
R.P. McEachran and P.A. Fraser, Proc. Phys. Co. 86:369 (1965).
T. M. Miller and B. Bederson, Advances in Atomic and Molecular Physics, Vol. 13, eds. D.R. Bates and B. Bederson (Academic Press, New York) p. 1 (1977).
S.N. Nahar and J.M. Wadehra, Phys. Rev. A 35:4533 (1987).
I.C. Percival and M.J. Seaton, Proc. Camb. Phil. Soc. 53:654 (1957).
A. Salmona and M.J. Seaton, Proc. Phys. Soc. London, 77:617 (1961).
K.P. Sarkar, M. Basu and A.S. Ghosh, These proceedings, 1988.
K.P. Sarkar, B.C. Saha and A.S. Ghosh, Phys. Rev. A 8:236 (1973).
I. Sloan, Proc. Roy. Soc. A 281:151 (1964).
K. Smith, The Calculation of Atomic Collision Processes, (Wiley-Inter-Science, New York), (1971).
T.S. Stein, M.S. Dababneh, W.E. Kauppila, C.K. Kwan and Y.J. Wan, in Atomic Physics with Positrons ed. J.W. Humberston and E.A.G. Armour, (Plenum, New York, 1988).
T.S. Stein, R.D. Gomez, Y.F. Hsieh, W.E. Kauppila, C.K. Kwan and Y.J. Wan, Phys. Rev. Lett. 55:488 (1985).
P.M. Stone, Phys. Rev. 141:137 (1966).
A. Temkin, Phys. Rev. 107:1004 (1957).
A. Temkin, Phys. Rev. 116:358 (1959).

A. Temkin and J.C. Lamkin, Phys. Rev. 121:788 (1961).

J. M. Wadehra, Can. J. Phys. 60:601 (1982).

H.R. Walters, J. Phys. B 9:227 (1976).

S.J. Ward, M. Horbatsch, R.P. McEachran and A.D. Stauffer, in press J. Phys. B (1988).

PRECISION MEASUREMENT OF THE TRIPLET POSITRONIUM DECAY RATE IN GASES

D.W. Gidley, C.I. Westbrook, R.S. Conti and A. Rich

Randall Laboratory of Physics

University of Michigan, Ann Arbor, Michigan 48109, USA

INTRODUCTION

In a recent publication[1], we presented a new 200 ppm measurement of the vacuum decay rate, λ_T of orthopositronium (o-Ps) formed in a gas. Our result, λ_T = 7.0516 ± 0.0013 μs^{-1}, represents a factor of four improvement over previous measurements. Our result is in substantial agreement with experimental results, the most recent of which are[2-5]: 7.056 ± 0.007 μs^{-1}, 7.045 ± 0.006 μs^{-1}, 7.051 ± 0.005 μs^{-1}, and 7.050 ± 0.013 μs^{-1}. These values are 1-2.5 standard deviations above the present theoretical value[6,7] and our new measurement exceeds theory by 10 experimental standard deviations.

In this paper, we briefly review the present theoretical and experimental situation. We will discuss our method for measuring λ_T by formation of o-Ps in gases and present new results for λ_T measured in neon (with a small admixture of either isobutane or neopentane. Appropriate for this workshop we will present a detailed discussion of systematic effects that are specifically related to the technique of Ps formation in a gas and how we handled these systematics at the present level of accuracy. We also consider some suggested future experiments to rigorously and directly investigate their effect.

THEORY

The theoretical value for the decay rate of o-Ps may be expressed as the sum of decay rates into three photons (λ_3), five photons (λ_5), etc:

$$\lambda_T = \lambda_3 + \lambda_5 + \ldots .$$

The contribution of λ_5 has been calculated[8] to be $\lambda_5/\lambda_3 \sim 10^{-6}$, and is thus negligible. The leading term is:

$$\lambda_3 = \frac{\alpha^2 mc^2 2(\pi^2 - 9)}{\hbar \quad 9\pi} [1 + A(\alpha/\pi) + \frac{1}{3}\alpha^2 \ln \alpha + B(\alpha/\pi)^2 + \ldots]$$

The two most recent calculations give A = 10.266 ± 0.011[6] and A = 10.282 ± 0.003[7]. The coefficient B is still uncalculated. If B = 1, its contribution to λ_3 is about 5 ppm or 3.9×10^{-5} μs^-. Taking A from Ref. 7, one obtains through order $\alpha^2 \ln\alpha$, $\lambda_3 = 7.03830 \pm 0.00007$ μs^{-1}. A positive value of this uncalculated coefficient of B = 340 ± 33 would bring theory into agreement with our new measurement. We note here that some theorists have argued that it may be more appropriate to write the second order term as a coefficient times α^2 rather than $(\alpha/\pi)^2$ (see Ref. 9). In this case the second order coefficient is obviously a factor of ten smaller.

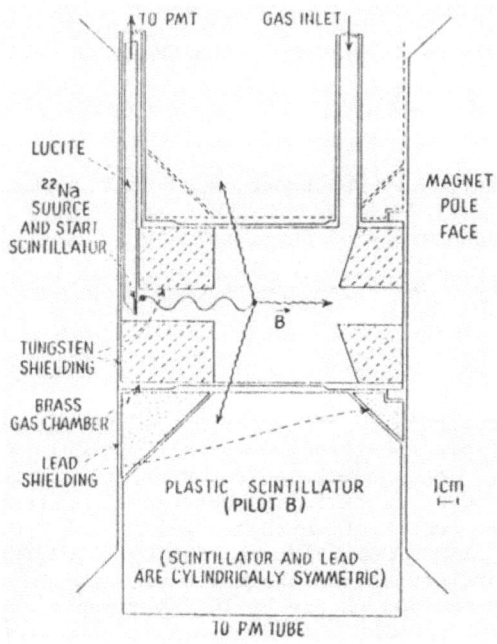

Fig. 1. Ps formation chamber and detector arrangement

EXPERIMENT

The experimental technique used in our new measurement is discussed in greater detail in Ref. 1, but we present an overview here for completeness. Positrons from a 10-µCi radioactive ^{22}Na source pass through a 0.1 mm thick scintillator and form Ps in a gas in a magnetic field of 6.8 kG (Fig. 1). The scintillator is coupled via a 45 cm Lucite light pipe to an Amperex XP2020 photomultiplier tube (PMT). The pulse from this PMT provides a start signal to a digital timing system. The magnetic field causes all positrons with a forward momentum component to spiral into the chamber. The positrons then slow down in the gas to energies of order 10 eV and some fraction of these (roughly 25%) form Ps. The magnetic field mixes the m = 0 singlet and triplet magnetic substates of Ps resulting in perturbed triplet (and singlet) states that decay with lifetimes of 13 ns (0.12 ns) at 6.8 kG. The m = ± 1 triplet states are unperturbed and continue to decay with a rate λ_T. When Ps decays, the annihilation γ rays are detected by an annular plastic scintillator coupled via light pipes to four Hamamatsu R1250 PMT s. These provide the stop timing signal. One data run consists of a histogram of start-stop delay times. The shape of the histogrammed time spectrum, to sufficient accuracy for this work, has the following form for t > 150 ns:

$$N(t) = (ae^{-\lambda t} + b) e^{-Rt} \tag{1}$$

Here, λ is the decay rate of o-Ps at a particular gas density and R is the total stop rate. The magnetically perturbed triplet state has decayed away sufficiently by 150 ns to be excluded from Eq. 1. The spectrum was fitted by the use of a maximum-likelihood routine to extract the parameters a, b and λ. We measured λ over the pressure range of 100 torr to 1200 torr (pressure is measured with a Baratron capacitive manometer). With a positron start rate of $\sim$ 75 kcps and an o-Ps rate of $\sim$ 400 cps, (a/b) in Eq. 1 ranged from 5-20. Ten days of running permit measurement of λ at the $\pm$ 0.002 μs^{-1} level. The vacuum decay rate is then determined by extrapolating λ to zero density.

UPDATED RESULTS

We reported in Ref. 1 measurements of λ_T in three gases; isobutane, $(CH(CH_3)_3)$ (100 - 1200 torr); neopentane, $(C(CH_3)_4)$ (100 - 800 torr); and nitrogen, (200 - 1400 torr) with a 20 torr admixture of isobutane to quench low energy positrons. We have since performed a new measurement in neon (200 - 1200 torr) using small admixtures of either isobutane or neopentane. All of these results are summarized in Table 1 and shown in Fig. 2.

Table 1. Extrapolated values of λ_T derived from above figure.

Isobutane	7.0522 ± 0.0013	s^{-1}
Nitrogen	7.0493 ± 0.0018	s^{-1}
Neopentane	7.0543 ± 0.0026	s^{-1}
Neon	7.0505 ± 0.0026	s^{-1}

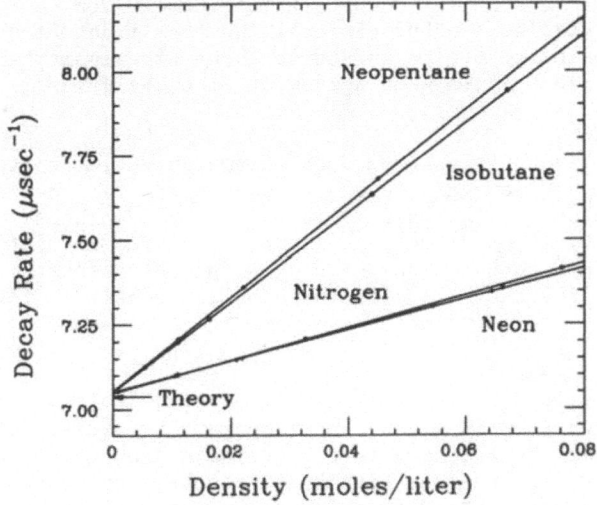

Fig. 2. Plot showing extrapolations of λ in 4 gases.

We now turn to a discussion of systematic effects in this experiment. The electronics have been rigorously checked and calibrated (see Ref. 1). The simultaneous use of two independent, and very different digital timing systems (with and without electronic noise rejection schemes) and the fact that they measure the same value for λ, renders essentially negligible the possibility of any electronics based systematic error. We consider that by far the most important systematic errors are related to physical effects in the gases. The most important of such effects include: 1) any non-linearities in λ vs density, 2) formation of long-lived excited states of Ps, and, 3) incomplete thermalization of Ps that could result in time dependent collisional quenching ("pickoff") of Ps. There are other effects related to possible systematic errors in the isobutane and neopentane virial coefficients, or due to incomplete mixing of the quench gas in the nitrogen and neon runs, but these are directly ruled out by the agreement of the extrapolations in the different gases. In fact this agreement is considered to be the single most important systematic test we have performed because it excludes a wide range of gas or pressure-dependent effects (including 1-3 above) which if present, would produce differences in the extrapolated decay rate. Thus we consider it unlikely that the effects mentioned above could produce the observed 1900 ppm discrepancy with the theory to order $\alpha^2\ell n\alpha$. However, at the 200 ppm level, we must consider more closely these gas related effects and future improvement in λ_T measurements may require dedicated experiments to explore these effects (to be discussed).

Non-linearities in λ vs. Density

To check for non-linearities in the decay rate extrapolation, we compare our measured slopes, $m = d\lambda/dn$, with other measurements in high densities of noble gases and nitrogen. Many such experiments[10] have been carried out at densities of order a few amagat and higher (1 amagat = 0.044615 moles/1 and is the number density corresponding to 1 atmosphere of ideal gas at 0°C). By using a large range of densities these experiments were able to measure the slopes of the lines (i.e. the pickoff rates) to fairly high precision, although they were not able to measure the value of the intercept accurately. Coleman et al.[11] reported a measurement in Ne gas at densities between 7 and 39 amagat and found that the slope of the straight line fit was: $m = 0.198 \pm 0.006$ μs^{-1}/amagat. This value was found by assuming that the zero density intercept was 7.24 μs^{-1}. We have reanalyzed the data of Ref. 11 making the intercept either 7.04 or 7.05 μs^{-1} (it does not matter which at this level of precision). We then find: $m = 0.204 \pm 0.002$ μs^{-1}/amagat. (The source of the factor of three discrepancy in the error is unknown). This result is in good agreement with the value reported here: $m = 0.206 \pm 0.004$ μs^{-1}/amagat.

In nitrogen at densities from 7 to 46 amagat, Coleman et al.[12] found $m = 0.214 \pm 0.001$ μs^{-1}/amagat. Again we have corrected the zero density intercept to 7.05 μs^{-1}. The value measured in this work is: $m = 0.2146 \pm 0.0016$ μs^{-1}/amagat, in excellent agreement. Thus the consistency between our low density data and previous high density measurements demonstrates that λ is indeed linear over a wide range of gas densities for neon and nitrogen.

Ps Excited States

When positrons stop in a gas, one expects that excited state positronium, Ps*, is formed in addition to the ground state atoms. Excited state yields (in the n = 2 states) as high as 6% of the stopped positron rate have been observed in experiments[13] in which a low energy beam of positrons is incident on a low density gas cell target. There are 16 states in the n = 2 level: The 12 P states all decay by Lyman α emission to the ground state in 3.2 ns, the singlet S has an annihilation lifetime of about 1 ns, and the triplet S states annihilate with a lifetime of 1.1 μs. Clearly even a small amount of 2^3S states, which, because of collisions, could have any lifetime shorter then 1.1 μs, could affect the measurement of the ground state triplet lifetime at the present level of accuracy. A yield of 6%, which must be multiplied by 3/16 to obtain a 2^3S yield of 1%, is likely to be an over estimate for our apparatus because in Ref. 13, the gas was rarefied enough to allow the atoms to pass out of a gas cell and into a region where a detector observed the Lyman α decay. In our work, Ps atoms formed at low energies ($\sim$ 10 eV) have mean free collision times below 1 ns even in the lowest gas densities. Thus, excited state atoms whose kinetic energies are larger than their binding energies (1.7 eV for n = 2) undergo many collisions during which they can dissociate, and if 1% of the positrons do form Ps in the 2^3S_1 state, only a small fraction will survive dissociation. This point is even more important for Ps* in states higher than n = 2. Nevertheless, one must carefully examine the possibility of Ps* surviving dissociation and causing systematic errors in the fitted decay rate.

It is perhaps worthwhile to mention that searches for Ps* in dense gas environments[14,15] have failed. (Dense here means 1 amagat or higher). In these experiments a β^+-source was placed in a gas chamber containing an atmosphere or more of gas. An ultraviolet sensitive photomultiplier together with an interference filter or monochromator was used to search for the Lyman α decay of the 2P state. As an example, in Ref. 14, no Lyman signal was observed at the level of 0.05% of the number of stopped positrons. The first demonstration of Ps* formation in gases was that of Ref. 13, but it was recognized before this demonstration that Ps* formation should be substantial[15.5] and that some mechanism such as dissociation or radiationless deexcitation must have been quenching Ps* quickly compared to even the 3.2 ns 2P state lifetime. Thus, as a result of the above searches and an argument based on dissociation, we assume that the rate of 2^3S_1 state formation after dissociation and quenching is no more than 0.1% of the positron stopping rate. This would correspond to about 0.5% of the ground state formation rate, which is still sufficiently high to produce a 1000 ppm shift in the fitted value of λ_T if the Ps* lifetime is pathologically "just right", i.e. 50-100 ns. Direct analysis of the fitted lifetime spectrum indicates that there are no such components with an intensity or lifetime which could cause a 500 ppm shift in λ_T. However, we expect collisional deexcitation to significantly lower this upper limit as we will now discuss.

The most likely fate of the 2S positronium states which do not dissociate is collisional "quenching" into a nearby 2P state. In an electric field, E, a 2S state will get an admixture (in first order) of 2P state with an amplitude of order:

$$\eta = \frac{eE < 2P|r|2S >}{h\nu}$$

Where $h\nu$ is the energy splitting between 2S and 2P levels ($\nu \sim$ 10 GHz) and the numerator is just the electric dipole matrix element between the 2S state and one of the 2P states ($\sim 1 \times 10^{-8}$ e-cm). The field necessary to produce $\eta \sim 1$ is about 10^4 V/cm. Typical atomic electric fields are about 5 orders of magnitude larger than this so one expects significant mixing between 2S and 2P states during collisions. Measurements on excited state H atoms tend to support this assertion. In collisions with nitrogen molecules the deexcitation cross section was found to be $\sigma \sim 10^{-14}$ cm^2 for H atoms travelling at velocities of $\sim 10^6$ cm/s[16]. and $\sigma \sim 10^{-15}$cm^2 at velocities of 7×10^7 cm/s[17].

For Ps* with n > 2, the above arguments can only be extended to the n = 3,4 and 5 states. The radiative lifetimes of the 3P, 4P and 5P states are short enough that one expects any angular momentum state to be collisionally mixed with the short lived P states quickly compared to 140 ns. This argument does not work for higher n levels because even the P states in these levels have long radiative decay lifetimes. The 6P state lifetime, for example, is 80 ns. However, there are additional effects in polyatomic gases which will in all likelihood cause highly excited Ps* to disappear quickly if it is formed at all and can survive dissociation. An analogous process of collisional, radiationless de-excitation of Na atoms is observed in discharge tubes when small admixtures of polyatomic gases such as C_6H_6 and N_2 are admitted. Cross sections of order 10^{-15} cm^2 are deduced from these observations[18]. It is generally understood that long range dipole-dipole interactions are responsible for the radiationless de-excitation and the same mechanism will cause de-excitation to the ground state with a similar or even greater cross section[19]. This mechanism, in addition to the low probability of highly excited Ps surviving dissociation, removes any concern that Ps* in high n levels can live longer than about 1 ns at the pressures used in this experiment.

Thus, lacking direct measurements of Ps* quenching, we conclude that on the basis of the above indirect evidence, there are no Ps* states of sufficient intensity and lifetime to produce a shift as large as 100 ppm in λ_T. There is an additional quenching mechanism associated with the 6.8 kG magnetic field which helps quench the 2^3S_1 state. Curry[20] calculated Zeeman and motional Stark shifts in the n = 2 level due to Ps moving through magnetic fields of up to 7 kG. Curry finds that the vacuum lifetime of the longest lived n = 2 Ps state with energy 0.1 eV transverse to a 6.5 kG field is 20 ns. This effect by itself rules out all but the shortest lifetimes of n = 2 Ps which could affect the measurement. A 20 ns lifetime is easy to search for and rule out by observing the change in the fitted lifetime as the fit is stepped out over 200 ns.

There is clearly a need for direct Ps* quenching experiments if one is to definitively resolve the effect of excited states. We have attempted to observe quenching of the 2^3S_1 state using the microwave transition experiment which recently[21] produced measurements of the 3 allowed 2S-2P transitions in Ps. The apparatus was backfilled with approximately 1 millitorr of gas and microwaves were applied to drive the 2^3S_1-2 P_2 transition (8.6 Ghz). A decrease in the resonance signal (microwaves on - microwaves off) indicated collisional quenching of the $2 S_1$ state with a cross section between 1 and 4×10^{-14} cm^2 for Ps* atoms with an average energy of 1.7 eV. One problem with the experiment was an uncertainty about the gas composition that was causing the quenching. Nitrogen gas was bled into the vacuum system initially, but outgassing (probably CO) constituted a significant fraction of the total pressure during the several hour runs. The result, though crude, is comparable to the H atom quenching experiments mentioned above[16,17].

Ps thermalization

The last gas dependent systematic effect that we will consider is associated with the possibility that the velocity distribution of Ps changes, i.e. "thermalizes" on a time scale similar to λ^{-1}. This would mean that λ in Eq. 1 could be decreasing throughout the lifetime spectrum and thus the fitted decay rate would be artifically high, especially at low pressures where thermalization would be slower. In fact we observe a trend consistent with thermalization in the fitted decay rate at the lowest pressures used for each gas. As the start channel of the fitting program is successively stepped out beyond t = 150 ns, the fitted decay rate decreases by roughly 1000 ppm before asymptotically approaching a constant value at the at 175-275 ns. depending on the gas (see Fig. 3). Upon doubling or tripling of the gas pressure, and hence of the Ps-gas-molecule collision rate, the

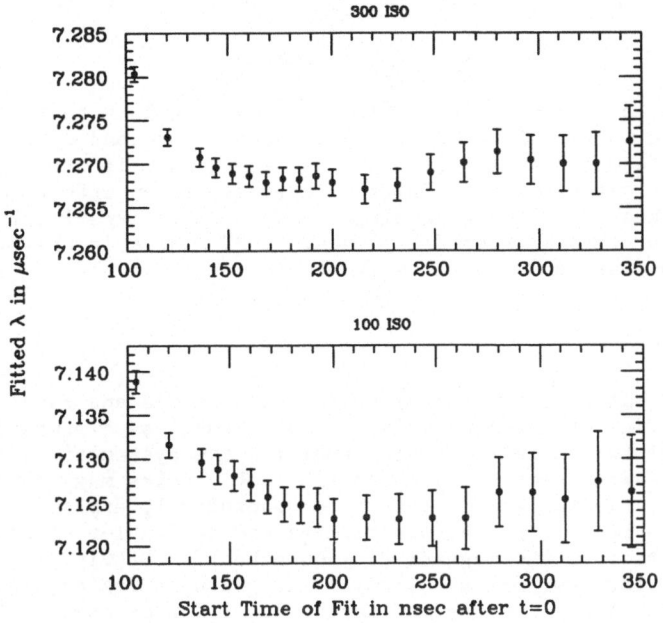

Fig. 3. Fitted decay rates at 100 and 300 torr of isobutance

effect disappears and λ is observed to be constant beyond 150 ns. We conclude that Ps thermalization effects are negligible at all but the lowest gas pressures used and that, at these low pressures, we have correctly accounted for such effects by selecting the asymptotic value of λ. We therefore expect no pressure-dependent shift in the extrapolated value of λ_T.

The thermalization interpretation is, however, not totally satisfying under closer scrutiny. We would anticipate the polyatomic isobutane and neopentane gases with their vibrational and rotational modes to thermalize Ps more quickly. In a classical sense, the typical mass of the atom struck by the Ps atom should be effectively that of hydrogen in these gases. However the Ps thermalization rate is no more than a factor of two slower in nitrogen and neon, not a factor of 14 or 20. Only in argon, with atomic mass of 40, did the rate of thermalization appear to be so slow that we abandoned a precision measurement in low pressures (400 torr or lower). Ps thermalization in argon, however, is certainly not 40 times slower than in isobutane.

It is possible that the effects discussed above are not entirely due to thermalization. Other lifetime experiments have failed to observe temperature dependent pickoff. Fox, Canter and Fishbein[22] did a lifetime experiment in high pressures of helium gas at varying temperatures. At a density of 200 amagat, they observed that the o-Ps pickoff rate at 77°K and 300°K was 17.0×0.5 μs^{-1}/amagat, and that the rates at the two temperatures were consistent within their own uncertainties. Writing the pickoff rate as:

$$\lambda_{pickoff} = n\sigma v,$$

and assuming that the Ps had thermalized, one can conclude that the quantity σv is constant to within 3% for energies between 0.0064 and 0.025 eV. This means that σ in He obeys at 1/v law at those energies. If other gases do this too, one can set stringent limits on thermalization effects at low densities because at the lowest densities used in this experiment the pickoff rate is only 1% of the observed decay rate. A 3% change in the pickoff is only a 300 ppm variation in the observed decay rate.

FUTURE EXPERIMENTS

We believe that direct measurements of Ps-molecule and Ps-atom collisions (together with Ps*-atom or molecule collisions) as discussed earlier would be very helpful in definitively resolving the systematic effects discussed in the previous sections. Moreover, such studies may provide an interesting challenge for atomic theory to quantatively explain. Experimentally, it may be possible to perform time-dependent Doppler broadening studies of Ps thermalization on the magnetically quenched m = 0 triplet state (which decays primarily into two 511 keV γ rays). A group in Peking has developed this method[23] and studied Ps thermalization in silica aerogels. More temperature dependent measurements like those in Ref. 22 at high densities (where pickoff dominates) and over a wide temperature range in a variety of gases would be very useful in ascertaining the velocity dependence of pickoff. Eventually, the recent work in forming beams of Ps (and presumably Ps*) in gas-cells[24] and on surfaces[25] may be helpful in understanding Ps collisions with atoms and molecules.

In order to effectively bypass the question of Ps thermalization, and Ps* effects we are presently endeavouring to extract values of λ_T for He, Ne, Ar and N_2 gases using the high density data presented in Refs. 11 and 12 and only one precisely measured value of λ at or near 1 amagat (our maximum chamber pressure is about 1300 torr). Thermalization should certainly be negligible at 1 amagat or above. We estimate that an accuracy of ± 0.003 μs^{-1} can be obtained for each gas. The values of λ_T derived in this way should be quite independent of our previous values since the measurements of λ mainly determined the slope, m and had little direct effect on the intercept λ_T. Thus, they will provide a strong systematic confirmation of our present value of λ_T if agreement is found.

To provide an additional rigorous check of our present measurement we have begun a new experiment (designed to reach an accuracy of ± 0.007 μs^{-1}) using a slow positron beam with Ps formation in an evacuated cavity (similar to that described in Ref. 5).

We are grateful to P.W. Zitzewitz, G.W. Ford, G.P. Lepage, J. Sapirstein, G. Adkins and members of the Michigan positron group for helpful discussions. We also thank M. Charlton and P. Coleman for supplying us with the high density decay rate results from Refs. 11 and 12. This work is supported by the National Science Foundation under Grant No. PHY-8403817.

REFERENCES

1. C.I. Westbrook, D.W. Gidley, R.S. Conti and A. Rich, Phys. Rev. Lett. 58:1328 (1987).

2. D.W. Gidley, A. Rich, P.W. Zitzewitz and D.A.L. Paul, Phys. Rev. Lett. 40:737 (1978).

3. T.C. Griffith, G.R. Heyland, K.S. Lines and T.R. Twomey, J. Phys. B 11:743 (1978).

4. D.W. Gidley, A. Rich, E. Sweetman and D. West, Phys. Rev. Lett. 49:525 (1982).

5. D.W. Gidley and P.W. Zitzewitz, Phys. Lett. 69A:97 (1978).

6. W.G. Caswell and G.P. Lepage, Phys. Rev. A20:36 (1979).

7. G.S. Adkins, Ann. Phys. (N.Y.) 146:78 (1983).

8. G.S. Adkins and F.R. Brown, Phys. Rev. A28:1164(1983); G.P. Lepage, P.B. Mackenzie, K.H. Streng and P.M. Zerwas, Phys. Rev. A28:3090 (1983).

9. G.P. Lepage and G.S. Adkins, private communcation. This modification was suggested since bound state corrections of order α^2 will occur as well as radiative corrections of order $(\alpha/\pi)^2$.

10. M. Charlton, Rep. Prog. Phys. 48:737 (1985).

11. P.G. Coleman, T.C. Griffith, G.R. Heyland and T.L. Killeen, J. Phys. B8:1734 (1975).

12. P.G. Coleman, T.C. Griffith, G.R. Heyland and T.L. Killeen, Atomic Physics 4, ed. G. ZuPutlitz, E.W. Weber and A. Winnacker (New York Plenum) p.355.

13. G. Laricchia, M. Charlton, G. Clark and T.C. Griffith, Phys. Lett. 109A:97 (1985).

14. M. Leventhal, Proceedings of the National Academy of Sciences, 66:6 (1970).

15. B.G. Duff and F.F. Heymann, Proc. R. Soc. A272:363 (1963) and L.W. Fagg, Nucl. Instrum. Methods, 85:53 (1970).

16. W.L. Fite, R.T. Brackman, D.G. Hummer and R.F, Stebbings. Phys. Rev. 116:363 (1959).

17. F.W. Byron, R.V. Krotkov, J.A. Medeiros, Phys. Rev. Lett. 24:83 (1970).

18. For a wide range of excited state atomic quenching cross sections, see H.S.W. Massey, E.H.S. Burhop and M.B. Gilbody, Electronic and Ionic Impact Phenomena, (Clarendeon, Oxford, 1971), Vol. 3, pp 1656-1696.

19. G.W. Ford, private communication.

20. Stephen M. Curry, Phys. Rev. A7:447 (1973).

21. S. Hatamian, R.S. Conti and A. Rich, Phys. Rev. Lett. 58:1833 (1987).

22. R.A. Fox, K.F. Canter and M. Fishbein, Phys. Rev. A15:1340 (1977).

23. Tianbao Chang et al. submitted for publication.

24. B.L. Brown, paper in these proceedings and references therein.

25. D.W. Gidley, R. Mayer, W.E. Frieze and K.G. Lynn, Phys. Rev. Lett. 58:595 (1987), and references therein.

THE BROOKHAVEN POSITRON-HYDROGEN SCATTERING EXPERIMENT:

MOTIVATION AND GENERAL SCOPE

M.S. Lubell

Department of Physics
City College of the City University of New York
New York, NY 10031

INTRODUCTION

In this paper, I will emphasize the general motivation for the Brookhaven National Laboratory (BNL) e^+-H project by drawing appropriate analogies to e^--H scattering, with special attention given to the recent results of spin-tagged experiments carried out first at Yale University and later at City College. I will also describe the general layout of the BNL experiment as originally envisioned, and in this context I will discuss some of the results of a hybrid beam transport design study performed largely by F.J. Mulligan.[3] Finally, I will present the updated design plans for the experiment along with some comments about the general scope of the project.

In my discussion, I will omit a great many of the details of the experiment, since the specific design features, the measurement, and the data reduction methods will be explained by G. Sinapius in the succeeding paper. I should make it clear that Dr. Sinapius and I are presenting the material on behalf of the entire e^+-H collaboration comprising the following people and institutions:

AT&T Bell Laboratories: A.P. Mills, Jr.
University of Bielefeld: W. Raith, A. Schwab, and G. Sinapius
Brandeis University: K.F. Canter and S. Berko
Brookhaven National Laboratory: K.G. Lynn
City College of CUNY: Xuan Liu, M.S. Lubell, and L.O. Roellig
St. Patrick's College, Maynooth: F.J. Mulligan and J. Slevin.

GENERAL MOTIVATION

In its function as a target, the hydrogen atom has served as a principal theoretical testing ground for the physics of atomic collisions. By virtue of simplicity and the precision with which hydrogenic wavefunctions are known, the electron-hydrogen collision problem, for example, has been the subject of more than two hundred theoretical papers during the last two decades. As the starting point for the many-body problem with long-range forces, however, the electron-hydrogen system is still not amenable to solution in closed analytic form, since it is a three-body problem, or perhaps more appropriately, a two-and-a-half-body problem, if the proton is accorded the status of a spectator particle.

The electron-hydrogen system is further complicated by the indistinguishability of two of the three collision partners, a situation mimicked in nuclear physics by proton-deuteron scattering. Unlike its nuclear analog, however, the electron-hydrogen problem must contend with the infinite number of bound states supported by the Coulomb potential describing the atomic target, a complication that has defied the development of a single, reliable, simplified approach with applicability over a wide range of kinematic conditions. In fact, experiments using polarized beams[2] or angular correlation techniques,[3] two methods that provide more detailed access to the nature of the collision than is ordinarily available, have recently revealed substantial difficulties with many of the calculational techniques previously regarded as acceptably accurate.

By contrast with electron-atom scattering, positron-atom scattering is often viewed as a complementary but yet distinct collision process. For the case of an atomic hydrogen target, however, the two processes can be regarded as inseparable partners under certain conditions. Consider first the scattering amplitude for an electron-hydrogen collision. It is well known[4] that can be written in the form

$$= - \frac{m}{2\pi h^2} \langle \Psi_f | T | \Psi_i \rangle, \tag{1}$$

where m is the mass of the electron, h is Planck's constant, T is the transition operator, and Ψ_i and Ψ_f are the initial and final state wave functions respectively, both of which must be antisymmetrized in accordance with the requirements of Fermi-Dirac statistics.

Using $\vec{k}$ to denote the wave vector of the incident (initial) free electron state, u to denote the wave function of the initial atomic state, η to denote the spinor of the initial free-electron state, and χ to denote the spinor of the initial atomic-electron state, and using primes to denote the corresponding final state quantities, the wavefunctions Ψ_i and Ψ_f can be expressed in antisymmetrized forms as

$$\Psi_i = \sqrt{2} \ [e^{i\vec{k}\cdot\vec{r}_1} u(\vec{r}_2)\eta(1)\chi(2) - e^{i\vec{k}\cdot\vec{r}_2} u(\vec{r}_1)\eta(2)\chi(1)] \tag{2}$$

and

$$\Psi_f = \sqrt{2} \ [e^{i\vec{k}'\cdot\vec{r}_1} u'(\vec{r}_2)\eta'(1)\chi'(2) - e^{i\vec{k}'\cdot\vec{r}_2} u'(\vec{r}_1)\eta'(2)\chi'(1)], \tag{3}$$

where $\vec{r}$ is the spatial coordinate and the indices 1 and 2 specify the individual electrons, and where for simplicity we have used the notation for a final bound state atomic wavefunction. Substituting Eqs. (2) and (3) into Eq. (1) we find that can be written as

$$= \frac{1}{2} \{f(\vec{k},\vec{k}') \ [\langle \eta'(1)|\eta(1)\rangle\langle\chi'(2)|\chi(2)\rangle + \langle\eta'(2)|\eta(2)\rangle\langle\chi'(1)|\chi(1)\rangle]$$
$$- g(\vec{k},\vec{k}')[\langle\eta'(1)|\chi(1)\rangle\langle\chi'(2)|\eta(2)\rangle + \langle\eta'(2)|\chi(2)\rangle\langle\chi'(1)|\eta(1)\rangle]\}, \tag{4}$$

where the direct scattering amplitude $f(\vec{k},\vec{k}')$ is given by

$$f(\vec{k},\vec{k}') = - \frac{m}{2\pi h^2} \langle e^{i\vec{k}'\cdot\vec{r}_1} u'(\vec{r}_2)|T| \ e^{i\vec{k}\cdot\vec{r}_1} u(\vec{r}_2)\rangle, \tag{5}$$

and the exchange scattering amplitude $g(\vec{k},\vec{k}')$ is given by

$$g(\vec{k},\vec{k}') = - \frac{m}{2\pi h^2} \langle e^{i\vec{k}'\cdot\vec{r}_2} u'(\vec{r}_1)|T|e^{i\vec{k}\cdot\vec{r}_1} u(\vec{r}_2)\rangle \tag{6}$$

with the indices 1 and 2 being interchangable in Eqs. (5) and (6).

If we examine Eqs. (5) and (6), we see that for fixed $\check{R}$ and u, the amplitudes f and g differ only in the specification of their final states. To some extent, therefore, f and g simply represent two separate scattering channels, in much the same way that elastic scattering and any particular atomic excitation represent two separate channels. Final state specification of course also depends upon the direction of $\check{R}$ (scattering angle), as well as its magnitude, although such a continuously varying kinematic label is not ordinarily classified under the heading of scattering channel.

From a general perspective it is probably more appropriate to classify all these different labels simply as variations in the specification of the state vectors. Indeed it is by such variations that theoretical methods are usually examined for their predictive capabilities. In other words, for a given transition operator, corresponding to a particular electron-atom collision system, cross sections are calculated and measurements are carried out for different incident energies, different initial and final target states, different scattering angles, and different angular momentum configurations.

It is not unreasonable to ask whether this is the only way that theoretical methods can or should be tested. Consider for example, the alternative case where state vectors are held fixed and the transition operator is allowed to vary in a well controlled fashion. Ordinarily such conditions are hard to achieve experimentally. However, with the increasing sophistication of polarized beams technology and with the increasing intensity of low-energy positron beams, we now appear to be on the threshold of providing the necessary experimental environment for such studies. Indeed, as computers become more sophisticated and calculational techniques become more elaborate, benchmark experiments that provide detailed, controlled variations in the T operator as well as in the state vectors themselves will become more and more essential.

The combination of electron-hydrogen and positron-hydrogen scattering experiments hold the promise for providing such benchmark capabilities. Through the use of spin-tagging techniques in which initial and final state spinors are completely specified, it is possible in principle to separate f from g completely for electron scattering, as can be seen from Eq. (4). For example, if we consider the process

$$e\uparrow + H\downarrow \rightarrow e\uparrow + H\downarrow \tag{7}$$

for which $\langle \eta'|\eta\rangle = \langle \chi'|\chi\rangle = 1$ and $\langle \eta'|\chi\rangle = \langle \chi'|\eta\rangle = 0$, Eq. (4) reduces to

$$= f(\check{R},\check{R}'). \tag{8}$$

Variation of $\check{R}$, $\check{R}'$, u, and u' of course permits $|f|^2$ and hence $|\ |^2$ to be measured for different state vector specifications in the usual fashion. Now, however, if $\check{R}$, $\check{R}'$, u, and u' are held fixed and a positron is substituted for an electron, T will be varied in the required controlled fashion with the state vectors remaining constant.

Thus we see the intimate relationship between e^+-H and e^--H collisions in establishing benchmark tests of calculational methods. Unfortunately, the tools required for carrying out the necessary experiments are not yet fully developed, and hence the initial studies, which will explore only angle-integrated ionization and positronium formation, must be more modest in their immediate goals. Nonetheless, the first e^+-H experiments are certainly prerequisites for future more ambitious measurements of differential elastic scattering cross sections.

It should be noted that e^+-H scattering possesses a richness beyond its relationship to e^--H scattering. For example, the dynamics of positronium formation has its own theoretical intrinsic interest as well as a strong relationship to the understanding of the 511 keV and 1.81 MeV cosmic γ-ray emissions from the near galactic center.[5-8] Energy loss processes, including both excitation and ionization for positrons passing through hydrogen gas[8] are also of considerable astrophysical interest. Therefore the first e^+-H experiments may be regarded as fundamental in their own right and not merely precursors to the more complex studies that will follow.

STATUS OF SPIN-TAGGED ELECTRON-HYDROGEN EXPERIMENTS

In the early 1970's shortly after the development of the Fano-type[9] polarized electron sources, we began a study of electron-hydrogen scattering with polarized beams. Using a crossed beams configuration in which the spins of the incident free electron and target hydrogen electron were prepared either parallel (↑↑) or antiparallel (↑↓), we measured the cross section asymmetries

$$A_{90^\circ} (1s \to 1s) = \frac{d\sigma(\uparrow\downarrow) - d\sigma(\uparrow\uparrow)}{d\sigma(\uparrow\downarrow) + d\sigma(\uparrow\uparrow)} \tag{9}$$

for 90° elastic scattering and

$$A_I = \frac{\sigma(\uparrow\downarrow) - \sigma(\uparrow\uparrow)}{\sigma(\uparrow\downarrow) + \sigma(\uparrow\uparrow)} \tag{10}$$

for total impact ionization. Expressed in terms of the appropriate amplitudes f and g, the quantities $A_{90^\circ}(1s\to1s)$ and A_I can be written as[2]

$$A_{90^\circ}(1s\to1s) = \frac{\text{Re }(f^*g)}{d\bar{\sigma}\ d\Omega} = \frac{|f||g|\cos\theta}{d\bar{\sigma}\ d\Omega} \tag{11}$$

and

$$A_I = \sigma_I^{int}/\bar{\sigma}_I, \tag{12}$$

where θ is the relative phase of f and g, $d\sigma/d\Omega$ is the spin averaged differential cross section given by

$$\frac{d\bar{\sigma}}{d\Omega} = \frac{1}{4} |f+g|^2 + \frac{3}{4} |f-g|^2, \tag{13}$$

$\bar{\sigma}_I$ is the total spin-averaged ionization cross section given by

$$\bar{\sigma}_I = \frac{1}{2} \int_0^E dE_3' \quad \frac{k_3'k_2'}{k_3} (\frac{1}{4}|f+g|^2 + \frac{3}{4}|f-g|^2)\ d\hat{k}_3'd\hat{k}_2', \tag{14}$$

and σ_I^{int} is the total interference cross section given by

$$\sigma_I^{int} = \frac{1}{2} \int_0^E dE_3' \quad \frac{k_3'k_2'}{k_3} \text{Re }[f^*(k_3',k_2')g(k_3',k_2')]d\hat{k}_3'd\hat{k}_2'. \tag{15}$$

Here $\hat{k}_3$, $\hat{k}_3'$, and $\hat{k}_2'$ are, respectively, the momenta of the incident, scattered, and ejected electrons, E_3' is the energy of the scattered electron, and $E + \frac{1}{2}$ is the energy of the incident electron, all quantitites being given in atomic units.

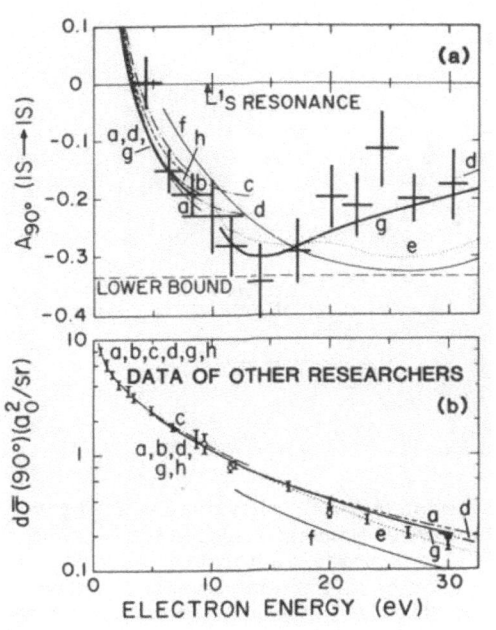

Fig. 1. (a) Measured values of A_{90° (1s→1s) as a function of incident electron energy as presented in Ref. 2. The vertical error bars are one standard deviation uncertainties, and the horizontal error bars indicate the energy spread of the beam. Theoretical curves are obtained from information in the following references using procedures given in Ref. 2: a, Ref. 10; b, Refs. 11-13; c, Ref. 14; d, Ref. 15; e, Ref. 16; f, Ref. 17; g, Refs. 18-21; h, Ref. 22; (b) Measurments of $d\bar{\sigma}(90^\circ)$ taken from Refs. 16, 23, and 24, solid bars, and from Refs. 25 and 26, open circles. Theoretical curves are from the same references as in (a).

The results of the first series of measurements of A_{90°(1s→1s) and A_I are shown in Figs. 1(a) and 2(a) respectively together with a variety of theoretical predictions, by no means exhaustive in number. In Figs. 1(b) and 2(b) are shown for comparison measurements of the respective spin-averaged cross sections and the respective theoretical calculations. The contrasting disagreement between experiment and theory for the asymmetries and the agreement between experiment and theory for the spin averaged cross sections is quite evident.

Fig. 2. (a) Measured values of A_I as a function of incident electron energy as presented in Ref. 2. The vertical error bars are one standard deviation uncertainties, and the horizontal error bars indicate the energy spread of the beam. Theoretical curves are obtained from information in the following references using procedures given in Ref. 2: a, Ref. 27; b, Ref. 28; c, Ref. 29; d, Ref. 30; e, Ref. 28; f, Ref. 31; g and h, Ref. 27; i, Ref. 30; j, Ref. 32; k, Ref. 33; l, Ref. 34; (b) Measurements of σ_I from Refs. 35 and 36. The vertical error bars indicate the spread of the measurements. Theoretical curves are from the same references as in (a).

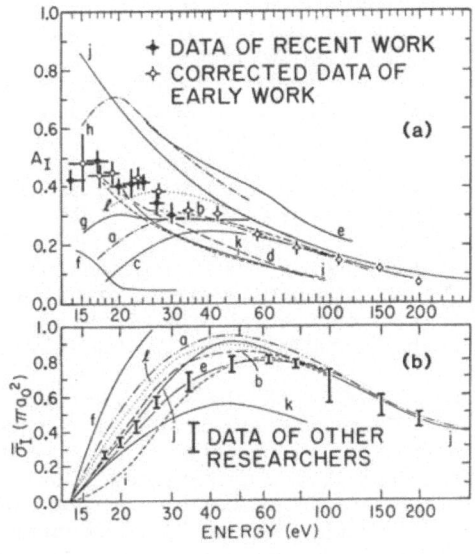

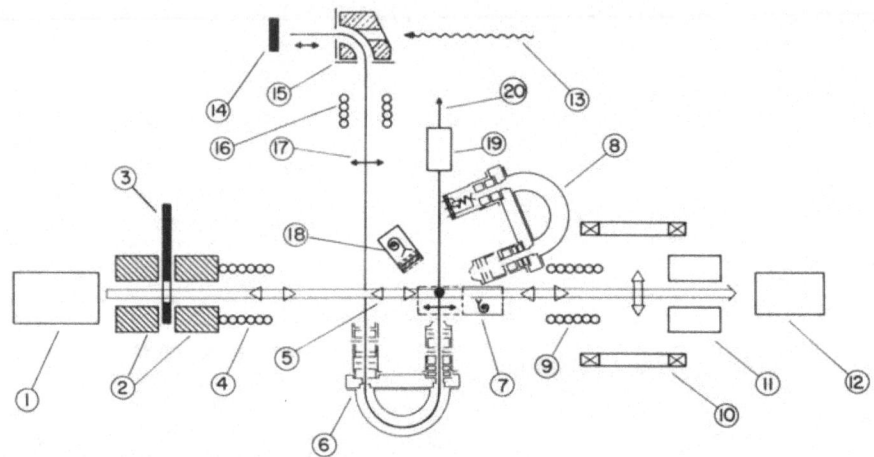

Fig. 3. Layout of polarized e-H experiment at CCNY showing (1) rf hydrogen source (Ref. 38); (2) hexapole magnetic high-field state
selector; (3) beam chopper; (4) solenoidal adiabatic spin precessor; (5) direction of the electronic polarization vector of hydrogen atoms; (6) molybdenum hemispherical electron monochromator;
(7) channel multiplier ion detector; (8) molybdenum hemispherical
electron spectrometer; (9) and (10) adiabatic spin precessors;
(11) Stern-Gerlach polarimeter; (12) quadrupole mass analyzer;
(13) circularly polarized 787-nm laser light for photoemission;
(14) GaAs crystal mounted on sapphire block; (15) copper spherical
90° bender; (16) solenoidal spin precessor; (17) direction of polarization vector of electrons; (18) channel multiplier Lyman-α
photon detector; (19) movable Faraday cup; (20) electron beam
exiting to Mott polarimeter.

New measurements of A_I and $A_\theta(1s \rightarrow 1s)$ for $20° < \theta < 105°$ are currently
being carried out with an apparatus recently constructed at CCNY.[37] Shown
schematically in Fig. 3, the experiment relies on a GaAs polarized
electron source,.[39] an rf atomic hydrogen source,[38] high-field atomic state
selection,[40] and a hemispherical electron monochromator-spectrometer pair.
Preliminary results in the near ionization threshold region are presented
in Fig. 4 for the measured asymmetry Δ_I, which is related to the physical
asymmetry A_I by the expression $\Delta_I = P_e P_H (1-F_2) |\cos\alpha| A_I$, where P_e and P_H
are the electron and atomic hydrogen polarizations, respectively, F_2 is
the fraction of events attributable to molecular hydrogen, and α is the
angle between the electron and hydrogen polarization vectors. When the
ionization measurements are finally completed they should shed significant
light on the validity of the Wannier model[42-45] of threshold ionization and
the competing Coulomb dipole model.[46-50]

Although the measurements of $A_\theta(1s \rightarrow 1s)$ and A_I do not provide the
full separation of f and g required for the benchmark experiments
previously suggested, they do represent the first steps toward the more
sophisticated spin-tagged studies needed. In fact, the first such studies
for electron scattering from mercury and xenon have recently been
reported,[53] and the experimental configuration described in that report
can easily be adapted to electron-hydrogen scattering. The required
combination of e^+-H and spin-tagged e^--H scattering experiments should
thus become a reality in the not too distant future.

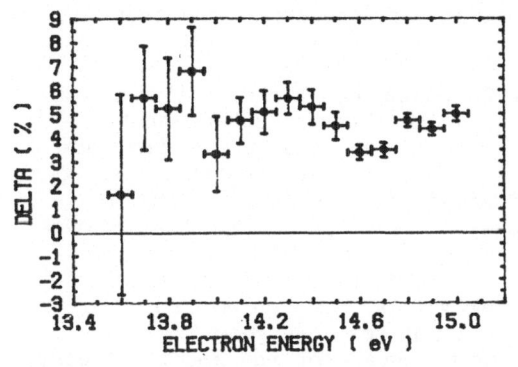

Fig. 4. Preliminary data from polarized e^--H experiment at CCNY showing Δ_I as a function of incident electron energy in the near ionization threshold region. The vertical error bars represent one standard deviation uncertainties, and the horizontal error bars indicate the energy spread of the beam.

DESCRIPTION OF THE BROOKHAVEN e^+-H EXPERIMENT

It is largely the development of the high-intensity low-energy positron beam[52] at the High-Flux Beam Reactor (HFBR) at Brookhaven National Laboratory (BNL) that makes it possible to plan a comprehensive program of positron-hydrogen scattering experiments. The characteristics of the existing HFBR positron beam that are relevant to gas scattering experiments are summarized in Table 1 along with the requirements of the e^+-H experiment. Note that the maximum intensity in the table is a factor of five higher than that reported in Ref. 52, which was published more than one year ago. In fact, with the use of a frozen neon moderator,[53] a modification to be incorporated into the BNL source later this year, the intensity is expected to rise yet another factor of ten to 2×10^9 e^+/s, making it at least four orders of magnitudes more intense than that produced by conventional ^{22}Na sources.

Table 1. Positron Beam Characteristics

Characteristic	Existing HFBR Beam	e^+-H Experimental Requirements	EBE Hybrid Beam Design
Energy (eV)	10-2000	1-500	1-500
Energy Width (meV)	75	<100	<100
Maximum Intensity (e^+/s)	2×10^8		4×10^7
Half Life (h)	12.8		12.8
Time Available(%)[a]	75		75
Average Intensity(e^+/s)[b]	3.8×10^7	$\geq 10^6$	8×10^6
Emittance(mrad cm eV$^{3/2}$)[c]	330	33	33
Magnetic Field (G)	≥ 20	<0.1	<0.1

[a]For 72-hour run with 24-hour reloading time for positron source.
[b]For 72-hour run with 75% time-availability.
[c]At stated magnetic field.

Although potentially suitable for positron-atom studies, the BNL beam is not without its drawbacks. Its diameter is approximately 1 cm, and at an energy of 400 eV its divergence (half cone angle) is 2°, giving it an emittance of 350 mrad cm eV$^{3/2}$, which is at least an order of magnitude larger than that required for low-energy crossed positron-atom beam studies. Moreover the BNL transport optics is purely magnetic, with the positron source itself residing in a magnetic field of 20-50 G, a configuration that inflates the generalized emittance[40] of the beam even further.

In order to circumvent these difficulties we carried out a design study[2] that employed the technique of brightness enhancement remoderation[54-59] to lower the phase space of the beam and at the same time to effect a transition from magnetostatic to electrostatic transport optics. We further removed the moderated positron source from its magnetic environment and replaced its extraction optics with that of the modified Soa gun geometry,[60] thereby creating a hybrid beam transport system that contains a long magnetostatic section sandwiched between two short electrostatic sections. (The removal of the source from the magnetic field and the use of the modified Soa gun geometry are consistent with the frozen neon moderator upgrade of the source already in progress.) The layout of the hybrid beam line and its relationship to the e[+]-H experiment, as well as to two other experiments using the HFBR e[+] beam, is shown in Fig. 5.

In order to carry out the design study we used the SLAC ray tracing code developed by W.B. Hermannsfeldt,[63] which first solves Poisson's equation at an array of mesh points for a given set of boundary conditions and after calculating the potential at each mesh point then determines the electric vector at each point. The code allows the use of either cylindrical or rectangular coordinates and, although not germane to our application, contains provision for including the effects of space charge. It became clear quite rapidly that optimum injection into and ejection out of the magnetostatic solenoidal transport optics dictated the use of magnetic termination plates at each end of the solenoid. Using the computer code POISSON,[62] we modeled a solenoid with 1-cm thick soft iron terminations plates having beam holes 1.2 cm in diameter. We proceeded to investigate various tuning conditions for the modified Soa gun and launched the extreme trajectories into the solenoid with the use of the SLAC/Hermannsfeldt code.

Close analysis showed that for optimal tuning of the modified Soa gun, a solenoid length could be found for which a 2 keV transport energy and a 50 G magnetic field would result in a spot size radius of < 0.75 mm on a thin film transmission remoderator. With the invariant emittance ε given by

$$\varepsilon = r \; \frac{Et}{E} \; \sqrt{E} = r\sqrt{E_t} \tag{16}$$

for a transverse escape energy E_t, and with E_t having a typical value[55] of 0.2 eV, the emittance of the remoderated beam in the design study was 33 mrad cm eV$^{3/2}$, a value appropriate for low-energy crossed beams scattering experiments. The design study showed, moreover, that the spot size radius on the remoderator varied by no more than 5 percent as the potential on the remoderator was varied from 0 to +500 V. Thus if the interaction region, where the positron and hydrogen beams cross, is to be held at ground potential, and the positron beam energy is to be varied by variation of the potential on the remoderator, a kinematic range of 1 to 500 eV should be easily accessible with the hybrid beam transport design.

The beam characteristics of the design study are shown in the final column of Table 1, in which it is assumed that the efficiency of the transmission remoderator[54,58,59] is 0.2. Comparison with the minimum requirements of an e[+]-H crossed beams experiment, also given in Table 1, shows that the EBE hybrid system combined with the technique of brightness enhancement remoderation should provide a positron beam of the quality necessary to begin studies of positron-hydrogen collisions.

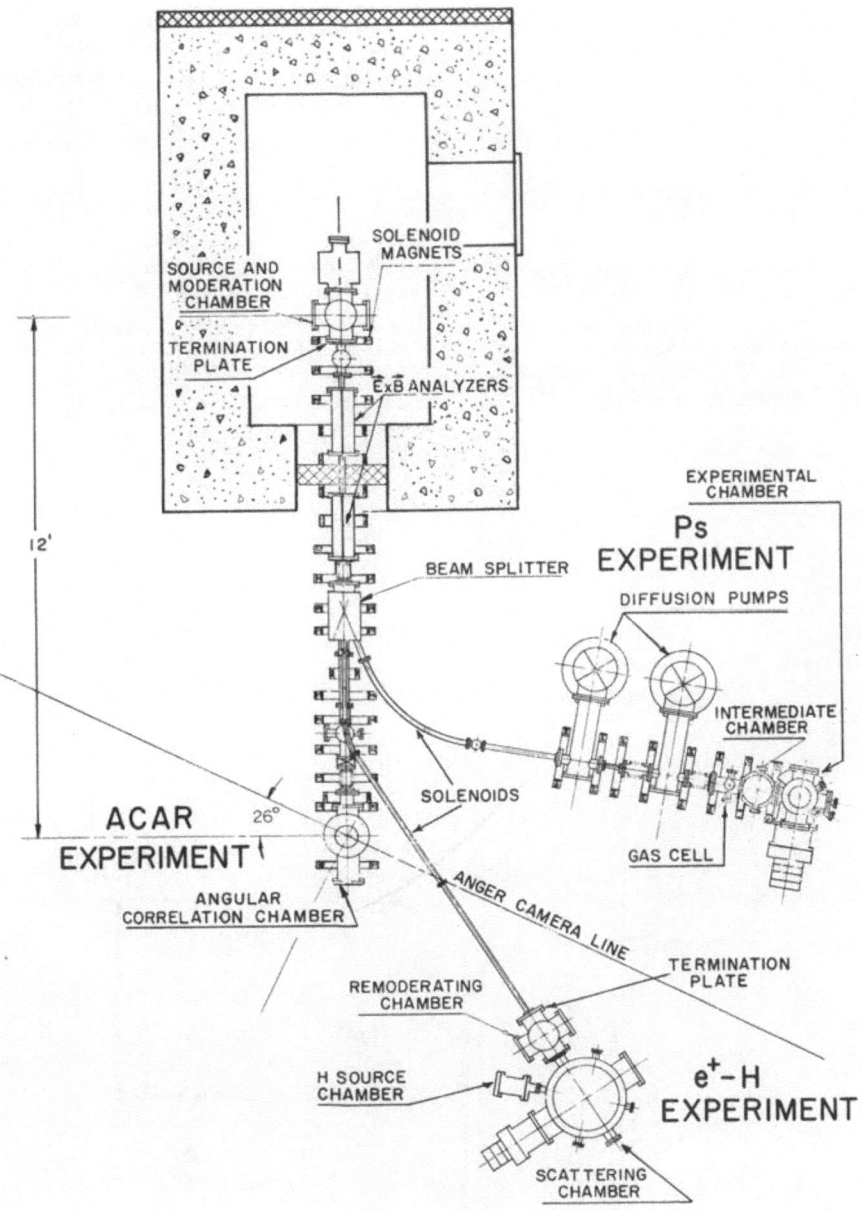

Fig. 5. Scale drawing of low-energy, high-intensity positron facility at
the Brookhaven High-Flux Beam Reactor. The locations of the
existing ACAR (angular correlation of annihilation radiation) and
Ps experiments are shown together with the proposed location of
the e^+-H experiment. The concept of the hybrid beam line is
illustrated by the presence of the two termination plates at the
beginning and the end of the solenoidal magnetic transport
section that connects the source and moderation chamber (located
inside the concrete block shielding house) to the remoderation
chamber (located immediately upstream of the e^+-H scattering
chamber).

For completeness, it should be noted that the final electrostatic transport section in the design study utilized the modified Soa gun geometry for beam extraction from the remoderator foil and conventional tube lenses for transport to the interaction region, the last lens elements comprising a "zoom lens" configuration. For the calculations of the final electrostatic section, we employed standard matrix transfer techniques using the notes of C. E. Kuyatt for guidance.[63]

As an added comment about the hybrid beam transport, we point out that while our design study strictly applies only to a straight solenoid, the electron optical properties of the existing BNL beam are well understood and can be incorporated easily into a final design. In this context we also note that for the 5-meter radius of curvature assumed for the bent solenoid illustrated in Fig. 5, the chosen conditions of 2 keV transport energy and 50 G magnetic field satisfy the requirements of both the guiding-center approximation and the preservation of relative phase of cyclotron orbits.[55,64-66]

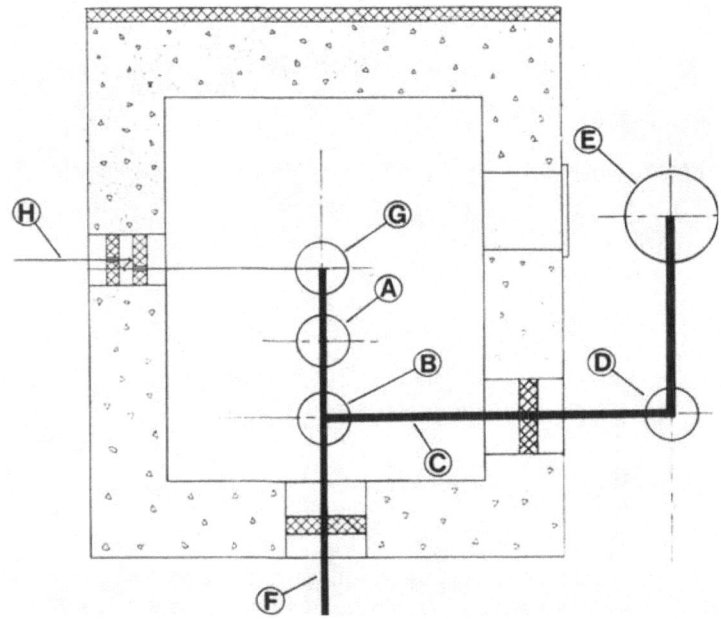

Fig. 6. Schematic layout of the full electrostatic beam line option for the e⁺-H experiment, illustrating the future figuration of the Brookhaven high intensity positron facility. The following elements are shown: A, source and moderation chamber with modified Soa gun geometry (Ref. 60); B, electrostatic switchyard; C, transverse electrostatic beam line; D, remoderator and 90° electrostatic bender; E, e⁺-H scattering chamber; F, ACAR and Ps beam line; G, future isotope-separation chamber; H, laser for isotope separation.

Within the last few months it has become clear that for greater accessibility the BNL positron beam will be moved from its present location on the operations level of the HFBR to a new location in another laboratory building. In connection with the move, the concrete block shielding house will be enlarged, new experimental locations will be developed, and a new beam transport system will be designed. As presently envisioned, the layout of the new e^+-H experiment and associated beam line will be as shown in Fig. 6. For this new configuration, the beam transport system for the e^+-H experiment will be entirely electrostatic. As in the case of the hybrid transport design, thin film transmission brightness enhancement remoderation will be employed, as indicated in Fig. 6. The final beam characteristics are expected to be the same as those given in the final column of Table 1.

The BNL e^+-H experiments will employ an rf atomic hydrogen source similar to the one used in the spin-tagged e^--H studies. Based upon data obtained at CCNY and elsewhere, the anticipatated hydrogen beam density will be between approximately 10^{12} atoms/cm^3 and the dissociation fraction will be approximately 0.8. As already indicated, the experiments will be carried out in a crossed beams configuration, the details of which will be discussed by Dr. G. Sinapius. In general, the positron measurements will be normalized to electron measurements with the use of a GaAs photoemission electron beam originating at the site of the transmission remoderator. By contrast with a thermionic electron beam, which has a typical energy width of >300 meV and a large angular divergence at the cathode surface, a GaAs photoemission beam is characterized by an intrinsic energy spread of <130 meV and a comparative narrow angular divergence, properties that more closely approximate those of the remoderated positron beam. Thus the electron normalization should be enhanced considerably in reliability.

SCOPE OF THE e^+-H EXPERIMENT

As presently envisioned, the e^+-H program will begin with the study of positron impact ionization with and without positronium formation, as discussed in the succeeding paper by G. Sinapius. Contingent upon the experience gained in these studies, the project will continue with differential elastic scattering measurements. Although at this time it is premature to discuss the details of such future measurements and as a consequence unnecessary to examine the precise status of their theoretical basis, it is fair to say in the context of the background material presented at the beginning of this paper that such measurements will likely gain in their importance as the sophistication of computer programs improve and the technology of positron beams and polarized electron beams also improve.

ACKNOWLEDGEMENT

Support for this project is provided in part by the U. S. National Science Foundation (Grant No. PHY-8603166) and by the Research Foundation of the City University of New York PSC-CUNY No. 667353.

REFERENCES

1. F. J. Mulligan and M. S. Lubell, Bull. Am. Phys. Soc. 32:1279 (1987); to be submitted to Rev. Sci. Instrum.
2. G. D. Fletcher, M. J. Alguard, T. J. Gay, V. W. Hughes, P. F. Wainwright, M. S. Lubell, and W. Raith, Phys. Rev. A 31:2854 (1986).

3. J. Slevin, Rep. Prog. Phys. 47:461 (1984).
4. J. Kessler, "Polarized Electrons," 2nd Ed., Springer Verlag, Berlin (1985), Ch. 4.
5. W. R. Webber, V. Schonfelder, and R. Diehl, Nature 323:692 (1986).
6. M. Leventhal, C. J. McCallum, and P. D. Stang, Astrophys. J. 225:L11 (1978); M. Leventhal, C. J. McCallum, A. F. Huters, and P. D. Stang, Astrophys. J. 260:L1 (1982).
7. W. A. Mahoney, J. C. Ling, W. A. Wheaton, and S. A. Jacobson, Astrophys. J. 286:578 (1984); G. H. Share, R. L. Kinzer, J. D. Kurfess, D. J. Forrest, E. L. Chupp, and E. Rieger, Astrophys. J. 292:L61 (1985).
8. R. W. Bussard, R. Ramaty, and R. J. Drachman, Astrophys. J. 228:928 (1979).
9. See for example P. F. Wainwright, M. J. Alguard, G. Baum, and M. S. Lubell, Rev. Sci. Instrum. 49: 571 (1978).
10. W. C. Fon, P. G. Burke, and A. E. Kingston, J. Phys. B 11:521 (1978).
11. C. Schwartz, Phys. Rev. 124:1468 (1961).
12. D. Register and R. T. Poe, Phys. Lett. 51A:431 (1975).
13. R. L. Armstead, Phys. Rev. 171:91 (1968).
14. A. Temkin and J. C. Lamkin, Phys. Rev. 121:788 (1961).
15. S. Geltman, Phys. Rev. 119:1283 (1960).
16. J. Callaway and J. F. Williams, Phys. Rev. A 12:2312 (1975).
17. G. Khayrallah, in "Eleventh International Conference on the Physics of Electronic and Atomic Collisions, Abstracts of Contributed Papers," K. Takayanagi and N. Oda, eds. North Holland, Amsterdam, 1979, p.114.
18. P. G. Burke and H. M. Schey, Phys. Rev. 126:147 (1962); 126:163 (1962).
19. P. G. Burke and K. Smith, Rev. Mod. Phys. 34:458 (1962).
20. P. G. Burke, H. M. Schey, and K. Smith, Phys. Rev. 129:1258 (1963).
21. B. L. Scott, Phys. Rev. 140:A699 (1965).
22. P. G. Burke, D. F. Gallaher, and S. Geltman, J. Phys. B 2:1142 (1969).
23. J. F. Williams, J. Phys. B 7:L56 (1974).
24. J. F. Williams, J. Phys. B 8:1683, 2191 (1975).
25. P. J. O. Teubner, C. R. Lloyd and E. Weigold, Phys. Rev. A 9:2552 (1974).
26. C. R. Lloyd, P. J. O. Teubner, E. Weigold, and B. R. Lewis, Phys. Rev. A 10:175 (1974).
27. M. R. H. Rudge and M. J. Seaton, Proc. R. Soc. London Ser. A 283:262 (1965).
28. R. K. Peterkop, Zh. Eksp. Teor. Fiz. 41:1938 (1961) [Sov. Phys. -- JETP 14:1377 (1962)].
29. S. Geltman, M. R. H. Rudge, and M. J. Seaton, Proc. Phys. Soc. London 81:375 (1963).
30. J. E. Golden and J. H. McGuire, Phys. Rev. Lett. 32:1218 (1974).
31. M. R. H. Rudge and S. B. Schwartz, Proc. Phys. Soc. London 88:563 (1966).
32. V. I. Ochkur, Zh. Eksp. Teor. Fiz. 47:1746 (1964) [Sov. Phys. -- JETP 20:1175 (1965)].
33. D. F. Gallaher, J. Phys. B 7:362 (1974).
34. M. R. H. Rudge, J. Phys. B 11:L149 (1978).
35. W. L. Fite and R. T. Brackman, Phys. Rev. 112:1141 (1958).
36. E. W. Rothe, L. L. Marino, R. H. Neynaber, and S. M. Trujillo, Phys. Rev. 125:582 (1962).
37. D. Crowe, F. C. Tang, A. Vasilakis, M. S. Lubell, K. Rubin, F. J. Mulligan, J. Slevin, and M. Eminyan, Bull. Am. Phys. Soc. 32:1283 (1987).
38. J. Slevin and W. Stirling, Rev. Sci. Instrum. 52:1780 (1981).
39. F. C. Tang, M. S. Lubell, K. Rubin, A. Vasilakis, M. Eminyan, and J. Slevin, Rev. Sci. Instrum. 57:3004 (1986), and references therein.
40. V. W. Hughes, R. L. Long, Jr., M. S. Lubell, M. Posner, and W. Raith, Phys. Rev. A 5:195 (1972).

41. G. H. Wannier, Phys. Rev. 90:817 (1953).
42. A. R. P. Rau, Phys. Rev. A 4:207 (1971).
43. R. Peterkop, J. Phys. B 4:513 (1971).
44. H. Klar and W. Schlecht, J. Phys. B 9:1699 (1976).
45. C. H. Greene and A. R. P. Rau, J. Phys. B 16:99 (1983).
46. A. Temkin, Phys. Rev. Lett. 16:835 (1966).
47. A. Temkin and Y. Hahn, Phys. Rev. A 10:708 (1974).
48. A. Temkin, J. Phys. B 7:L450 (1974).
49. A. Temkin, Phys. Rev. Lett. 49:365 (1982).
50. A. Temkin, IEEE Trans. Nucl. Sci. NS-30:1106 (1983); Phys. Rev. A 30:2737 (1984).
51. O. Berger and J. Kessler, J. Phys. B 19:3539 (1986).
52. K. G. Lynn, A. P. Mills, Jr., L. O. Roellig, and M. Weber, in: "Electronic and Atomic Collisions," D. C. Lorentz, W. E. Meyerhof, and J. R. Peterson, eds., Elsevier, Amsterdam, 1986, p.227.
53. E. M. Gullikson and A. P. Mills, Jr., Phys. Rev. Lett. 57:376 (1986).
54. A. P. Mills, Jr., Appl. Phys. 23:189 (1980).
55. K. F. Canter and A. P. Mills, Jr., Can. J. Phys. 60:551 (1982).
56. K. G. Lynn and A. Wachs, Appl. Phys. A 29:93 (1982).
57. D. M. Chen, K. G. Lynn, R. Pareja, and B. Nielsen, Phys. Rev. B 31:4123 (1985).
58. W. E. Frieze, D. W. Gidley, and K. G. Lynn, Phys. Rev. B 31:5628 (1985).
59. K. F. Canter, T. Horsky, P. H. Lippel, W. S. Crane, and A. P. Mills, Jr., in: "Proceedings of the Third International Workshop on Positron (Electron)-Gas Scattering," W. E. Kauppila, T. S. Stein, and J. M. Wadehra, eds., World Scientific, Singapore, 1986, p.202.
60. K. F. Canter, P. H. Lippel, W. S. Crane, and A. P. Mills, Jr., in: "Positron Studies of Solids, Surfaces, and Atoms," A. P. Mills, Jr., W. S. Crane, and K. F. Canter, eds., World Scientific, Singapore, 1986, p.199.
61. W. B. Hermannsfeldt, SLAC Report No. 226.
62. The POISSON code was originally written by R. F. Holsinger of Lawrence Livermore National Laboratory and was subsequently modified at CERN.
63. C. E. Kuyatt, Lecture Notes, unpublished. For an abbreviated version, see J. H. Moore, C. C. Davis, and M. A. Coplan, "Building Scientific Apparatus," Addison Wesley, London, 1983, Ch. 5.
64. W. E. Kauppila, T. S. Stein, G. Jesion, M. S. Dobabneh, and V. Pol, Rev. Sci. Instrum. 48:822 (1977).
65. J. D. Jackson, "Classical Electrodynamics," 2nd Ed., John Wiley & Sons, New York, 1975, p.584.
66. K. F. Canter, private communication.

THE BROOKHAVEN POSITRON-HYDROGEN SCATTERING EXPERIMENT:

EXPERIMENTAL SET-UP

Günther Sinapius

Fakultät für Physik

Universität Bielefeld, D-4800 Bielefeld, Fed.Rep. of Germany

INTRODUCTION

The aim of this experiment is to study the positron impact ionisation of atomic hydrogen with and without positronium formation. It will be run by a collaboration of members of the Brookhaven Consortium, the City College of CUNY, St. Patrick's College at Maynooth and the University of Bielefeld.[1] The Bielefeld positron group will provide the scattering chamber and the experimental set-up to measure the cross sections.

The apparatus that is now being tested at Bielefeld will be modified and is planned to be shipped to Brookhaven in the Spring of 1988. It will go on the projected electrostatic beam-line and the measurement procedure will differ from that currently employed at Bielefeld.[2]

EXPERIMENTAL SET-UP

A sketch of the apparatus as it is to be linked to the electrostatic beam-line behind the transmission remoderator and a 90-degree spherical deflector is shown in Fig. 1. At the gate valve next to the deflector we expect a positron beam of about 10^7 e$^+$/s with a phase-space of r × θ = 5 mrad cm at 40 eV. Because of the 10^{-6} Torr base pressure in the scattering region there will be a differential pumping stage to meet the vacuum requirements for the positron beam-line and the remoderator.

The hydrogen beam is produced by a Slevin radio-frequency atomic hydrogen beam source.[3] A grid structure surrounds the region where the positron and hydrogen beams cross. All positrons scattered into angles less than 30° pass the interaction region. When the beam intensity is low they are counted on the micro-channel-plate (MCP) mounted downwards in the beam line; for higher intensity the MCP is operated as a current amplifier. In a later stage of the experiment the positron beam will be transported further through a 90° deflector in order to reduce background from the annihilation radiation. The grids at the interaction region are biased to facilitate the extraction of the ions. They are transported to a channel-electron-multiplier (CEM) shielded against the Lyman-α radiation from the hydrogen source. A mass spectrometer allows the identification of H$^+$ and H$_2^+$ ions.

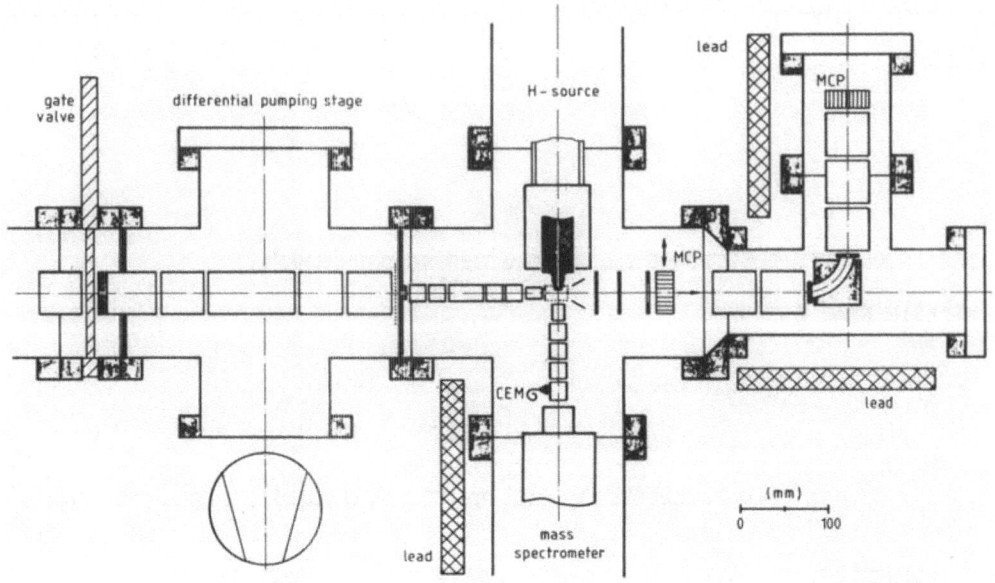

Fig. 1.   Sketch of the experimental set-up to measure the positron
impact ionisation of atomic hydrogen.

MEASUREMENTS

At low positron intensities the time correlation between detected
positrons and ions identifies ionisation without positronium formation.
This method has been successfully applied for the study of positron impact
ionisation of He and H_2[4,5]. There are, however, three additional
difficulties:

1.   There will be two different types of ions: H^+ and H_2^+. If
     no positronium is formed they can be identified from the time-
     of-flight (TOF) spectrum. Cross sections for ionisation with
     and without positronium formation are deduced from measurements
     with the hydrogen discharge tube turned on and off.

2.   The degree of dissociation of the H_2 might change in time.
     Frequent control measurements at a given energy will keep
     track of the degree of dissociation.

3.   Only those positrons that scatter into the forward direction
     after ionisation are detected. Measurements with electrons
     will tell us how important scattering through large angles is.
     A magnetic guiding field for the positrons in this part of the
     apparatus might be employed.

The TOF method cannot be applied at high positron intensities, when
the current is measured. In that case, however, the mass spectrometer
allows the measurement of the sum of both ionising processes independently
for H and H_2. With the data taken at low positron beam intensity, ionisation with and without positronium formation can be deduced. With a target
thickness of 10^{12} cm^{-2} and a cross section of 5×10^{-17} cm^2 we expect an
event rate of 5×10^{-5} ions/e$^+$.

In all stages of the experiment measurements of the electron impact ionisation cross sections and comparison with values from the literature offer the possibility of checking the performance of the apparatus and provide a normalisation for the positron data.

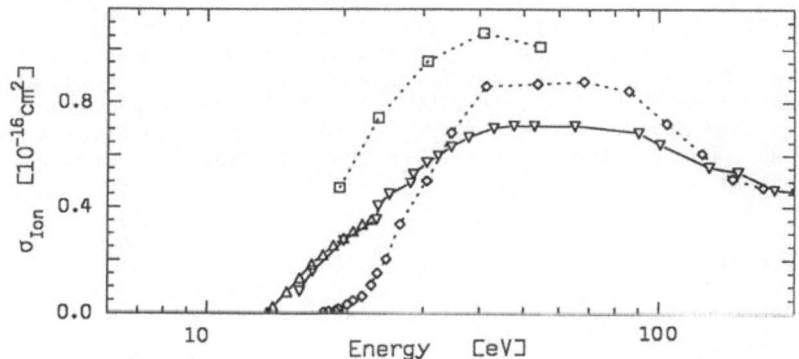

Fig. 2. Calculations for the e^+ – H ionisation cross section ☐– DW1[6] and ◇ –CTMC[7] compared with experimental electron data: Δ[8], ∇[9] .

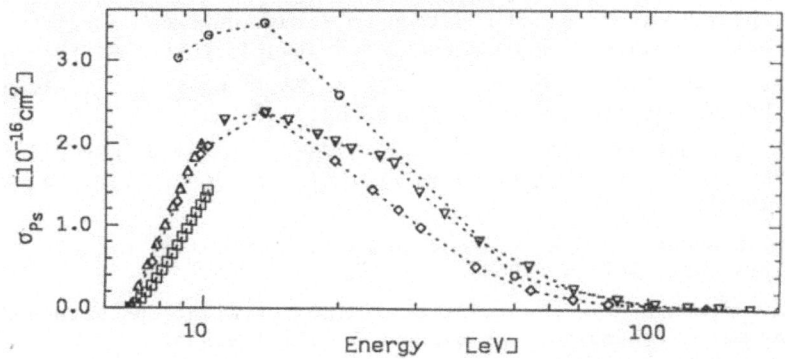

Fig. 3. Theoretical results for the e^+ – H positronium formation cross section, ☐ –CS[10], ◇ –DW[11], Δ–KV[12], ∇–CTMC[7], O–TM[13].

THEORETICAL PREDICTIONS

There are not many calculations for the positron impact ionisation of atomic hydrogen without positronium formation. In Fig. 2. results from a distorted-wave polarized-orbital method (DW)[6] and from a classical trajectory Monte Carlo method (CTMC)[7] are displayed together with experimental data for electron impact ionisation.[8,9] We choose that version of the distorted-wave (DW1) calculation which the authors estimate as more reliable. In the case of He a similar treatment (a plane wave for the outgoing positron) gave good agreement with the experimental data.[4] The calculations give different results. Both, however, exceed the corresponding electron cross sections above 30 eV.

A variety of calculations have been performed to describe positronium formation in e^+ - H collisions. Fig. 3 shows some results obtained with different approaches: coupled-static approximation (CS)[10], polarized-orbital distorted-wave calculation (DW1)[11], Kohn variational method (KV)[12], classical trajectory Monte Carlo (CTMC)[7], and quantum mechanical T-matrix above 100 eV. Major differences between the theoretical predictions arise at lower energies.

ACKNOWLEDGEMENTS

This text is the result of many discussions with Prof. W. Raith, A. Schwab, G. Spicher and W. Sperber.

REFERENCES

1. M.S. Lubell, The Brookhaven Positron-Hydrogen Scattering Experiment, this conference.

2. G. Spicher, A. Glasker, W. Raith, G. Sinapius and W. Sperber, Ionisation of Atomic Hydrogen by Positron Impact, poster at this conference.

3. J. Slevin and W. Stirling, Radio Frequency Atomic Hydrogen Beam Source, Rev. Sci. Instrum. 52:1780 (1981).

4. D. Fromme, G. Kruse, W. Raith and G. Sinapius, Partial-Cross Section Measurements for Ionisation of Helium by Positron Impact, Phys. Rev. Lett. 57:3031 (1986).

5. D. Fromme, G. Kruse, W. Raith and G. Sinapius, Measurements of the Impact Ionisation and Positronium Formation Cross Sections for Positron Scattering on Molecular Hydrogen, poster at this conference.

6. A.S. Ghosh, P.S. Majumdar and M.Basu, Positron-Impact Ionisation of Hydrogen Atoms, Can. J. Phys. 63:621 (1985).

7. A.E. Wetmore and R.E. Olson, Ionisation of H and He^+ by Electrons and Positrons Colliding at Near-Threshold Energies, Phys. Rev. A34:2822 (1986).

8. J.W. McGowan and E.M. Clarke, Ionisation of H(1s) near Threshold, Phys. Rev. 167: 3 (1968).

9. W.L. Fite and R.T. Brackman, Collisions of Electrons with Hydrogen Atoms: I Ionisation, Phys. Rev. 122:1141 (1958).

10. M.A. Abdel-Raouf, J.W. Darewych, R.P. McEachran and A.D. Stauffer, Application of the Coupled-Static Approximation to Positron-Hydrogen Inelastic Scattering, Phys. Lett. 100A:353 (1984).

11. P. Khan and A.S. Ghosh, Positronium Formation in Positron-Hydrogen Scattering, Phys. Rev. A27:1904 (1983).

12. J.W. Humberston, Positronium - Its Formation and Interaction with Simple Systems, Adv. At. Mol. Phys. 22:1 (1986).

13. E. Ficocelli Varracchio and M.D. Girardeau, Ideal-Space Treatment of the $e^+ + H \quad Ps + H^+$ Process, J. Phys. B: At. Mol. Phys. 16:1097 (1983).

SYNTHESIS OF ANTIHYDROGEN

H. Poth[*]

Kernforschungszentrum Karlsruhe

Institut für Kernphysik, Karlsruhe, Fed.Rep. Germany

INTRODUCTION

An antihydrogen atom consists of an antiproton nucleus and an orbiting positron, but although the constituents are well known, atomic antimatter has been neither produced nor observed. High-precision spectroscopy may provide deeper information about the properties of the constituents, in particular those of the antiproton. In the first instance, however, a comparison of the properties of antihydrogen with the corresponding ones in hydrogen allows the validity of the CPT theorem to be tested in a compound system of hadrons and leptons.

The properties of antihydrogen which would be measured in the laboratory could be different from those of hydrogen, even without breaking the CPT theorem, if the antiatom were to be subjected to a long-range force exerted with different strengths by ordinary matter and antimatter respectively.

The final fate of antihydrogen when it collides with normal matter is that it decomposes into positronium and an antiprotonic atom. The positronium decays into gammas, and the antiproton annihilates with a proton or a neutron. The resulting products − pions and gammas − do not carry any specific characteristics which would allow us to conclude that an antihydrogen-hydrogen annihilation took place. Since the force between antihydrogen and matter atoms appears to be attractive, there may exist bound or resonant states. For the hydrogen-antihydrogen system this seems to be excluded because of the strength of the annihilation. However, it could be possible if the partner of the antihydrogen is a heavier normal-matter atom.

The interaction of positrons with matter is well understood, but this is not the case for the antinucleon-nucleon system. The strong force between antinucleons and nucleons can be partially constructed from the nucleon-nucleon force. However, the annihilation is so far described in only an empirical way. For instance, the spin dependence of the antiproton-proton annihilation potential is not yet known since, as no polarized antiproton beams exist, this aspect cannot be studied in sufficient detail. The formation of antihydrogen provides us with the possibility to align the antiproton

---

* Visitor at CERN, Geneva, Switzerland.

spin with respect to the positron spin, and in this way to produce polarized beams of antiprotons.

The synthesis of antihydrogen acquaints us with a technology which certainly can be extended to heavier antimatter systems. Antideuterium and antihelium may become accessible once one has learned to produce antihydrogen.

Antihydrogen can be produced in several ways. Here, only the radiative capture of positrons by antiprotons is considered. The cross-section depends inversely on the energy difference between both particles. In order to produce antihydrogen at a reasonable rate, a large number of quasi-monoenergetic antiprotons and positrons of equal velocity are needed. Antiprotons are created through pair production in high-energy proton collisions with nuclear targets. For the purpose considered here, positrons can be obtained from radioactive sources or from electron-, proton-, or photon-induced pair creation.

In the following, the production and accumulation of antiprotons and positrons is discussed, followed by a section on antihydrogen formation and a brief outline of possible experiments.

There might be occasionally an overlap with the following article, given by A. Rich. This reflects the many discussions that are taking place within our joint attempt to form antihydrogen.

STORAGE OF THE CONSTITUENTS

For the production of antihydrogen at a rate of several kilohertz it will be absolutely necessary to accumulate its constituents and to cool and recirculate them. The accumulation and storage of positrons as well as of antiprotons at relativistic energies is facilitated by the long lifetime of the beam. Also, the intensity limits are much less severe at high energies. At low energy the beam would blow up owing to repeated Coulomb scattering from residual gas particles. In these collisions, momentum is transferred from the longitudinal to the transverse motion. Since the focusing forces which act on the transverse motion are much weaker than the bending forces of the dipoles, particles will already be lost when only a small fraction of the longitudinal momentum shows up in the transverse direction. At very low energy, particles can again be stored for long times in ultrahigh-vacuum systems, namely when transverse and longitudinal momenta are of the same order.

ANTIPROTON PRODUCTION AND ACCUMULATION

Antiprotons are at present produced in four laboratories: at CERN (Geneva), at FNAL (Batavia), at BNL (Upton) and at KEK (Tsukuba). However, it is only at CERN and at FNAL that anitprotons are accumulated and also stored.

At CERN pulses of antiprotons are produced by protons which are accelerated in the proton synchrotron (PS) to 26 GeV/c and directed onto a target with generally a repetition rate of about 0.4 Hz. The spectrum of antiprotons produced in such collisions is shown in Figure 1. Antiprotons are collected at their production maximum (3.5 GeV/c antiproton momentum) and injected into the Antiproton Accumulator (AA). Once injected the antiproton bunch is cooled in phase space and then displaced in orbit and added to the previously injected antiprotons (stack). Phase-space cooling is needed in order to provide space for the next pulse. The stack also is continuously cooled. The accumulation procedure is illustrated in Figure 2. The antiproton accumulation rate achieved was typically 5×10^9 h^{-1},

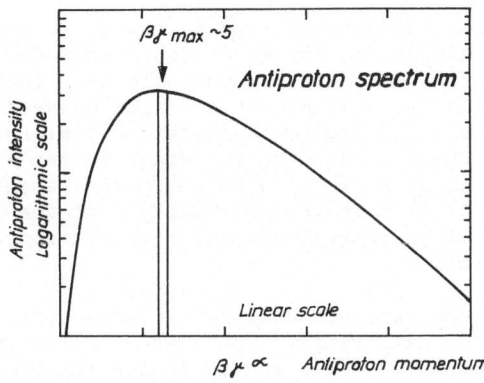

Fig. 1. The momentum spectrum for antiprotons produced in proton-
         nucleus collisions.

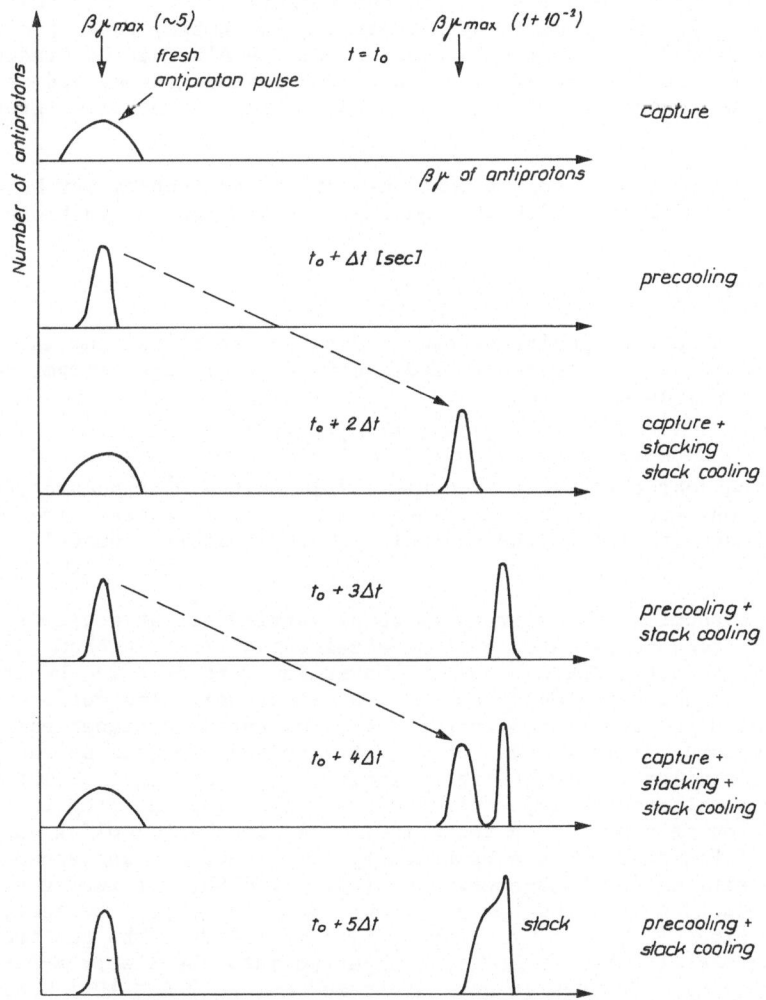

Fig. 2. Schematic illustration of antiproton production and
         accumulation (longitudinal velocity spectrum).

and the highest antiproton intensity stored in the AA is 5×10^{11} .[1] At present a new ring, the Antiproton Collector (AC), is under construction; this will be inserted in the accumulation chain between the antiproton production target and the AA. It will accept a larger phase space for the primary antiproton spectrum and provide a faster initial cooling. The precooled antiprotons will then be added to the stack in the AA and further cooled. With the new scheme it is hoped to increase the antiproton accumulation rate at least tenfold[1]. It seems unlikely that the maximum number of stored antiprotons will be increased much, since it appears to be close to the beam stability limit.

At FNAL the antiprotons are produced by 120 GeV/c protons accelerated in the proton synchrotron (main ring). The higher proton energy shifts the production maximum to about 8.9 GeV/c and increases the production cross-section. Antiprotons are collected at this momentum in the debuncher and after precooling they are transferred to the accumulator for stacking and further cooling. This procedure is similar to the new accumulation scheme at CERN. The design accumulation rate is 1.1×10^{11} per hour, and the aim is to produce a stack of up to 4×10^{11} antiprotons in the accumulator[2].

At BNL and KEK there exist only secondary antiproton beams; their intensity is too low for them to be interesting for antihydrogen production. At BNL it is planned to construct a booster for the Alternating Gradient Synchrotron (AGS) in the framework of the heavy-ion programme, and it was recently thought to accumulate antiprotons by making use of this future facility[3].

In the more distant future a powerful antiproton factory may be provided at a new hadron facility[4] which will have an intense proton synchrotron.

POSITRON PRODUCTION

Unlike antiprotons, positrons can be obtained at almost any laboratory, at least from radioactive sources. Radioactive β^+ emitters deliver positrons continuously at a rate of

$$R_e = 3.7 \times 10^{10} \ C_i \ f_\beta \ s^{-1} \qquad (1)$$

where C_i is the source activity in curie and f_β is the β^+ branching ratio. Sources emit positrons continuously into the full solid angle. The energy of the positrons ranges from the end-point energy (mostly around 1 MeV) down to almost zero.

Pulsed production of positrons is being carried out at accelerators. Positrons are produced by intense linac-accelerated electron beams showering-off in a metal target. The achievable conversion rate at a few hundred MeV is about 1% (one collected positron per 100 electrons). The positrons are collected behind the conversion target and accelerated to higher energies in a linear structure. This reduces beam emittance and momentum spread, and the effect is similar to precooling of antiprotons. At highly relativistic energies the positrons are injected into a storage ring in which they cool, through emission of synchrotron radiation. This inherent cooling eases accumulation. Figure 3 shows schematically the production and accumulation process of positrons for high-energy physics. At CERN, for instance, positrons are produced by the 200 MeV electron beam, of 5 A peak current, of the LEP Injector Linac (LIL) at a repetition rate of 100 Hz. The positrons are collected, accelerated to 600 MeV, and injected into the Electron-Positron Accummulator (EPA). The design goal is to accumulate 2.4×10^{11} positrons in EPA and cool them in momentum spread and in emittance by synchrotron radiation[5]. In order to make use of them for antihydrogen production, they

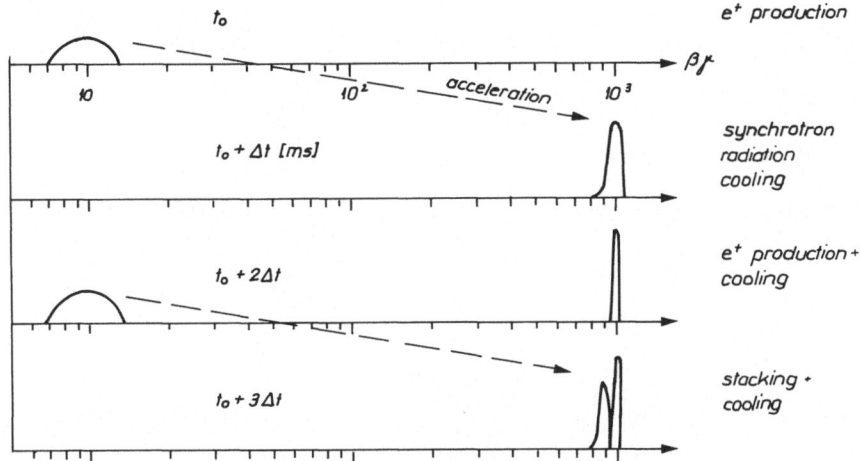

Fig. 3. Schematic illustration of positron production and accumulation at high energies.

would have to be decelerated and cooled again at low energies.

In contrast to antiprotons, positrons can be efficiently thermalized in appropriate moderators[6-8], which makes their accumulation at low energy competitive (cheaper and simpler) with high-energy schemes. Radioactive β^+ emitters can be used as primary positron sources. Positrons, collected over a large solid angle, are moderated. Since the positrons are emitted continuously, accumulation is not easy. Possibilities for accumulating slow positrons, emitted continuously from a radioactive source, are discussed in the talk of A. Rich. Pulsed production of slow positrons is being done at electron linacs[9-11]. In the following we will consider how such positrons can possibly be accumulated at low energy.

For this exercise let us consider slow positrons produced in pulses of 20 ns width at a repetition rate of 1 kHz and containing about 10^6 particles. Similar numbers were obtained at Livermore[9] and recently also at Giessen[11]. The spatial extension of such bunches of slow positrons is of the order of centimetres. Let us now consider a storage device consisting of a solenoid to guarantee radial confinement, and end-electrodes to provide electric potential wells (Figure 4). The first pulse of thermal positrons is injected by keeping the front electrode potential U_L on ground. Once the pulse is in the trap it is rapidly increased to a small positive value (e.g. twice the energy width of the positrons). The rear electrode is kept unchanged at high positive potential U_R. The next pulse is accelerated so that it just can pass the small front ridge. Again, as soon as it is in the trap, the potential of the front electrode is increased by a small step. The accumulation procedure is illustrated in Figure 5. In this way pulses of positrons are successively stacked energy-wise, one on top of the other, and the confining potential increases after each pulse. With a step size of 2 V, one would have accumulated 10^9 positrons in 1 s. However, their energy would range from almost zero to 2 keV.

Now one can play the trick of remoderating them, which can be done with high efficiency[8]. For this purpose the positrons are extracted by lowering the potential of one end-electrode and applying a high, positive potential

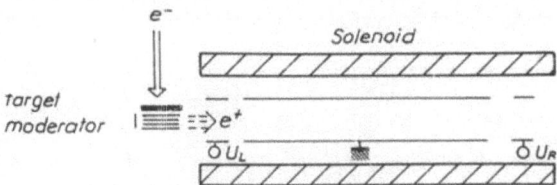

Figure 4. An arrangement for trapping pulsed positrons.

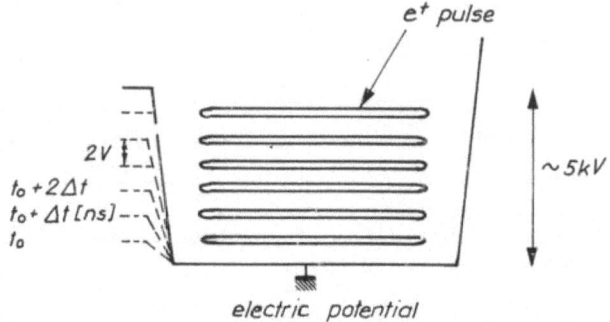

Figure 5. Stacking of positrons at low energy.

to the central electrode and the other end-electrode. The remoderated positrons are transferred to another trap, where they are stored in a potential
well of a few volts. The accumulation in the first trap is resumed while
the positrons in the second trap are being cooled. Cooling is envisaged in
order to provide phase space for the next pulse, which would come 1 s later.
Of course, one could again stack the positrons energy-wise in the second
trap. However, phase-space cooling might be fast enough to allow stacking
to be done without having to increase the energy very much. In this way
one would not need a second remoderation. There are several ways of cooling the positrons in the second trap: resistive cooling, electron cooling,
or synchrotron radiation. Resistive cooling in a harmonic trap is a standard technique for the achievement of cold electron gases[12], and cooling
times below 1 s were achieved. Electron cooling at such low energy has not
yet been done, but it should be possible as long as cold electrons can be
provided. Cooling the positrons in the second trap allows the longitudinal
trapping potential to be reduced. The next pulse is then put on top of the
first one, but the separation energy is only a few volts. The second trap
would therefore serve as a real accumulator, providing a stack of cold positrons at the end. With such a collection technique (Figure 6) it should be
possible to achieve an accumulation rate of 10^8 s^{-1} (allowing for a factor
of 10 losses) and stored positron intensities similar to those in relativistic damping rings.

ANTIHYDROGEN FORMATION PROCESSES

Antihydrogen can be produced in different reactions[13]:

$e^+ + \bar{p} \rightarrow \bar{H} + h\nu$ spontaneous radiative capture (2)

$n \cdot h\nu + e^+ + \bar{p} \rightarrow \bar{H} + h\nu + n \cdot h\nu$ induced radiative capture   (3)

$e^+ + e^+ + \bar{p} \rightarrow \bar{H}^+ (+ h\nu), \bar{H} + e^+$ three-body formation   (4)

$e^+ + \bar{p} + \bar{p} \rightarrow \bar{H}_{\bar{2}} (+ h\nu), \bar{H} + \bar{p}$ three-body formation   (5)

$Ps + \bar{p} \rightarrow \bar{H} + e^-$ charge exchange   (6)

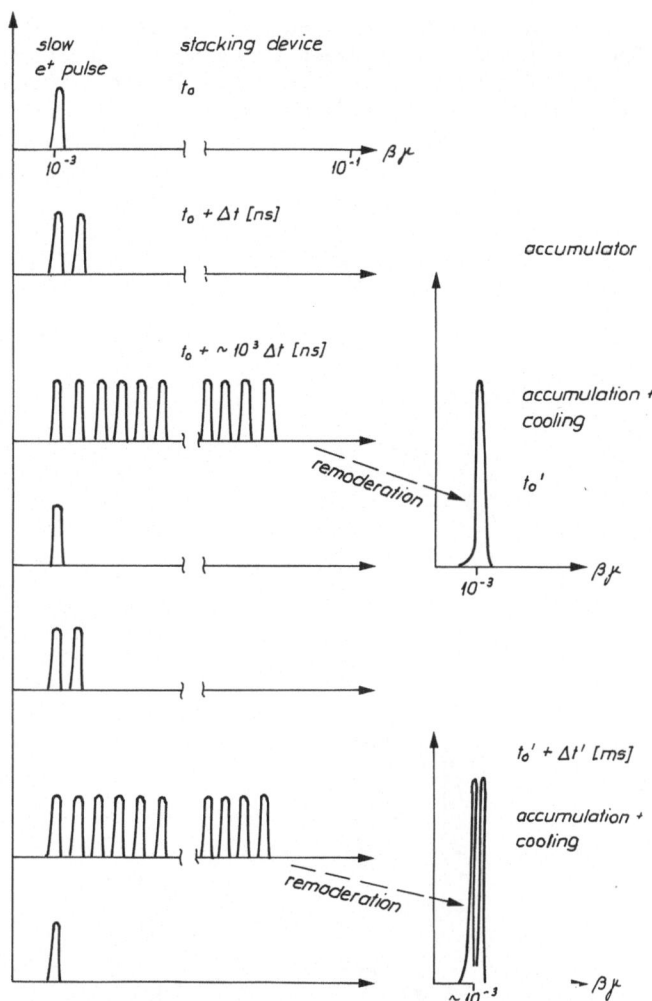

Figure 6. Schematic illustration of positron accumulation at low energy.

$$p + A \rightarrow \bar{H} + \bar{H} + \text{anything} \qquad\qquad \text{direct production} \qquad (7)$$

$$\bar{\Lambda}, \bar{n} \rightarrow (\bar{p}e^+) + \nu_e \qquad\qquad \text{antibaryon decay.} \qquad (8)$$

Here we consider reactions (2) and (3) only. The production of antihydrogen through the interaction of antiprotons with positronium (Ps), reaction (6), is described in another paper presented at this conference[14]. A discussion of various methods of producing antihydrogen can also be found elsewhere in the literature[15].

FORMATION SCHEMES FOR RADIATIVE CAPTURE

For the formation of antihydrogen through radiative capture, positrons and antiprotons of equal velocity are required. Antiprotons are accumulated at $\beta\gamma$ values of 3.9 (CERN) and 9.5 (FNAL) (β: particle speed in unit of speed of light; $\gamma = 1/\sqrt{1 - \beta^2}$). At present, accumulation schemes for positrons exist only at very relativistic velocities, $\beta\gamma > 1000$. However, as argued above, it should be possible to accumulate positrons at thermal energies also. One has the choice between bringing the positrons to a βγ-value between 0.1 and 9.5 (existing storage rings for antiprotons) or decelerating antiprotons to thermal energies (Figure 7).

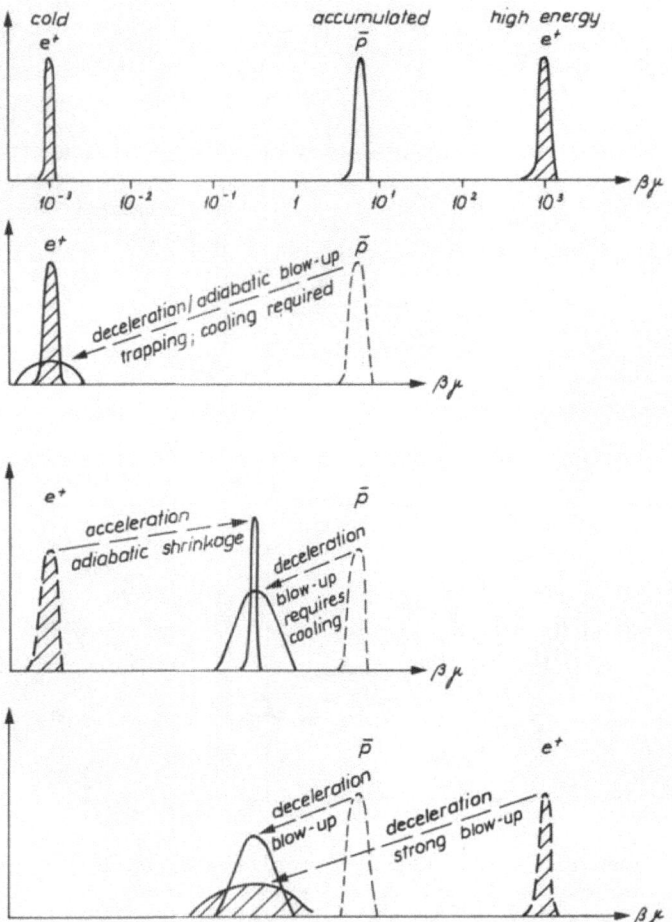

Figure 7. Possibilities for matching antiprotons and positrons in velocity.

At CERN, antiprotons are decelerated in the PS by radio frequency (RF) from the accumulator energy down to about 175 MeV (600 MeV/c), and, further on, in the low-energy antiproton ring LEAR[16] to a minimum of 5 MeV (100 MeV/c). Deceleration blows up the phase space of the antiproton beam proportionally to βγ. Hence the antiproton beam has to be cooled at certain stages of deceleration so as to avoid beam losses. In LEAR this can be done at injection energy and at several lower energies, including 5 MeV, by stochastic cooling[16]. Very soon electron cooling will also be available in LEAR at any energy up to 75 MeV[17].

MERGED BEAM SINGLE-PASS SCHEME

Recently it was proposed to cool the antiproton beam with the electrons at 50 MeV to a very low temperature (velocity spread), and to merge it in one straight section of the ring with a positron beam derived from a moderated source in order to study the feasibility of antihydrogen production in a single-pass experiment[18]. The antihydrogen production rate was expected to be sufficient for testing the principle of the method as well as for the first artificial production of antimatter atoms. However, for experimentation with antihydrogen, higher rates are needed.

The major limitations for antihydrogen production come from the low available density of its constituents. Therefore, both species have to be used repeatedly for the formation process, which means that those which did not interact to produce antihydrogen have to be recuperated[13]. This is possible when dealing with interactions between charged particles as in reactions (2) to (5), because they can easily be stored and recirculated. If all other loss processes can be suppressed, positrons and antiprotons can be entirely coverted into antihydrogen. Antiprotons and positrons can be stored and recirculated in traps and storage rings. These possibilities are considered next.

FORMATION IN TRAPS

The formation of antihydrogen in traps was discussed some time ago[19]: it requires the slowing down of antiprotons to very low energy (a few keV). When starting from a few MeV this can be achieved by RF deceleration or by energy degradation in a foil. The last possibility is very lossy since antiprotons diappear owing to annihilation in the foil. However, it has recently been successfully applied in order to capture a few hundred antiprotons in a Penning trap[20]. For RF deceleration, the trapping efficiency depends very much on the beam quality. Another problem is that a high-intensity antiproton stack cannot be decelerated all at once with high efficiency, owing to beam stability requirements. This means that low-intensity bunches have first to be decelerated and cooled, and then stacked in the trap. Yet for the measurement of the gravitational mass of the antiproton it is planned to decelerate them in a radio-frequency quadrupole (RFQ) to 20 keV, capture them in a trap at around 3 keV, and then cool tham down resistively to temperatures below 1 keV[21]. There are certainly methods for slowing down antiprotons and storing them at thermal temperatures, but detailed studies will have to show that intensity and spatial densities can be achieved.

For antihydrogen formation it should be possible to store positrons and antiprotons at the same time in a RF trap[19] or in combined traps[22]. Once formed therein, antihydrogen would, however, escape in any direction if no means were found to store the neutral atom. It would be difficult to distinguish an antihydrogen annihilation from an ordinary antiproton annihilation if no clear signature were to be found. This is different for antihydrogen formation in flight, which is considered next.

DOUBLE-RING SCHEME

Instead of slowing down antiprotons, then trapping and mixing them with positrons at thermal energies, both species can be merged at relativistic velocities in overlapping rings. Therefore positrons accumulated and cooled in high-energy damping rings would have to be decelerated to a $\beta\gamma$ value a around unity, or accelerated from trapping energies to this value. As in the case of antiprotons, deceleration leads to beam blow-up if no cooling takes place. Fast cooling of positrons with $\beta\gamma$ around unity is probably impossible, so they would have to be remoderated and thermalized. Hence one would not gain from accumulation and storage at highly relativistic energies. Thus the accumulation and storage of positrons at thermal energies is the most appropriate process. They have to be accelerated and injected into a small storage ring overlapping with the antiproton ring. Such a scheme would be applicable at CERN and at FNAL. Acceleration of thermal positrons to the considered velocities can be done electrostatically, which guarantees the achievement of monochromaticity. Also the transverse beam temperatures should remain approximately constant.

The recirculation of positrons for the formation of antihydrogen has already been discussed earlier[23]. The basic limitation in that scheme was, however, the achievable positron density, since the authors did not consider positron accumulation at low energy in order to increase it.

A storage ring for positrons of energies around 1 MeV would be very simple compared with the storage ring for antiprotons of the same velocity. For the positrons, an electrostatic beam transport structure might even be practicable. The lifetime of the positrons therein depends critically on the achievable vacuum, because Coulomb scattering can transfer energy from the longitudinal to the transverse motion and heat up the beam, which ulti- mately leads to losses as pointed out above. For instance, in a 10 m ring with a vacuum of 10^{-12} Torr the positrons would pass through the equivalent of about 3 ng/cm^2 of hydrogen in one second, which probably deteriorates the beam considerably. Cooling techniques might be applied to compensate for the transverse heating. This has to be investigated in more detail.

ANTIHYDROGEN FORMATION RATES FOR A DOUBLE-RING SCHEME

Spontaneous capture

The luminosity L for two merged rings multiplied by the capture cross- section yields the spontaneous capture rate. Taking into account the finite temperature of beams, we obtain[13]

$$R = L\sigma = N_{\bar{p}} N_e k \ell \gamma^{-2} A (C_{\bar{p}} C_e A_{\bar{p}} A_e)^{-1} \alpha_r, \qquad (9)$$

where $N_{\bar{p}}(N_e)$ is the number of antiprotons (positrons) circulating in k bunches in a ring of circumference $C_{\bar{p}}(C_e)$. The quantities $A_{\bar{p}}$ and A_e define the area cross-section of the particle beams, A the total overlap area, and ℓ the length of the overlap region. The value of the recombination coeffici- ent α_r can be taken as 2×10^{-12} cm^3 s^{-1}. It is reasonable to assume that the positron beam can be made as small as the antiproton beam, hence $A_e = A_{\bar{p}} = A$.

In the following we consider two concrete examples, namely: anti- hydrogen production at 0.3 GeV/c in LEAR ($C_{\bar{p}}$ = 78 m, γ = 1, ℓ = 2 m, $N_{\bar{p}}$ = 4×10^{10}); and in the FNAL accumulator at 4 GeV/c ($C_{\bar{p}}$ = 474 m, γ = 4.3, ℓ = 15 m, $N_{\bar{p}}$ = 4×10^{11}). Beam cross-sections of 0.1 cm^2, C_e = 3ℓ, α_r = 2×10^{-12} cm^3 s^{-1}, k = 1, and a stored positron intensity of 10^8 are assumed Inserting these values into Equation (9) one arrives at about

$$R = 3400 \text{ s}^{-1} \qquad (10)$$

for LEAR; the rate for the other cases would be a factor of 10 lower.

This assumes that the positrons can be stored and recirculated for times that are comparable to accumulation times. As argued in Ref. 13, in a dedicated and optimized arrangement it might be possible to achieve even

$$R = 2 \times 10^{-4} N_e \beta \text{ s}^{-1} . \qquad (11)$$

Induced capture

Positron capture by antiprotons can be stimulated through irradiation of the merged beams with light of an appropriate wavelength[24]. From Eqn. (33) of Ref. 24 we find, for a typical arrangement, an enhancement factor G of

$$G = 2Pn^5(\hbar c/13.6 \text{ eV})^3/(a^2 mc\Delta_\parallel \Delta_\perp), \qquad (12)$$

where P is the radiation power, a is the radius of the photon beam, and $\Delta_\parallel$ and $\Delta_\perp$ are the longitudinal and transverse velocity spread of the positrons in the particle rest frame. In Ref. 24, capture into the n = 2 state of antihydrogen was considered. It would yield a gain factor of about 100 with commercially available lasers. The benefit of the induced capture depends very much on the laser and defines the operation regime. If only pulsed lasers can be used in order to achieve the required radiation power, the antiproton and positron beams have to be matched in time structure and repetition rate, otherwise the duty cycle is so low that this process becomes unimportant. Given the flexibility of choice for the particle beam energy and final capture state, it should be possible to reach a duty cycle close to unity. In this case the hydrogen formation rate via induced capture becomes

$$R_{ind} = GR . \qquad (13)$$

Under these optimal conditions the antiproton beam lifetime versus recombination decreases to about one day, which means that nearly all antiprotons could be converted into antihydrogen. The scheme of forming antihydrogen in a merged beam arrangement is jointly pursued by the CERN-Heidelberg-Karlsruhe groups[19], by the group from the University of Michigan[25], and by K.G. Lynn, from Brookhaven National Laboratory.

EXPERIMENTS

Obviously there are a number of interesting experiments that one would like to do once antihydrogen is formed in sufficient quantities. Possible experiments with antihydrogen formed in flight have been discussed previously[26,27]. In summary, this concerns the high-precision spectroscopy of the antihydrogen atom, in particular the measurement of the Lamb shift and the hyperfine structure (HFS) splitting. A precise measurement of these quantities allows us to make a validity test of the CPT theorem. Here, a particular role is played by the ground-state HFS splitting, a quantity which is extremely well measured for hydrogen. The theoretical calculations are far behind this precision. A comparison of the HFS splitting between hydrogen and antihydrogen also contains information on the structure of the nucleon.

The interaction between neutral matter and antimatter atoms has for some time attracted the interest of theorists[28] because there was the suspicion that metastable molecules could be formed. Although this idea seems to be ruled out on the basis of theoretical investigations, the interaction of antihydrogen with heavier normal atoms is still of considerable interest. However, such studies require very low energy antihydrogen atoms.

Finally, the formation of antihydrogen has one interesting application for medium-energy and high-energy physics, which lies in the possibility to polarize antiprotons by passing through a particular state of antihydrogen 8,29,30.

CONCLUSION

Undoubtedly antihydrogen will be synthesized within the next decade and its properties will be measured. The precision will depend crucially on the achievable production rates. In order to maximize the rates, it seems to be inevitable that the constituents of antihydrogen will have to be accumulated and recirculated. This might be achieved by storing them in a ring at relativistic energies and producing antihydrogen with fast merged beams, or by storing them in traps with antihydrogen emerging at thermal energies. Both alternatives have advantages and disadvantages which, however, were not discussed in detail here.

ACKNOWLEDGEMENTS

The author would like to thank K. Lynn, A. Rich, B. Seligman and A. Wolf for helpful discussions.

REFERENCES

1. E. Jones, report CERN/PS/86-30 (1986).
2. G. Rapidis, "Proc. First Workshop on Antimatter Physics at Low Energy, Batavia, 1986", B.E. Bonner and L.S. Pinsky, eds. Fermi National Acc. Lab., Batavia (1986), p.83.
3. "Proc. 1986 Summer Workshop on Antiproton Beams in the 2.10 GeV/c Range, Upton, 1986", D. Lazarus, ed., BNL 52082, Upton (1987); Y.Y. Lee and D.I. Lowenstein, "A conceptual design for a very low energy anti-proton source", Brookhaven Nat. Lab. Report AGS/AD/Techn, Note 269 (1986).
4. See for instance: "Proc. Int. Conf. on a European Hardron Facility, Mainz, 1986"., Th. Walcher, ed. North-Holland, Amsterdam (1986), also Nucl. Phys. B279:(1987).
5. LEP Design Report, Vol. "The LEP injector chain", report CERN-LEP/TH/ 83-29 (1983).
6. A. Vehanen, K.G. Lynn, P.J. Schultz and M. Eldrup, Appl. Phys. A32:2572 (1983).
7. A.P. Mills, Jr. in: "Positron Scattering in Gases", J.W. Humberston and M.R.C.McDowell, ed. Plenum Press, Inc. New York (1981), p. 121.
8. R.S. Conti and A. Rich, "Proc. Workshop on the Design of a Low-Energy Antimatter Facility, Madison, 1985", D. Cline ed. World Scientific Singapore, (1986), p. 97.
9. R.H. Howell, R.A. Alvarez and M. Stanek, Appl. Phys. Lett. 40:751 (1982).
10. M. Begemann, G. Graff, H. Herminghaus, H. Kalinowsky and R. Ley, Nucl. Instrum. Methods 201:287 (1982); G. Graff, R. Ley, A. Osipowicz, G. Werth and J. Ahrens, Appl. Phys. A33:59 (1984).
11. F. Ebel, W. Faust, C. Hahn, S. Langer, H. Schneider and A,Singe, "Slow positron set-up at the Giessen 65 MeV Linac", this conference.
12. L.S. Brown and G. Gabrielse, Rev. Mod. Phys. 58:233 (1986).
13. H. Poth, Appl. Phsy. A43:287 (1987).
14. F.M. Jacobsen, L.H. Andersen, B.I. Deutch, P. Hvelplund, H. Knudsen, M. Charlton, G. Laricchia and M. Holzscheiter, "On antihydrogen production", this conference.
15. A. Wolf, "Proc. 8th European Symposium on Nucleon-Antinucleon Inter-actions (Antiproton '86), Thessalonki, 1986, S. Charalambus, C. Papastefanou and P. Pavlopoulos, eds., World Scientific, Singapore, (1987), p. 123.
16. D. Allen, E. Asseo, S. Baird, J. Bengtsson, M. Chanel, J. Chevallier, R. Gianni, P. Lefevre, F. Lenardon, R. Ley, D. Manglunki, E. Martensson, J.L. Mury, C. Mazeline, D. Mohl, G. Molinari, J.C. Perrier, T. Petters-son, P. Smith, G. Tranquille and H. Vestergard, preprint CERN/PS 87-

  26 (1987): "Performance update on LEAR", contribution to the Particle Accelerator Conference, Washington, 1987.

17. C. Habfast, H. Poth, B. Seligmann, A. Wolf, H. Haseroth, C.E. Hill and J.-L. Vallet, "Proc. 3rd LEAR Workshop on Physics with Cooled Low-Energy Antiprotons in the ACOL Era, Tignes, 1985", U. Gastaldi, R. Klapisch, J.-M.Richard and J. Tran Thanh Van, eds., Editions Frontieres, Gif-sur-Yvette (1985), p. 129.

18. J. Berger, P. Blatt, C. Habfast, H.Haseroth, P. Hauck, Ch. Hill, R. Neumann, H. Pilkuhn, H. Poth, G. zu Putlitz, B. Seligmann, A. Winnacker and A. Wolf, (CERN-Heidelberg-Karlsruhe Collab.), "Feasibility Study for Antihydrogen Production at LEAR", proposal CERN/PSCC/85-45, PSCC/P86, CERN/PSCC/86-21, Add. 1; and CERN/PSCC/86-37, Add. 2.

19. G. Torelli, Proc. 5th Antiproton Symposium, Bressanone, 1980, M. Cresti, ed. CLEUP, Padua (1980), p. 43.

20. G. Gabrielse, X. Fei, S.L. Rolston, R. Tjodker, T.A. Trainor, H. Kalinowsky, J. Haus and W.P. Kells, Phys. Rev. Lett. 57:2505 (1986).

21. N. Beverini, J.H. Billen, B.E. Bonner, L. Bracci, R.E. Brown, L.J. Campbell, D.A. Church, K.R. Crandall, D.J. Ernst, A.L. Ford, T, Goldman, D.B. Holtkamp, M.H. Holzscheiter, S.D. Howe, R.J. Hughes, M.V. Hynes, N. Jarmie, R.A. Kenefick, N.S.P. King, V. Lagomassimo, G. Manuzio, M.M. Nieto, A. Picklesimer, J. Reading, W. Saylor, E.R. Siciliano, J.E. Stovell, P.C. Tandy, R.M. Thaler, G. Torelli, T.P. Wangler, M. Weiss and F.C. Witteborn, Los Alamos report LAUR-86-260 (1986).

22. G. Gabrielse, Penning traps, masses and antiprotons, to appear in "Proc. Int. School of Physics with Low-Energy Antiprotons: Fundamental Symmetries, Erice, 1986". To be published.

23. H. Herr, D. Mohl, and A. Winnacker, in: "Proc.2nd Workshop on Physics with Cooled Low-Energy Antiprotons at LEAR, ERice, 1982", U.Gastaldi, and R. Klapisch, eds. Plenum Press, Inc. New York (1983), p. 659.

24. R. Neumann, H. Poth, A. Wolf and A. Winnacker, Z. Phys. A313:253 (1983).

25. A. Rich, R. Conti, W. Frieze, D.W. Gidley, M. Skalsey, T. Steiger, J. Vautlouse, H. Griffin, W. Zheng and P.W. Zitzewitz, "Antihydrogen: Production and Applications", this Conference.

26. H. Poth, in: "Proc. 2nd Conf. on the Intersection between Particle and Nuclear Physics, Lake Louise, 1986", D.F. Gessaman, ed. AIP Conf. Proc. No. 150, AIP, New York (1986), p. 580.

27. R. Neumann, Possible experiments with antihydrogen, to appear in "Proc. Int. School of Physics with Low-Energy Antiprotons: Fundamental Symmetries, Erice, 1986". To be published.

28. H. Chojnacki and S. Roszak, Acta Phys. Pol. A67:811 (1985) and references therein.

29. K. Imai, "Proc. 6th Int. Symposium on Polarization Phenomena in Nuclear Physics, Osaka, 1985", M. Kudu et al. eds. Supplement to J. Phys. Soc. Japan 55:302 (1986).

30. H. Poth, Polarized antiprotons from antihydrogen, to appear in "Proc. Int. School of Physics with Low-Energy Antiprotons: Fundamental Symmetries, Erice, 1986". To be published.

ANTIHYDROGEN: PRODUCTION AND APPLICATIONS

A. Rich, R. Conti, W. Frieze, D.W. Gidley, H. Griffin,[a]
M. Skalsey, T. Steiger, J. Van House, W. Zheng,[b] and
P.W. Zitzewitz[c]

Department of Physics, Univ. of Michigan, Ann Arbor, MI 48109

(a) Department of Chemistry, University of Michigan
(b) Shanghai Institute of Nuclear Research, Academia Sinica
     P.O. Box 8204, Shanghai, China
(c) Department of Natural Sciences, University of Michigan-
     Dearborn, Dearborn, MI 48128

1. INTRODUCTION

The formation of antihydrogen ($\bar{H}$) is of interest for a variety of
reasons. Properties of the $\bar{H}$ such as the electronic energy levels, fine
structure, Lamb shift, and hyperfine structure can be measured and compared
to the corresponding quantities in hydrogen as tests of CPT invariance.
Novel investigations of the interactions of $\bar{H}$ with atoms and with gravita-
tion can be undertaken. Finally, applications such as the production of
polarized antiprotons or the storage of macroscopic quantities of $\bar{H}$ can also
be pursued.

Several methods for producing $\bar{H}$ have been proposed.[1-8] All of these
share certain general features. Copious numbers of antiprotons ($\bar{p}$) are
available only from p − $\bar{p}$ pair production at high projectile energies
(> 10 GeV). These must be cooled and decelerated to usable energies.
Because $\bar{H}$ formation occurs only slowly in all the proposed methods, the antiprotons must be conserved by recirculating them in traps or storage rings.
In addition, sources of cold positrons (e^+) or positronium (Ps), more intense
than any now used, are necessary to achieve usable rates of $\bar{H}$ production.

Figure 1 summarizes the possible methods for slowing and cooling the
$\bar{p}$, converting $\bar{p}$ to $\bar{H}$, and slowing and cooling the $\bar{H}$. Three levels of deceleration and cooling are considered. First, $\bar{p}$ are slowed from their 4 GeV
formation energy in an antiproton accumulator while being stochastically
cooled before they are cooled again in a low energy storage ring, such as
the Low Energy Antiproton Ring (LEAR) at CERN. This is energy level I.
Electron cooling[7,8] can be used to reduce the effective temperature to an
energy spread of kT = 0.1 eV.

Two methods have been proposed for the next deceleration step to $\sim$ 3 keV
(level II). Energy loss by means of collisions in a solid and subsequent
trapping have already been demonstrated.[9] This method is at a disadvantage
in the long run due to the small fraction of the incident $\bar{p}$, 10^{-4}, that can
be trapped. While this fraction can surely be increased substantially, it

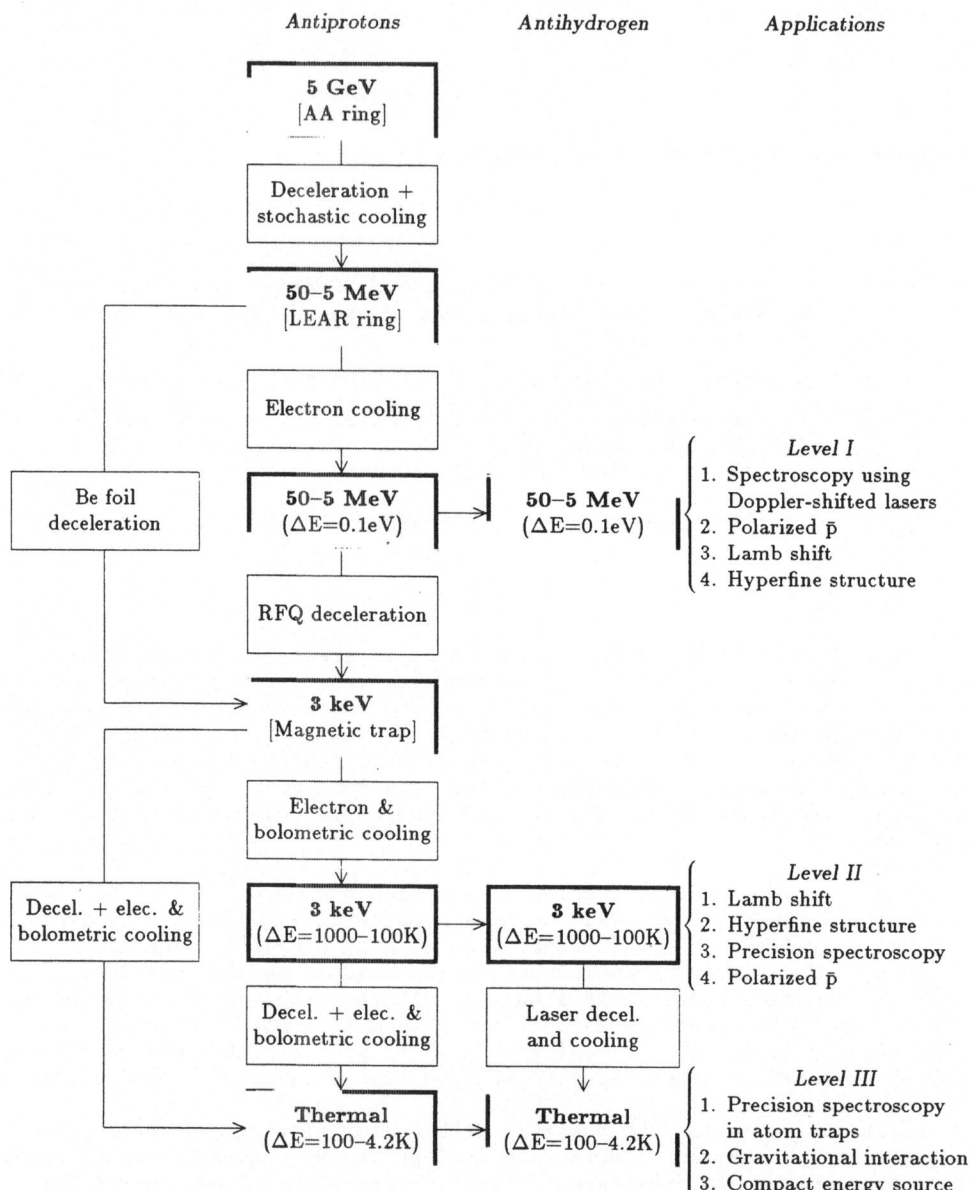

	Antiprotons	*Antihydrogen*	*Applications*

Fig. 1. Antihydrogen Production and Applications.
Main column shows antiproton production, deceleration
and cooling methods (light boxes) as well as storage
energies (heavy boxes). Second column shows antihydrogen
energies. Antihydrogen formation methods for the three
energy levels are described in the text.

is unlikely that the fraction of $\bar{p}$ trapped using this technique can ever approach unity. The other suggested deceleration method is to use an RFQ (radiofrequency quadrupole).[10] The RFQ should be more efficient, but it has yet to be operated as a decelerator. Once in a 3 keV trap or storage ring, the antiprotons can be electron or bolometrically cooled such that monochromatic antiprotons are stored in the trap at $\sim$ 3 keV. Finally, the $\bar{p}$ can be both cooled and decelerated until the velocity of the center of mass is zero (level III).[11]

The three energy levels are convenient points of departure for discussion of the conversion of the $\bar{p}$ to $\bar{H}$. Five $\bar{p} \to \bar{H}$ conversion processes are considered.

a) $e^+ + \bar{p} \to \bar{H} + h\nu$

b) $e^+ + \bar{p} + nh\nu \to \bar{H} + (n + 1)h\nu$

c) $e^+ + \bar{p} + e^+ \to \bar{H} + e^+$

d) $e^+ + \bar{p} + e^- \to \bar{H} + e^-$

e) $Ps + \bar{p} \to \bar{H} + e^-$.

In all of these processes the $\bar{H}$ emerges with momentum in essentially the same direction as the incident $\bar{p}$ momentum. Thus, at energy levels I and II the $\bar{H}$ will be produced in a directed beam while at level III the $\bar{H}$ will be distributed isotropically. Process (a), photocapture, requires a cold, high density, e^+ beam moving at the same velocity as the $\bar{p}$ for levels I and II, while at level III the average $\bar{p}$ velocity is zero. At sufficiently high positron density the 3-body process (c) will dominate. The $\bar{H}$ formation rate increases with lower relative positron $\bar{p}$ velocities. For example, for $n_e = 10^{10}$ cm^{-3}, process (c) will dominate over process (a) at a positron temperature below 1000 K.[12] If a high density of e^- is present then process (d) can dominate. We note that a high density of e^- is more easily achieved than the same density of e^+.

In process (b), stimulated photocapture, the presence of photons from a laser enhances the $\bar{H}$ production rate. The near relativistic velocities (β = 0.3) in energy level II with their resulting Doppler shifts allow visible lasers to be used to induce capture. For example, capture into the n = 2 $\bar{H}$ state can be accomplished using 506 nm light. The ultraviolet lasers (365 nm) necessary to do this in energy levels II and III exist, but the high power densities that are required are more difficult to achieve.

The charge exchange reaction, process (e), requires positronium (Ps) to have low velocities in the $\bar{p}$ frame (< 1 eV) in order to produce $\bar{H}$ efficiently. This requirement, coupled with the short lifetime of Ps (140 ns in the longest lived ground state 1^3S_1), limits the usefulness of process (e) in energy level I. In levels II and III the short Ps lifetime and resulting loss of e^+ offsets much of the advantage of its large cross section (3 × 10^{-16} cm^2) in comparison to processes in which e^+ may be reused. Applications of antihydrogen formed at the three energy levels are listed in Figure 1.

Antihydrogen is formed on Level I as a well collimated 50 MeV beam with $\sim$ 0.1 eV energy spread. This configuration is well suited for making fast beam measurements of $\bar{H}$ such as electronic structure, fine structure, hyperfine structure and Lamb shift,[13] especially when these measurements require high frequency light that can result from the Doppler shift of a visible laser. Particularly appropriate to energy level I is the production [14] of polarized $\bar{p}$ via $\bar{H}$ formation. The e^+ within the $\bar{H}$ would be either polarized prior to $\bar{H}$ formation or polarized by optical pumping after $\bar{H}$ formation. The polarization would then be transferred to the $\bar{p}$ via the hyperfine interaction and the e^+ stripped off by passage through a thin foil, leaving polarized $\bar{p}$ s.

Level II antihydrogen would be produced as a directed beam at $\sim$ 3 keV
with an energy spread of 0.1 eV or smaller. The experiments that could be
done with these $\bar{H}$ are essentially the same as those for level I. The disadvantage is that UV lasers are required for some experiments. On the other
hand, the lower beam velocity allows the experiments to be physically more
compact.

Level III $\bar{H}$ would be produced as an isotropic cloud at rest in the lab
frame with energy spread of 100 K to 4 K, depending on the degree of cooling
achieved. This isotropic nature of the $\bar{H}$ velocity makes it difficult to use
in spectroscopic, fine, and hyperfine structure measurements, as well as
polarized $\bar{p}$ production. However, the low temperatures are necessary to make
any measurements of the gravitational interactions of antimatter[15] or to
accumulate macroscopic amounts of antimatter in a trap.[16] Antihydrogen for
these experiments can alternately be produced at level II and might be decelerated and trapped by laser cooling techniques. So far, however, laser cooling
techniques have been used only with Na and Cs atoms. Extension to normal
hydrogen has not been accomplished.

II. ANTIHYDROGEN FORMATION WITH FAST, MERGED BEAMS

The attempt to form $\bar{H}$ using the fast (level I) merged beams techniques
is being undertaken by the following group:

Antiproton beam

H. Haseroth, C.E. Hill and J.L. Vallet
CERN, Geneva, Switzerland

C. Habfast, H. Poth, B. Seligman and A. Wolf
Kernforschungszentrum Karlsruhe, Inst. für Kernphysik,
Fed. Rep. Germany.

P. Blatt, R. Neumann, A. Winnacker and G. zu Putlitz
Physik, Inst. der Universität, Heidelberg, Fed. Rep.
Germany.

Positron beam

The authors of this article, based at the University
of Michigan, (Ann Arbor and Dearborn, Michigan)

K.G. Lynn, Brookhaven National Laboratory.

The preceding article by Dr. Poth describes $\bar{p}$ formation and properties
of LEAR as well as overlapping some of the $\bar{H}$ formation technique discussed
in this section.

A. Methods

The formation rate of antihydrogen, $R_{\bar{H}}$, using an antiproton storage ring
such as LEAR and a merged positron beam of equal cross section is given by[17]

$$R_{\bar{H}} = N_{\bar{p}} n_e \alpha \eta \gamma^{-2} \qquad (1)$$

where $N_{\bar{p}}$ is the number of stored antiprotons, n_e is the positron density in
the overlap region, α is the recombination rate coefficient, η is the fraction of the $\bar{p}$ ring overlapped by the e^+ beam, and γ is the relativistic
factor that transforms from the rest frame to the laboratory frame. The rate
coefficient α is calculated assuming that the positrons have a velocity spread
parallel to the beam momentum, $\Delta_{\parallel}$, much smaller than the spread perpendicular
to the beam, $\Delta_{\perp}$. This is the so-called flattened velocity distribution.

Defining the transverse energy (temperature) of the positrons, $T \equiv \frac{1}{2}m\Delta_\perp^2$, the rate coefficient is approximately[18]

$$\alpha = \frac{r_e^2 c^2}{\Delta_\perp} \left\{ 0.28 + 0.07 \left(\frac{T_\perp}{E_o} \right)^{1/3} - 0.125 \ln \left(\frac{T_\perp}{E_o} \right) \right\} \quad (2)$$

Table 1. Summary of expected antihydrogen production rates for four production methods using relativistic merged beams

Capture Method	Number of e^+ passes	Bunched e^+	$R_{\bar{H}}(s^{-1})$
Spontaneous	1	no	0.01
Stimulated	1	yes	1
Spontaneous	10^6	yes	10^4
Stimulated	10^6	yes	10^6

where r_e is the classical radius of the electron (2.8×10^{-13} cm), and E_o is the binding energy of antihydrogen (13.6 eV). Equation 2 is in good agreement with experimental rates obtained for $e^- + p \to H + h\nu$ radiative capture experiments.[19]

For the LEAR facility operated at a $\bar{p}$ energy of 50 MeV, the expected parameters are $N_{\bar{p}} = 1 \times 10^{11}$, $\eta = 0.04$, $\beta = 0.32$ ($\gamma = 1.06$). If the positron beam has $T_\perp = 0.1$ eV ($\Delta_\perp = 1.9 \times 10^7$ cm/s) and if $n_e = 1$ cm^{-3}, then we have $\alpha = 3.4 \times 10^{-12}$ cm^3/s and $R_{\bar{H}} = 0.01$ s^{-1}. This value of $R_{\bar{H}}$ is large enough to demonstrate and study the merged beam $\bar{H}$ formation process, but it is too small to perform most of the $\bar{H}$ research mentioned in the introduction. Schemes for much more copious $\bar{H}$ formation, using merged beams with the positrons in a multi-pass rather than a one-pass mode, will be discussed shortly (see also Table 1).

The photocapture process can be enhanced via the stimulated emission of photons, $e^+ + \bar{p} + nh\nu \to \bar{H} + (n + 1)h\nu$. Neumann et al (ref. 2) show that an increase in the $\bar{H}$ production rate by a factor $G_f^{max} = 110$ can be achieved if the positron distribution is sufficiently flattened ($\Delta_\perp/\Delta_\parallel \approx 30$) and if the intensity of the stimulating light is 18 MW/cm^2. A photon with wavelength 506 nm in the laboratory frame produces transitions into the $n = 2$ level. The required light intensity can be obtained only with pulsed lasers. The use of the positrons is optimized if the beam is pulsed to assure that the positrons are present only when the laser is on. If positrons can be so compressed in time while maintaining their time-averaged density at 1 cm^{-3} then the antihydrogen formation rate would be increased by the full gain factor G_f^{max} to become $R_{\bar{H}} = 1$ s^{-1}. The next section discusses methods to accumulate and bunch positrons to achieve this goal.

With a pulsed positron beam positron recirculation becomes possible, and, as we will show, would yield $R_{\bar{H}} \gg 1$ s^{-1}. For a 10 ns long positron pulse repeated every 10 ms the instantaneous positron density is 10^6 times higher than the average density. Consider recirculating the e^+ in a storage ring 1 m in circumference, having a 10 ns period. If the positrons would retain their small energy spread (0.1 eV) for 10 ms, then the $\bar{H}$ production rate can be enhanced over the spontaneous recombination rate by the same factor of 10 . Thus $10^4 \bar{H}$ s^{-1} would be produced. If the positron storage ring is not able to maintain the small energy distribution then active cooling using cold electrons may be possible.

Laser stimulated photocapture can be incorporated into the recirculating positrons if laser pulses at the required intensity can be repeated with a 10 ns period. This would require recirculating photons at 10 ns intervals for the 10 ms time between laser pulses. If these requirements can be met, an antihydrogen production rate $\bar{H} = 10^6$ s^{-1} could be achieved. These estimates are summarized in Table 1.

B. Technical Details

1. Positron Source

The positron beam that we propose to use in preliminary $\bar{H}$ formation experiments will originate from a radioactive ^{22}Na source. This nuclide offers the best compromise between long half-life (2.6 y) and high specific activity. A ^{22}Na source can be produced that is sufficiently thin ($\sim$ 10 mg/cm^2) to enable most of the positrons to escape the source material and be available for moderation. We plan to use an encapsulated ^{22}Na source of activity in excess of 1 Ci that will have an active area of about 2 cm in diameter. A 3.2 mg/cm^2 Ni end window covering the source deposit will allow the emitted positrons to pass through with only 15% attenuation.

Activation of the ^{22}Na isotope will be done at the Los Alamos LAMPF 800 MeV proton accelerator.[20] A disk of high-purity Al metal is placed in the beam dump of the accelerator. The ^{22}Na is produced by many reactions [e.g. ^{27}Al(p, pnα) ^{22}Na] during irradiation times that are typically several months. The first 120 g Al trial disk was activated to 0.5 Ci ^{22}Na activity. It is being used to test procedures for extracting the ^{22}Na isotope and depositing the source.

The ^{22}Na will be separated from the bulk Al target and then purified by ion-exchange chromatography. Several separation techniques are presently being investigated using the facilities of the University of Michigan Phoenix Memorial Laboratory. The purified source will then be deposited on to a tungsten backing and hermetically sealed. The specific activity obtained during the deposition process is measured using a Si surface-barrier detector to determine the positron flux and a NaI scintillation detector to measure the γ flux. Preliminary experiments have been conducted using the short-lived (15h) ^{24}Na isotope as a tracer. Stable deposits as small as 1 mm in diameter have been achieved using this method.

2. Positron Beam Requirements

In order to form antihydrogen with acceptably large efficiencies by the photocapture process using merged $\bar{p}$ and e^+ beams, the velocities, diameters and angular divergences of the two beams must be closely matched. Thus the characteristics of the positron beam are determined by the properties of the LEAR $\bar{p}$ beam. In this section we summarize these characteristics and describe the design parameters of a positron beam that would satisfy the requirements for a demonstration of one-pass $\bar{H}$ production ($R_{\bar{H}} \sim 10^{-2}$ s^{-1}). As stated earlier because our actual program involves a stored e^+ beam, the discussion that follows is presented only to illustrate the technical considerations involved in design of the final beam.

The stored, electron-cooled, antiproton beam circulating in LEAR at 300 MeV/c will eventually have an emittance of < πmm.mrad, a relative transverse momentum spread $\sim 10^{-3}$, and a longitudinal momentum spread as small as 3×10^{-5}. The beam diameter is roughly 1 mm. The requirements of a velocity-matched, merged positron beam are an energy of 26 keV, transverse energy spread below 0.1 eV, longitudinal energy spread 1 eV, and diameter 1 mm. The required emittance of the 26 keV positron beam is thus ∿ 1 mm.mrad, where emittance is written as the product of the radius and angular half-width. The intrinsic energy distribution of positrons emitted from a moderator (or remoderator) is significantly smaller than 1 eV. Thus longitudinal energy spread of the positron beam caused by effects such as a lack of stability of the acceleration voltage supply must be controlled to achieve maximum $\bar{H}$ formation rates, especially when the stimulated photocapture method is used. The projected emittance of a state-of-the-art 26 keV positron beam using a 0.1 Ci source with diameter 3 mm is of order 2.5 mm.mrad. Increasing the active diameter to 20 mm to accommodate 10 Ci ^{22}Na results in an emittance of 20 mm.mrad. A single stage of brightness enhancement that reduces the emittance by a factor of 20 is thus needed to obtain the required emittance of 1 mm.mrad.

The above positron beam would have four sections: (1) positron moderator and electrostatic lenses; (2) magnetic field transition region; (3) remoderator and pre-accelerator region; and (4) final acceleration to 26 keV beam energy. The first section is designed to accept a 20 mm diameter radioactive source and a transmission moderator producing ∿ 400 mm.mrad emittance at 100 eV. We note that a moderator having a 0.15 eV FWHM transverse energy spread would meet this requirement. The positrons are accelerated to 5 keV and focused to a 10 mm diameter spot with transverse energy of 1 eV FWHM. The second section of the beam is designed to bring the positrons into a 100G axial magnetic field that is needed to guide the final beam through the interaction region within LEAR. The converging beam produced by the first section together with the focusing effect of the magnetic field gradient reduces the beam diameter to 1 mm on the remoderator. At the same time the transverse energy increases to 100 eV. The function of the remoderator is to reduce transverse energy spread from 100 eV back to 0.15 eV without appreciably increasing the beam diameter. In order to maintain this energy width, the magnetic field must be constant from the remoderator to the interaction region. After the remoderator, the beam is accelerated to 26 keV for merging with the LEAR beam.

In summary, the design presented above will yield $R_e = 1 \times 10^8$ e$^+$ s^{-1} using a 10 Ci ^{22}Na source, 10^- moderation and 0.25 remoderation efficiency.[21] Such a beam would have $n_e = R_e/A\beta c = 1$ cm^{-3} for A = 0.8 mm^2 [Although such moderation efficiencies have been achieved,[22] if not surpassed by a factor of 3-7,[23] it is clear that all positron transmission losses must be carefully minimized along with the emittance.] We next discuss e$^+$ storage techniques that would permit a great improvement on this minimum rate.

3. Accumulation Methods

Substantial gains in $\bar{H}$ production rates are possible if laser-induced recombination is used, and even higher gains are possible if the positrons can be made to pass through the interaction region many times. In the case of the laser, improvements are possible only if the positrons are accumulated and sent into the interaction region in a short pulse that is comparable in length and repetition rate to the laser pulse. Similarly, a pulse of duration less than the period of a recirculator is necessary to recycle all of the positrons. Thus it is clear that the closely related topics of accumulation and e$^+$ storage are of great importance to the eventual success of an $\bar{H}$ production project.

We outline here several aspects of the problem of accumulating positrons in a trap from a continuous beam. A trap will be defined as a configuration of electric and/or magnetic fields that confines charged particles in all three dimensions. Typically an axial magnetic field provides radial confine-

ment. If the fields are increased at the ends a magnetic "bottle" is formed.
Alternatively, axial electric fields can be used in the "Penning trap"
configuration. Finally, trapping can also be achieved in a Paul or RF
quadrupole trap.

The problem, then, is to collect positrons in a trap over a time interval of 10-100 ms. Trapping methods that use conservative fields obey the
Liouville theorem and trade reduction in time uncertainty for an increase in
the spread in energy, position, or angular divergence of the beam. Once the
positrons are in the trap, we wish to cool them to recover the small size,
divergence, and energy width that they had prior to trapping. This reduction
in phase space also allows a new e^+ population to be injected into the trap
without detrapping the previous population. Finally, we must extract all the
accumulated positrons in a bunch of $\sim$ 10 ns duration, the period of a e^+
storage ring or length of a laser pulse.

There are three general methods that can be used to provide the cooling
required to compress in time the e^+ randomly emitted from a radioactive source
while still retaining their initial compact distribution in phase space.
These three are (1) collisional cooling wherein the positrons collide with
particles of a cooler medium; (2) radiative cooling, the energy loss due to
acceleration of e^+ in a trap; and (3) stochastic cooling, which requires knowledge of the time of arrival of e^+ (time tagging) permits forces to be applied
that compress the e^+ bunch in time.

Several media for collisional energy loss are possible. A neutral gas
can effectively cool the e^+, but Ps formation is possible, resulting in the
annihilation of the e^+ before extraction from the gas. This technique would
also require several stages of differential pumping. Free electrons can cool
e^+ without excessive loss due to recombination ($e^- + e^+ \rightarrow Ps + h\nu$). However,
space charge limitations on the electron density restrict the cooling rate
that can be achieved. Positive ions can be added to neutralize the e^- space
charge but the plasma will partially recombine and increase the Ps formation
rate. The negative work function of e^+ in many metals allows remoderators[21]
to cool positrons while the high e^- density prohibits Ps formation. Less
than 30% of the initial positrons are retained in the moderated beam. Radiative cooling is a relatively slow process. For example, synchrotron radiation
in a 6 Tesla Penning trap cools with time constants $\sim$ 0.1 s. Thus long
confinement times that are independent of cooling are required. If the positron source is already pulsed[24] the arrival time information can be used to
stack successive pulses. Alternatively a "tagging" signal[25] from a randomly
occuring e^+ source also allows stacking of single positrons in the same
manner.

One specific method to obtain a positron bunch 10 ns long with a repetition rate of 100 Hz from a random source of slow positrons is a double remode-
rator-Penning trap configuration that we will call the "double ΔV" accumulator. In this scheme remoderation of the positrons provides the cooling
that allows the e^+ arrival time to be compressed. A positron is emitted from
a moderator with a few eV of kinetic energy. An axial magnetic field confines
the e^+ to near the axis. It travels across the first trap, reflects from the
potential of a grid V_{g1}, and returns toward the moderator. During this first
transit of the trap a cylinder potential V_{c1} is lowered sufficiently rapidly
(e.g. 5V/10 ns) so that the e^+ cannot reach the moderator. The total energy
of this e^+ will continue to drop in subsequent passes through the trap while
new e^+ are added at the top of the trap. When the increase in the trap depth
reaches 5 kV (i.e. 1000 trap periods or 10 µs) grid voltage V_{g1} is suddenly
reduced, allowing all of the e^+ to pass through the mesh and strike the
remoderator during one trap period (10 ns). Up to 30% of these will be
emitted from the remoderator with the original energy spread. A second stage
then compresses 1000 of these 10 ns bunches (with a 100 kHz repetition rate)

to a single 10 ns bunch with a 100 Hz repetition rate again followed by remoderation to restore the original energy spread. The chief difference in this second trap is that the potential V_{c2} does not decrease in a continuous ramp as V_{c1}, but rather in discrete 5 volt steps only when the bunches of e^+ from the previous stage are within the second trap. We estimate that 10% of the e^+ emitted from the original moderator would be obtained in 10 ns, 100 Hz bunches.

An alternative scheme is to replace the second stage trap with an electron cooling trap. The electron cooling trap would be opened to admit the remoderated e^+ bunch from stage 1. A dense beam of e^- is slightly slower than and moving at matched velocities with the e^+ during their passage in one direction. During the interval between successive pulses from stage 1, the e^+ are collisionally cooled to a sufficiently low energy so that they do not escape the trap when a new bunch of e^+ is introduced. This scheme has the advantage that the 30% remoderation efficiency for the second stage is circumvented and that maximum accumulation time is extended beyond the 10 ms available with the double ΔV trap. The accumulation time could be of order one second.

A third accumulation method involving cyclotron cooling in a 6 T field for the purpose of using $\bar{H}$ formation process (c) is being developed.[4]

III. OTHER $\bar{H}$ PRODUCTION SCHEMES

In a previous section we have detailed schemes to produce $\bar{H}$ at 50 MeV using LEAR and an intense ^{22}Na source. In Table 1 we gave expected $\bar{H}$ formation rates for different merged beam schemes. By way of comparison we discuss three other methods of producing $\bar{H}$ that have been proposed. We compare the expected $\bar{H}$ production rates and difficulties that must be addressed to produce the $\bar{H}$.

One suggested scheme[5] involves slowing $\bar{p}$ s from LEAR with an RFQ to a level II (4 keV) trap and using the positronium charge exchange reaction (e) to form $\bar{H}$. Positronium at $\sim$ 1 eV would be produced by stopping positrons from a 50 Ci ^{64}Cu source on the inner surface of the $\bar{p}$ trap. This would result in a positronium density n $\sim$ 2 cm^{-3}. For 10^6 $\bar{p}$ in the trap the antihydrogen production rate would be $R_{\bar{H}}$ = 2/min. The $\bar{H}$ would exit the trap on either end with essentially the momentum of the original $\bar{p}$ at formation. The major advantage of this scheme is the simplicity of providing the Ps to the interaction region. Some negative considerations are related to the difficulty of fabricating and transporting the necessary ^{64}Cu ($T_{1/2}$ = 12.7hr) source from a reactor once a day, the unproven schemes to decelerate and trap 10^6 $\bar{p}$ at 4 keV, and the difficulty of greatly improving the $\bar{H}$ production rate. The next paper in these proceedings contains an update of this method.

A second proposed method of $\bar{H}$ production[6] is a variation on the first. In this case the $\bar{p}$ s trapped in LEAR are decelerated to a momentum of 14 MeV/c. The authors of ref. 6 estimate there will be 10^3 antiprotons circulating in the ring at this momentum. Positrons from a 0.5 Ci ^{22}Na source are moderated with 10% efficiency in an electric field-assisted moderator (high purity Si) and sent into a Ps charge-exchange interaction region in LEAR itself. The expected $\bar{H}$ production rate is $R_{\bar{H}}$ = $(10^{-3} - 10^{-4})$ s^{-1}. This method has the advantage of being a relatively straightforward scheme to produce the first $\bar{H}$. However, except by increasing by a factor of 20 the ^{22}Na source strength, no improvement in the initially small value of $R_{\bar{H}}$ is forseeable. Further, $\bar{p}$ deceleration in LEAR to 14 MeV/c and the 10% moderation efficiency are both speculative, and the storage lifetime of the antiproton beam at low momenta is likely to be very short.

A recently proposed scheme[4] is to decelerate $\bar{p}$ s by energy loss in a Be foil and to trap them in a 3 keV trap as already accomplished.[9] The 10^4 $\bar{p}$ that are expected to be trapped could then be cooled by electron cooling to < 0.1 eV and transferred to another trap where they circulate at $\sim$ 1 eV. A sufficient density of cold positrons in a nested trap produces a very high $\bar{H}$ production rate. Kells et al[4] have suggested that at sufficiently low positron temperatures the three body reaction (c) of Table 1 will dominate. In particular if the calculations[12] used in reference 4 are extrapolated to 4 K then the $\bar{H}$ production rate would be

$$R_{\bar{H}} = (10^4)(10^{10} \text{ cm}^{-3})(1 \text{ cm}^3 \text{ s}^{-1}) = 10^{14} \text{ s}^{-1}$$

for a positron density of $n_e = 10^{10} \text{ cm}^{-3}$. This high positron density would be achieved by trapping e^+ produced from a pulsed linac at 10^6 e^+/pulse. A total of 10^4 pulses would be accumulated and cooled to 4 K by cyclotron radiative cooling. The tremendous $\bar{H}$ production rate of 10^{14} s^{-1} of course will only be maintained as long as antiprotons exist in the trap. Thus $R_{\bar{H}}$ is no longer a useful figure of merit. The conversion of $\bar{p}$ to $\bar{H}$ is total, resulting in 10^4 $\bar{H}$ in the trap with which to perform experiments. Questions involved with this scheme are whether the equations can be extrapolated by more than two orders of magnitude in temperature beyond their stated range and the unproven nature of cooling large numbers of positrons to the required temperatures and densities. The inherent inefficiency of slowing the $\bar{p}$ from LEAR by energy degradation in Be foils could be circumvented if the RFQ decelerator works.

IV. CONCLUSIONS

The goal of producing antihydrogen is an important one with many ramifications, both fundamental and applied. The initial steps towards antihydrogen production are now being taken by several groups and, as might be expected, a number of different routes are under consideration. In this paper we have described in some detail the particular approach to the problem that our group is taking. In addition we have tried to review, albeit somewhat more briefly, other possible approaches.

Comparison of the various $\bar{H}$ proposals involves considerations related to ease of production, characteristics of the $\bar{H}$ beam such as energy and angular divergence, and eventual use of the $\bar{H}$. We feel that in terms of feasibility of production of useful quantities of $\bar{H}$ the merged beam approach is perhaps the most practical. We say this because, apart from electron cooling, the existing $\bar{p}$ beam may be used parasitically as is for merged beam $\bar{H}$ formation so that the entire burden of the research falls on producing an adequate e^+ beam and on laser enhancement if this proves to be necessary. Both of these activities can be carried out essentially independently of LEAR. In addition loss of $\bar{p}$ using the merged beam method is minimal so that $\bar{H}$ formation rates may be highest using this method. Finally, $\bar{H}$ formed in a beam is suitable for a range of fast-beam (level I) experiments as discussed above. Production of the required positron beam is now proceeding at Michigan, and we note from the viewpoint of research described at this conference that even partial progress towards the goals envisioned will result in e^+ beams greatly improved over those that currently exist. Such beams would facilitate a number of improved versions of ongoing experiments as well as giving the possibility of initiating entirely new areas of e^+ - Ps research.

V. ACKNOWLEDGEMENTS

We would like to thank G.W. Ford, W.P. Kells, K.G. Lynn, H. Poth and A.Wolf, for helpful discussions. Support for the research described in this paper has been provided by Richard Wood and Company and by the National Science Foundation under grant PHY 8403817.

VI. REFERENCES

1. H. Herr, D. Mohl and A. Winnacker, in Physics at LEAR with Low-Energy Cooled Antiprotons, Eds. U. Gastaldi and R. Klapisch, (Plenum, New York, 1982) pp. 659ff.

2. R. Neumann, H. Poth, A. Winnacker and A. Wolf, Z. Phys. A A313:253 (1983).

3. A. Wolf, talk at Antiproton '86.

4. G. Gabrielse, L. Haarsma, X. Fei, S.L. Rolston, T.A. Trainor, H Kalinowsky, W.P. Kells, post deadline contribution to the American Physical Society Division of Atomic, Molecular, and Optical Physics Meeting, May 18-20, 1987.

5. B.I. Deutch, A.S. Jensen, A. Miranda and G.C. Oades, in Proceedings of the First Workshop on Antimatter Physics at Low Energies, (FNAL, Batavia, 1986), pp. 371ff.

6. J.W. Humberston, M. Charlton, F.M. Jacobsen and B.I. Deutch, J. Phys. B20:L25 (1987).

7. G.I. Budker and A.N. Skrinsky, Sov. Phys. Ups. 21:277 (1978).

8. H. Poth, Review of Electron Cooling Experiments, Proc. Workshop on Electron Cooling and Related Applications (ECOOL84), Ed. H. Poth, p.45.

9. G. Gabrielse, X. Fei, K. Helmerson, S.L. Rolston, R. Tjoelker, T.A. Trainor, H. Kalinowsky, J. Haus and W. Kells, Phys. Rev. Lett. 57:2504 (1986).

10. M.H. Holzscheiter, in Low Energy Antimatter, Ed. D.B. Cline, (World Scientific, Singapore, 1986), pp. 120ff.

11. N. Beverini, L. Bracci, V. Lagomarsino, G. Monuzio, R. Prodi and G. Torelli, in Physics at LEAR with Low-Energy Cooled Antiprotons, Eds. U. Gastaldi and R. Klapisch, (Plenum, New York, 1982), pp. 649ff.

12. D.R. Bates, A.E. Kingston and R.W.P. McWhirter, J. Roy. Phys. Soc. 267A:297(1962).

13. H. Poth, in Proceedings of the Second Conference on the Intersections between Particle and Nuclear Physics, Lake Louise 1986, Ed. D.F. Ceesamon, (AIP Conf. Proc. No. 150, New York, 1976), pp. 580ff.

14. K. Imai, in Proceedings 6th International Symposium on Polarization Phenomena in Nuclear Physics, Osaka, Japan (1985), pp. 302ff; R.S. Conti and A. Rich, in Low Energy Antimatter, Ed. D.B. Cline, (World Scientific, Singapore, 1986), pp. 97ff.

15. T. Goldman and M. Nieto, in Physics with Antiprotons at LEAR in the ACOL Era, Eds. U. Gastaldi, R. Klapisch, J.M. Richard, J. Tran Thanh Van (Editions Frontieres, Singapore, 1985) pp. 639ff. and N. Beverini, L. Bracci, V. Lagomarsino, G. Monuzio, ibid, pp. 649ff.

16. R. Forward, J. Prop. and Power 1:370 (1985) and B.N. Cassenti, ibid p. 143.

17. M. Bell and J.S. Bell, Part. Accl. 12:49 (1982).

18. M. Bell and J.S. Bell, Part. Accl. 11:233 (1981).

19. G.I. Budker, et al., Part. Accel. 7:197(1976) and H. Poth, in CERN Accelerator School, Oxford, 1985 CERN 87-03 (1987), pp. 534ff.

20. The sources are procurred for us by Dr. Kelvin Lynn of Brookhaven National Laboratory.

21. P.J. Schulz, E.M. Gullikson and A.P. Mills, Jr., Phys. Rev. B34:442 (1986) and references therein.

22. J. Van House and P.W. Zitzewitz, Phys. Rev. A29:96 (1984).

23. A.P. Mills, Jr., and E.M. Gullikson, Appl. Phys. Lett. 49:1121 (1986).

24. A.P. Mills, Jr., Appl. Phys. 22:273 (1980).

25. J. Van House, A. Rich and P.W. Zitzewitz, Origins of Life 14:413 (1984).

26. R.H. Howell et al., in Positron Scattering in Gases, Eds. J.W. Humberston and M.R.C. McDowell, (Plenum, New York, 1984) pp. 155ff.

ON ANTIHYDROGEN PRODUCTION

F.M. Jacobsen, L.H. Andersen, B.I. Deutch, P. Hvelplund
and H. Knudsen

Institute of Physics, University of Århus
DK-8000 Århus C, Denmark

M. Charlton and G. Laricchia

Department of Physics and Astronomy, University College
London, Gower Street, London WC1E 6BT, U.K.

M. Holzscheiter

Los Alamos National Laboratory, Los Alamos
NM 87454, USA

ABSTRACT

In this paper we present our plans for production of anti-hydrogen, $\bar{H}$, in collisions of antiprotons, $\bar{p}$, with positronium, Ps, atoms. MeV energy $\bar{p}$ extracted from the Low Energy Antiproton Ring, LEAR, at CERN and cooled down to about 2.5 keV are injected into a RFQ race-track ion-trap. Low energy positrons are electrostatically focussed onto an Al foil inserted in one of the trap electrodes whereby a large fraction of them intersect the $\bar{p}$ in the trap as Ps. By using known technology this method will result in the formation of a few $\bar{H}$ per sec. Ways of improving this rate of $\bar{H}$ formation will be discussed and furthermore some of the more important experiments to precede the $\bar{H}$ production will be described.

1. INTRODUCTION

Recently there has been much interest in the possibility of producing antihydrogen, $\bar{H}$, in a laboratory[1-6], in particular, since the Low Energy Antiproton Ring, LEAR, at CERN has made available sufficient stored antiproton, $\bar{p}$, current[7] for its formation. The creation of $\bar{H}$ was first discussed by Budker and Skrinskij[8]. Today three different schemes are under consideration for $\bar{H}$ formation namely 1) in the LEAR at high energy by injecting a velocity matched beam of positrons into one of the straight sections of LEAR where $\bar{H}$ can be formed by radiative combination of e^+ and $\bar{p}$. The capture rate can be enhanced by a factor of 100 if it is stimulated by a laser[9], 2) by ternary combinations in a trap at cryogenic temperatures of simultaneously stored e^+ and $\bar{p}$[6] and 3) by $\bar{p}$ collisions with positronium, Ps, in a Radio Frequency Quadropole, RFQ, race-track ion-trap[10]. These various $\bar{H}$ formation processes are summarized below

$$\bar{p} + e^+ \longrightarrow \bar{H} + v \qquad\qquad (1)$$
$$\bar{p} + e^+ + v \longrightarrow \bar{H} + v + v \qquad\qquad (2)$$
$$\bar{p} + e^+ + e^+ \longrightarrow \bar{H} + e^+ \qquad\qquad (3)$$
$$\bar{p} + Ps \longrightarrow \bar{H} + e^- \qquad\qquad (4)$$

It is not possible to state which of the three methods (1 and 2 considered to be similar) that are the most effective, much depends upon the kind of experiments which one wishes to perform with $\bar{H}$. Concerning the formation processes themselves it is needless to state that all of the methods involve very exciting physics.

In principle the reactions 1 - 3 have the advantage over 4 in that no e^+ need to be lost in the former processes due to annihilation. Characteristics of methods 1 and 2 are a very small cross-section and very energetic $\bar{H}$ results. The ternary reaction results in very low energy $\bar{H}$. This method may yield a very high production rate of $\bar{H}$, $R_{\bar{H}}$, at lower temperatures. (proportional to $T^{-4.5}$).

The main advantages of forming $\bar{H}$ in $\bar{p}$ - Ps collisions are that its cross-section[4,5,11,12] is large (of geometric size), a fairly collimated beam of $\bar{H}$ can be produced in the energy range 0 - 20 keV and with modest improvements over the current state-of-the-art very decent production rates are achievable. In fact the rate of the formation of $\bar{H}$ when measured per hour (for all of the proposed schemes) may be limited by the number of $\bar{p}$ which can be extracted from LEAR.

In the following we shall only be concerned with $\bar{H}$ formation as a result of $\bar{p}$ - Ps collisions. For the other methods of producing $\bar{H}$ the reader should consult the papers given by Poth and Rich et.al. (included in these proceeding) and Galbrielse[6].

2.   FORMATION OF $\bar{H}$ IN $\bar{p}$ - Ps COLLISIONS

A general outline of the present research program can be found in ref. 10 and we shall here only give a brief account of the procedure. Fig. 1 displays a schematic layout of the experimental arrangement. MeV $\bar{p}$ extracted from LEAR and cooled to about 2.5 keV are injected into the RFQ race-track ion-trap of a design originally developed by Church[13]. During the $\bar{p}$ injection, a quarter of the trap will be turned off. After the injection has been completed this section is turned on before the particles reach this part of the trap. Estimates show that as many as 10^8 $\bar{p}$ can be injected into the trap[10,14] yielding an intensity of $1.4 \cdot 10^{14}$ sec^{-1} at any cross-section of the trap. Low energy positrons are electrostatically focussed onto an Al foil inserted in one of the trap electrodes (see the lower part of Fig. 1) whereby a large fraction of them intersects the $\bar{p}$ in the trap as Ps. Due to momentum conservation the $\bar{H}$ atoms formed in the $\bar{p}$ - Ps collisions will be emitted in the direction of the incident $\bar{p}$ beam resulting in a fairly collimated beam of $\bar{H}$.

2.1 THE Ps GAS TARGET

It is known that Al(111) after being exposed to about 500L of O_2 converts those e^+ which diffuse to the surface within their lifetime[2] into Ps with an efficiency very close to 1[15]. A study of a clean Al(111)

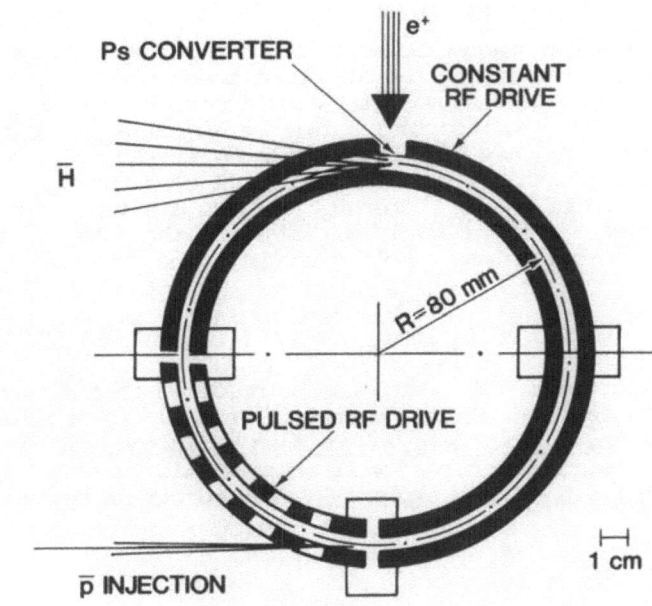

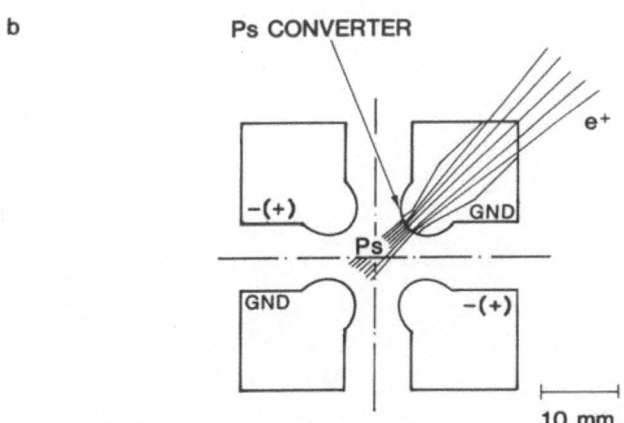

Fig. 1 A schematic layout of the experimental arrangement for
the formation of antihydrogen (a). In the lower part of
the figure (b) it is shown how the e$^+$ are injected onto
the Ps converter positioned on one of ground electrodes
of the RFQ ion - trap.

surface[16] has shown that the Ps atoms are emitted with an average kinetic
energy close to the temperature of the sample when 2 keV e$^+$ are injected
into the crystal. By assuming that the O_2 coverage of the Al(111) surface
does not alter the mean kinetic energy of Ps then ortho-Ps travels 1 cm
within its vacuum lifetime of about 140 nsec. In passing we mention that
the Al foil is only one of several possible candidates for the Ps
converter. An alternative solution could be to use silicon which also
converts e$^+$ into Ps with an efficiency close to 1 and possibly with a
somewhat lower emission velocity[17].

The average number of Ps atoms in the trap, N_{Ps}, can be expressed as

$${}^\bullet N_{Ps} = 0.5 \cdot 0.75 \cdot 1.4 \cdot 10^{-7} I_{e^+} \qquad (5)$$

where we have assumed an efficiency of 0.5 for the Ps convertor. I_{e^+} represents the intensity of the slow positrons on the convertor and the factor 0.75 is included as only the fraction of Ps that is in the triplet state can be used. The important parameter is not N_{Ps} but the number per cm^2 of Ps atoms, n_{Ps}, as seen by the $\bar{p}$ beam. If the positron beam hits the Ps convertor within a spot that is small compared to the flight distance of triplet Ps, L_{Ps}, then n_{Ps} becomes a strong function of the distance, r, from the convertor. If we assume that the Ps atoms are emitted isotropically into a half plane with the speed v_{Ps} then n_{Ps} is given as

$$n_{Ps} = (0.5 \cdot 0.75 \cdot I_{e^+})/(v_{Ps} r \pi) \tan^{-1}((L_{Ps}/r)^2 - 1)^{0.5} \qquad (6)$$

For the $\bar{H}$ experiment ^{81}Rb (which can be produced to a β^+ strength of 1 Ci at CERN[18,19]) can be used as a source of fast e^+. By assuming that the β^+ can be converted to low energy e^+ with an efficiency of 10^{-3} then, for the present design, the $\bar{p}$ beam will see a static number of Ps scattering centers of $n_{Ps} = 5 - 10$ Ps atoms/cm^2. This number is believed to be a

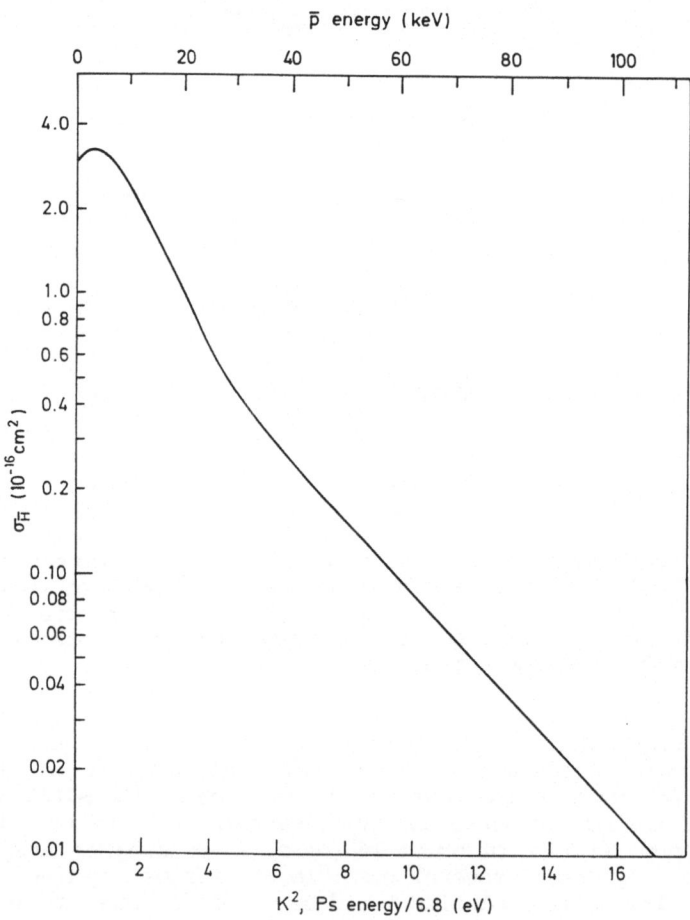

Fig. 2 The cross-section for formation of ground state H in p-Ps collisions (from ref. 5).

conservative estimate as we have ignored any enhancement caused by Ps atoms which may be re-injected into the $\bar{p}$ beam when they ricochet from one of the other trap electrodes. If the electrodes in the vicinity of the Ps convertor are coated with oxygen (or any other non-metallic atoms) it is expected that Ps can survive many encounters with the walls of the trap[20,21].

2.2 CROSS-SECTION FOR $\bar{H}$ FORMATION AND PRODUCTION RATE

There exist several theoretical results for the cross-section for the formation of $\bar{H}$ in $\bar{p}$ - Ps collisions[4,5,11,12]. In the most accurate of these, the cross-section for the formation of ground state $\bar{H}$ was obtained from very accurate variational calculations of the Ps formation cross-section in e^+ - H collisions using CPT arguments[5]. These results are shown in Fig. 2. As observed this cross-section is appreciable in the range 0 - 20 keV of $\bar{p}$ kinetic energies with a maximum of about 3 $\AA^2$ at 2.5 keV. Also the formation of $\bar{H}$ into the n=2 states have been calculated (see Darewych in these proceedings) yielding a cross-section which is at least as great as that for the ground state. Furthermore, included in these proceeding (see Ermolaev et.al) is a calculation of the total cross-section, $\sigma_{\bar{H}}$, for $\bar{H}$ formation in $\bar{p}$ - Ps collisions indicating a value of 10-20 $\AA^2$ for the range of $\bar{p}$ energies 2 - 10 keV. In the latter paper the cross-section for disintegration of Ps was also calculated and the results show that this cross-section first becomes comparable to $\sigma_{\bar{H}}$ at a $\bar{p}$ energy of 20 keV.

By assuming $\sigma_{\bar{H}} = 10$ $\AA^2$ at a $\bar{p}$ energy equal to 2.5 keV and a $\bar{p}$ intensity $I_{\bar{p}} = 1.4 \cdot 10^{14}$ sec^{-1} then the rate of $\bar{H}$ formation can be expressed as

$$R_{\bar{H}} = \sigma_{\bar{H}} n_{Ps} I_{\bar{p}} = 1\text{-}2 \text{ sec}^{-1} \qquad (7)$$

This production rate of $\bar{H}$ is large enough for its detection and for some first generation experiments.

2.3 THE e^+ BEAM

The beam optics under construction for guiding the positrons to the Ps convertor will be electrostatic to facilitate interface with the ion trap. The optics (see fig. 3) consist of a modified Soa immersion lens positron gun in which the β^+ moderator acts as the cathode. This part of the beam is identical to that used by Canter[22]. The positrons leave the gun with an energy of 400 eV and are further accelerated to 3 keV before they enter into a cylindrical mirror analyser to be deflected 90° to get the beam out of sight of the primary positron source background. The beam is further transported to the Ps convertor by means of an Einzel lens. Depending on the technical details of the interface with the ion - trap and the choice of the type of the Ps convertor to be employed an extra accelerator lens may be inserted after the Einzel lens.

Due to the lifetime of the $\bar{p}$ in the ion - trap and, in particular, due to backgound pressure problems in the test experiments (see below) the whole apparatus is being constructed of stainless steel capable of being baked at 900°C (without becoming magnetic) in order to drive out H from the bulk of these materials. The system will be pumped by a Ti sublimation- and an ion pump with a turbo acting as a roughing pump. The latter pump will be isolated from the apparatus during the final stage of the pumping. This design secures a base pressure in the 10^{-12} torr regime for the whole apparatus.

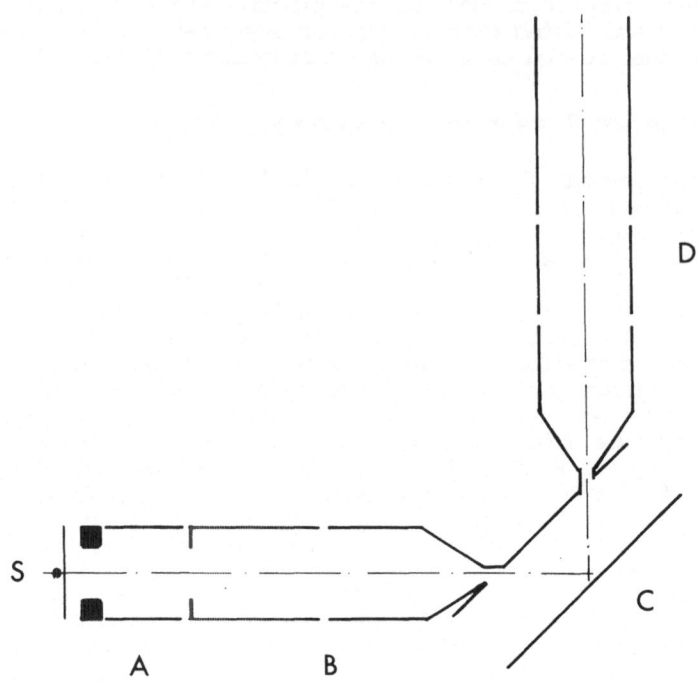

Fig. 3 shows the electrostatic e$^+$ beam. S: β$^+$ source, A: e$^+$
extraction optics, B: accelerator gap, C: cylindrical
mirror analyser, D: Einzel lens.

3. IMPROVEMENTS

There are several ways in which R_H can be increased over the value
given above. The most obvious ones are i) an increase of the source
strength, ii) improvement of the fast to low energy e$^+$ conversion
efficiency and iii) enhancement of the effective Ps density as seen by
the p̄ beam by confining the Ps atoms to a smaller area/volume by reducing
the mechanical dimensions of the Ps gas cell (see Sect. 2.1).

The main effort will be to develop more efficient positron
moderators. A standard UHV magnetic e$^+$ beam has been constructed with the
main purpose of developing field assisted moderators. The idea of
electric field assisted moderators stems from Lynn and McKee[23] . A recent
theoretical study indicates that such moderators may yield moderator
efficiencies of the order of 0.1[24] . In ref. 25 the silicon - silicide
system is discussed by Beling et.al. (in these proceedings). In the
selection of possible systems to be used as field assisted moderators the
following parameters are important 1) irradiation damage caused by the
slowing down of the β$^+$ particles, 2) the e$^+$ work function in the
moderator and 3) the interface between the metal contact and the bulk
material as well as the e$^+$ work function of the contact itself.

The effect of irradiation damage including charge up can be minimized by selecting low band gap semi-conductors. We mention that perhaps
one should not pay too much attention in the initial stage of the
development of field assisted moderators to possible effect of ir-

radiation damage. High band gap semi-conductors and e.g. ionic crystals may have other attractive properties and as long as they can be used for a sufficient length of time (hours) they could provide workable systems. Furthermore, the effect of charge up may in fact be an advantage as it may cause an increase in the internal field above the average value at the emitting surface.

Concerning positron work functions in insulators and semi-conductors not much is known today. As there exist no evidences at all of e^+ trapping into neutral vacancy type defects in these kind of materials it seems reasonable to assume that the e^+ work function is somewhat positive in most of these systems. If so it will in most cases be necessary to apply a very strong field in order to heat up the e^+ above thermal energies so they can overcome the potential barrier at the emitting surface. As the electric field strength needed to heat up the e^+ sufficiently may be an order of magnitude greater than that necessary for achieving the "saturation" drift velocity electric breakdown may cause a problem. Electric breakdown along the surface of the moderator can possibly be eliminated if the electric field in the main part of the moderator is adjusted to yield only the "saturation" drift velocity and then to increase strongly close to the emitting surface. Such field distributions can be constructed for MIS systems (due to the effect of band bending) and in semi-conductors by ion implantation.

Investigations of these two types of systems are currently taking place at Århus. In order to avoid problems with the contact at the emitting surface, due to interface defects and/or the e^+ work function itself, this contact will be made as a fine grid with approximately 50 % opening area.

Finally, it will soon be investigated how much the cross-section of the ion - trap, in the vicinity of the Ps convertor, can be reduced without causing loss of the antiprotons in the trap. If this cross-section can be reduced to 0.1x0.1 cm^2 and a field assisted moderator can be constructed to yield an efficiency of a few %, R_H increases to become in excess of 10^3 sec^{-1}.

4. OTHER EXPERIMENTS

With the development of a Ps gas target other new types of collision systems can be studied. Theoretical aspects of Ps collisions[26,27] are covered by Drachman and Peach in these proceedings. The Ps gas target is different from other normal gas targets since the two particles in Ps are of equal mass with the result that the centers of mass and charge coincide. Thus, there exist no mean static electrostatic interaction between Ps and other collision partners but only the interaction caused by the non localized exchange force. Also both particles in the neutral Ps can absorb energy, a property which influences the close-collision part of the ionization contribution and the second Born terms like the 'Thomas peak' in charge-exchange collisions[28].

Both because of the very exciting physics involved and with the purpose of optimizing the experimental set-up, the following collisions will be studied

$$e^{-*} + Ps \longrightarrow e^+ + e^- + e^- \qquad (8)$$
$$e^{-*} + Ps \longrightarrow Ps^- + e^- \qquad (9)$$
$$p + Ps \longrightarrow H + e^+ \qquad (10)$$
$$H^- + Ps \longrightarrow HPs + e^- \qquad (11)$$

These collisions and others will be studied in a 'crossed beam' arrangement; currently an ion source capable of delivering mA of p, H⁻ and e⁻ is under construction. These collision systems have been discussed by Hvelplund[29] and by use of a 70 mCi ^{22}Na as a β^+ emitter, product rates of the order of 10-50 sec⁻¹ are achievable.

The ionization of Ps by e⁻ impact (8) can also be used to characterize the density profile of the Ps gas target by sweeping an electron beam across the Ps gas. At e⁻ energies greater than ~ 500 eV reliable theoretical results exist for the cross-section of reaction (8)[27]. For the ortho- to para-Ps conversion by e⁻ impact (9) theory predicts[30] a cross-section of the order of 5-10 Å^2 at impact energies below the first excitation threshold. For the charge exchange process (11) no theoretical information exists for its cross-section, however, the bound state HPs has been studied[31] and is found to be stable with a binding energy of 1.06 eV.

ACKNOWLEDGEMENTS

The authors wish to thank C.D. Beling, K. Canter, J. Chevallier, P.G. Coleman, J. Darewych, M. Dunn, A. Ermolaev, J.W. Humberston, K.G. Lynn, B. Nielsen, G. Peach, R.I. Simpson and G. Sinapius for usefull discussions. Special thanks is due to Morten Eldrup. This work has been supported by The Danish Natural Science Research Council(DK), The Carlsberg Foundation(DK), The Lennard Olsson Basic Research Fund (DK) and The Science and Engineering Research Council(UK). MC wishes to thank The Royal Society(UK) and FMJ to thank the organizer committee for the invitation to speak at the workshop.

REFERENCES

1. H. Poth, in these proceedings.
2. A. Rich, R. Conti, W. Frieze, D.W. Gidley, M. Skalsey, T. Steiger and J. Van House, in these proceedings.
3. H. Herr, D. Mohl and A. Winnacker, Proc. Workshop on Physics at LEAR with Low-Energy-Cooled Antiprotons, Erice, 1982, Eds. V. Gastaldi and R. Klapisch (Plenum Publ. Co., NY, 1982) p. 659.
4. B.I. Deutch, A.S. Jensen, A. Miranda and G.C. Oades, Proc. 1st. workshop on Antimatter Physics at Low Energies, Fermilab, 1986 (FNAL, Batavia, IL,1986) p. 371.
5. J.W. Humberston, M. Charlton, F.M. Jacobsen and B.I. Deutch, J. Phys. B 20, L25 (1987).
6. G. Galbrielse, Workshop and Symp. on The Physics of Low-Energy Stored and Trapped Particles, 1987 (AFI, Stockholm 14-18 June, 1987) to be published.
7. P. Lefevre, D. Mohl and D.J. Simon, in ref.4 p. 185.
8. G.I. Budker and A.N. Skrinskii, Sov. Phys. Usp. 21, 277 (1978).
9. R. Neumann, H. Poth, A. Winnacker and A. Wolf, Z. Phys. A 313 253 (1983).
10. B.I. Deutch, F.M. Jacobsen, L.H. Andersen, P. Hvelplund, H. Knudsen, M. Holzscheiter, M. Charlton and G. Laricchia, in Ref. 6, to be published
11. J.W. Darewych, private communication and to be published, see also in these proceedings.
12. A.M. Ermolaev, B.H. Bransden and C.R. Mandel, in these proceedings.
13. D.A. Church, J. Appl. Phys. 40, 3127 (1969).
14. W. Poul, in Ref. 6, to be published.
15. K.G. Lynn and H. Lutz, Phys. Rev. B 22, 4143 (1980).

16. A.P. Mills and L. Pfeiffer, Phys. Rev. B $\underline{32}$, 53 (1985).
17. K.G. Lynn, private communication.
18. G. Sinapius, G. Spicher and H.L. Ravn, J. Phys. E $\underline{19}$, 987 (1986).
19 G. Sinapius and H. Ravn, in these proceedings.
20. R. Paulin and G. Ambrosio, J. de Physique $\underline{29}$, 263 (1968).
21. W. Brandt and R, Paulin, Phys. Rev. Lett. $\underline{21}$, 193 (1968).
22. K. Canter, T. Horsky, W.S. Lane and A.P. Mills, "Positron (Electron)-Gas Scattering" (Eds. W.E. Kauppila, T.S. Stein and J.M. Wadehra, World Scientific, Singapore, 1986) p. 202.
23. K.G. Lynn and B.T.A. McKee, Appl. Phys. $\underline{19}$, 124 (1979).
24. C.D. Beling, R.I. Simpson, M. Charlton, F.M. Jacobsen, T.C. Griffith, P. Moriarty and S. Fung, Appl. Phys. A $\underline{42}$, 111 (1987).
25. C.D. Beling, R.I. Simpson and M. Charlton, in these proceedings.
26. R.J. Drachman, in these proceedings.
27. G. Peach, in these proceedings.
28. E. Horsdal-Pedersen, C.L. Cocke and M. Stockli, Phys. Rev. Lett. $\underline{50}$, 1910 (1983).
29. P. Hvelplund, in Ref. 6, to be published.
30. S.J. Ward, J.W. Humberston and M.R.C. McDowell, J. Phys. B $\underline{20}$, 127 (1987).
31. Y.K. Ho, Phys. Rev. A $\underline{34}$, 609 (1986).

μ^+ CHARGE EXCHANGE, MUONIUM FORMATION AND DEPOLARIZATON IN GASES

Donald G. Fleming and Masayoshi Senba

TRIUMF and Department of Chemistry
University of British Columbia
Vancouver, B.C., V6T 1Y6, Canada

INTRODUCTION

The muon, like the electron, can be found in two charge states, μ^+ and μ^-, and can be produced at 'MeV' energies and <u>100%</u> (longitudinally) <u>polarized</u> from $\pi \to \mu\nu$ decay.[1] Most of this initial energy is lost in ionization processes as the muon slows down, during which no loss in polarization occurs. At 'keV' energies, when the muon velocity is comparable to orbital electron velocities in the moderator, additional mechanisms begin to dominate the energy loss process. Negative muons are captured into highly excited "mesic" orbits and because of this are often referred to as "heavy electrons" (m_μ = 206 m_e). They are not considered further here. <u>Positive</u> muons suffer an entirely different fate. At 'keV' energies, the μ^+ undergoes charge exchange with the moderator, producing the muonium atom (Mu = μ^+e^-) and returning the free muon in a series of charge exchange cycles, which can be thought of in complete analogy with those for protons. Subsequent thermalization at 'eV' energies is accomplished by elastic and inelastic scattering processes. The same features underlie the Ore concept of positronium (Ps) formation in gases[1], but the μ^+ in matter acts much more like a light proton (m_μ = 1/9 m_p) than a heavy positron, a theme which is emphasized throughout this paper.

Although much of the focus these days is on multiply charged ions, it is still important to pursue charge exchange studies of singly charged ions, particularly those of point charges. As a light proton, the positive muon provides a unique opportunity to explore <u>intrinsic</u> mass effects, if any, on the charge exchange process. In this regard the Mu atom can simply be thought of as an isotope of the H atom. At high energies, "velocity scaling" in the charge exchange process is expected to be correct, at least for total cross sections.[2,3] In general, though, there is very little data to test this concept. In the few cases where isotopic H^+ and D^+ charge exchange cross sections have been compared, velocity scaling seems to work for the total, but not for the differential cross sections.[3] A wider mass variation though is clearly desirable since the proton and deuteron differ by only a factor of two. Comparative measurements, for example, of total electron capture cross sections for energetic He^+ ions in a variety of gases with those of the corresponding H^+ ions at four times the energy[4,5] give very different results, but it must be noted that He$^+$ is already a one electron system. At low energies, in the 'eV' regime, velocity scaling is expected to break down, although in a merged-beam study of H$^+$-H and H$^+$-D charge transfer collisions it seems to hold very well[6].

The polarization of the muon provides a unique handle on the energy dependence of both charge exchange and elastic and inelastic scattering cross sections as the μ^+ thermalizes in the gas. The main mechanism for depolarization is formation of the Mu atom itself. Since the μ^+ is initially 100% polarized, but the electron is not, muonium can form equally in "ortho", $|\alpha_\mu\alpha_e\rangle$, or "para", $|\alpha_\mu\beta_e\rangle$ spin states, designated herein as o-Mu and p-Mu, in analogy with positronium formation. In zero or weak magnetic fields, the ortho state is an eigenstate of the hyperfine interaction, $|F,M_F=1,1\rangle$, but the para state is not, oscillating then between the M=0 substates of $|1,0\rangle$ and $|0,0\rangle$ at the hyperfine frequency, ν_o = 4463 MHz. At low pressures and low energies, where the time between collisions, τ_c, is $\gtrsim 1/\nu_o$, complete depolarization can result as the μ^+ cycles between these two spin states; conversely, at higher pressure, $\tau_c \ll 1/\nu_o$ and little or no depolarization occurs. Present results confirm one's expectation that the muon results are particularly sensitive to changes in cross sections in the eV regime. As such, comparisons with experiment utilizing velocity scaled proton cross sections should provide insight into the viability of mass scaling in low energy charge exchange.

EXPERIMENTAL

The μSR Technique

Spin polarized muons enter the gas target of interest with kinetic energies $\gtrsim$ 3 MeV. The energy loss process that the μ^+ then undergoes can roughly be divided into three stages[7,8]. Most of the incident energy is lost in the first stage, where (Bethe-Bloch) ionization processes dominate until a kinetic energy $\lesssim$ 30 keV is attained. Then follows a regime of cyclic charge exchange with the moderator 'M', μ^+ + M $\rightleftharpoons$ Mu, in which muonium is produced, with an electron capture cross section σ_C and the free μ^+ is returned, via an electron loss collision with cross section σ_L. There are ~100 charge exchange cycles occurring in competition with both elastic and inelastic scattering events. At kinetic energies $\lesssim$ 30 eV, depending crucially on the ionization potential (IP) of M and the magnitudes of the cross sections involved, the muon emerges as either the Mu atom or retains its identity as a free muon, subsequent thermalization to k_BT being accomplished by both elastic and inelastic scattering processes. At observation times ($\gtrsim$ 20 ns after a μ^+ enters the target) one can expect to observe thermalized muons in two distinct environments: diamagnetic, with polarization P_D, or as the paramagnetic muonium atom, with polarization P_{Mu}. A third possibility, the formation of muonated free radicals[9] will not be considered here. In addition, pressure-dependent depolarization can result in an apparant "missing" or "lost" fraction, P_L, such that $P_D + P_{Mu} + P_L = 1$, the initial beam polarization.

Regardless of its environment, the μ^+ ultimately decays in the parity-violating process $\mu^+ \rightarrow e^+\bar{\nu}_\mu\nu_e$, emitting the positron preferentially along the muon spin axis. The decay positron is detected in coincidence with a data gate that is triggered by a single stopped muon; the time difference between muon and positron is, on average, just the mean life of the muon, τ_μ = 2.2 μs. This time-differential process is repeated many times until a histogram of ~10^6 events has been accumulated[7],[9-12]. In a weak transverse magnetic field, muons bound in the paramagnetic Mu atom and those in diamagnetic environments precess with distinctly different Larmor frequencies ($\gamma_{Mu} = 1.03 \times \gamma_D = 1.39$ MHz G^{-1}), manifest as modulations of the decay histograms, and referred to as Muon Spin Rotation (μSR) "signals" $S(t)$. Examples (labelled "asymmetry") are shown in Fig. 1 for room-temperature H_2 gas at ~3 atm pressure at magnetic fields of 7.7 G (top) and 318 G (bottom), respectively. These can be thought of in analogy with free induction decay signals in magnetic resonance, and have the general functional form

$$S(t) = \sum_i A_i e^{-\lambda_i t} \cos(\omega_i t + \phi_i) \tag{1}$$

where A_i, ω_i, ϕ_i and λ_i are the initial amplitudes, precession frequencies, phases and transverse relaxation times ($\lambda = 1/T_2$) for muons observed in the i'th environment, respectively. In the weak (7.7 G) field of Fig. 1 (top), the signal is dominated by fast Mu precession (A_{Mu}), that from slowly precessing diamagnetic muons appearing only as an underlying curvature in the data; whereas in the bottom spectrum at 318 G, Mu precession is averaged out, both by the experimental time resolution (~2 ns) and by the coarse binning used, so that only the diamagnetic contribution (A_D) from Eqn. (1) is relevant.

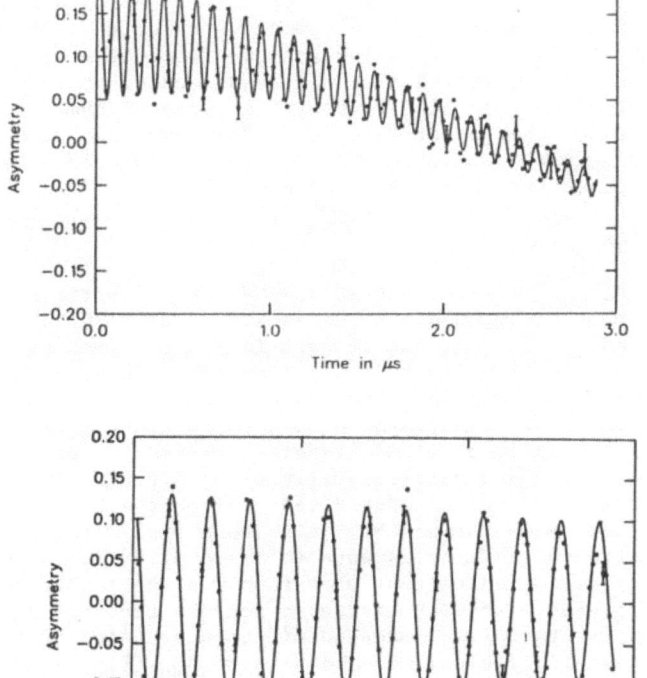

Fig. 1. μSR Asymmetry spectra of ~3 atm H_2 at 7.7 Gauss (top) and 318 Gauss (bottom), showing also a fit of Eqn. (1) to the data.

Of the parameters in Eqn. (1), we are primarily concerned in this paper with the <u>initial</u> amplitudes of paramagnetic muonium (A_{Mu}) and of muons in diamagnetic environments (A_D), since these (indirectly) reveal the nature of the collision processes occuring <u>during</u> the slowing down of the muon; the relaxation rate λ, on the other hand, is a measure of muonium reaction rates <u>at</u> thermal energies[13,14]. The muon polarizations (P_{Mu} and $\overline{P}_D$) and <u>relative</u> fractions (f_{Mu} and f_D) are defined experimentally by

$$P_{Mu} = 2A_{Mu}/A_{max}, \quad P_D = A_D/A_{max} \tag{2a}$$

$$f_{Mu} = P_{Mu}/(P_{Mu}+P_D), \quad f_D = P_D/(P_{Mu}+P_D) \tag{2b}$$

where A_{max} is the maximum possible amplitude for the experimental conditions, typically measured in an Al plate ($A_{max} \sim 0.3$). The total observed polarization is $P_{tot} = P_D + P_{Mu}$, which is < unity. It is assumed that there is an equal chance of forming o-Mu and p-Mu, although at very low pressures this may not be true[15]. The factor of 2 in Eqn. (2) then accounts for the non-observation of p-Mu, since the rapid time scale for the (μ^+-e^-) hyperfine interaction (0.2 ns) is much faster than the experimental time resolution ($\sim$2 ns). It is noted that p-Ps is also largely unobservable in positron experiments, due in this case to its very fast annihilation lifetime. In the discussion to follow, the neutral fractions f_{Mu} are compared with the corresponding results found for positrons thermalizing in gases, f_{Ps}. It should be pointed out that, in the definition in Eqn. (2), A_D is meant to correspond to diamagnetic environments <u>in</u> the gas, and as such care must be taken that no "wall signals" contribute to A_D (free μ^+ which scatter into the walls of the taret vessel). This is accomplished by a careful tuning of the μ^+ beam as well as by running several "air" runs over a range of charge densities comparable to those for the gas of interest. Spin-exchange with paramagnetic O_2 efficiently converts all o-Mu to p-Mu states, thereby leading to 100% depolarization of the muon ensemble[14].

Experimental Results

The experiments were carried out at the TRIUMF cyclotron, a meson facility adjacent to the campus of the Univerity of British Columbia. Surface (4.1 MeV) muons were brought to rest in a gas target positioned between a pair of Helmholtz coils of 1.5 m diameter. These coils provided a magnetic field transverse to the muon spin in the range $\sim$1 G to $\sim$300 G, thus enabling the measurement of the μSR signals described earlier (Fig. 1). The gases or liquid vapors employed in this study were all obtained commercially and were generally of high purity. Those materials obtained as gases were taken directly from the bottle without further purification, except He and Ne, which were passed through a LN_2 cold trap; those obtained as liquids were degassed prior to usage in order to remove any dissolved oxygen. For each run the target vessel was filled to a given pressure and μSR histogram spectra recorded by detecting positron events in two independent "counter telescopes" at each of two different magnetic fields; $\sim$8 G to measure A_{Mu} (Fig. 1, top) and $\sim$300 G to measure A_D (Fig. 1, bottom).

Table I presents values taken from several references[7,11,12,16,17], as well as some new data for H_2, D_2 and Ar, for the muon polarizations and relative fractions for selected gases at different pressures. The most complete set of data, up to $\sim$6 atm pressure, have been obtained in the cases of Xe and Ar. Results for Xe[17] are shown below in Fig. 2 as the muonium amplitude; since A_D is consistent with zero (Table I), $P_{tot} = 2A_{Mu}$ in Xe. The P_{tot} vs. pressure curve for Ar is similar, although ~100% of the polarization is recovered at the highest pressures (Table I). It is noted, of the noble gases there is essentially no diamagnetic signal in Xe (and Kr), whereas just the opposite situation prevails in He (and Ne), with Ar being an intermediate case. In contrast to the polarization themselves, the relative fractions are largely independent of pressure, except in some of the heavier alkane gases[16], notably hexane in Table I.

Table I. Percent μSR Amplitudes for Different Pressures in Selected Gases

Gas	Pressure(torr)	P_D	P_{Mu}	P_L	f_D	f_{Mu}
H_2	1025	25.0±0.6	31.7±1.0	43.3±1.2	44±2	56±2
	1510	29.0±0.6	38.5±1.1	32.5±1.2	43±2	57±2
	2000	33.0±0.9	43.1±1.5	23.9±1.7	43±2	57±2
	2500	37.3±0.6	42.0±1.1	20.7±1.2	47±3	53±3
D_2	1000	27.0±0.4	34.9±0.5	38.1±0.6	44±1	56±1
	1800	31.9±0.6	43.6±1.1	24.5±1.2	42±2	58±2
	2500	35.6±0.6	45.5±2.8	18.9±2.9	44±4	56±4
He	900	75.0±3.0	≲3.0	≲25.0±4.0	≳96	≲4
	1300	87.0±2.0	≲2.0	≲13.0±3.0	≳98	≲2
	1500	88.0±2.0	≲2.0	≲12.0±3.0	≳98	≲2
N_2	455	9.3±2.3	56.7±3.5	34.0±4.2	14±3	86±3
	760	12.0±1.5	78.1±4.4	9.9±4.6	13±2	87±2
	1010	15.4±1.0	78.6±1.9	6.0±2.1	16±1	84±1
	1820	17.6±0.5	83.4±2.0	−1.0±2.1	17±2	83±2
Ar	380	14.0±2.7	38.0±5.0	48.0±6.0	27±8	73±8
	760	17.0±1.0	55.0±1.7	2.0±2.0	24±3	76±3
	1520	24.0±1.0	62.5±2.2	13.5±2.4	28±3	72±3
	2130	23.5±0.6	72.0±2.5	4.5±2.6	25±3	75±3
	4560	28.5±1.0	71.0±2.5	−0.5±2.7	28±4	72±4
H_2O	740	3.6±1.1	47.0±1.7	49.3±2.0	7±2	93±2
	990	6.5±2.5	57.5±4.8	36.0±5.4	10±4	90±4
	1520	8.2±0.9	71.6±3.1	20.3±3.3	10±1	90±1
	1900	9.4±0.8	78.0±1.8	12.6±2.0	11±1	89±1
CH_4	800	9±1	76±2	15±2	12±2	89±2
	2280	13.2±0.6	87.8±2.4	1.0±2.5	13±3	87±3
C_6H_{14}	225	12.6±1.5	78.2±2.1	9.2±2.6	14±2	86±2
	430	24.7±1.4	72.0±1.7	3.3±2.2	25±2	75±2
	705	24.3±2.8	75.6±2.0	.05±3.4	25±2	75±2
	1010	24.8±1.3	75.5±1.8	−0.3±2.2	25±1	75±1
CCl_4	125	17.5±1.9	18.5±2.8	64.1±3.3	49±4	51±4
	250	30.0±1.2	26.5±2.5	43.5±2.8	53±3	47±3
	380	33.0±2.0	27.2±2.2	39.8±3.1	53±2	47±2
	760	35.1±3.9	32.1±3.1	32.8±5.0	52±4	48±4
Xe	325	≲4.0	45.5±6.0	≲54.5±7.0	≲6	≳92
	760	≲3.0	69.5±7.0	≲30.5±8.0	≲5	≳96
	1520	≲2.0	76.4±3.4	≲23.6±4.0	≲3	≳97
	2280	≲2.0	76.5±3.5	≲23.5±4.0	≲3	≳97
	4650	≲2.0	83.4±4.1	≲16.6±5.0	≲2	≳98

A MODEL FOR MUON CHARGE EXCHANGE

In studies of μ^+ charge exchange and muonium formation, the experimental observable, the polarization, plays a central role. At high energies, even from 10's of MeV to 10's of keV, the dominant energy loss process is ionization, which, being a pure Coulomb interaction, has no effect on the μ^+ polarization. As mentioned, when the <u>velocity</u> of the μ^+ becomes comparable to orbital electron velocities in the moderator, charge exchange begins to play an increasingly important role, with the cross section σ_C for electron capture peaking at energies roughly given by the "Massey criterion" of adiabatic energy transfer. Although σ_C for

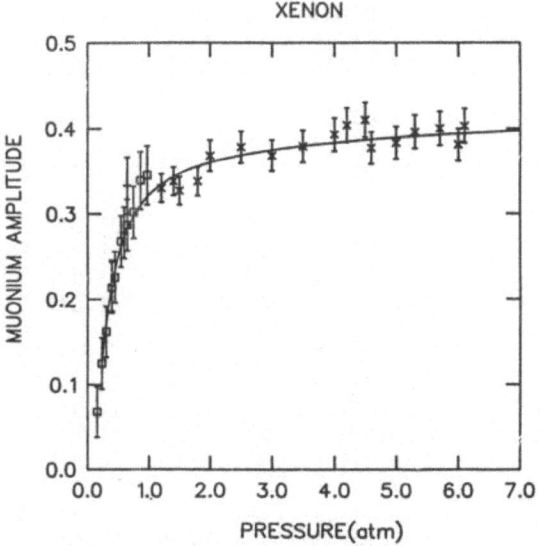

Fig. 2. The observed Mu amplitude (A_{Mu}) in Xe gas as a function of pressure (in atm). Since $A_D \approx 0$ (Table I), $P_{tot} \approx 2A_{Mu}$ in this case, giving rise to an apparent missing fraction, $P_L \sim 0.2$, at the highest pressures. The solid curve is a slightly modified fit to the data of the density matrix calculations of Turner[15].

inner shell capture will be of some importance at high energies in the slowing down process for the μ^+, as it is for the proton[18], the largest cross sections are for "valence" shell electrons; important in the case of He, for example, from ~30 keV to ~30 eV (or from ~300 KeV to ~300 eV for proton charge exchange). It is in this charge exchange regime where muon depolarization occurs via the formation and subsequent electron loss collisions of p-Mu, shown for Xe as a function of pressure in Fig. 2.

An early model[7] simply related $P_{tot} = P_{Mu} + P_D$ to the total time spent in the charge exchange regime as a time average of the $\mu^+ - e^-$ hyperfine interaction, $\langle \cos \omega_o t \rangle = P_{tot}$, where $\omega_o = 2\pi\nu_o = 2.8 \times 10^{10}$ rad s^{-1}. This gives acceptable fits to the observed pressure dependence seen in P_{tot}[7],[11]. A much more sophisticated approach has been developed by Turner[15], using a density matrix formalism to investigate the muon spin dynamics from a solution of the Boltzmann equation. In this formalism, the electron capture and loss cross section (rates) are treated as free parameters to be fit to the experimental data, a recent example of which is shown in Fig. 2. Note the large apparent missing fraction in Xe, $P_L \approx 0.2$, even at 6 atm pressure. An even bigger lost fraction, $P_L \approx 0.4$, is seen in CCl_4 at an asymptotic "high" pressure of only 1.2 atm (Table I and ref. 17). The fact that recovery of the polarization is incomplete at moderate pressures in these particular cases needs to be studied further at much higher pressures. Such work is in progress. Present results though are likely an indication of just how inefficient these heavy (spherical) molecules are at slowing down the muon.

In order to explore the effects of elastic (and inelastic) scattering on the muon polarization, a formalism has been developed[8] which explicitly includes these cross sections, as well as those for electron gain (σ_C) and electron loss (σ_L). Such a treatment is perfectly general, and could just as well be applied to, for example, proton (or positron) charge exchange, with the muon polarizaton emerging then as a special cse. The approach is similar in philisophy to that of Allison[4] and also to the Monte-Carlo calculations of Malcome-Lawes in addressing the question of hot tritium reactivity[19]. It is outlined below and applied to the case of muons stopping in He, utilizing proton cross sections and assuming velocity scaling in the case of charge exchange. The central problem is to calculate the energy loss in a given charge exchange cycle and the time dependence of this process, since it is this quantity that is directly related to the loss of muon polarization.

The Neutral Fraction, $f_0(E)$

If there were no competing processes, the energy loss per charge exchange cycle would be trivial, essentially given by the ionization potential of the moderating gas (I_M). The actual situation is more complex though, involving both elastic and inelastic (ionization and excitation) scattering processes as well, so that in general, after N consecutive collisions of a given type k, the final energy E_N can be written in the form[8]

$$E_N = EF_k^N - \frac{I_k(1-F_k^N)}{(1-F_k)} \tag{3}$$

where F_k is a (weakly) energy dependent kinematic factor which depends on both the masses and scattering angles of colliding partners and I_k is essentially the endothermicity for the process of interest. By differentiating Eqn. (3) wrt N (N→0), the energy loss per collision can be shown to have the form

$$(dE/dN)_k = \ln F_k [E + I_k/(1-F_k)] \tag{4}$$

For example, in the case of elastic scattering, $I_k = 0$, and Eqn. (4) may be integrated between some initial and final energy to give an expression for N, $N = (1/\ln F)\ln(E_f/E_i) \approx (M/2m)\ln(E_i/E_f)$, where the latter approximation is a familiar expression arising from the assumption of isotropic scattering for a light particle of mass m incident on a heavy target of mass M. It can be noted that Eqn. (4) is valid also for a composite collision process (e.g. cyclic charge exchange), provided that $F(E)$ and $I(E)$ are defined appropriately.[8] The situation in a given charge exchange cycle (cy) is pictured in Fig. 3. Note that a complete cycle involves <u>four</u> segments: electron capture (σ_C) between E_0 and E_1; elastic and inelastic energy loss of Mu between E_1 and E_2; electron loss (σ_L) between E_2 and E_3; and elastic and inelastic energy loss of the μ^+ between E_3 and E_4.

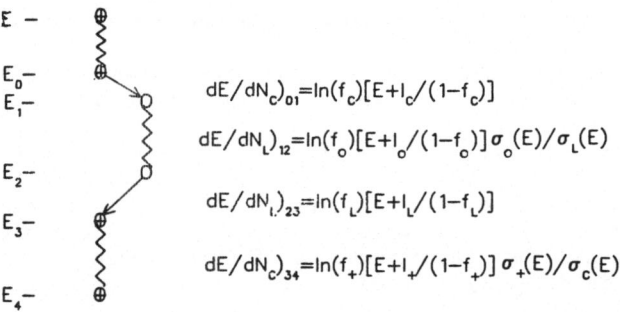

$$dE/dN_C)_{01} = \ln(f_c)[E + I_c/(1-f_c)]$$

$$dE/dN_L)_{12} = \ln(f_o)[E + I_o/(1-f_o)]\sigma_o(E)/\sigma_L(E)$$

$$dE/dN_L)_{23} = \ln(f_L)[E + I_L/(1-f_L)]$$

$$dE/dN_C)_{34} = \ln(f_+)[E + I_+/(1-f_+)]\sigma_+(E)/\sigma_c(E)$$

Fig. 3. Energy loss rates per charge exchange collision for four different portions of the cycle. Charge transfer (0,1; 2,3) is assumed to be instantaneous and is represented by the arrows; competing processes (1,2; 3,4) are represeted by wavy lines.

As defined in Figure 3, there are several contributions to the energy loss per collision for a given cycle at some average E, the sum of which can be represented by

$$(dE/dN)_{cy} = (dE/dN_C)_{01} + (dE/dN_L)_{12} + (dE/dN_C)_{23} + (dE/dN_C)_{34} \quad (5)$$

Since $(dN/dt)_i = (dN/dE)_i \cdot (dE/dt)_i = nV_{rel}(E)\sigma_i(E)$, where n is the number density and $V_{rel}(E)$ is the relative velocity at energy E, the average time to complete a cycle at energy E can be found from the sum of capture and loss contributions,

$$(dt/dN)_{cy} = (dt/dN_L)_{1,2} + (dt/dN_C)_{3,4} = \frac{\sigma_C(E) + \sigma_L(E)}{\sigma_C(E) \cdot \sigma_L(E)} \times \frac{1}{nV_{rel}(E)} \quad (6)$$

In Eqn. (6) it is assumed that the specific charge exchange transitions $0 \to 1$ and $2 \to 3$ (Fig. 3) are instantaneous. Also, additional charge exchange processes, for example forming Mu^-, are neglected. It is noted that Eqn. (6) is just the sum of mean free paths. From Eqns. (5) and (6), the energy loss rate during cyclic charge exchange can be expressed by

$$(dE/dt)_{cy} = nV_{rel}(E)\{A_+(E)[\ln(F_C)(E+I_C/(1-F_C))\sigma_C + \ln(F_+)(E+I_+/(1-F_+))\sigma_+] + A_0(E)[\ln(F_L)(E+I_L/(1-F_L))\sigma_L + \ln(F_0)(E+I_0/(1-F_0))\sigma_0]\} \quad (7)$$

where here '+' and '0' represent the charged particle (μ^+) and neutral (Mu), respectively, and $A_+(E)$ and $A_0(E)$ are ratios of charge exchange cross sections, defined by,

$$A_+(E) = \sigma_L(E)/(\sigma_C(E)+\sigma_L(E)) \quad (8a)$$

$$A_0(E) = \sigma_C(E)/(\sigma_C(E)+\sigma_L(E)) \quad (8b)$$

with σ_0 and σ_+ representing the total cross sections but <u>excluding</u> charge exchange (i.e. in a <u>given</u> charge state); e.g.,

$$\sigma_0(E) = \sum_k \sigma_{0k}(E) = \sigma_{0E\ell}(E) + \sigma_{0I}(E) + \sigma_{0X}(E) \quad (9)$$

for Elastic, Ionization and eXcitation contributions, respectively (see also Fig. 3).

The energy of the last charge exchange cycle and the position of energy thresholds are crucial in determining whether an incident muon actually thermalizes as a bare muon or as a muonium atom. In the case of muonium formation ($\mu^+ + M \underset{\sigma_C}{\longrightarrow} Mu + M^+$), the threshold energy, although slightly modified by kinematic factors, in given essentially by the difference in energy between I_M and that of Mu itself; in He, e.g. this value is ~11.1 eV. In the case of electron loss ($Mu+M \underset{\sigma_L}{\longrightarrow} \mu^+ + \bar{e} + M$), the threshold is just I_{Mu} (13.6 eV), or, in the case of formation of a bound M^- ion, $I_{Mu} - A_M$, where A_M is the electron affinity of the moderator (typically $\lesssim 1$ eV). For example, in Fig. 3, if the charge capture threshold fell between E_0 and E_2, the final product would be a Mu atom. Statistically, if the energy at which the last charge exchange collision takes place near E is sufficiency random, the fraction of <u>neutral</u> species (i.e., Mu) expected below the cut-off at E is approximately given by the ratio $E_0 - E_2/E_0 - E_4$, which is reminiscent of the Ore model prediction for positronium formation[1,12,20]. This ratio is given by (Fig. 3),

$$f_0(E) = [(dE/dN_C)_{01} + (dE/dN_L)_{12}]/(dE/dN)_{cy} \quad (10)$$

If charge exchange is the dominant energy loss mechanism, then $f_0(E)$ is largely <u>independent</u> of $\sigma_C(E)$ and $\sigma_L(E)$. On the other hand, if elastic and inelastic scattering mechanisms dominate and if it can be assumed that $\sigma_0 \sim \sigma_+$ and kinematic factors can be ignored, then $f_0(E)$ is essentially given by the value of $A_0(E)$ from Eqn. (8b), an expression which can also be derived from the "steady state" assumption[4,7,8] for the time dependence of $f_0(E)$. This is also then the fraction of <u>time</u> spent as muonium since the charge transfer collisions are always assumed to be instantaneous. It can be pointed out that, for most moderators, $I_{Mu} \gtrsim I_M$, in contrast to positronium, and hence Mu formation is usually exothermic and can proceed right down to thermal energies so that both Eqns. (8) and Eqn. (10) would predict 100% Mu formation. Helium and Neon are in fact notable exceptions.

The Muon Polarization, $\langle P \rangle$

Every time Mu is formed in the para (M=0) spin state, the muon spin polarization is reduced by the factor $\cos(\omega_o \tau_i)$, where τ_i is the residence time of muonium during the i'th muonium formation (in Ref. 7 this was treated as an integrated time over all cycles). As noted earlier, since the captured electron is not polarized, it is assumed that muonium is formed with equal probabilities in ortho and para spin states. If there are N muonium formations during cyclic charge exchange, the observed polarization after thermalization is obtained by the average polarization over all possible sequences of parallel and anti-parallel states through the charge exchange regime,

$$\langle P \rangle = 1/2^N \sum_{\{N_1 N_2 N_3 \ldots N_N\}} [\cos(\omega_o \tau_1)]^{N_1} [\cos(\omega_o \tau_2)]^{N_2} \ldots [\cos(\omega_o \tau_N)]^{N_N} \quad (11)$$

where $N_i = 0$ and 1 designate the parallel and anti-parallel states, respectively, and the sum is taken over all possible sequences of states $\{N_1, N_2, \ldots N_N\}$ and the initial muon polarization is taken to be unity. For example, the sequence $\{0, 0, \ldots 0\}$ denotes the case where all the muonium formation is in the M=1 state, resulting in no depolarization. Conversely the sequence $\{1, 1, \ldots 1\}$ results in the maximum depolarization. This summation can be readily evaluated and the polarization as a function of density (n) is,

$$\langle P(n) \rangle = \prod_i [1 + \cos(\omega_o \tau_i)]/2 = \prod_i [\cos(\omega_o \tau_i/2)]^2 \quad (12)$$

The task now is to evaluate τ_i, the residence time for muonium in the i'th cycle. Again referring to Fig. 3, suppose a muon undergoes an electron capture collision at E_0, which produces a muonium atom at E_1, at t=0. The probability that the particle at E_1 does <u>not</u> undergo any electron loss collision between t=0 and t=t is given by the familiar "attentuation" equation,

$$S(0,t) = \exp(-\int_0^t n \, \sigma_L(E)V(E)dt) = \exp(-N_L(0,t)) \quad (13)$$

If there are no competing processes, $\sigma_L(E)$ and $V(E)$ are idependent of time until the next electron loss collision takes place and the mean free time of the neutral species can be calculated trivially from the condition that the number of electron loss collisions is unity ($N_L(0,t)=1$) by E_2, which gives the familiar result $\tau = 1/n\sigma_L(E)V(E)$. In the presence of competing processes, however, the cross sections and velocities are now time dependent. In this case, in terms of <u>energy</u>, the survival probability between E_1 (t=0) and E_2 (t=t) can be written as,

$$S(E_1, E_2) = e^{-N_L(E_1, E_2)} = \exp(-\int_{E_1}^{E_2} \frac{\sigma_L(E)dE}{\sigma_0(E) \cdot \ln F_0(E)[E+I_0/(1-F_0(E))]}) \quad (14)$$

The <u>mean</u> energy E_2 at which the next electron loss collision takes place is obtained from the condition $N_L(E_1, E_2) = 1$. Since $dt/dE = dt/dN \cdot dN/dE$, and recalling the form of (dN_L/dE) in Fig. 3 and integrating over energy, one obtains an expression for the (mean) residence time of muonium in the neutral charge state (in the i'th cycle) as

$$\tau_0(E_1, E_2) = \frac{1}{n} \int_{E_1}^{E_2} \frac{dE}{\ln F_0(E)[E+I_0/(1-F_0(E))] \cdot v(E)\sigma_0(E)} \quad (15)$$

It is noted that the energy E_2 in Eqn. (15) implicitly depends on $\sigma_L(E)$.

To calculate the polarization, the result from Eqn. (15) is inserted in to Eqn. (12), and the process repeated over many cycles. At the end of each cycle, the charge state of the muon is monitored and at the end of these many cycles the polarization can be recorded as $\langle P_\mu \rangle$ for the μ^+ or $\langle P_{Mu} \rangle$ for the muonium atom. Note that the pressure dependence is contained in the number density, n.

Muons in Helium. A Case Study

Muonium fractions and muon spin polarizations have been calculated in He as a function of energy and pressure with charge exchange, elastic, and ionization processes for both neutral and positive species taken into account (excitation cross sections are small and have been ignored). The cross sections for muon (muonium) have been obtained from proton (hydrogen) results by a simple scaling scheme based on the following assumptions: (i) the cross sections for both charge exchange and ionization processes are the same at the same relative velocity (i.e., velocity scaling holds), and (ii) the elastic cross sections are equal at the same energy in the center of the mass system. For electron capture the data of Stedeford and Hasted[21] have been used, recorded also in the compilation by Tawara[22]. For the electron loss cross sections we have used the data of Roussel et al.[23] The ionization cross sections for μ^+ and Mu are taken from the corresponding p and H data of Solov'ev et al.[24]. For the elastic scattering cross sections, the measurements of Ruzic and Cohen[25] at low energies, where collisions are dominated by elastic scattering, have been used. These calculations are denoted by (He-1) in the figures to follow. For comparison, calculations are also carried out for the case where only charge exchange collisions are included (He-0). Further calculations as well as discussion of the treatment of the angular distributions, kinematic factors, etc., are reported by Senba[8].

In the present calculations, the statistical fluctuation of the residence times are fully taken into account while the competing processes are treated by a mean value approximation implied by Eqns. (5)-(7). The muon polarization P_{tot} and the neutral fraction f_0 are calculated in the following way. Using the initial energy of the positive muon (E), the energy at which the first electron capture collision takes place (E_0) is found from an equation of the form of (14) with the requisite number of collisions generated by $N_C(E, E_0) = -\ln(1-R_0)$ where R_0 is a random number ranging uniformly from 0 to 1 (see Fig. 3). The quantity $N_C(E, E_0)$ found in this way has an exponential distribution: the probability (density) that $N_C(E, E_0)$ takes a value x is given by $\exp(-x)$ and the mean value of x is one. This procedure simulates the statistical distribution of $N_C(E, E_0)$ and therefore of E_0 also. In terms of E_0, from Eqn. (3), the energy of the first muonium formation is calculated by $E_1 = F_C(E_0)E_0 - I_C(E_0)$.[8] Similarly, from Eqn. (14) the energy at which the first electron loss occurs (E_2) can be generated by $N_L(E_1, E_2) = -\ln(1-R_2)$ through another random number R_2. The muonium residence time between E_1 and E_2 is then calculated from Eqn. (15). The procedure is repeated until the muonium (positive muon) energy reaches the electron loss (capture) threshold, both ~12 eV in He.

The polarization retained in this particular muon is calculated by Eqn. (12) and the average polarizations retained in the positive and neutral final products, $\langle P_\mu \rangle$ and $\langle P_{Mu} \rangle$, have been calculated at a given pressure for a sampling of 10000 muons. The total polarizations as a function of pressure are expressed in terms of these values in order to compare with those observed experimentally at thermal energy. The results for $P_{tot}(n)$ are shown in Fig. 4, where they are compared also with recent experimental data (Table I). It can be noted that P_{tot} in He is dominated by contributions from the μ^+, and hence $\langle P_\mu \rangle$. Considering the fact that the experimental errors are $\gtrsim \pm 10\%$ (earlier values reported for He in Ref. 7 are considerably in error, likely due to errors in the determination of wall signals), the absolute agreement with the experimental polarization is quite good and the trend with pressure is well reproduced. The importance of including competing processes, particularly elastic scattering, is shown by the comparison between the (He-1) and (He-0) calculations. While the trend with pressure in the

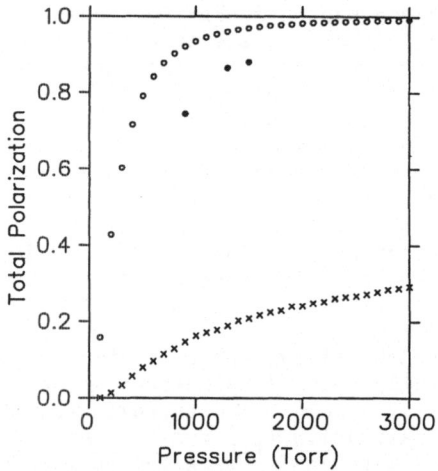

Fig. 4. The total muon polarization in He, P_{tot}, as a function of moderator pressure in torr. The open circles are for the (He-1) calculation which includes competing processes while the crosses are the (He-0) calculation for pure charge exchange. The filled circles are experimental results.

later calculation, which includes only charge exchange, is again well reproduced, as it is in the density matrix calculations of Ref. 15, the absolute magnitude of the polarization is grossly in error. The additional energy loss due to competing processes leads to a reduced number of muonium formations and to a higher energy (~200 eV) for the last muonium formation, both of which, in turn, lead to an enhanced polarization. In this case the polarization is insensitive to the details of the charge exchange cross sections at low energies.

The neutral fraction $f_0(E)$ is calculated at several energies during cyclic charge exchange, also by a Monte Carlo sampling of individual events. Results are given in Fig. 5. In contrast to the polarization, $f_0(E)$ is sensitive to a subtle balance between charge exchange and its competing processes down to low energies, depending on the choice of elastic cross sections used. In order to compare with experiment, for the (He-1) calculation, the relevant energy E in Fig. 5 is the electron loss threshold, 13.6 eV. This indicates that a sizable neutral fraction ($f_{Mu} > 0.3$) is expected in gas phase He; but, experimentally, little or

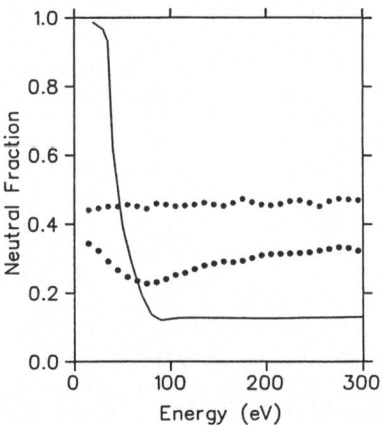

Fig. 5. The neutral fraction $f_0(E)$ expected for Mu in He from cyclic charge exchange. The solid line results from the "steady state" assumption of Eqn. (8b). The upper and lower points are (He-0) and (He-1), respectively.

no muonium signal has been observed, $f_{Mu} \lesssim 0.03$ (Table I). Possible causes for this discrepancy are discussed below. The (He-0) calculation assumes charge exchange dominates the energy loss, and consequently, in concert with the polarization data in Fig. 4, gives much more Mu at all energies. Figure 5 also shows the neutral fraction expected from the "steady state" assumption (Eqn. (8b)), which actually approaches unity as $E \rightarrow 0$ because $\sigma_L(E) = 0$ at threshold, in complete disagreement with experiment. The calculated values for $f_0(E)$ in Fig. 5 are in fact likely to be most correct for $E \gtrsim 100$ eV, since the polarizations are largely determined by these energies and agree quite well with the data (Fig. 4).

DISCUSSION

The μSR Results

The polarization data of Table I (P_{tot}) are plotted as a function of valence electron density (Pressure (atm) x number of valence electrons (Z_e)) in Fig. 6, assuming that molecules can be simply treated as a sum of their constituent atoms. This parameter is consistent with our contention that it is the outer electrons which play the dominant role in determining the amount of muonium seen. The figure illustrates the statement made earlier that $P_L \rightarrow 0$ at moderate pressures, with the exception of Xe (and CCl_4). It also illustrates the concept developed above that charge exchange is not the only process responsible for energy loss during the charge exchange regime, otherwise one would expect all the data to lie more closely on the same universal curve. While the trend with increasing $P \cdot Z_e$ is the same in all cases[11], there are significant differences even between molecules of the same "valency" (H_2, He; Ar, Xe and H_2O).

Consider the differences between He and H_2 (D_2). Because of an anticipated similarity in elastic cross sections, kinematic factors could be expected to dictate that D_2 and He would both be less efficient than H_2 in moderating the muon, thus predicting a lower polarization. The fact that P_{tot} is the same for H_2 and D_2 (as are the relative fractions) and considerably less than that seen for He at the same density, suggests that it is predominately the second (endothermicity) term in Eqn. (3) that is determining the energy loss: for a given charge exchange cycle, this is essentially just I_M, 24.5 eV for He but only 15.4 eV for $H_2(D_2)$. A related point is that Mu can form at lower energies in H_2/D_2 than in He, thereby suffering more depolarization. On the other hand, the fact that H_2, N_2 and Ar, with essentially the same IP, exhibit considerably different recoveries of the polarization in comparison with He indicates that elastic scattering is surely important. For the molecular gases, ro-vibrational inelastic scattering should also be considered with isotopic differences also anticipated[26]. Again one's naive expectation is that, in general, such a contribution should enhance the moderating efficiency of molecules compared to atoms, giving rise to a larger polarization for H_2, (D_2) than in He. The conclusion that H_2 (D_2) is a relatively inefficient moderator compared to He is consistent with recent studies elsewhere of energy moderation of ~2 eV hot tritium atoms in these same systems[27]. It may simply be that pronounced forward scattering on light molecules effects an inefficient energy transfer.

Table II lists the relative fractions f_D and f_{Mu} (Eqn. (2)) for selected gases. The IP for each gas is also given as is an estimate of the neutral fraction f_H expected from velocity scaled experimental proton charge exchange cross sections[7,12,22] down to thermal energies. As mentioned earlier, for all gases with an IP <13.6 eV (Mu), even from Eqn. (8b), this fraction should be 100%, since $\sigma_C \gg \sigma_L$ at eV energies, consistent also with the f_{max} prediction from the Ore model[1,12]. For those gases whose IP is >13.6 eV, the fraction of neutral H (Mu) atoms expected depends crucially on the charge state of the muon at threshold, which, as discussed above, depends in turn on the magnitude of the appropriate elastic and inelastic cross sections at higher energies. For Ar, H_2 and N_2, at energies of a few eV above the electron capture threshold (~2 eV), σ_C is again appreciably greater than σ_L and f_H ~ 90(±5)% is expected in each case. The situation is less clear for He (and Ne). The calculations cited above (Fig. 5) suggest $f_{Mu} > 30\%$ in He, but values of ≲10% can also be expected if elastic scattering were to dominate at higher energies.[7,8]

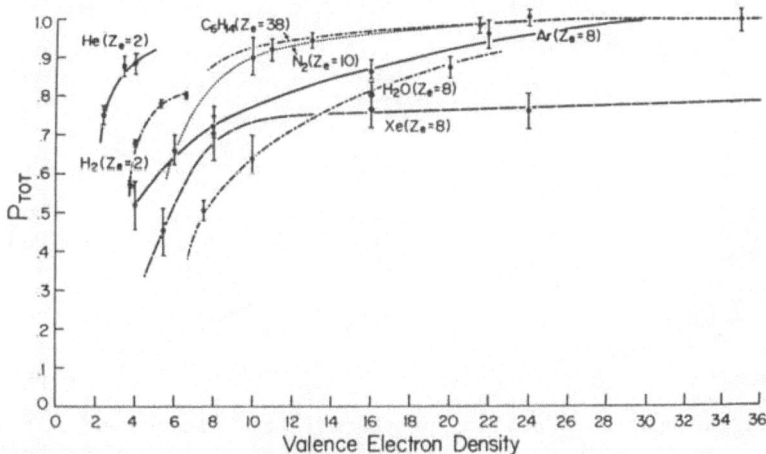

Fig. 6. The total muon polarization $P_{tot} = P_D + P_{Mu}$ for selected gases
from the data of Table I, vs. the Valence Electron Denisty
(P(atm).Ze(no. of valence electrons)). Molecules are treated
simply as sum of constituent atoms. The lines are drawn just to
guide the eye, but they exhibit the expected density dependence
(Fig. 4 and Refs. 7, 8, 11, 15, 17).

Table II. Ionization Potentials, and Charge Fractions for Positive Muons and Protons Thermalizing in Gases

Gas	IP (eV)	$f_D(\%)^a$	$f_{Mu}(\%)^a$	$f_H(\%)^b$
He	24.5	100±5	0±5[c]	≳30
Ne	21.6	95±5	5±5[c]	≳30
Ar	15.8	26±4	74±4	85
Kr	14.0	0±5[d]	100±5	100
Xe	12.1	0±5[d]	100±5	100
H_2	15.4	39±4	61±4	90
N_2	15.6	16±4	84±4	90
NH_3	10.2	9±4	91±4	100
CO_2	13.8	10±3[e]	90±5[e]	100
H_2O	12.6	11±2	89±2	100
CH_3OH	10.8	16±2	84±2	100
CCl_4	11.5	53±3	47±3	100
$CHCl_3$	9.3	28±2	72±2	100
CH_2Cl_2	11.4	21±3	79±3	100
CH_4	12.6	12±2	88±2	100
C_2H_6	11.5	19±2	81±2	100
C_3H_8	11.1	21±2	79±2	100
C_4H_{10}	10.6	21±2	79±2	100
C_5H_{12}	10.4	15±2	85±2	100
C_6H_{14}	10.2	25±2	76±2	100
$Si(CH_3)_4$	9.9	20±2	80±2	100

[a] Relative fraction from Eqn. (2), taken from Refs. 7,11,12,16 and Table I. All obtained at pressures ≳ 1 atm.
[b] Corresponding neutral fraction for thermalized H atoms, estimated from proton charge exchange cross sections[22], as explained in the text.
[c] Trace impurities could be responsible for a small amount of muonium.
[d] Wall signals could obscure a small diamagnetic fraction.
[e] Preliminary values obtained at only one pressure.

In comparing the experimental values for f_{Mu} with the "expected" values estimated as f_H, in Table II, several interesting points emerge. Only in the case of Kr and Xe, with IP's ≲13.6 eV, is ~100% Mu formation actually seen, as expected. In all other gases with low IP's, with the exception of CCl_4 (and possibly $CHCl_3$)[11], f_{Mu} ~ 90%. This difference from 100%, manifest in a corresponding non-zero value of ~10% for f_D, has been interpreted as evidence for hot (Mu^*) atom reactivity[12],[16], in the energy range below the 13.6 eV threshold for electron loss. In the larger alkanes, some loss of Mu could be expected from intramolecular electron loss[5,16,28]. However, in H_2, hot atom reactivity likely dominates ($Mu^*+H_2 \rightarrow MuH+H$), with f_D much larger than seen in N_2, which has a comparable IP and expectation for f_H. Nitrogen has a much stronger chemical bond than H_2, so it should have a much reduced Mu^* reactivity, and the value found for f_{Mu} is indeed in excellent agreement with f_H. The neutral fraction f_{Mu} for Ar also agrees well with the charge exchange prediction from f_H; in the absence of any dimer formation[12], Mu reactivity is not possible. It is noted that the estimate of f_H as the fraction of neutrals expected from proton charge exchange ignores the possibility of hot atom (H^*) reactivity, which could be appreciable, based on current comparison between Mu^* and T^* reactions[12,16].

In principle, the noble gases should allow the most straightforward interpretation and, particularly for Ar, Kr and Xe, the large amounts of Mu seen are as expected. Qualitatively, with their large IP's, we would expect less Mu in He and Ne, but the essentially 0% seen appears inconsistent with the model calculation discussed above. This disagreement suggests that either velocity scaling breaks down at <u>low</u> energies, say well below 100 eV, or that there are additional (unknown) mechanisms for loss of polarization. The latter possibility seems unlikely though in the case of He or Ne, but such a mechanism (e.g., Mu* reactivity of p-Mu following cyclic charge exchange)[12] may provide an explanation for both the reduced amount of Mu seen and the large missing fraction in CCl₄ (Table I). Another possibility is that the proton charge exchange and/or scattering cross sections are simply incorrect at low energies, and certainly in some cases there is a wide scatter in reported values[22].

Finally, the nature of the diamagnetic environment for muons stopping in gases can be briefly commented on. There are three possibilities: bare muons (μ^+), muon molecular ions (e.g., $HeMu^+$) and muonium molecules (e.g., MuH), all of which would be unresolved in a typical μSR experiment. In the noble gases (He, Ne, Ar) we have established[12,29] that the diamagnetic environment is due to molecular ions, with no evidence in favor of bare muons. The mechanism for the formation of these ions is not firmly established, but they are most likely formed via termolecular collision processes of epithermal μ^+ at energies well below the charge exchange threshold[12]. Muon molecular ion formation in molecular gases is not yet established, but the energies for appreciable μ^+ capture are low enough (<1 eV) that it seems unlikely such a process can effectively compete with Mu* reactivity.

Comparison with Positronium Formation

Table III compares the neutral fractions for total positronium formation (f_{Ps}) with the corresponding values for muonium formation (Table II), for some selected gsses, all at ~1 atm pressures. The f_{Ps} values are taken from several sources, cited in Ref. 12. Table III gives the IP for each gas again as well as the Ore predictions[1,12] for f_{Ps}.

TABLE III. Positronium and Muonium Fractions in Gases

Gas	IP(eV)	$f_{Ps}(exp)^a$	$f_{Ps}(Ore)^b$	$f_{Mu}{}^c$
He	24.5	.24	.14-.28	<.05
Ne	21.6	.26	.09-.32	<.10
Ar	15.8	.33	.16-.43	.74
Kr	14.0	.19	.20-.49	1.0
Xe	12.1	.07	.26-.56	1.0
H_2	15.4	.30	.18-.44	.61
N_2	15.6	.19	0.0-.44	.84
CO_2	13.8	.39	0.0-.49	.90
H_2O	12.6	–	.07-.54	.89
NH_3	10.2	–	.22-.66	.91
CH_4	12.6	.38	.21-.54	.88
C_2H_6	11.5	.41	.26-.59	.81

[a]Experimental values for total fraction of Ps seen at ~1 atm.
[b]Prediction of the simple Ore Gap model for the limits $f_{min}-f_{max}$.
[c]The experimental values of f_{Mu} from TRIUMF (Table II).

Not surprisingly, one can tell at a glance that there is little correlation between the experimental values for f_{Mu} and f_{Ps}, with much more Mu seen in every case except in He and Ne, where there is much less. In first order, this simply reflects the large IP of Mu (13.6 eV) compared to Ps (6.8 eV); positronium formation at thermal energies is rarely (if ever) an exothermic process in gases, whereas muonium formation most frequently is. With the notable exception of particularly Xe, where the amount of Ps seen experimentally is far below even the minimum amount expected, the experimental results for all the other gases of Table III are well within the Ore limits (to the authors' knowledge, those for H_2O and NH_3 in the gas phase have not been reported). Although the Ore predictions in Table III are based on the simplest interpretation of that model, a uniform distribution of positron energies, more extended versions do not appreciably alter the fact that there is far too little Ps seen in Xe[20]. This discrepancy is most likely due to the long slowing down time for e^+ in such gases, thus enhancing the liklihood of annhiliation. As pointed out above, Xe is also an inefficient moderator for muonium, thus giving rise to an unusually slow recovery of the muon polarization (Fig. 2).

In addition to the differences in the values themselves for f_{Mu} and f_{Ps}, there is a different dependence seen on density (pressure) in a variety of gases. The results given in Table III are all for gases at ~1 atm pressure. Up to ~30 atm for some of the rare gases and up to ~3 atm for some molecular gases, no pressure dependence in f_{Mu} has been seen for any of the gases in Table III. As noted previously, in some of the heavier alkanes, there is a tendency for f_{Mu} to decrease (f_D increases) in the pressure range up to ~1 atm[16]. This situation is in contrast to that prevailing in positronium studies, where there is a general tendency for f_{Ps} to increase noticeably with an increase in pressure[12,30] up to gas densities of about 50 amagats (~50 atm pressure at 300 K), by a factor of about 1.5 relative to the ~1 atm values quoted in Table III. At higher densities, either the Ps yield exhibits saturation or begins to decrease, depending markedly on temperature in some cases. Only in the rare gases, He, Ne and Ar are the values for f_{Ps} found to be pressure (and temperature) independent, these results alone being consistent with the muonium ones. It is worth pointing out though that the muon results to date have generally been obtained over a much more restricted range of (low) pressures than the positron data and that the possibility of temperature-dependent muonium amplitudes has not really been investigated.

A number of suggestions have recently been advanced to explain the dependence of Ps yields on pressure and temperature in gases, including fragmentation processes, the formation of e^+ complexes, and Ps formation in a "spur" caused by ionizing radiation as the positron thermalizes in the gas[31,32]. The latter model seems to give the most reasonable account of the data[31]. In its simplest form, the spur model predicts that the fraction of Ps formed is governed by an equation of the form of

$$f_{Ps}(spur) \approx 1 - e^{-r_c/r_t} \tag{16}$$

where r_c is the Onsager escape radius and r_t is the average distance between the e^+ and e^- in the spur; since the latter will decrease with increasing density while the former is largely density independent, the spur model predicts that f_{Ps} will increase with increasing pressure in the gas, in accord with observations. It is noted that such a trend is inconsistent with any appreciable hot atom reactivity for positronium[12].

In general, one can conclude that in gases at low pressures (~1 atm) Ps formation due to radiation induced spurs is negligible and it is in this pressure region (Table III) where the Ore Gap model is likely to enjoy its greatest success. Jacobsen[31] has concluded that at pressures $\gtrsim 10$ atm, Ps formation is likely dominated by e^+ and e^- recombination in the spur. The present muonium results in ~1 atm gases (and higher in the noble gases) demonstrate that spur processes are of negligible consequence and f_{Mu} is dominated by charge exchange and hot atom chemistry. In this respect, it is of considerable interest to extend the present studies of muonium formation and reactivity in gases to much higher pressures, through their critical points. This would allow a systematic exploration of the possibility that additional mechanisms, including spurs, could contribute to f_{Mu} (f_D) in regions of very high density, as the liquid phase is approached. Such experiments are in progress, motivated in large part by the fact that μSR results in the condensed phase are markedly different from those discussed herein in the gas phase[12]. A spur model has been advanced to account for the condensed phase μSR results[32,33] for which data on high pressure fluids is a crucial missing link.

REFERENCES

1. H.S.W. Massey, E.H.S. Burhop and H.B. Gilbody, "Electronic and Ionic Impact Phenomena", Vol. 5, "Slow Positron and Muon Collisions", Oxford-Claredon Press, 1974.
2. B.H. Bransden, Rep. Prog. Phys. 35, 949 (1972); H. Tawara and A. Russek, Rev. Mod. Phys. 45, 178 (1973).
3. E. Rille, R.E. Olson, J.L. Peacher, D.M. Blankenship, T.J. Krale, E. Redd and J.T. Park, Phys. Rev. Letts. 49, 1819 (1982).
4. S.K. Allison, Rev. Mod. Phys. 30, 1137 (1958).
5. S.L. Varghese, G. Bissinger, J.M. Joyce and R. Loubert, Phys. Rev. 31A, 2202 (1985).
6. J.H. Newman, J.D. Cogan, D.L. Ziegler, D.E. Nitz, R.D. Rundel, K.A. Smith and R.F. Stebbings, Phys. Rev. 25A, 2976 (1982).
7. D.G. Fleming, R.J. Mikula and D.M. Garner, Phys. Rev. 26A, 2527 (1982).
8. M. Senba, J. Phys. B, to be submitted; M. Senba et al., contributed paper, XV ICPEAC, Brighton, July, 1987.
9. E. Roduner, Prog. Reaction Kinetics 14, 1 (1986).
10. R.H. Heffner and D.G. Fleming, Phys. Today, Dec. 1984, p. 2.
11. D.J. Arseneau, D.M. Garner, M. Senba and D.G. Fleming, J. Phys. Chem. 88, 3688 (1984); D.J. Arseneau, M.Sc. Thesis, Univ. of British Columbia, 1984.
12. D.G. Fleming, Radiat. Phys. Chem. 28, No. 1, 115 (1986).
13. D.M. Garner, D.G. Fleming and R.J. Mikula, Chem. Phys. Letts. 121, 80 (1985); I.D. Reid et al., J. Chem. Phys., in press (1987).
14. M. Senba, D.M. Garner, D.J. Arseneau and D.G. Fleming, Hyp. Int. 17-19, 703 (1984); D.G. Fleming, R.J. Mikula and D.M. Garner, J. Chem. Phys. 73, 2751 (1980).
15. R.E. Turner and M. Senba, J. Chem. Phys. 84, 3776 (1986); Phys. Rev. 29A, 2541 (1984).
16. D.G. Fleming, M. Senba, D.J. Arseneau, I.D. Reid and D.M.Garner, Can. J. Chem. 64, 57 (1986).
17. M. Senba, R.E. Turner, D.J. Arseneau, D.M. Garner, L.Y. Lee, I.D. Reid and D.G. Fleming, Hyp. Int. 32, 795 (1986).
18. M.E. Rudd, R.D. Dubois, L.H. Toburen, C.A. Ratcliffe and T.V. Goffe, Phys. Rev. 28A, 3244 (1983); G. Lapicki and F.D. McDaniel, Phys. Rev. 22A, 1896 (1980).

19. D.J. Malcolm-Lawes, in "Hot Atom Chemistry: recent trends and applications", T. Matsuura, ed., Kodansha Ltd., Tokyo, 1984, p. 39.
20. D.M. Schraeder and R.E. Svetic, Can. J. Phys. 60, 517 (1982).
21. J.B.H. Stedeford and F.B. Hasted, Proc. Roy. Soc., A227, 466 (1955).
22. H. Tawara, At. Data Nucl. Data Tables 22, 491 (1978).
23. F. Roussel, P. Pradel and G. Spiess, Phys. Rev. A16, 1854 (1977).
24. E.S. Solov'ev, R.N. Il'in, V.A. Oparin and N.V. Fedorenko, Soviet Phys. JETP (Eng. Trans.) 15, 459 (1962).
25. D.N. Ruzic and S.A. Cohen, J. Chem. Phys. 83, 5527 (1985).
26. G. Bischof, V. Hermann, J. Krutein and F. Linder, J. Phys. B At. Mol. Phys. 15, 249 (1982); D.W. Davies and S.J. Till, Mol. Phys. 39, 757 (1980); G.D. Billing, Chem. Phys. 30, 387 (1978).
27. S. Aronowitz, T. Scattergood, J. Flores and S. Chang, J. Phys. Chem. 90, 1806 (1986).
28. G. Bissinger, J.M. Joyce, G. Lapicki, R. Loubert and S.L. Varghese, Phys. Rev. Letts. 49, 318 (1982).
29. D.G. Fleming, R.J. Mikula, M. Senba, D.M. Garner and D.J. Arseneau, Chem. Phys. 82, 75 (1983); D.J. Arseneau, Ph.D. Thesis, in progress.
30. G.R. Heyland, M. Charlton, T.C. Griffith and G. Clark, Chem. Phys. 95, 157 (1985); G.L. Wright, M. Charlton, G. Clark, T.C. Griffith and G.R. Heyland, J. Phys. B16, 4065 (1983).
31. F.M. Jacobsen, "Proc. of NATO Workshop on e⁺ Scattering in Gases", Plenum Press, New York, 1983, p. 85; F.M. Jacobsen, Chem. Phys. 101, 259 (1986).
32. F.M. Jacobsen, Hyp. Int. 31, 501 (1986).
33. O.E. Mogensen and P.W. Percival, Rad. Phys. Chem. 28 No. 1, 85 (1986); P.W. Percival, J.C. Brodovitch and K.E. Newman, Hyp. Int. 17-19, 721 (1984).

SLOW MUON PHYSICS

Dale R. Harshman

A T & T Bell Laboratories

600 Mountain Avenue, Murray Hill, New Jersey 07974, U.S.A.

INTRODUCTION

Recently, slow positive muons (μ^+) have been observed to be emitted from solid rare-gas moderators exposed to a 4.2 MeV incident μ^+ beam. Energy spectra obtained from the time of flight data indicate a maximum at $\sim$ 5 eV with a tail extending to higher energies. The data suggest a "hot muon" emission mechanism, implying a long diffusion length for low-energy μ^+ in these solids. Of the targets measured, argon was observed to produce the highest yield ($\sim 10^{-5}$ slow μ^+ per incident μ^+), providing a useful flux for further experimentation. The discussion presented here centers around these results and their future implications for slow μ^+, Mu (μ^+e^-) and Mu$^-$ ($\mu^+e^-e^-$) beams. The applications of such beams will also be discussed.

It would be very useful to have beams of slow positive muons (μ^+) analogous to the slow positron beams now employed in a variety of solid state and atomic physics experiments.[1] One possible method of producing such a beam is by moderating the energetic (4.2 MeV) μ^+ that are available from stopped pion decay at a number of accelerators.[2] The first successful moderator[3] is single crystal LiF which has an efficiency of a few parts in 10^7 for converting fast μ^+ into slow μ^+ and into muonium negative ions ($\mu^+e^-e^-$ or Mu$^-$). Results of a recent experiment[4] indicate that solid rare gases are more efficient muon moderators (by about two orders of magnitude) and could be used to make a practical slow μ^+ beam.

The choice of LiF and the rare gas solids as possible slow μ^+ moderators was made by drawing guidance from previously reported positron results. A large e$^+$ emission probability is now attributed to the emission of "hot positrons", i.e. slowly thermalizing positrons with energies below the $\sim$ 10 eV inelastic thresholds for positronium and electron-hole pair formation.[5-7] Energetic positive muons slow down rapidly in condensed media by ionization until the velocity of the μ^+ becomes comparable to that of the valence electrons.[8] At lower velocities, the cross sections for electronic processes are significantly reduced.[9] Because of this, there should be no analogous sharp thresholds for muonium (μ^+ e$^-$ or Mu) and electron-hole pair formation for slow muons in insulators. At $\sim$ 10 eV the energy loss should therefore be dominated by phonon excitation, resulting in a comparatively small energy loss per unit distance, dE/dx. Low energy muons thus have a relatively long diffusion length in insulators and should be able to escape

with high efficiency. Unlike the alkali halides, rare-gas solids do not
support optical phonons (FCC lattice), possibly making dE/dx even smaller
in the low energy thermalization region. The present work is motivated by
the positron results mentioned, the interest in understanding better the
energy loss mechanisms involved at near thermal velocities, and the desire
to obtain an efficient slow μ^+ moderator.

EXPERIMENTAL TECHNIQUE

The apparatus used to study the energy spectrum of muons emerging from
the surface of possible moderators is shown in Fig. 1. It consists of a
high-vacuum scattering chamber, an electrostatic extraction lens, and a
magnetic spectrometer. Energetic (4.2 MeV, $\Delta E \approx 20\%$, 350,000/s) positive muons
from the M20(B) secondary channel at TRIUMF were incident on a cold target
assembly. A scintillation counter upstream of the target was used to detect
the incident muons. The incident beam momentum was tuned so that the stopping
distribution was centered at the downstream surface of the target. An electrostatic immersion lens, immediately downstream from the target, collected
and accelerated to 10 keV any low-energy charged particles that emerged from
the target. The extracted particles were injected into a mangetic (dipole-
quadrupole-quadrupole) spectrometer which momentum-selected and focused the
10 keV particles on to a 4 cm diameter channelplate detector. The basic
measurement was a time of flight (TOF) between the incident μ^+ counter and

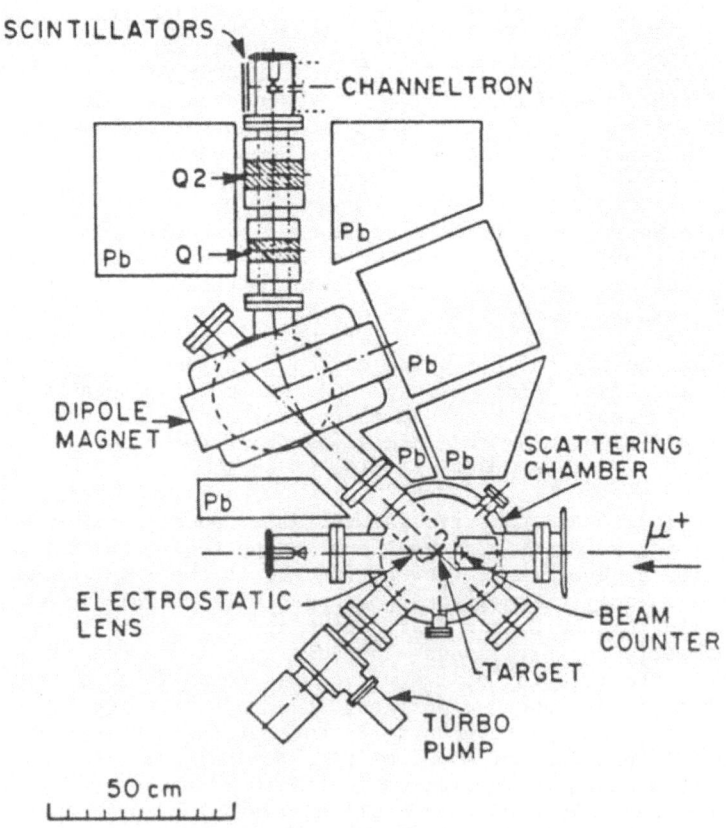

Figure 1. Diagram of the apparatus

the channelplate detector. The gap ($\sim$3.5 mm wide) between the target surface and the immersion lens was forward biased at two different voltages;
+ 500 V and + 9.45 V. The higher value was used in determining the total
yield, while the lower one allowed a better determination of the energy
spectrum.

Several improvements have been made to the apparatus since the previous
work on LiF was reported.[3] In particular, the vacuum has been improved to
provide a routine operating range down to 5×10^{-9} Torr, the quadrupoles
were redesigned to give a larger acceptance, and the large diameter (4 cm)
channelplate was introduced. Monte Carlo calculations, assuming uniform
distributions in energy (50 eV wide), emission angle, and origin position
on the target surface, give the acceptance of the beam transport as $\sim$ 80%.
Assuming the detection efficiency of the channelplate to be $\approx$ 50% for 10 keV
muons, the experimental efficiency is thus estimated to be about 40%. The
time-of-flight (TOF) spectrum was measured between single events in the
incident muon counter and the channelplate detector. Because of the high
incident μ^+ rates, the TOF was reversed in the electronics (i.e. the start
was delayed with respect to the stop), allowing faster data accumulation
times. The experimental time resolution, determined in the same manner as
previously reported,[3] was found to be about 3.5 ns FWHM, with no long time
tails observed.

The cold target consisted of a rectangular stainless steel helium
reservoir having two 0.025 mm thick stainless steel windows through which
the muons pass. The windows are separated by 3 mm and have a usable projected area of 25 by 25 mm^2, approximately three times smaller than the
incident beam spot. The target was cooled by flowing ^{4}He gas or liquid via
a transfer line. The gases were condensed at two different rates; one with
1.5×10^{-5} mbar partial pressure for 15 seconds and another at 1.5×10^{-4}
mbar partial pressure for a period of 1.5 seconds, both corresponding to
about 150 atomic layers.[10] To condense argon, krypton and xenon only the
cold helium vapor was required, whereas for neon, it was necessary to have
liquid in the helium reservoir.

EXPERIMENTAL RESULTS

The TOF data for solid argon and krypton are shown in Fig. 2. The data
peaked at early times were taken with the target biased at + 500 volts with
respect to the electrostatic lens, and the data peaked at later times were
taken with a + 9.45 volt bias. A small flat background, determined from the
integrated rate over a 100 ns interval between 75 and 175 ns, has been subtracted for all of the spectra reported here. We note that the full width
at the base of the 9.45 volt bias data is about 60-70 ns which is equal to
the time of flight of a μ^+ of zero initial kinetic energy through the gap
region between the target and the lens. Any thermal diffusion process would
be signified by a $1/\sqrt{t}$ dependence in the late time TOF,[3] which was not
observed. By integrating the 500 volt bias data over the region between
350 ns and 400 ns, the slow μ^+ yield for solid argon, Fig. 2(a), was found
to be y = $(1.8 \pm 0.2) \times 10^{-6}$ per incident fast μ^+. By taking the experimental efficiencies into account, one obtains a yield y $\sim 10^{-5}$ slow μ^+
per incident μ^+. For deposition times on the order of a few minutes the
yield was observed to be a factor of $\sim$ 2 smaller, but the shape of the TOF
flight spectrum was indistinguishable (within statistics) from Fig. 2(a).
The yield was also observed to decrease as a function of time with a $\sim$ 30
minute time constant. The smaller amplitude observed in N(t) for krypton,
Fig. 2(b), at short times, as compared to that observed for argon, is
consistent with a weighting of the krypton energy distributions to lower
energies and the larger ionization potential of argon. The yield for solid
krypton was found to be y = $(8.9 \pm 0.4) \times 10^-$ per incident fast μ^+.

Solid xenon and neon were also measured and the data were found to be similar in shape to that observed for argon and krypton. In the case of xenon, the yield was found to be $(3.4 \pm 0.3) \times 10^{-7}$ per incident fast μ^+, while the yield for neon was $(1.2 \pm 0.3) \times 10^{-6}$ per incident fast μ^+. The yield for neon might have been substantially reduced by the presence of liquid helium in the reservoir and the associated large variations in stopping density. Indeed, the comparatively large band gap energy, smaller positron scattering cross section,[11] and the increasing trend on the slow μ^+ yield in going from xenon to argon would lead one to expect neon to have about twice the yield of the argon.

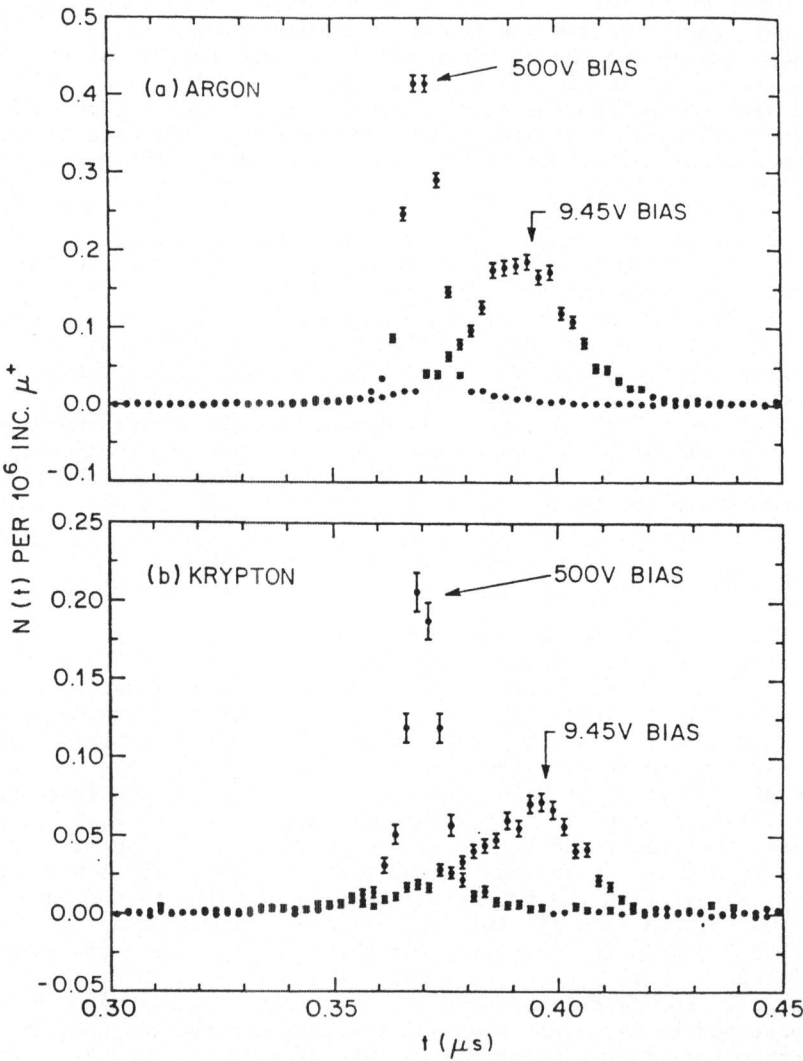

Figure 2.   Slow μ^+ time of flight spectra for (a) argon and (b) krypton, with the accelerating bias set at 500 volts (data peaked at early times) and 9.45 volts (data peaked at late times). The errors shown are purely statistical.

Energy distributions corresponding to the TOF spectra were calculated for argon and krypton by applying a Jacobian transformation $N(E_\perp) = -(dE_\perp/dt)^{-1} \cdot N(t)$ to the 9.45 volt bias data. Here, $E_\perp$ is the kinetic energy associated with the perpendicular component of velocity. The resulting distributions indicate a maximum in $N(E_\perp)$ at low energies (~ 5 eV), with a high-energy tail falling off monotonically. Unfortunately, the exact shape of the energy distributions cannot be established at this time, due to the imprecise determination of time-zero and the uncertainty in the gap measurement. Our best estimates of the energy spectra (corresponding to a 3.5 mm gap distance) are shown in Fig. 3. If one includes the systematics associated with the error in the gap distance, one finds that the position of the maximum in $N(E_\perp)$ can vary in the range 0 - 10 eV.

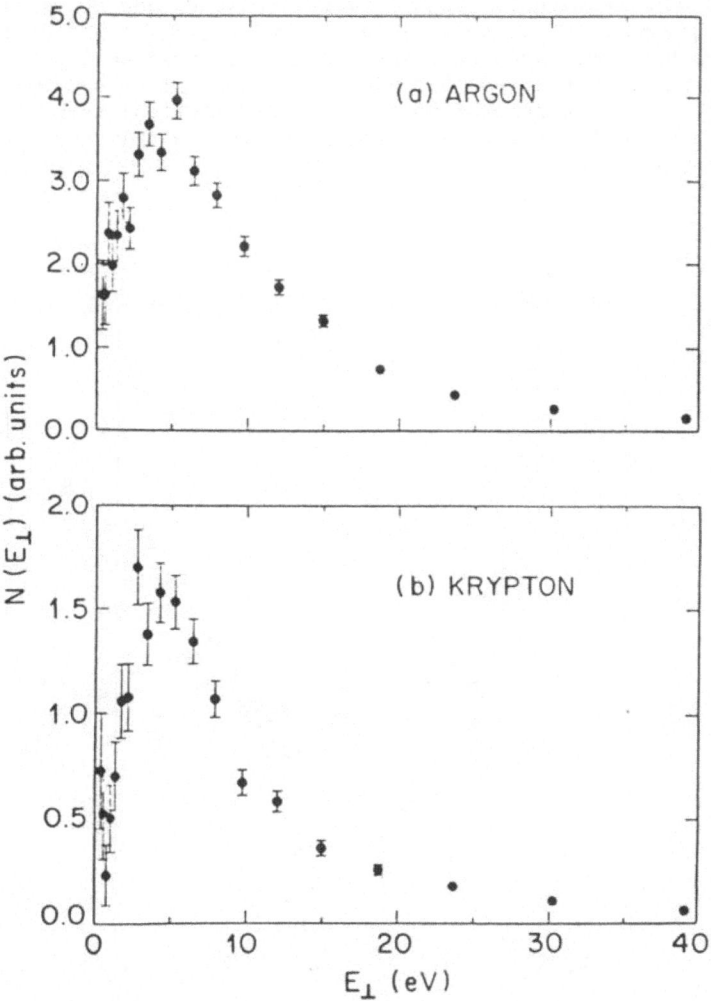

Figure 3.   Energy spectra estimated from the TOF data (9.45 volt bias) for (a) argon and (b) krypton. The errors in $N(E_\perp)$ shown are the transformed statistical errors. As mentioned, the distributions pictured here are subject to some uncertainty due to the imprecise gap measurement.

As mentioned, the slow μ^+ emission probability for argon depends on the rate at which the rare-gases are deposited and the age of the target. It is likely that contamination by background gases such as H_2 and CO is greater for the longer deposition time, and increases the density of trap sites. As well, the observed reduction in the yield with time after deposition is probably caused by surface contamination. An improved vacuum and a more stable cryogenic target would therefore be required before the true yield and energy spectrum could be determined.

Although no direct measurement of the polarization of the emitted μ^+ beam has been made to date, one would expect the initial polarization (i.e. $\sim$ 100%) to be largely retained if one accepts the hot muon representation.

Further possible candidates for slow μ^+ moderators are liquid He and high surface area targets such as ^{4}He-coated silica powder.[12,13] Since the electron affinities of liquid He and solid Ne are negative,[10] the affinity of Mu$^-$ could also be negative, possibly making such moderators efficient sources of slow Mu$^-$. Transverse-field SR measurements[12,13] of the muonium fraction have been made as a function of surface coverage for helium adsorbed on fine fine silica powder (at 6.0 ± 0.1 K). The relative muonium asymmetry (which is directly proportional to the muonium fraction) is plotted against the specific volume V_S of the ^{4}He adsorbed on to the silica surface in Fig.4.

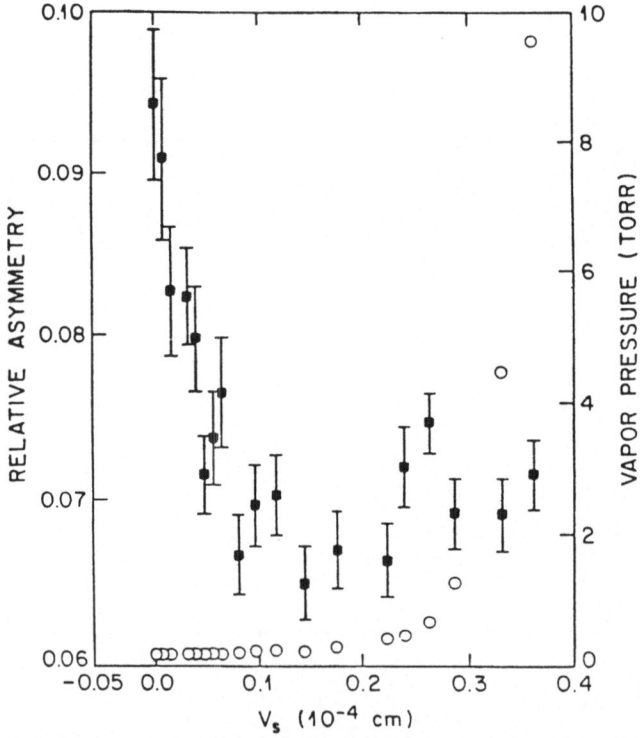

Figure 4.   Transverse field muonium asymmetry He coverage (closed squares) for SiO$_2$ powder. The corresponding vapor pressure data are represented by the open circles.

These data clearly show the muonium fraction to decrease with increasing surface coverage, suggesting that the charge exchange cross section is significant at the helium-silica interface. Unfortunately, it is impossible to say at this time what role the helium atoms play in the charge exchange process. One possibility is that the helium atoms are relatively passive and only serve to cover up areas on the surface, thereby impairing surface muonium formation. There is, however, an alternative possibility which casts the helium atoms in a more active role, where they might act to dissociate the muonium atoms at the surface, or perhaps facilitate the formation of Mu^- ions. Since electrons are known to be stabilized on helium films, the formation of Mu^- ions seems a likely possibility. In either case, the muons would be left in a diamagnetic state. Since the Mu^- affinity could be negative for such a surface, this type of moderator could produce slow Mu^- with an efficiency of a few percent.

REMODERATION OF SLOW POSITIVE MUONS

An argon yield of $y \approx 2 \times 10^{-5}$ corresponds to a hot μ^+ diffusion length of $\lambda = \Delta R \cdot y \approx 35 \text{Å}$ in solid argon, where $\Delta R \approx 175$ μm is the μ^+ range width. Using the number density $n = 2.66 \times 10^{22}$ cm^{-3} for the argon atoms and our inferred value of λ, we calculate a collision cross section of

$$\sigma_\mu = (n\lambda)^{-1} \sqrt{2M_{Ar}/3M_\mu} \approx 2 \times 10^{-15} \text{ cm}^2.$$

Here we estimate the number of collisions necessary to lose a significant amount of energy to be $\sim 2M_{Ar}/M_\mu$. Owing to the fact that under ideal conditions y may be larger, and since we are not accounting for Mu formation during thermalization, the σ_μ calculated here is an upper limit. At these low energies, σ_μ should be roughly the same as the zero-energy limit of the e^+ cross section, $\sigma_\mu(0)$. Extrapolating the measurements of Ref. 14 to zero momentum we find $\sigma_\mu(0) \approx 1.2 \times 10^{-15}$ cm^2, in rough agreement with our estimate for σ_μ. It is interesting that the much smaller zero-energy e^+ cross section[15] for Ne would imply a slow muon yield about five times larger than the Ar value. Since muons with tens of keV kinetic energy would have a range comparable to the estimated diffusion length, they could be brought to a sharp focus and remoderated[16] to obtain a brighter beam of slow muons with little loss of intensity. A remoderation chamber, with electrostatic injection and extraction optics, is currently being developed and will be situated at the end of the existing spectrometer shown in Fig. 1. The remoderated beam will be extracted from the incident surface (reflection geometry), and by tuning the incident beam energy in the range from 0 to 10 kV and the angle of incidence, one can investigate the angular distribution of the remoderated beam, the hot μ^+ diffusion length λ, and the branching ratios for μ^+, Mu, and Mu^- emission from several moderators.

POSSIBLE APPLICATIONS OF SLOW POSITIVE MUONS

If a moderator, such as that described here, were placed near the pion production target of a high flux facility, the slow μ^+ rate could be enhanced nearly one-hundred fold due to the increased efficiency for collecting the fast μ^+. The resulting beam of $\sim 10^3$ slow μ^+/s would have applications including surface diffusion studies,[14,15] electron density measurements near surfaces, microbeam formation,[16] muonium beam formation, and atomic physics experiments. Specifically, such a beam will allow one to investigate:

a) The quantum diffusion of a light interstitial in two dimensions.

b) The electron density at or near a surface, measured as a function of the implantation depth of the slow μ^+.

(c) Distortions in the muonium hyperfine interaction for adsorbed muonium atoms.

(d) Thermalization processes below 100 eV.

(e) Catalytic reactions on well characterized surfaces, utilizing muonium as a "light chemical isotope" of hydrogen.

(f) Fundamental problems in QED — A slow μ^+ beam could facilitate precise measurements of the muonium Lamb-Shift and is likely necessary for the production of "muium" ($\mu^+\mu^-$) in vacuum.

Of these, the eventual production of $\mu^+\mu^-$ in vacuum is of particular importance since it would provide a unique opportunity for QED studies, owing to the various decay channels available. Of course, one would also require an intense source of slow μ^- which might be accomplished through muonic atom (i.e., μ^- − H or μ^- − He) formation. The formation of muonic helium has already been observed[18] in a 14-atm. helium with 2% xenon admixture, but perhaps the yield could be enhanced using high surface area cryogenic targets.

ACKNOWLEDGEMENTS

The author would like to acknowledge the many helpful discussions with A.P. Mills, Jr., and the TRIUMF staff for technical support. Research at TRIUMF is supported by the Natural Sciences and Engineering Council of Canada and, through TRIUMF, by the Canadian National Research Council.

REFERENCES

1. See for example Positron Scattering in Gases, J.W. Humberston and M.R.C. McDowell, eds. (Plenum, New York, 1983).

2. T. Bowen, Physics Today 38:7,22 (1985).

3. D.R. Harshman, J.B. Warren, J.L. Beveridge, K.R. Kendall, R.F. Kiefl, C.J. Oram, A.P. Mills, Jr., W.S. Crane, A.S. Rupaal and J.H. Turner, Phys. Rev. Lett. 56:2850 (1986); D.R. Harshman, Invited Talk, 21 October 1985, Lawrence Livermore National Laboratory, Livermore, California, USA

4. D.R. Harshman, A.P. Mills, Jr., J.L. Beveridge, K.R. Kendal, G.D. Morris, M. Senba, J.B. Warren, A.S. Rupaal and J.H. Turner, Phys. Rev. Lett. (submitted)

5. A.P. Mills, Jr. and W.S. Crane, Phys. Rev. Lett. 53:2165 (1984).

6. E. Gullikson and A.P. Mills, Jr., Phys. Rev. Lett. 57:376(1986).

7. K.G. Lynn and B. Nielson, Phys. Rev. Lett. 58:81 (1987).

8. J.H. Brewer, K.M. Crowe, F.N. Gygax and A. Schenck, in Muon Physics, V.W. Hughes and C.S. Wu eds. (Academic, New York, 1975) Vol. III.

9. J.K.Berkowitz and J.C. Zorn, Phys. Rev. A29:611 (1974); H.H. Anderson and J.F. Ziegler, Hydrogen Stopping Powers and Ranges in All Elements, (Pergamon, New York, 1977); S.H. Overbury, P.F. Dittner, S. Datz and R.S. Thoe, Radiat. Eff. 41:219 (1979); E. Fermi and E. Teller, Phys. Rev. 72:399 (1944); P.M. Echenique, R.M. Neiminen and R.H. Ritchie, Solid State Commun. 37:799 (1981).

10. Rare Gas Solids, M.L. Klein and J.A. Venables, eds. (Academic, London, 1976). Rare-gas solids grown from the vapor phase are typically polycrystalline.

11. W.E. Kaupilla and T.S. Stein, in Ref. 1, p. 15.

12. D.R. Harshman, Ph.D. Thesis, Univ. of British Columbia (1986), available from University Microfilms, Ann Arbor, MI.

13. D.R. Harshman, Hyperfine Interactions, 32:847 (1986).

14. W.E. Kauppila, T.S. Stein and G. Jesion, Phys. Rev. Lett. 36:580 (1976).

15. R.P. McEachran, A.G. Ryman and A.D. Stauffer, J. Phys. B. At. Mol. Phys. 11:551 (1978).

16. A.P. Mills, Jr., Appl. Phys. 23:189 (1980).

17. K. Nagamine, Invited talk at IIIrd LAMPF II Workshop (Los Alamos, July 1983); A Ohsaki, T. Watanabe, K. Nagamine and I Iguchi, Abstract A62, IXth Int. Conf. on Atomic Phys. (Seattle, Wa., USA, July 1984); Q. Ma, X. Cheng, Z. Liu, Y. Liu and T. Watanabe, ibid, Abstract A63.

18. P.A. Souder, T.W. Crane, V.W. Hughes, D.C. Lu, H. Orth, H.W. Reist, M.H. Yam, G. zu Putlitz, Phys. Rev. A22:33 (1980).

THE INCLUSION OF THE CONTRIBUTIONS TO LOW ENERGY e^+H_2 SCATTERING AND
ANNIHILATION FROM THE LOWEST $\sum_u^+$ AND Π_u WAVES

E.A.G. Armour, D.J. Baker[+] and M. Plummer

Department of Mathematics
University of Nottingham
Nottingham NG7 2RD, England

We have shown[1] that the inclusion in the Kohn trial function of
Hylleraas-type functions containing the positron-electron distance as a
linear factor greatly increases the contribution to the total cross section
from the lowest partial wave below incident energies of about 2.5 eV.
However, this increase is insufficient to account for the experimental
value above about 1.5 eV.

Thus for incident energies above about 1.5 eV there must be appreciable
contributions to the total cross section from higher partial waves. In
positron or electron atom scattering the next partial wave is the P-wave.
However, the non-spherical hydrogen molecule splits the P-wave up into two
waves, the lowest partial wave of $\sum_u^+$ symmetry and the lowest partial wave
of Π_u symmetry.

For sufficiently low incident energies the phase shifts for these and
higher partial waves can be obtained from the Born approximation[2] using the
asymptotic form of the positron-hydrogen potential

$$V(r_3,\theta_3) \underset{r_3 \to \infty}{\sim} \frac{QP_2(\cos\theta_3)}{r_3^3} - \frac{\alpha_0}{2r_3^4} - \frac{\alpha_2 P_2(\cos\theta_3)}{2r_3^4} \tag{1}$$

where Q is the quadrupole moment of the hydrogen molecule and

$$\alpha_0 = \frac{1}{3}(\alpha_{11} + 2\alpha_\perp) \tag{2}$$

$$\alpha_2 = \frac{2}{3}(\alpha_{11} - \alpha_\perp) \tag{3}$$

α_{11} and $\alpha_\perp$ are the parallel and perpendicular polarisabilities,
respectively, of the hydrogen molecule. r_3, θ_3, ϕ_3 are the spherical polar
coordinates of the positron with respect to the midpoint of the line

joining the nuclei as origin and Z-axis directed along the intermolecular axis.

Outside the low energy region in which the Born approximation is applicable, the detailed interaction of the positron and the hydrogen molecule must be taken into account. We have done this by carrying out variational calculations using the generalised Kohn method as in the case of the lowest partial wave[1].

The calculations for the lowest $\sum_u^+$ wave are an extension of earlier calculations[3]. The trial function used is of the same form as for the lowest partial wave except the open-channel function is now appropriate to the lowest $\sum_u^+$ wave and the short-range correlation functions are chosen to be of $\sum_u^+$ symmetry. Thus Hylleraas-type functions are included in the basis set, as in the case of the lowest partial wave.

The trial function, Ψ, for the Π_u wave is of the form

$$\Psi = \Lambda(c,\lambda_3,\mu_3,\phi_3;\tau,a)\Psi_G + \sum_{i=1}^{M} g_i \chi_i \Psi_G \tag{4}$$

where $\Lambda(c,\lambda_3,\mu_3,\phi_3;\tau,a)$ is an open channel function appropriate to the lowest Π_u wave. The short-range correlation functions $\{\chi_i\}$ are of the form

$$\chi_i = (\lambda_1^{a_i}\lambda_2^{b_i}\mu_1^{c_i}\mu_2^{d_i}[M_{13}\cos(\phi_1-\phi_3)]^{p_i}[M_1\cos\phi_1]^{q_i}$$

$$+ \lambda_1^{b_i}\lambda_2^{a_i}\mu_1^{d_i}\mu_2^{c_i}[M_{23}\cos(\phi_2-\phi_3)]^{p_i}[M_2\cos\phi_2]^{q_i})$$

$$\times \exp[-\beta(\lambda_1+\lambda_2)]N\lambda_3^{r_i}\mu_3^{s_i}[M_3\cos\phi_3]^{t_i}e^{-\alpha\lambda_3} \tag{5}$$

where

$$M_i = [(\lambda_i^2-1)(1-\mu_i^2)]^{\frac{1}{2}} \tag{6}$$

and

$$M_{ij} = M_i M_j \tag{7}$$

a_i, b_i, c_i, d_i, p_i, q_i, r_i, s_i and t_i are non-negative integers and α, β and N are constants. The variables in Λ and $\{\chi_i\}$ are as described in the case of the lowest partial wave[1]. Three sets of values are used for p_i, q_i and t_i: $p_i = 0$ $q_i = 0$ $t_i = 1$, $p_i = 1$ $q_i = 0$ $t_i = 1$ and $p_i = 0$ $q_i = 1$ $t_i = 0$. Over 70 short-range correlation functions are used in the calculation.

Results show that the use of Hylleraas type functions significantly increase the calculated phase shifts and total cross sections for the $\sum_u^+$

wave. Long range polarisation effects, though improved, were not adequately taken into account, as shown by comparison with the Born approximation at low energies, and we are currently modifying the calculation to deal with this. Similar conclusions may be drawn for the Π_u wave calculation at low energies. However, results for this partial wave indicate that taking polarisation effects into account together with the inclusion of Hylleraas type functions in the calculation should, with the contributions from the Σ_g^+ and Σ_u^+ waves, account for the experimental results up to the positronium formation threshold. The contribution to Z_{eff}, the effective number of electrons available to the positron for annihilation, was very small for these partial waves compared to the Σ_g^+ wave contribution[1], but rose at higher energies. The Hylleraas type functions increased the contribution from the Σ_u^+ wave by a factor of between 1.5 and 2 across the energy range. However, it is still insignificant except at the upper end of the energy range.

† Present address: Rutherford Appleton Laboratory, Chilton, Didcot, Oxfordshire OX11 OQX, England.

REFERENCES

1.   E.A.G. Armour and D.J. Baker, Abstracts of the XVth ICPEAC (1987).

2.   N.F. Mott and H.S.W. Massey, The Theory of Atomic Collisions, Third Edition (Oxford University Press), p. 89.

3.   E.A. G. Armour, Proc. 3rd Int. Workshop on Positron (Electron) – Gas Scattering (Detroit) 1985 (World Scientific, Singapore) Invited paper p. 85.

POSITRONIUM FORMATION IN POSITRON-HYDROGEN SCATTERING AT INTERMEDIATE ENERGIES

D. Basu, Madhumita Basu and A.S. Ghosh

Indian Association for the Cultivation of Science

Theoretical Physics Department, Calcutta-32, INDIA

Recently, experiments[1-3] have been performed to measure the positronium (Ps) formation cross section in positron-atom and positron-molecule scattering. The results of the University College London group[1] differ drastically from those of the Texas group[2]. Moreover, the most recent findings of the Texas group[3] show that their results in the case of e^+- He scattering differ from first Born (FBA) predictions appreciably even at an incident positron energy of 250 eV. These observations have stimulated considerable theoretical activities[4-6] along this line.

The validity and suitability of the FBA for charge transfer processes is not assured even at high energies. Therefore, a discrepancy between the prediction of the FBA results and the measured data is not unlikely. In the case of e^+- H scattering, the second order contribution is required to establish the behaviour of the charge transfer cross section even at high energies. This suggests that the effect of double scattering may be important for electron capture by positrons. Moreover, the exotic positronium atom may also be formed in the continuum.

Here, in the present article, we investigate the influence of the second order contribution in the formation of Ps in positron-atom scattering. In particular, we have performed a model calculation for the formation of Ps in e^+- H scattering. The first order matrix element has been obtained by solving the coupled static equations. The second order matrix element has been evaluated by retaining one eigenstate (2s) and two pseudo-states (2p, 3s) as intermediate ones. The pseudo-state is an admixture of higher excited and continuum states. Therefore, the effects of the continuum have been partially taken into account by this model. Preliminary results are very encouraging.

REFERENCES

1. M. Charlton, G. Clark, T.C. Griffith and G.R. Heyland, J. Phys. B16:L465 (1983).

2. L.S. Fornari, L.M. Diana and P.G. Coleman, Phys. Rev. Lett. 51:2276 (1983).

3. L.M. Diana, P.G. Coleman, D.L. Brooks, P.K. Pendleto and D.M. Norman, (Private communication).

4. J.W. Humberston in Positron (Electron)-Gas Scattering edited by W.E. Kauppila, T.S. Stein and J.M. Wadehra (World Scientific, Singapore) (1986) p. 35.

5. P. Khan and A.S. Ghosh, Phys. Rev. A 27:1904 (1983).

6. P. Khan and A.S. Ghosh, Phys. Rev. A 28:2180 (1983).

EPITHERMAL POSITRON EFFECTS IN SURFACE MEASUREMENTS

D.T. Britton, P.C. Rice-Evans and J.H. Evans[*]

Physics Department, Rolay Holloway College
University of London
*Harwell Laboratory, UK Atomic Energy Auth., Oxon OX11 ORA

ABSTRACT: Positrons annihilating at a molybdenum surface have been studied with a beam apparatus. A new extended method of analysis has been employed to model the annihilation spectra as function of incident positron energy. The results indicate that epithermal emission can be accounted for by elastic scattering of positrons near the surface.

INTRODUCTION

When positrons are implanted into solids at low energies a large fraction may be reemitted before being fully reduced to thermal energies. As they pass backwards through the surface these epithermal positrons can pick up an electron and emerge as epithermal positronium (Howell et al. 1986). Indeed for a dirty surface this is usually the only form of positronium produced as more than 1 molecular overlayer suppresses thermal Ps production (Huomo et al. 1987).

Usually in positron solid-state profiling experiments epithermal emission is regarded as an unwanted complication, particularly in back-diffusion measurements. In their measurement on positron diffusion in molybdenum the Helsinki group found it necessary to ignore all data points up to an incident energy of 4 kV (Huomo et al. 1987). In this paper we suggest that the epithermal fraction itself also yields valuable information. This has important implications in a number of applications, for example in the development of moderators for slow positron and Ps production.

The experiment was performed with the Xenophon positron beam apparatus (Britton et al. 1985). The Doppler-broadening of the annihilation radiation from the target was measured with a germanium detector placed behind the target. The line-height parameter (F) is defined conventionally (Rice-Evans et al. 1978). For statistical reasons we find it sensible to use the running integrated difference (RID) method to determine the maximum change (Sr) in F with respect to a reference line-shape (the data for 12 keV incident positron energy).

POSITRON IMPLANTATION AND DIFFUSION

Monoenergetic positrons implanted into a surface have a stopping profile which is a derivative of a Gaussian (Makhov 1960, Vehanen et al. 1987):

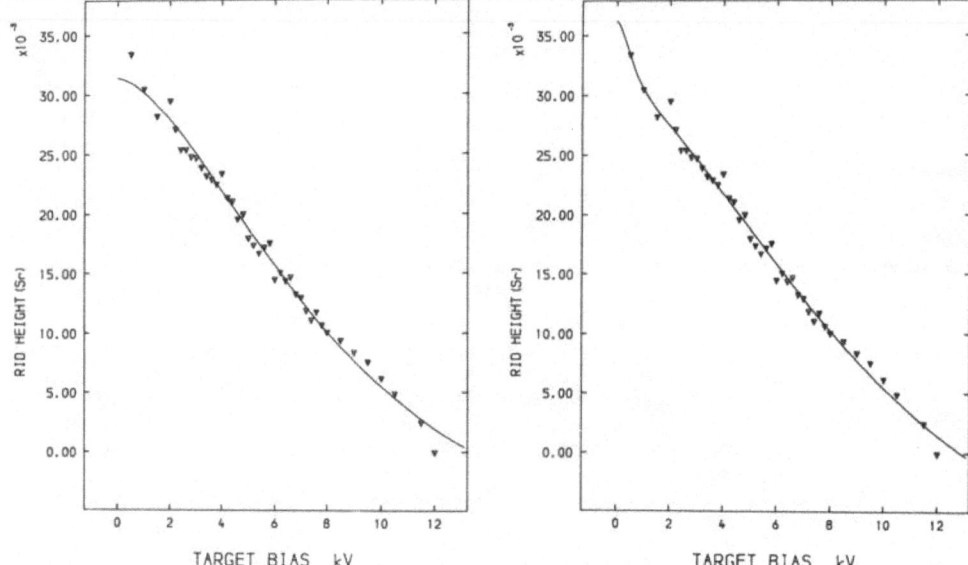

Fig.1.   Fit to data with no allow-
         ance for epithermal
         emission.

Fig. 2.   Improved fit by including
          allowance for epithermal
          emission.

$$P(z,E) = -\frac{d}{dz}e^{-\frac{z^m}{z_0^m}}$$

where

$$z_0 = \frac{\alpha}{\rho}E^n \ ,$$

ρ is the material density, E is the positron energy in keV, and z is the distance from the surface. If 'stopping' is taken to mean thermalisation then a positron has the probability of diffusing back to the surface given by the Laplace transform of the stopping profile,

$$J = \int_0^\infty P(z,E)e^{-z/L_+}dz$$

The one-dimensional positron diffusion length is related to the diffusion coefficient by $L_+ = \sqrt{\tau D_+}$ where τ is the free lifetime.

The positron lineshape parameter can then be fitted to a simple two state model for surface and bulk terms. In fitting to our data the values of α = 4.5μgcm^{-3}, m = 2 and n = 1.6 were used in the profile (Vehanen et al. 1987).

Incorporating an extra probability into the analysis and a third lineshape parameter value corresponding to the sum of all states occuring as a result of epithermal emission

$$F = J_{ep}F_{ep} + (1 - J_{ep})(J F_s + (1 - J)F_b)$$

where the epithermal probability is given by

$$J_{ep} = \int_0^\infty P(z,E)e^{-z/l_+}dz$$

if epithermal emission is due to elastic scattering with a mean scattering length l_+. This is not an unreasonable assumption if it is considered that the dominant effect in the implantation profile is multiple inelastic electron scattering at higher energies (Nieminen, 1983).

Fitting to lineshape parameter measurements on Mo yields a diffusion length of 1300 Å without considering the epithermal fraction. Including this extra term gives a longer diffusion length of 1600 Å and a scattering length of 20 Å. Close examination of the fits shows, incidently that the curves coincide in the range 4-10 kV with the epithermal model giving closer agreement with the data below 4 kV.

CONCLUDING REMARKS

This analysis indicates that epithermal emission of positrons is dominated by simple elastic scattering of positrons. The Makhovian profile is a valid model for the positron stopping profile if a 'stopped' positron is an epithermal positron with insufficient energy to excite an electron state.

The scattering length l_+ is equivalent to the mean free path of a hot positron in the material. As such, any structure smaller than this is unlikely to be resolved in a profiling experiment. This also gives an indication of the thickness of the overlayer which can be seen from line-shape parameter profiling. A layer less than the scattering length would only contribute to a 'surface state' and would not be seen independently. Doppler-broadening experiments are therefore useful in profiling measure-ments with dirty surfaces with coverages up to about 10 monolayers.

If ballistic positrons could be delivered to within the scattering length of the exit face of a moderator then epithermal emission would occur. For a moderator in backscatter mode this would imply a dense material with a long scattering length. Alternatively a composite of two materials, a dense substrate to stop the positrons with a lighter, low density (possibly non-metallic) coating would work. Such moderators, using MgO, were in use 15 years ago before the modern negative workfunc-tion moderators W and Cu became popular. With the right material a secondary reflection moderator could be used to give a well defined positronium beam.

REFERENCES

Britton, D.T., Rice-Evans, P. and Evans, J.H., Nuc.Inst.Meth., B12, 426, 1985.

Coleman, C.F., Appl.Phys. 19, 87, 1976.

Howell, R.H., Rosenberg, I.J. and Fluss, M.J., Phys.Rev.Lett. 34, 3069, 1986.

Huomo, H., Vehanen, A., Bentzon, M.D., Hautojärvi, P., Phys.Rev. B35, 8252, 1987.

Makhov, A.F., Sov.Phys.Solid State, 2, 1942, 1960.

Rice-Evans, P., Chaglar, I. and El Khangi, F., Phil.Mag. 38, 543, 1978.

Vehanen, A., Saarinen, K., Hautojärvi, P., Huomo, H., Phys.Rev.B 1987 in press.

A PROPOSED SOURCE OF ATOMIC HYDROGEN SUITABLE FOR THE MEASUREMENT OF

TOTAL AND PARTIAL SCATTERING CROSS SECTIONS

M. Charlton, N. Zafar, G. Laricchia, A.C.H. Smith and
S.J.B. Corrigan

Department of Physics and Astronomy, University College
London, Gower Street, London WC1E 6BT

Whilst extensive theoretical work on e^+-H scattering has been reported[1,2], there have been no experimental measurements on this system since it is difficult to produce atomic hydrogen in large quantities in a form suitable for scattering applications. With the advent of intense e^+ beams it is feasible to perform experiments using sources of H which have been applied to the scattering of electrons and protons[3]. Measurements of various differential scattering and total ionisation cross sections are planned at the Brookhaven beam[4]. In this contribution we wish to draw attention to another type of H source which may produce area densities $\simeq 10^{15}$ atoms cm^{-2} such that many cross sections can be measured using moderate intensity e^+ beams.

The design is based upon a d.c. operated Woods[5] tube as used recently by Schwab et al[6]. In that experiment, which studied high energy ($\simeq$ 5MeV) p^+ - H interactions, the proton beam was passed through the central region of the discharge tube itself where the dissociation fraction is highest ($\geq$ 95%). This is unsuitable for low energy ($\leq$ 10eV) e^+ beams as the tube may charge up and since the electric field across the discharge would change the energy of the e^+. To overcome this we have considered the effect of placing a small protrusion off the central portion of the discharge through which a e^+ beam could pass and such that a high density of H atoms could be maintained (see figure 1).

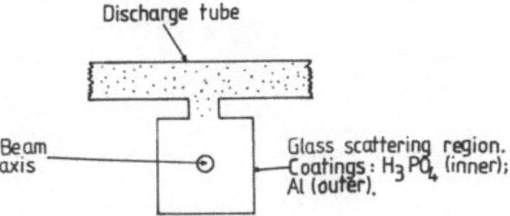

Figure 1: Schematic illustration of discharge tube and coated protrusion. Final details await experimental tests though the tube itself will be $\approx$ 1-1.5m long, at a pressure 0.06-0.6 mbar and immersed in a cooling arrangement held at 77K.

To achieve this it is necessary to find a material which can be used to coat the discharge tube which, (i) inhibits the recombination of H atoms and (ii) will prevent charging up. It has been known for some time that ortho-phosphoric acid ($H_3 PO_4$) satisfies the first criterion. Recent work at U.C.L. has shown that it also satisfies point (ii). Several experiments have been performed on pyrex tubes internally coated with this material. Firstly, passing a beam through an earthed $H_3 PO_4$ coated tube had no effect upon its energy spread as measured using a retarding grid arrangement. This, however, did not preclude that the beam may have undergone an energy shift in the tube which would not have been detected. Thus the tube was inserted into a time-of-flight arrangement such that the e^+ energy could be monitored over the entire flight path. No energy shift was detected for 10eV e^+. A beam of $\geq$ 3000 e^+s^{-1} was transported through this arrangement for several days and no effects of charging up were noted. A prototype discharge tube is nearing completion and testing will begin soon.

References

1. A.S. Ghosh, N.C. Sil and P. Mandal, Positron-atom and positron-molecule collisions, Phys. Rep. <u>87</u> (1982) 313
2. J.W. Humberston, Positronium - its formation and interaction with simple systems, Adv. At. Mol. Phys. <u>22</u> (1986) 1
3. J. Slevin and W. Stirling, Radio frequency atomic hydrogen beam source, Rev. Sci. Inst. <u>52</u> (1981) 1780.
4. Contributions by M.S. Lubell and G. Sinapius. This volume.
5. R.W. Wood, Spontaneous incandescence of substances in atomic hydrogen gas Proc. Roy. Soc. A. <u>97</u> (1920) 1.
6. W. Schwab, G.B. Baptista, E. Justiniano, R. Schuch, H. Vogt and E.W. Weber, Measurement of the total cross sections for electron capture of 2.0 - 7.5 MeV H^+ in H, H_2 and He, J. Phys. B. <u>20</u> (1987) 2825
7. eg. A. Ding, J. Karlau and J. Weise, Production of H-atom and O-atom beams by a cooled microwave discharge source. Rev. Sci. Inst. <u>48</u> (1977) 1002

ANGULAR CORRELATION STUDIES OF

POSITRON AND POSITRONIUM ANNIHILATION IN GASES

P.G. Coleman,* S.Rayner,* and R.N. West*
M. Charlton[+] and F. Jacobsen[+a]

*School of Mathematics and Physics
University of East Anglia, Norwich NR47TJ, U.K.

[+]Department of Physics and Astronomy
University College London, Gower Street, London WC1E6BT, U.K.

A series of angular correlation spectra for positrons annihilating in the noble gases and in gas mixtures has been obtained using the University of East Anglia two-dimensional angular correlation spectrometer.[1]   Beta positrons from a sodium-22 source are guided by a 0.8T magnetic field to the interaction region, which is first evacuated and then filled with gas at 1 bar pressure at room temperature. Gamma cameras placed at 3m on either side of the sample chamber record the two-photon annihilations of free positrons and positronium (Ps) atoms in a 9mm-diameter volume of gas at the centre of the cell. Events recorded include annihilation from the triplet (m = 0) - singlet state, which at 0.8T is 91% mixed and has a mean lifetime of 9.7ns. In contrast, a negligible number of two-photon decays are rec - orded from (triplet) o-Ps atoms quenched on the (distant) walls of the gas cell or through collisions with gas atoms within the viewed volume.

Typical spectra for helium and xenon are illustrated in Figure 1. The calibration is 0.5 mrad per channel and the system resolution,which is the measured FWHM of the Ps peak for a sample of quartz mounted in the evacu - ated sample chamber, is **5 channels (i.e. 2.5 mrad)**. The widths of the gas spectra increase monotonically with Z; this is attributed to the progressively less efficient slowing-down of the mixed-state Ps atoms prior to annihilation. In the case of helium the Ps atoms are slowed down through momentum transfer collisions with the light gas atoms to a final mean energy low enough to enable visual identification of a narrow component on the spectrum. This is not seen in any of the other spectra, and as such no direct method for obtaining Ps formation fractions in the heavier noble gases is offered by these measurements. These fractions have long been a focus of interest because of their unexpectedly low values.[2] However, some information may be gained from a full fitting analysis of the spectra; because of their symmetry they can be integrated to yield equivalent one - dimensional spectra from which, after deconvolution of the resolution function, effective momentum distributions may be derived.

Spectra were also recorded for mixtures of krypton and xenon with helium, in an attempt to identify Ps formation fractions in the heavy gases.

a) Present address: Institute of Physics, Aarhus University,
       DK-8000 Aarhus C, Denmark

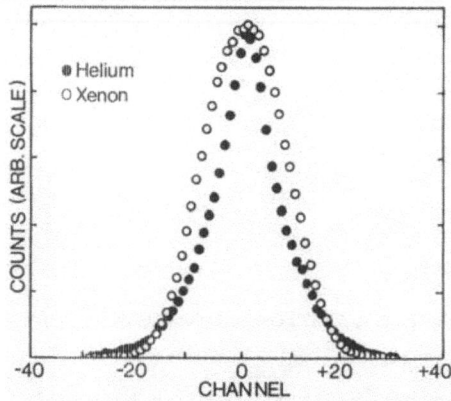

Fig. 1. Central sections through the two-
dimensional angular correlation
spectra for helium and xenon.

Because the Ore gaps of the mixed gases do not overlap, inelastic collisions
with krypton and xenon swamp Ps formation collisions in the helium Ore gap,
so that the helium essentially acts only as a moderator for the mixed-state
Ps formed in the heavier gases.

The derivation of effective momentum distributions should prove particularly fruitful in a comparative study of Ps slowing down in ^{3}He and ^{4}He.
In both these gases the Ps components may be resolved and the final Ps momentum distributions deduced with some confidence. Then, assuming that ^{3}He
and ^{4}He share a common Ps scattering cross section Q(E), and that the momentum transfer interactions in the two gases differ only as a result of the
different atomic masses, information on Q(E) can be deduced.

Full analyses of the noble gas series, and the ^{3}He-^{4}He spectra, are
currently in progress.

References

1. R.N. West, J. Mayers, and P.A. Walters
   J.Phys.E. 14, 478 (1981).

2. P.G. Coleman, T.C. Griffith, G.R. Heyland and T.L. Killeen
   J.Phys.B. 8, L185 (1975).

FORMATION OF ANTIHYDROGEN IN EXCITED STATES IN

ANTIPROTON-POSITRONIUM COLLISIONS

J.W. Darewych [+]

Department of Physics and Astronomy
University College London
Gower Street, London WC1E 6BT, U.K.

The possibility of producing antihydrogen, $\bar{H}$, in the laboratory has been discussed recently in the literature. (Humberston et al 1987 and references cited therein). The proposal is to collide slow (<100keV) antiprotons, $\bar{p}$, with positronium atoms, Ps, to produce antihydrogen through the reaction $\bar{p}$ + Ps → $\bar{H}$ + e$^-$. Humberston et al (1987) have determined the formation cross section for $\bar{H}$ in its ground state from the previously obtained, accurate calculations of the positronium formation cross section σ(e$^+$ + H(1s) → Ps(1s)+p) in low energy positron - hydrogen collisions (Humberston 1986 and references cited therein). These cross sections are related in the well known way:

$$\sigma(\bar{p}+Ps \rightarrow \bar{H}+e^-) = \sigma(p+Ps \rightarrow H+e^+) = k_e^2 \, \sigma(e^++H \rightarrow Ps+p)/k_{Ps}^2$$

Here k_e and k_{Ps} are the wavenumbers of e$^+$ and Ps respectively. In this way Humberston et al (1987) show that the cross section for the formation of antihydrogen in its ground state is substantial, being about $3\mathring{A}^2$ for antiproton energies <10keV.

We estimate the cross section for the production of antihydrogen in its lower excited states (2s, 2p, 3s, ...) using the Born approximation (BA), which is known to work reasonably well for the ground state (Humberston et al 1987). These excited state cross sections are shown to be of comparable size to the ground state ones, which increases the feasibility of antihydrogen formation in the laboratory. Partial s, p and d cross sections in the BA are presented for the various processes.

References

Humberston J.W. 1986 Adv. At. Mol. Phys. 22 1-36.

Humberston J.W., Charlton M., Jacobsen F.M. and Deutch B.I. 1987
        J. Phys. B: At. Mol. Phys. 20 L25-29.

+ Permanent address: Department of Physics, York University,
                        Downsview,
                        Toronto, M3J1P3
                        Canada.

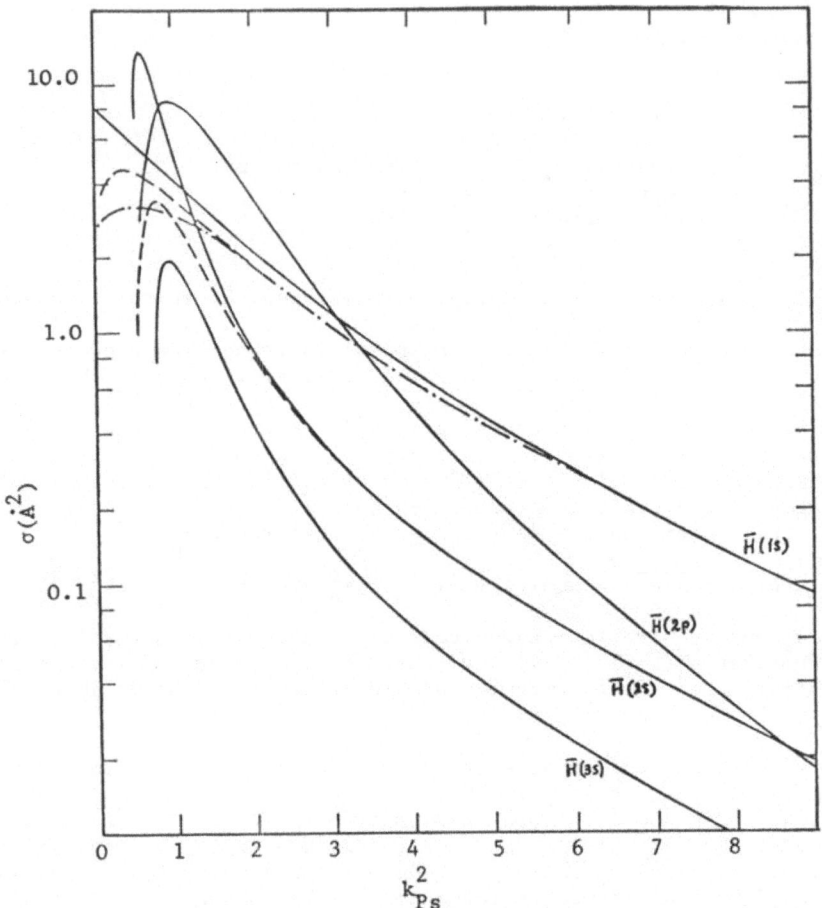

Fig. 1. Antihydrogen formation cross sections $\sigma(Ps(1s) + \bar{p} \to \bar{H}(n,\ell) + e^-)$. Solid curves: Born approximation. Dashed curves: s-wave-subtracted Born approximation. Dot-dashed curve: Humberston's (1987) 'exact' results.

THE EFFECT OF AN ELECTRIC FIELD ON FREE POSITRON ANNIHILATION IN ATOMIC AND MOLECULAR GASES

S.A. Davies, M. Charlton and T.C. Griffith

Department of Physics and Astronomy
University College London
Gower Street, London WC1E 6BT, U.K.

The conventional positron lifetime technique described by Charlton[1] has been used to investigate free positron annihilation in various atomic and molecular gases under the influence of a static electric field. A uniform field was generated using the enclosed ring system designed by Charlton and Curry[2], consisting of two Al end plates, one held at earth potential and one at $+V$, linked by a resistive chain of Al rings and insulating spacers. A 10 μCi ^{22}Na source was mounted centrally on the earthed end plate and the enclosure was mounted inside a pressure vessel capable of withstanding up to 40 atmospheres of gas.

The motivation behind this work was to test recent theoretical calculations[3] and to provide further insight into the diffusion and annihilation mechanisms of free positrons in gases, since such experiments ultimately yield information on positron momentum transfer cross sections in an energy region which is currently inaccessible to slow positron beams.

In a lifetime spectrum, the equilibrium region of free positron annihilation is characterised by a decaying exponential with annihilation rate $\bar{\lambda}_f$, defined by

$$\bar{\lambda}_f = \int_0^\infty y(v) \, \lambda_f(v) \, dv \Big/ \int_0^\infty y(v) \, dv \qquad (1)$$

where $y(v)$ is the equilibrium positron velocity distribution. Canter[4] has shown that the application of an electric field effectively "heats up" the free positrons, causing them to attain equilibrium at a higher energy than in the zero field case where they thermalise with the gas. An increase in positron energy results in a corresponding velocity increase; various authors, for example[3,4], have shown that $\lambda(v)$ is reduced as v is increased, so when a field is applied, $\lambda(v)$ will be less than in the zero field case, resulting in a decrease in $\bar{\lambda}_f$ according to (1).

Measurements were performed in room temperature He at densities of 3.5 and 36 amagats (1 amagat $\equiv 2.69 \times 10^{25}$ atoms (molecules) m^{-3}) to obtain $\bar{\lambda}_f$ at electric field strengths of $0 < \varepsilon_n < 11$ Vcm^{-1} amagat^{-1}. This quantity was then used to calculate Z_{eff}, the effective number of atomic (molecular) electrons available to a positron for annihilation in the gas according to

$$\bar{\lambda}_f = \pi \, r_o^2 \, cn \, \bar{Z}_{eff} \qquad (2)$$

where $\pi r_o^2 cn$ is the Dirac annihilation rate for a free electron gas; r_o being the classical electron radius, c the velocity of light and n the number of gas atoms (molecules).

At 3.5 amagats, agreement between the measured value of $\bar{Z}_{eff}$ as a function of ε_n and the theoretical calculation of Campeanu and Humberston[3] was very good (Figure 1), whereas at 36 amagat it can be seen that the minimum value of $\bar{Z}_{eff}$ was considerably higher than predicted theoretically. Other workers[5,6], have performed similar experiments in high density He which compare favourably with theory for $\varepsilon_n > 0$, but their zero field values of $\bar{Z}_{eff}$ were too low, so this agreement may be fortuitous and is probably unreliable. The existence of a density effect, although not predicted theoretically, was also present in the results obtained for Ar.

Similar measurements were performed in H_2, CH_4, CO_2 and N_2 at various densities. For these gases, no density effect was observed, and over the range of field strengths studied $\bar{Z}_{eff}$ was found to decrease linearly as ε_n was increased from zero. The results for these gases will be published later.

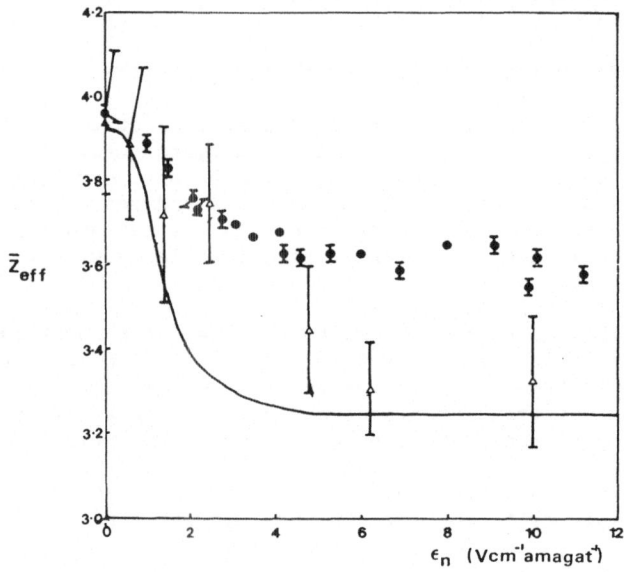

Figure 1. $\bar{Z}_{eff}$ versus ε_n for positrons in He.

REFERENCES

1. M. Charlton, Rep. Prog. Phys. 48:737 (1985).
2. M. Charlton and P.J. Curry, Nuovo Cimento 6:17 (1985).
3. R.I. Campeanu and J.W. Humberston, J. Phys.B 10:239 (1977).
4. K.F. Canter, Contemp. Phys. 13:457 (1972),
5. G.F. Lee, P.H.R. Orth and G. Jones, Phys. Lett. 28A:674 (1969).
6. C.Y. Leung and D.A.L. Paul, J. Phys.B 2:1278 (1969).

SLOW POSITRON SETUP AT THE GIESSEN 65 MeV LINAC

F. Ebel, W. Faust, C. Hahn, S. Langer,
H. Schneider, and A. Singe

Abt. Angew. Kernphysik, Strahlenzentrum der
Justus-Liebig-Universität Giessen
D-6300 Giessen, Leihgesterner Weg 217
W.-Germany

In our experiment the pulsed LINAC of the Giessen University,
Strahlenzentrum, is used to produce the slow positrons. The
LINAC is characterized by a maximum electron energy $E(e^-)$ of
65 MeV, an average current of $\bar{i}$ = 240 µA at 28 MeV, pulse
lengths T ranging from 6 ns to 2 µs, and a repetition rate ν
of max. 600 Hz.
The experimental arrangement (Fig. 1) consists of the water-
cooled tungsten electron-positron converter target, and the
tungsten moderator in a vacuum-chamber, a solenoid as slow
positron transport system, and different detector systems to
registrate the slow positrons in the experimental hall.

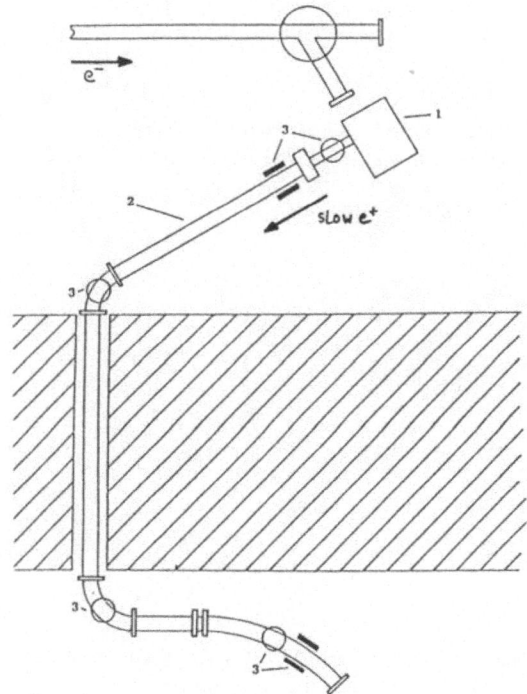

Fig. 1.

Schematic experimental setup

1: Vacuum chamber

2: Solenoid transport system

3: Helmholtz coils

A slow positron-production rate of $\simeq 10^8$ e$^+$/s was measured. The electron beam parameters were 37 MeV and 120 µA average current.
In Fig. 2 the energy distribution of the slow positron beam is shown. The beam current was measured by means of a channel plate-detector, using a retarding grid-method. For E(e$^+$) = 300 eV we obtained $\Delta E/E$ = 6.33 %.

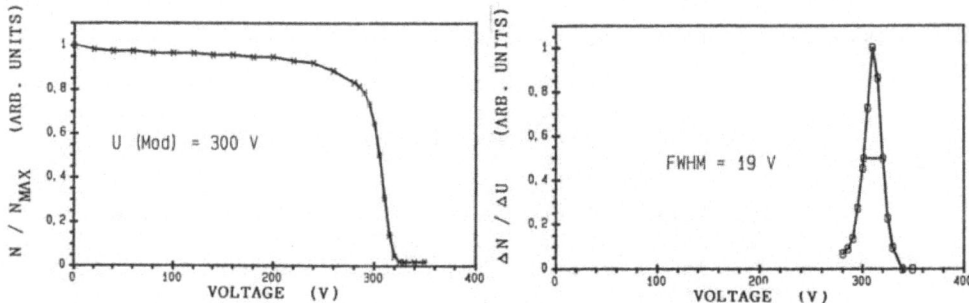

Fig. 2a.

Normalized number of slow positrons as a function of the repeller-grid voltage.

Fig 2b.

The distribution obtained by differentiation of the function in a).

In Fig. 3 the number of slow positrons at the end of the beam transport system, as a function of a diaphragm with various diameters ranging from 5 mm to 50 mm, is given.

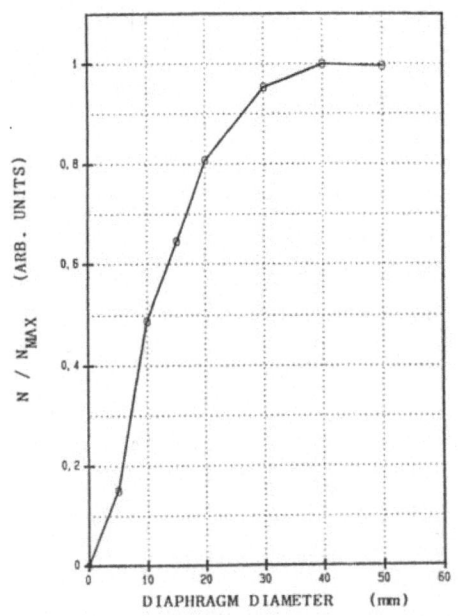

Fig. 3.

Number of slow positrons per second as a function of a diaphragm mounted right behind the 90° bend

We thank the Deutsche Forschungsgemeinschaft, Bonn-Bad Godes-berg, W. Germany, for the generous financial support.

INTERACTIONS BETWEEN $\bar{P}$ AND H AT INTERMEDIATE ANTIPROTON ENERGIES

A.M. Ermolaev

Department of Physics, Durham University, Science Laboratories

South Road, Durham, DH1 3LE, England

Cross sections for direct excitation and ionisation in slow $\bar{p}$ + H collisions (at few keV lab),

$$\bar{p} + H(1s) \rightarrow \bar{p} + H^*(n\ell m) \tag{1}$$

and

$$\bar{p} + H(1s) \rightarrow \bar{p} + p + e^- , \tag{2}$$

are currently of particular interest in view of the recent proposal (1) to produce antihydrogen, $\bar{H}$, under laboratory conditions. It has been pointed out (2) that cross sections for formation of $\bar{H}$ via the capture of positron by antiproton $\bar{p}$ in collisions between $\bar{p}$ and positronium, Ps, are likely to be larger by several orders of magnitude than the cross sections for radiative capture to $\bar{H}$ in the $\bar{p}$ + Ps collisions. The Born cross sections for formation of $\bar{H}$ via positron capture in $\bar{p}$ + Ps collisions computed by Darewych (3) and semiclassical and classical calculations (4) of the same process point at rather large (geometrical size) cross sections for the formation of $\bar{H}$ at low collision energies, thus giving additional support to the antihydrogen project. In these newly proposed experiments, the major source of background may be reactions between the trapped keV beam of antiprotons and the remnant hydrogen gas H_2 in the vacuum chamber. It is likely that this reaction can be estimated by considering a simpler process of interactions between antiproton and atomic hydrogen. For energies E > 100 keV lab (E is impact energy in the laboratory system where H is assumed to be at rest), cross sections for incident $\bar{p}$ are close to those for incident p. However this may not be the case for lower energies, hence the current interest in reactions (1) and (2) at E < 100 keV lab.

Reactions (1) and (2) are also of general interest due to the peculiarity of the $\bar{p}$ + H system with a 'negative coupling constant' regime. For E > 30 keV lab, this system was studied earlier by Reading and co-workers (5) using the semiclassical (impact parameter) method and expanding the electronic wavefunction $\Psi(\underline{r},t)$ in terms of a single-centre (SC) atomic orbital (AO) basis of some 68 pseudostates with $\ell \lesssim 3$. They found that ionisation cross sections for reaction (2) with both $\bar{p}$ and p projectiles, were close to each other if E $\gtrsim$ 60 keV lab. For the lowest energy 30 keV lab they treated, the difference was some 20 - 30 per cent. Their calculations based on the SC expansion did not converge fast enough for E $\lesssim$ 30 keV lab so that the low E region was omitted from the previous analysis.

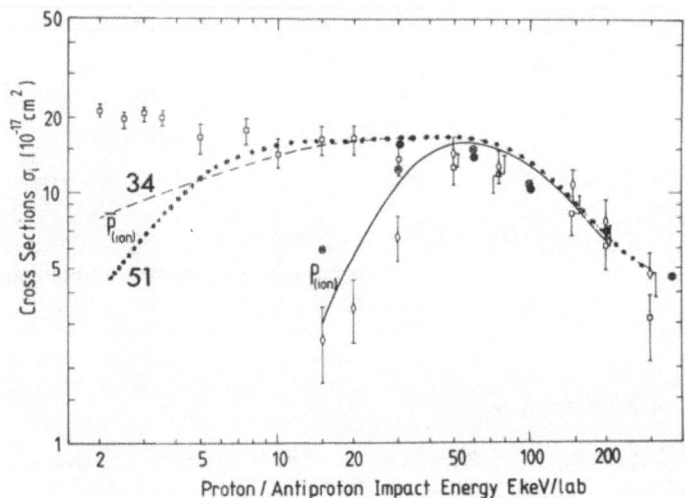

Figure 1. Computed cross sections for ionisation of atomic hydrogen by proton/antiproton impact (all cross sections are in units of 10^{-17} cm^2). **Proton projectile:** ———— , TC close-coupled, 70 states, Shakeshaft [6]; ⊗ , one-and-a-half centre expansion, 54 states, Reading et al. [7]; ⟨̷⟩ , CTMC, present work. **Antiproton projectile:** ••••• , 51, TC close-coupled, one pseudostate on p̄ and 50 states on H, present work; — — — , 34, TC close-coupled, four pseudostates on p̄ and 30 states on H, present work; ● , SC close-coupled, 68 states on H̄, Martir et al. [5]; ⊟̷ , CTMC, present work.

In the present work, the author has carried out impact parameter calculations of reactions (1) and (2) using a two-centre (TC) AO basis and numerical codes [8] developed earlier. The actual calculations were simplified considerably by employing an 'asymmetric basis' with a single (positive energy) $l = 0$ pseudostate centred on the antiproton, and a set of 50 states with $l \leqslant 3$, centred on the hydrogen target. This basis was somewhat smaller than that used in [5] but it had an important advantage of allowing the direct account of 'capture' into the antiproton continuum which effectively represents 'anticorrelation' existing between the motion of the atomic electron and incident p̄. In order to investigate the numerical convergence of such an asymmetric expansion, a second set of calculations has been carried out using a basis with four pseudostates (two states with $l = 0$ and two states with $l = 1$) on the antiproton centre. This second basis was tested for $E \lesssim 20$ keV lab where the proton-centred states with high positive eigen-energies are relatively unimportant. Consequently, the target-centred part of the original basis was reduced from 50 to 30 states so that the second set of calculations was carried out using the 34 state basis.

In addition to the semiclassical (impact parameter) calculations, this author has carried out two sets of classical trajectory Monte-Carlo (CTMC) calculations of reactions (1) and (2), for p̄ and p as projectiles. The classical cross sections were obtained using the CTMC code [9] with some 3000-6000 trajectory runs for each impact energy. In both methods, the full range of E treated was from 2 through 300 keV lab.

The computed cross sections are presented in Figure 1 and in tables 1 and 2. Both methods give ionisation cross sections which are remarkably different for p̄ and p at energies below 50 keV lab. This shows clearly that estimates of the antiproton cross sections which use the proton data become completely unreliable for E as low as 2 to 10 keV lab. On the other hand,

Table 1. Computed cross sections for ionisation of atomic hydrogen by proton/antiproton impact (all cross sections are in units of 10^{-17} cm^2).

E keV lab	$\bar{p}$ ai	$\bar{p}$ bi	E keV lab	$\bar{p}$ ai	$\bar{p}$ bi	p bc	p bi
2.0	3.5	22.5 ± 2.2	15.0	15.5	16.3 ± 2.2	33.5 ± 3.0	2.6 ± 0.9
2.5		20.1 ± 2.1	20.0	15.0	16.3 ± 2.1	34.7 ± 3.1	3.5 ± 1.0
3.0	9.0	21.9 ± 2.2	30.0	15.7	13.7 ± 2.0	28.6 ± 2.8	6.7 ± 1.4
3.5		20.7 ± 2.1	50.0	10.6	9.6 ± 1.7	8.8 ± 2.8	14.9 ± 2.0
5.0	11.6	16.5 ± 2.2	75.0	10.9	10.2 ± 1.6	2.9 ± 0.9	12.8 ± 1.9
7.5	15.0	17.8 ± 2.2	145.0	9.5	8.2 ± 1.5	0.30 ± 0.25	11.1 ± 1.8
10.0	14.9	14.6 ± 2.0	200.0	7.0	6.1 ± 1.3	0.28 ± 0.25	7.9 ± 1.5
			300.0	5.0	4.8 ± 1.0		

a = TC close-coupled, using the 51 state AO basis.
b = CTMC.
i = ionisation.
c = capture.

Table 2. Computed cross sections for direct excitation of atomic hydrogen by antiproton impact *) (all cross sections are in units of 10^{-17} cm^2).

E keV lab	nl	$H^*(nl)$ 2s	2p	3s	3p	3d
2.0		0.43	1.25	0.33	0.44	0.56
5.0		1.58	3.80	0.66	0.49	1.57
7.5		1.51	4.55	0.31	0.75	1.38
10.0		1.38	5.13	0.16	0.89	1.31
15.0		1.29	5.88	0.17	0.94	1.23
20.0		1.11	6.85	0.12	1.11	0.86
30.0		1.16	7.78	0.14	1.28	0.58
50.0		0.99	8.54	0.10	1.40	0.36
75.0		0.70	7.47	0.08	1.44	0.27
145.0		0.53	6.68	0.06	1.16	0.09
200.0		0.37	6.50	0.04	1.00	0.07
300.0		0.33	5.85	0.04	0.94	0.05

*) using the 51 state AO basis, semiclassical method.

the agreement between semiclassical and classical results is good everywhere with an exception of the low-energy range 2 - 5 keV lab. This confirms that the difference between the $\bar{p}$ and p cross sections at $E \lesssim 50$ keV lab is due to the opposite charges of the projectiles rather than the models used.

As Figure 1 shows, the anticorrelation effects are particularly strong for E below 5 keV lab. For such low energies it is not sufficient to use, in semiclassical calculations, a basis with a single $\ell = 0$ pseudostate centred on $\bar{p}$ (as i.e. done in the present 51 state calculation). The additional projectile-centred pseudostates with $\ell = 0$, and 1, increase the ionisation cross section by a factor of two at 2 keV lab. It is also very significant that this brings the low energy semiclassical and classical ionisation cross sections in better correspondence with each other.

Table 2 shows that as in the case of proton-hydrogen collisions [6], direct excitation (1) is dominated by the 1s $\rightarrow$ 2p transition which attains a maximum of $\sim 8.5 \times 10^{-17}$ cm^2 at E ~ 50 keV lab. For E as low as 15 keV lab, cross sections for this transition as well as those for excitation of the 3p level, are larger by a factor of two than corresponding cross sections for excitation of H by proton impact. However, for higher energies, this agreement is better: it is within 20 per cent at E = 40 keV lab, and within 6 per cent at E = 75 keV lab.

The present work suggests that $\sigma(\text{ion}) \sim 1.0 - 1.5 \times 10^{-16}$ cm^2 is probably a reasonable estimate of cross sections for the antiproton impact ionisation of atomic hydrogen in the low-energy range 2 - 10 keV lab. This cross section must be compared with the estimate of 1.0 - 2.0 $\times 10^{-15}$ cm^2 for antihydrogen formation in p + Ps collisions at the same energy as shown in [4]. One may conclude that the reactions (1) and (2) are not entirely negligible in comparison with the basis reaction $\bar{p} + \text{Ps} \rightarrow \bar{H} + e^-$ for antihydrogen production.

This work was supported by a grant from SERC of U.K. The author is pleased to acknoweledge a discussion with Prof. J.F. Reading of his antiproton results and important comments of Dr M Charlton on the experimental requirments. The reported calculations were carried out on the Amdahl computers of the Universities of Durham and Newcastle-upon-Tyne, U.K.

References

1. Deutch, B.I., Jacobsen, F.M., Andersen, L.H., Hvelplund, P., Knudsen, H., Holzscheiter, M., Charlton, M., and Laricchia , G., in 'Workshop and Symposium on the Physics of Low-Energy Stored and Trapped Particles (AFI, Stockholm 14-18 June, 1987), to be published.
2. Humberston, J.W., Charlton, M., Jacobsen, F.M., and Deutch, I. 1987 J. Phys. B: At. Mol. Phys. 20 L25-29.
3. Darewych, J.W. 1987, this workshop.
4. Ermolaev, A.M., Bransden, B.H., and Mandal, C.R. 1987, this workshop.
5. Martir, M.H., Ford, A.L., Reading, J.F., and Becker, R.L. 1982 J. Phys. B: At. Mol. Phys. 15 1729-47.
6. Shakeshaft, R. 1978 Phys. Rev. A18 1930-34.
7. Reading, J.F., Ford, A.L., and Becker, R.L. 1981 J. Phys. B: At. Mol. Phys. 14 1995-2012.
8. Ermolaev, A.M. 1984 J. Phys. B: At. Mol. Phys. 17 1069-81.
9. Peach, G., Willis, S.L., and McDowell, M.R.C. 1985 J. Phys. B: At. Mol. Phys. 18 -3921-37.

FORMATION OF ANTIHYDROGEN AND DESTRUCTION OF POSITRONIUM IN $\bar{P}$ + Ps

COLLISIONS IN THE ANTIPROTON ENERGY RANGE 2 - 100 keV

A.M. Ermolaev, B.H. Bransden and C.R. Mandal

Department of Physics, Durham University, Science Laboratories

South Road, Durham DH1 3LE, England

We present cross sections for the formation of antihydrogen, $\bar{H}$, and the destruction of the positronium atom, Ps, in collisions between anti-proton $\bar{p}$ and the positronium target:

$$\bar{p} + Ps(1s) \rightarrow \bar{H}(n\ell m) + e^- \tag{1}$$
and
$$\bar{p} + Ps(1s) \rightarrow \bar{p} + e^+ + e^- , \tag{2}$$

in the energy range 2 < E < 100 keV lab (E is impact energy in the labora-tory system where the positronium target is assumed at rest). Reaction (1) is of particular interest because it can be used [1] for production of antihydrogen in a laboratory, provided that cross sections for (1) are not too small at impact energies of few keV (lab).

First estimates of cross sections for reaction (1) were obtained by considering the time-reverse, charge-conjugate, reaction of positronium formation:

$$e^+ + H(1s) \rightarrow p + Ps(n\ell m) . \tag{3}$$

This reaction was the subject of earlier investigations by Humberston and co-workers [2] in a series of accurate variational calculations. Using these results and detailed balancing, they obtained [3] reliable estimates of the partial cross section for the formation of $\bar{H}(1s)$ at E $\lesssim$ 10 keV lab. The largest cross section of $\sim 3.2 \times 10^{-16}$ cm^2 was found at E = 2 - 3 keV lab. However, this contribution is expected to be only a (small) part of the total cross section for antihydrogen formation (1).

In the present work, we treat the problem in two different ways: (i) in the semiclassical (impact parameter) formulation and (ii) by using the classical trajectory Monte-Carlo (CTMC) model.

The use of a semiclassical model for reaction (1) needs particular justification because one of the colliding partners (positronium) is a light particle, the ratio of m_e/m_{Ps} is of the order of unity. Earlier Martir et al. [4] showed that the impact parameter method could produce reasonably accurate cross sections for electron impact excitation of atomic hydrogen down to energies at least as low as 20 eV. This suggests that semi-classical treatment should be useful in the present case. However, there exists an additional formal difficulty in such a treatment of the rearange-

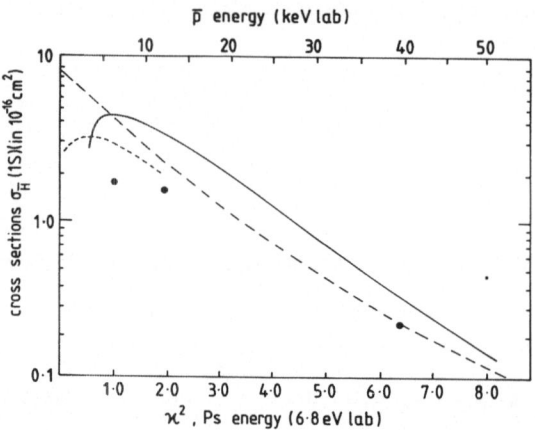

Figure 1. Computed cross sections for antihydrogen formation, eq. (1).
The partial cross sections for $\overline{\text{H}}$(1s) in units of 10^{-16} cm^2. ----- variational
calculations of Humberston et al. $\begin{pmatrix}3\end{pmatrix}$; ● - distorted wave approximation
of Shakeshaft and Wadera $\begin{pmatrix}5\end{pmatrix}$; — — — - first Born approximation of
Darewych $\begin{pmatrix}6\end{pmatrix}$; solid curve - semiclassical approximation, present work.

ment process (1) because the mass is different in the initial and final
channels, a feature which does not appear in the usual impact parameter
formulation for heavy particles. In the latter case, the momentum transfer
is fully taken into consideration by using plane-wave electronic translational factors (PWETFs) whereas in the present case this effect is difficult
to allow for correctly. It is well known that the neglect of the PWETFs
in the expansion of the wavefunction results in a gross overestimation of
the exchange matrix elements and, therefore, the rearrangement cross
sections at high E. On the other hand, the PWETFs are of lesser importance
if the collision takes place at low velocity and we expect the impact
parameter model to give a sensible estimate of the formation cross sections
at low energies.

Our impact parameter model used a two-centre (TC) atomic orbital (AO)
basis with 24 states and calculations were carried out with the help of
numerical codes $\begin{pmatrix}7\end{pmatrix}$. 7 states of the basis with $\ell \leqslant 2$ were in the positronium arrangement and 17 states with $\ell \leqslant 2$ were on $\overline{\text{p}}$. In Figure 1 our
partial cross sections for $\overline{\text{H}}$(1s) are compared with other data available,
including the accurate estimates of Humberston et al. $\begin{pmatrix}3\end{pmatrix}$. We see that
the impact parameter method overestimates by some 35 per cent and the
distorted wave approximation underestimates by 35 per cent the formation
cross sections at the maximum. We shall also note that the position of the
maximum predicted by both approximate methods is shifted towards higher E
comparing with the exact theory. The first Born cross sections of Darewych
$\begin{pmatrix}6\end{pmatrix}$ which are in agreement with earlier calculations of Omidvar $\begin{pmatrix}8\end{pmatrix}$, rise
steadily as E decreases and overestimate considerably at low E. For higher
E, the Born and distorted wave approximation converge numerically to the
same result whereas the impact parameter method overestimates. This is in
accord with the general discussion given above.

Figure 2 presents the total cross sections for the formation of anti-
hydrogen(numerical values of partial cross sections are in table 1). One can
see that in the general case the relation between the impact parameter and
Born rearrangement cross sections is similar to the case of $\overline{\text{H}}$(1s): the
Born cross sections are much higher than the impact parameter data at low
E, and the impact parameter cross sections are too large at high E.

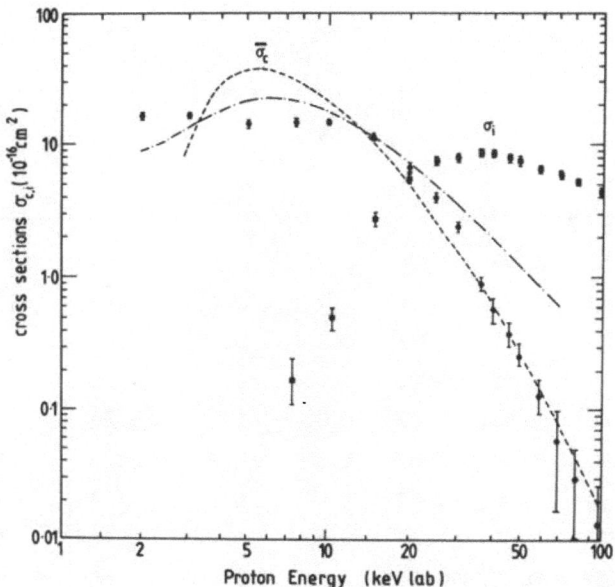

Figure 2. Computed total cross sections for antihydrogen formation $(\overline{\sigma}_c)$ and for the destruction of positronium (σ_i) in $\overline{p}$ + Ps collisions (in units of 10^{-16} cm^2). <u>Formation of antihydrogen:</u> —·—·— , impact parameter method, TC, close-coupled, 24 states, present work; ⬤ - CTMC, present work; -------- , Born approximation, Darewych [6] . <u>Destruction of positronium:</u> ▆ - CTMC, present work.

As an additional test, we have used the classical Monte-Carlo model for reactions (1) and (2). In this model, all conservation laws are satisfied exactly. The CTMC calculations used the computer code [9]. The results are displayed in Figure 2 and table 2. We found a good correspondence between CTMC and impact parameter cross sections for reaction (1) at low E, between 2 and 15 keV lab. For higher E, above 40 keV lab, the agreement between the CTMC and Born formation cross sections was excellent. Within an intermediate range, 15 < E < 40 keV lab, the correspondence between the Born and CTMC cross sections for (1) was fare.

The cross sections for the destruction of Ps are listed in table 2 and shown in Figure 2. Our CTMC model gives the cross sections below 0.5×10^{-16} cm^2 for E $\lesssim$ 10 keV lab. The maximum of $\sim 10 \times 10^{-16}$ cm^2 lies at about 40 keV lab.

Our main conclusion is that the total cross section for antihydrogen production, equation (1), is 10 – 20 $\times 10^{-16}$ cm^2 in the interval from 2 to 10 keV lab relevant to the antihydrogen project.

Finally we shall emphasise that the present semiclassical theory is not expected to provide very accurate cross sections for reaction (1). Nevertheless, we believe that it is good enough to be useful in obtaining an orientation in the present case, although more elaborate quantal calculations will be required to produce very accurate cross sections for antihydrogen formation at a later date.

This work was supported in part by grants from SERC of U.K. The authors thank Prof. J.W. Darewych for allowing to use his Born data prior to publication, Dr J.W. Humberston for sending us the unpublished data of Dr K. Omidvar, and Dr M. Charlton for his discussion of the experimental requirements. The reported calculations were carried out on the Amdahl

Table 1. Computed cross sections for the formation of $\overline{H}$ in collisions between $\overline{p}$ and Ps (semiclassical method, all cross sections are in units of 10^{-16} cm^2).

E keV lab	$\overline{H}(n\ell)$ 1s	2s	2p	3s	3p	3d	$\overline{\sigma}_c$ Total
2.0	0.157	2.73	4.99	0.989	0.104	0.118	9.09
3.0	0.387	5.24	6.41	0.782	0.433	0.233	13.5
4.0	1.84	7.00	7.31	0.548	0.740	0.508	17.9
5.0	4.35	7.75	7.40	0.790	1.08	0.760	22.1
7.5	4.00	6.68	5.90	0.86	1.72	1.41	20.6
10.0	4.03	4.30	3.63	1.20	1.94	1.73	16.8
15.0	3.22	2.08	2.42	1.67	1.62	1.21	12.2
25.0	1.50	0.57	1.57	0.42	0.714	0.269	5.04
50.0	0.165	0.278	0.55	0.103	0.139	0.060	1.30
75.0	0.036	0.168	0.22	0.016	0.030	0.022	0.49

Table 2. Computed cross sections for the formation of $\overline{H}$ and the destruction of Ps in collisions between $\overline{p}$ and Ps (Classical trajectory Monte-Carlo method, all cross sections are in units of 10^{-16} cm^2).

E keV lab	$\overline{\sigma}_c$	σ_i	E keV lab	$\overline{\sigma}_c$	σ_i
2.0	16.4 ± 0.7	0.29 ± 0.09	35.0	0.90 ± 0.10	8.78 ± 0.49
3.0	16.5 ± 0.7	0.29 ± 0.09	40.0	0.58 ± 0.13	8.75 ± 0.49
5.0	14.4 ± 0.6	0.29 ± 0.09	45.0	0.38 ± 0.10	8.17 ± 0.48
7.5	14.5 ± 0.6	0.18 ± 0.10	50.0	0.20 ± 0.05	7.45 ± 0.31
10.0	14.0 ± 0.6	0.50 ± 0.10	60.0	0.13 ± 0.04	6.62 ± 0.36
15.0	11.7 ± 0.6	2.80 ± 0.30	70.0	0.06 ± 0.04	5.91 ± 0.29
20.0	6.88 ± 0.43	5.66 ± 0.40	80.0	0.03 ± 0.02	5.32 ± 0.27
25.0	3.94 ± 0.34	7.70 ± 0.46	90.0	0.03 ± 0.02	4.65 ± 0.27
30.0	2.39 ± 0.26	8.25 ± 0.48	100.0	0.013 ± 0.011	4.37 ± 0.25

computers of the Universities of Durham and Newcastle-upon-Tyne.

References

1. Deutch, B.I., Jacobsen, F.M., Andersen, L.H., Hvelplund, P., Knudsen, H., Holzscheitler, M., Charlton, M., and Laricchia, G., in 'Workshop and Symposium on the Physics of Low-Energy Stored and Trapped Particles (AFI, Stockholm 14-18 June, 1987), to be published.
2. Humberston J.W. 1986 Adv. At. Mol. Phys. 22 1-36 and references therein.
3. Humberston, J.W., Charlton, M., Jacobsen, F.M., and Deutch, B.I. 1987 J. Phys. B: At. Mol. Phys. 20, L25-29.
4. Martir, M.H., Ford, A.L., Reading, J.F., and Becker, R.L. 1981 J. Phys. B: At. Mol. Phys. 15 1729-47.
5. Shakeshaft, R. and Wadera, J.M. 1980 Phys. Rev. A22 968-78.
6. Darewych, J.W. 1987, this workshop.
7. Ermolaev, A.M. 1984 J. Phys. B: At. Mol. Phys. 17 1069-81.
8. Omidvar, K. 1985, private communication.
9. Peach, G., Willis, S.L., and McDowell, M.R.C. 1985 J. Phys. B: At. Mol. Phys. 18 3921-37.

COMPARISON OF IMPACT-IONISATION AND CHARGE-TRANSFER CROSS SECTIONS FOR

POSITRON AND PROTON SCATTERING ON HELIUM

Dieter Fromme, Georg Kruse, Wilhelm Raith, and
Günther Sinapius

Fakultät für Physik, Universität Bielefeld
D-4800 Bielefeld, Federal Republik of Germany

Cross sections for impact ionisation of helium by positrons and protons as well as charge transfer cross sections for positron-helium and proton-helium scattering are usually compared at equal velocities of the incident projectiles (v_i) because the size of the cross sections depends on the duration of the interaction between projectile and target particle. The left parts of Figs. 1 and 2 show our measured values[1] of cross sections for impact ionisation and charge transfer and literature values[2-9] for the corresponding proton scattering processes. The charge transfer cross sections for positron and proton scattering on Helium intersect at $v_i = 4 \times 10^6$ m/s; at high velocities σ_{Ps} is significantly greater than σ_H.

The comparison of the positron and proton cross sections at equal velocities of the incident projectiles is questionable because of two reasons: (1) The velocities of the positronium and the hydrogen atom which are formed in the charge transfer processes differ significantly. (2) The range of the "post-collision interaction" exceeds that of the "pre-collision interaction". In the charge transfer processes the polarizabilities of the formed positronium and hydrogen atom exceed that of the helium atom by factors of 25 and 8, respectively. In the impact ionisation processes the charge of the helium nucleus is much less screened by remaining and out-going electron than by the two electrons of the helium atom before ionisation.

Thus a comparison of positron and proton cross sections at equal velocities of the particles after the scattering process, v_f, seems to be more appropriate than the comparison at equal velocities v_i. This comparison has been performed in the right parts of Figs. 1 and 2. In both cases positron and proton cross sections agree better than compared at equal velocities of the incident particles, v_i. This is most obvious in the case of the charge transfer processes. Thus we conclude that the size of the cross sections is indeed most substantially determined by the interaction between the helium ion and the outgoing particle(s).

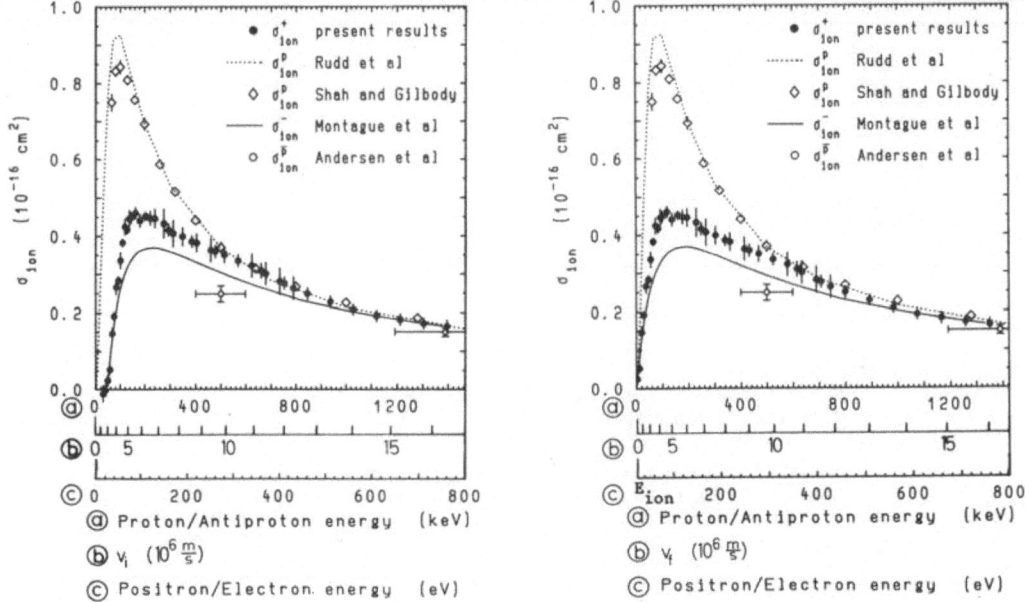

Fig. 1. Comparison of the partial cross sections for the impact ionisation of helium by positrons[1], protons[2,3], electrons[4] and antiprotons[5] at equal velocities of the incident (left) and outgoing (right) projectiles, respectively.

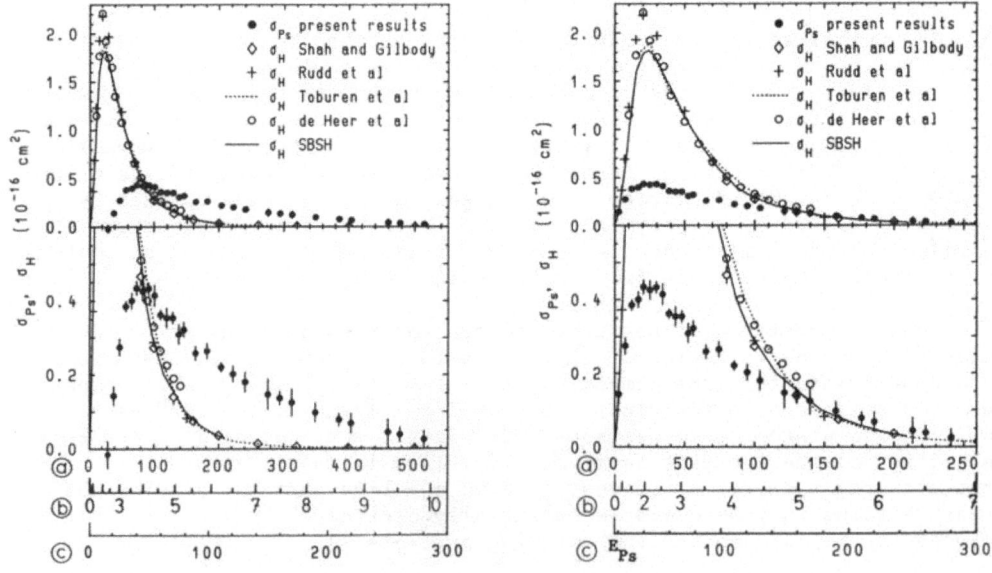

Fig. 2. Comparison of the charge transfer cross sections for positron[1] and proton [3,6-9] sacattering on helium at equal velocities of the incident projectiles (left) and outgoing positronium and hydrogen atom, respectively (right).

REFERENCES

1. D. Fromme, G. Druse, W. Raith, and G. Sinapius, Phys. Rev. Lett. 57, 3031 (1986).

2. M. E. Rudd, Y.-K. Kim, D. H. Madison, and J. W. Gallagher, Rev. Mod. Phys. 57, 965 (1985).

3. M. B. Shah and H. B. Gilbody, J. Phys. B 18, 899 (1985).

4. R. G. Montague, M. F. A. Harrison, and A. C. H. Smith, J. Phys. B 17, 3295 (1984).

5. L. H. Andersen, P. Hvelplund, H. Knudsen, S. P. Møller, K. Elsener, K.-G. Rensfelt, and E. Uggerhøj, Phys. Rev. Lett. 57, 2147 (1986).

6. M. E. Rudd, R. D. DuBois, L. H. Toburen, C. A. Ratcliffe, and T. V. Goffe, Phys. Rev. A 28, 3244 (1983).

7. L. H. Toburen, M. Y. Nakai, and R. A. Langley, Phys. Rev. 171, 114 (1986).

8. F. J. de Heer, J. Schutten, and H. Moustafa, Physca 32, 1766 (1966).

9. P. M. Stier and C. F. Barnett, Phys. Rev. 103, 896 (1956); J. B. H. Stedeford and J. B. Hasted, Proc. Roy. Soc. (London) A 227, 446 (1955).

MEASUREMENT OF THE IMPACT IONISATION AND POSITRONIUM FORMATION CROSS

SECTIONS FOR POSITRON SCATTERING ON MOLECULAR HYDROGEN

Dieter Fromme, Georg Kruse, Wilhelm Raith, and
Günther Sinapius

Fakultät für Physik, Universität Bielefeld
D-4800 Bielefeld, Federal Republic of Germany

We measured the impact ionisation and positronium-formation cross
sections for positron scattering on molecular hydrogen with the same
experimental set-up used for our measurements on helium.[1] As in the case
of helium the positron impact-ionisation cross section (not including
positronium formation) exceeds the corresponding electron cross section[2] in
the energy range around the respective maxima (Fig. 1). It is lower than
the electron cross section right above ionisation threshold. This is in
accordance with the Wannier threshold law, as described earlier.[3]

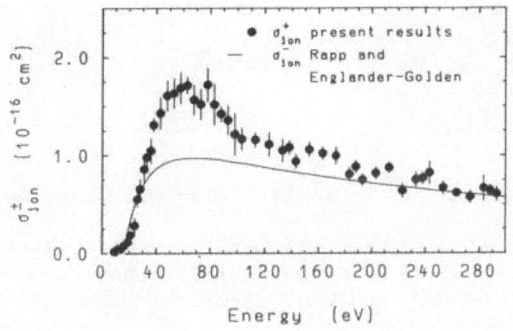

Fig. 1. Comparison of partial cross sections for impact ionisation of
molecular hydrogen by positrons (present results) and electrons.[2]

Our values on positronium formation in molecular hydrogen agree well
with the data of the Arlington group[4,5] (Fig. 2). The results of the
London group[6] are drastically lower than the other values in Fig. 2.
These observations apply to positronium formation in helium as well.

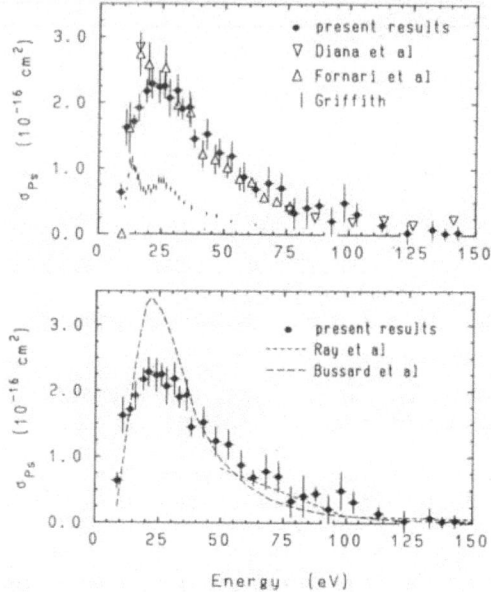

Fig. 2. Positronium-formation cross section of molecular hydrogen.
Top: comparison with other experimental results.[4-6]
Bottom: comparison with theoretical results.

Between the positronium formation and excitation thresholds the elastic
cross section can be obtained as the difference between the total and the
positronium formation cross sections. As in the case of helium[3] this yields
values of the elastic cross section which are distinctly lower than a "smooth"
extrapolation from below the positronium formation threshold. Furthermore,
the existence of a cusp at the positronium formation threshold is indicated.

REFERENCES

1. D. Fromme, G. Kruse, W. Raith, and G. Sinapius, Phys. Rev. Lett. 57,
    3031 (1986).
2. D. Rapp and P. Englander-Golden, J. Chem. Phys. 43, 1464 (1965).
3. R. I. Campeanu, D. Fromme, G. Kruse, R. P. McEachran, L. A. Parcell,
    W. Raith, G. Sinapius, and A. D. Stauffer, accepted for publication
    in J. Phys. B (April 1987).
4. L. S. Fornari, L. M. Diana, and P. G. Coleman, Phys. Rev. Lett. 51,
    2276 (1983).
5. L. M. Diana, P. G. Coleman, D. L. Brooks, P. K. Pendleton, and
    D. M. Norman, Phys. Rev. A 34, 2731 (1986).
6. T. C. Griffith, in Positron Scattering in Gases, J. W. Humberston,
    and M. R. C. McDowell ed., Plenum, New York, p. 53 (1984).
7. R. W. Bussard, R. Ramaty, and R. J. Drachman, Astrophys. J. 228, 928
    (1979).
8. A. Ray, P. P. Ray, and B. C. Saha, J. Phys. B 13, 4509 (1980).

AUTO-DISSOCIATING RESONANT STATES OF POSITRONIUM MOLECULES*

Y. K. Ho

Department of Physics and Astronomy
Baton Rouge, Louisiana 70803

This work reports a study of resonance phenomena in positronium molecules Ps_2, a system consists of two positronium atoms. A positronium molecule was first shown by Hylleraas and Ore[2] that these two Ps atoms form a bound system. The latest variational calculation[2] indicated that the binding energy of Ps_2 against dissociation into two positronium atoms is of 0.411 eV. The possible experimental studies of positronium molecules have been discussed by Mills.[3]

We have recently begun a theoretical study[4] of higher lying resonant states of positronium molecules. In some aspects, these high-lying states are similar to those in a positronium hydride, PsH. In PsH, it has been shown[5] that Rydberg series do exist as a result of the positron attaching to the H^- ion. Such Rydberg states, with the exception of the lowest S-wave state which also lies below the Ps + H threshold and becomes the ground state of PsH, would appear as resonances in Ps – H scattering. We would, therefore, expect such Rydberg series to also exist in the Ps_2 molecules as a result of the positron attaching to the Ps^- ion. The lowest state of the S-wave series also lies below the Ps–Ps scattering threshold and becomes the ground state of the positronium molecules. Higher members of the Rydberg states would lie in the Ps–Ps scattering continuum and appear as resonances in Ps–Ps scattering.

The method of complex-coordinate rotation (see for example a review by Ho[6]) is used to study resonance phenomena in Ps_2. The use of this method would provide parameters for both resonance energy positions and widths. We employ Hylleraas-type wave functions in which all six interparticle coordinates are used. A total of N = 500 terms are used in the present investigation. The resonance parameters, both resonance positions and total widths, are deduced from condition that the discreted complex eigenvalue is stabilized with respect to θ, where θ is the so called rotational angle of the complex transformation r → r exp(iθ). Using the condition

$$\partial|E|/\partial\theta = \text{minimum},$$

we have located six resonant complex eigenvalues. These are the lowest members of two Rydberg series, depending on the total spin states (singlet or triplet) of the two positrons, converging on the Ps^- threshold. The complex eigenvalues would appear as resonances in Ps–Ps scattering. The resonance positions expressed in eV are also shown in the Table.

Table 1. Auto-dissociating resonant states in Ps_2

E(Ry)	Γ(Ry)	E(eV)[a]	Γ(eV)
Series One			
−0.6855	0.0056	4.642	0.0762
−0.592	0.0080	5.551	0.109
−0.5625	0.0030	5.946	0.041
Series Two			
−0.626	0.016	5.089	0.216
−0.580	0.012	5.714	0.163
−0.553	0.016	6.082	0.216
Series Limit −0.5240[b]		6.476	

[a]Relative to the Ps–Ps scattering threshold.
[b]See Ref. (7) for example.

*The work was supported by the U. S. National Science Foundation Grant No. PHY-85-07133.

References

1. E. A. Hylleraas and A. Ore, Phys. Rev. 71, 493 (1947).
2. Y. K. Ho, Phys. Rev. A, 33, 1768 (1986).
3. A. P. Mills, Jr., in Positron Scattering in Gases, Eds. J. W. Humberston and M. R. C. McDowell (Plenum Press, New York, 1984) pp. 121.
4. Y. K. Ho, Phys. Rev. A 33, 3584 (1986).
5. R. J. Drachman, Phys. Rev. A 19, 1900 (1979).
6. Y. K. Ho, Phys. Reports, 99, 1 (1983).
7. Y. K. Ho, J. Phys. B, 16, 1503 (1983).

POSITRON-HYDROGEN RESONANCES ASSOCIATED WITH THE N=4 HYDROGEN

AND N=3 POSITRONIUM THRESHOLDS*

Y. K. Ho

Department of Physics and Astronomy
Louisiana State University
Baton Rouge, Louisiana 70803

There has been considerable interest to investigate resonances in e[+] - H scattering[1]. Several independent studies[2-4] have shown that S-wave resonances do exist associated with and below the hydrogen N=2 threshold. Resonances below the Ps(N=2) threshold have been investigated by Doolen[5] and recently by Ho and Greene[6]. It was found in Ref. 6 that the lowest resonance associated with the N=2 Ps threshold would lie at E=-0.150279 Ry. Such a finding contradicts with the result obtained by Doolen[5] that it would lie at E=-0.222 Ry. Resonances below the N=3 hydrogen threshold were also published in Ref. 6. Here we present calculations of resonances associated with the N=3 positronium and N=4 hydrogen thresholds. The method of complex-coordinate rotation[1] is used in the present investigation.

In this work, we use extensive Hylleraas-type wave functions of the form

$$\Psi = \sum_{klm} C_{klm} \exp[-\alpha(r_{1p}+r_{2p})] \; r_{12}^{k} \; r_{1p}^{l} \; r_{2p}^{m} , \qquad (1)$$

with $\omega \le k + l + m$ and ω is a positive integer or zero. In Eq. (1) r_{ij} represents the distance between particles i and j, with 1 and 2 denote the electrons and p the proton. A total of 969 terms ($\omega=16$) in the wave functions are used. Also, to overcome possible ill-conditioned natures of the Hylleraas-type wave functions, calculations are performed in quadruple precision arithmetic (about 30 significant digits) on an IBM 3084. Table 1 summarizes the S-wave resonances in positron-hydrogen scattering in these energy regions.

For resonances associated with the N=4 hydrogen threshold, we show three members of one resonance series, as well as the lowest member for another series. For resonance associated with the N=3 positronium threshold, we have found two members for one series. In addition, a shape resonance lying just above the Ps(N=3) threshold is located. The use of letters 'A' and 'B' in Table 1 is only for labeling purposes.

The present results are also shown in Figure 1. In the figure we connect resonance members that belong to the same series and the threshold to which each resonance converge.

Table 1. S-wave Resonances in e$^+$ - H scattering

E(Ry)	½ Γ(Ry)	series
Below H(N=4) threshold		
-0.067867	0.000048	A1
-0.064595	0.000022	A2
-0.063259	0.000010	A3
-0.063699	0.000050	B1
Below Ps(N=3) threshold		
-0.077063	0.0000476	A1
-0.0623	0.000025	A2
-0.05544[a]	0.00004	B1

[a]This is a shape resonance lying above the Ps(N=3) threshold of E=-0.055556 Ry.

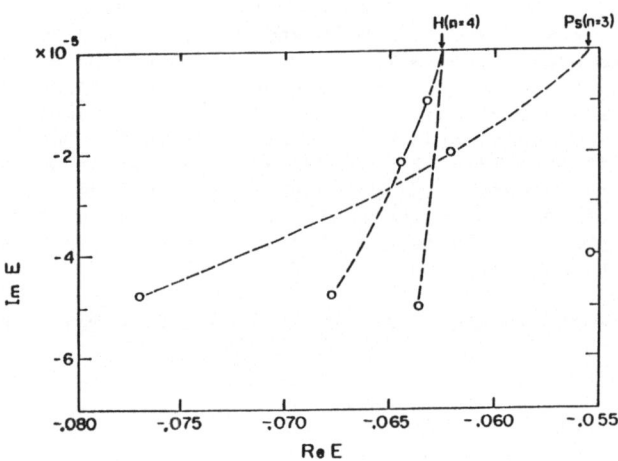

Figure 1. S-wave resonances in e$^+$ - H scattering in energy regions associated with the H(N=4) and Ps(N=3) thresholds. Resonance energy and half widths are in Rydbergs.

*This work was supported by U. S. National Science Foundation Grant No. PHY 85-07133.

References

1. For earlier references readers are referred to Y. K. Ho, Phys. Reports 99, 1 (1983).
2. L. T. Choo, M. C. Crocker, and J. Nuttall, J. Phys. B11, 1313 (1978).
3. G. D. Doolen, J. Nuttall, and C. Wherry, Phys. Rev. Lett. 40, 313 (1978).
4. E. Peliken and H. Klar, Z Phys. A 310, 153 (1983).
5. G. D. Doolen, Int. J. Quantum Chem. 14, 523 (1978).
6. Y. K. Ho and C. H. Greene, Phys. Rev. A, in press.

CALCULATION OF Ps DISTRIBUTION FROM SOLID SURFACE

Akira Ishii and Shigeru Shindo[*]

The Blackett Lab. Imperial College, London SW7 2BZ UK

[*]Dept. of Physics, Tokyo-Gakugei Univ. Koganei, Tokyo

In this paper we present calculations of the Ps distribution from a solid surface. The Ps distribution is important not only for surface science, but also for producing Ps beams. To obtain an intense monoenergetic Ps beam using a solid target, we should understand deeply the mechanism of Ps formation at surfaces.

Ps formation at surfaces is very different from that of positron-atom collisions, because we have no momentum conservation perpendicular to the surface and we have no Ps state in bulk. The equation for the Ps amplitude C_p and positron-electron joint amplitude $C_{k,q}$ is as follows [1-5];

$$(id/dt) \; C_p(t) = \sum_k V_{kq,p}{}^*(t) C_{k,q}(t) \tag{1}$$

$$(id/dt) \; C_{k,q}(t) = \sum_p V_{kq,p}(t) \; C_p(t) \tag{2}$$

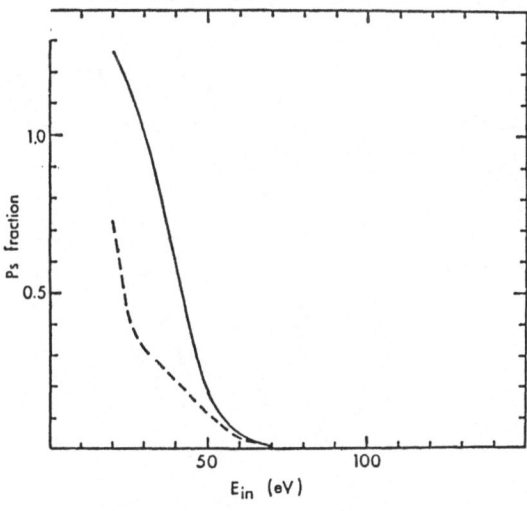

Fig. 1. [8]

Equation (1) is for forming Ps and (2) is for dissociating Ps. For high energy positrons, we only use (1) in the first Born approximation. But, for low energy positrons, we should couple Eq. (2) together with Eq. (1). In Figure 1, we show the Ps fraction for 6° glancing incidence. The solid line is the Born and dashed line is the 2nd order normalized Born [6]. The higher order correction is important in low energy incidence.

When positrons are implanted in a solid at energies greater than 1 keV most of the Ps is produced by positrons thermally diffusing backwards. In this case, we should do higher order calculations. As an example, we show the calculation of the Ps momentum distribution contour map for 2D–ACAR by Al [7]. The normalized Born is used in this calculation. As shown in Figure 2, the distribution of Ps if very broad and nearly isotropic. For transition metals having a narrow band in the vicinity of the Fermi surface, the

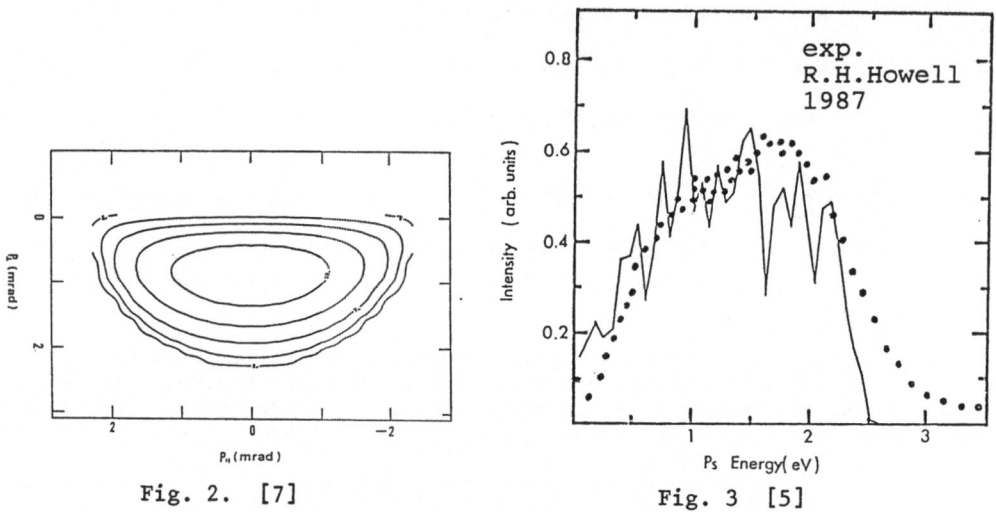

Fig. 2. [7]   Fig. 3 [5]

first Born and also the normalized Born is not a good approximation. In Figure 3, we show the energy distributions of Ps from a Ni(100) surface by solving numerically the integro-differential equation obtained by coupling (1) and (2) [5].

For Ps formation from surface scattering of energetic positrons, it is not necessary to include higher order corrections. But, there will be several processes to forming Ps; the Umklapp process and surface state

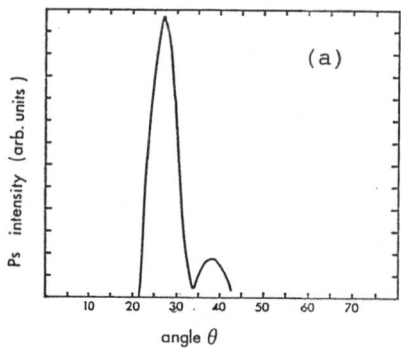
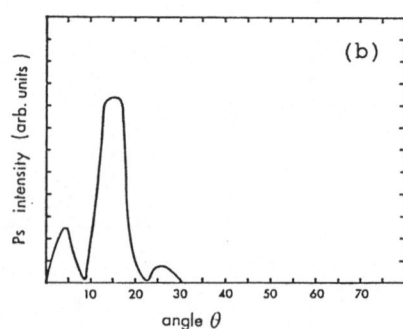

Fig. 4.

process [8]. In Figure 4, we show the ejection angle distributions of Ps for glancing angle incidence of 90 eV and 6° of Al(100). (a) is the ordin-ary process and (b) is the Umklapp process. The matrix element is Walker and Nieminen type [9].

This work is supported by Japan Society for the Promotion of Science and the Grant-in-Aide for the Science Research from the Ministry of Education of Japan.

REFERENCES

1.   A. Ishii, Surf. Sci. 147:277,295 (1984).

2.   A. Ishii, Phys. Rev. B36; in press.

3.   A. Ishii and S. Shindo, Phys. Rev. B 25:6521 (1987).

4.   A. Ishii and S. Shindo, Surf. Sci., in press.

5.   A. Ishii, to be published.

6.   S. Shindo and A. Ishii, Phys. Rev. B 35:8360 (1987).

7.   S.Shindo and A Ishii, Phys. Rev. B 36:(1) in press.

8.   A. Ishii, to be published.

9.   A.B. Walker and R.M. Nieminen, J. Phys. F16:L295 (1986).

ELASTIC SCATTERING OF FAST POSITRONS BY SODIUM AND POTASSIUM ATOMS

S.P. Khare and Vijayshri

Physics Department

Meerut University, Meerut-250005, India

So far no study seems to be available for the scattering of fast positrons by sodium and potassium atoms. Since these alkali atoms have high values of the dipole polarizability it is expected that polarization and absorption effects will play an important role in the scattering process. Hence, for these targets even in the intermediate energy range, appreciable differences between the cross sections for electrons and positrons are expected to exist.

In this paper we have investigated the differential cross section for the elastic scattering of positrons by sodium and potassium atoms for E > 100 eV. In this energy range one of the tractable methods of including polarization and absorption effects is to employ a modified Glauber approximation (MGA). However, a straight-forward application of MGA to Na and K, having large numbers of electrons, is a highly complicated problem. The evaluation of multiple scattering terms for these atoms become highly complicated. Nevertheless, Khare and Vijayshri (1983) showed that the values of the total cross section obtained for $e^{\pm}$ - Li scattering using full MGA were in good agreement with those obtained by MGA after neglecting all the multiple scattering terms (this model is referred to as the single particle scattering model - SPSM). This has encouraged us to employ SPSM to obtain DCS for the elastic scattering of positrons by Na and K for E > 100 eV. In the above energy range MGA-SPSM may not be a bad method.

In MGA the scattering amplitude is given by

$$f_{MGA} = f_G - f_{G2} + f_{B2}$$

where f_G, f_{G2} and f_{B2} are the Glauber amplitude, the second term of the Glauber series and the second Born term in the closure approximation, respectively. We have employed Hartree-Fock wave functions given by Clementi and Roetti (1974) to represent the ground state of Na and K. We have followed Khare and Vijayshri (1983) to obtain the value of the mean excitation energy . These values are 0.1401 a.u. and 0.1295 a.u. for Na and K, respectively.

Our present results are shown in Tables 1 and 2. A comparison of these cross sections with those for electrons shows that the former are smaller, which is according to expectations. Furthermore, the present cross sections do not show any rise in the backward direction as recently observed by Rao and Bharathi (1987) for e-Na and K scattering. This difference between electron and positron scattering may be real.

Table 1. Differential cross sections(in a_o^2) for the elastic scattering of positrons by sodium atoms in MGA-SPSM.

θ(deg.)/E(eV)	150	300	400
0	2.67(2)*	1.80(2)	1.54(2)
5	8.54(1)	5.56(1)	4.73(1)
10	3.40(1)	2.02(1)	1.59(1)
20	8.79	5.01	3.95
40	1.76	1.09	8.25(-1)
60	6.42(-1)	3.70(-1)	2.60(-1)
80	3.10(-1)	1.62(-1)	1.08(-1)
100	1.81(-1)	8.74(-2)	5.68(-2)
120	1.23(-1)	5.60(-2)	3.59(-2)
150	8.68(-2)	3.73(-2)	2.37(-2)
180	7.76(-2)	3.28(-2)	2.07(-2)

Table 2. Differential cross sections (in a_o^2) for the elastic scattering of positrons by potassium atoms in MGA-SPSM.

(deg.)/E(eV)	200	300	400
0	8.30(2)	6.55(2)	5.62(2)
5	2.41(2)	1.89(2)	1.60(2)
10	9.53(1)	7.09(1)	5.70(1)
20	2.65(1)	1.94(1)	1.51(1)
40	5.06	3.24	2.21
60	1.53	9.09(-1)	6.03(-1)
80	6.44(-1)	3.76(-1)	2.52(-1)
100	3.42(-1)	2.02(-1)	1.37(-1)
120	2.17(-1)	1.30(-1)	8.91(-2)
150	1.45(-1)	8.83(-2)	6.04(-2)
180	1.27(-1)	7.81(-2)	5.34(-2)

*The numbers within the brackets denote the power of 10 by which the corresponding entries are to be multiplied.

REFERENCES

1.   E. Clementi and C. Roetti, At. Data Nucl. Tables 14:3-4 (1974 .

2.   S.P. Khare and Vijayshri, J. Phys. B. 16:3621 (1983).

3.   M.V.V.S. Rao and S.M. Barathi, J. Phys. B 20:1081 (1987).

EXCITATION OF HELIUM BY POSITRON IMPACT AND A COMPARISON WITH ELECTRON DATA

R.P. McEachran*, L.A. Parcell†, A.D. Stauffer* and T. Zuo*

*Physics Department, York University, Toronto, Canada M3J 1P3
†School of Mathematics, Physics, Computing and Electronics,
 Macquarie University, Sydney, NSW 2109, Australia

We have calculated the excitation of the n^1P states of helium
(n=2,3,4) from the ground state by positron impact using a distorted wave
method (Parcell et al, 1983, 1987). As well as producing results for the
integrated and differential cross sections, we have obtained values for the
correlation and orientation parameters λ, χ, and O_{1-} (see, for example, Fon
et al, 1980, for a definition of these quantities).

The analysis of the partitioning of the positron-helium total cross
section (Campeanu et al, 1987) indicates that the sum of the integrated
cross sections for the excitation of the various bound states of helium by
positrons must exceed the equivalent sum in the case of excitation by electrons at energies greater than approximately 100eV. Since the cross sections for the excitation of the n^1P states contribute most to this sum, we
have also calculated the corresponding electron cross sections so that a
direct comparison can be made within a consistent set of calculations.

In our calculations we have used distorted waves produced by the static
plus polarisation potentials of the state in question. In the electron case
we have neglected the exchange of the scattered electron with the bound
electrons though in both cases the atomic wavefunctions are antisymmetrised.
Although we have used a method in which we calculate the difference between
the distorted wave t-matrix and the corresponding Born t-matrix we still
require a large number of partial wave contributions in order to obtain a
converged value for this difference. While of the order of fifty partial
waves was sufficient at energies below 100eV, at the higher energies several
hundred were required and we resorted to an extrapolation procedure to
produce sufficient partial wave contributions to ensure convergence of the
differential cross sections at large scattering angles.

Our results for the integrated cross sections are given in the table
below. In the case of electron excitation to the 2^1P and 3^1P states they
are in reasonable agreement with the ten-channel eikonal results of Flannery
and McCann (1975) which also neglects exchange of the scattered electron.
While our electron cross sections are always greater than the positron ones
at lower energies, at higher energies the positron cross sections are
larger. The energy at which this change occurs becomes larger the larger
the principle quantum number of the excited state. Thus at higher energies
our results display the expected qualitative behaviour through a quantita-

TABLE I

Integrated Excitation Cross Section (πa_0^2) from the Ground

State of Helium for Positron and Electron Impact

Incident Energy (eV)	Final State					
	2^1P		3^1P		4^1P	
	e+	e-	e+	e-	e+	e-
100.	0.1203	0.1357	0.0257	0.0368	0.0095	0.0162
150.	0.1106	0.1129	0.0253	0.0308	0.0096	0.0128
200.	0.0985	0.0973	0.0232	0.0259	0.0089	0.0108
250.	0.0881	0.0858	0.0211	0.0225	0.0082	0.0093
300.	0.0794	0.0768	0.0193	0.0200	0.0075	0.0083
400.	0.0669	0.0643	0.0164	0.0165	0.0064	0.0067
500.	0.0578	0.0556	0.0144	0.0142	0.0056	0.0055
600.	0.0508	0.0489	0.0127	0.0125	0.0050	0.0050
700.	0.0452	0.0436	0.0114	0.0111	0.0045	0.0045
800.	0.0416	0.0402	0.0104	0.0100	0.0042	0.0041
900.	0.0382	0.0370	0.0096	0.0092	0.0038	0.0037
1000.	0.0353	0.0342	0.0088	0.0085	0.0035	0.0034

tive evaluation is problematical because of uncertainties in the various cross sections, both experimental and theoretical.

The alignment and orientation parameters for both positron and electron excitation show little variation as a function of the principle quantum number of the excited state.

REFERENCES

Campeanu, R.I., Fromme, D., Kruse, G., McEachran, R.P., Parcell, L.A., Raith, W., Sinapius, G. and Stauffer, A.D., 1987, J. Phys. B. 20, 3557-70.

Flannery, M.R. and McCann, K.J., 1975, Phys. Rev. A 12, 846-55.

Fon, W.C., Berrington, K.A. and Kingston, A.E., 1980, J. Phys. B 13, 2309-25.

Parcell, L.A., McEachran, R.P. and Stauffer, A.D., 1983, J. Phys. B 16, 4249-57.

_____, 1987, J. Phys. B 20, 2307-15.

DIFFERENTIAL CROSS SECTIONS FOR ELECTRON AND POSITRON SCATTERING FROM ARGON

H. Nakanishi* and D. M. Schrader

Chemistry Department
Marquette University
Milwaukee, WI 53233, USA

Recent intense efforts to generate slow positron beams have led to many exciting developments in positron scattering experiments from atoms and molecules. Last year the first direct measurements were performed by Detroit group [1] for the differential cross sections of positrons elastically scattered from argon atoms at 100 and 200 eV. Since the differential cross sections provide more rigorous testing for theories and experiments, it may be of great interest to present the calculated differential cross sections of positrons and electrons from argon.

The scattering wave functions are calculated by solving a one-particle Schrodinger equation. The polarization effect from the target atom is included by the semiempirical polarization potential V_p developed previously [2] which has the functional form:

$$V_p = - \frac{\alpha_d}{2r^4}(1 - e_8^{r/p} \, e^{-r/p})^2.$$ (1)

Here α_d is the dipole polarizability of argon atom ($\alpha_d = 11.08a_0^3$, $e_8^{r/p}$ is the truncated exponential function after 8th power of r/p, and p is an adjustable parameter which is related the effective target radius r_0 by $r_0 = 10.617p$ [2].

The polarization potentials in the positronic and electronic systems are different even for the same target atom. The previous systematic studies [2] to account the difference show strong regularities among the effective target radii. The empirical method developed from these regularities enable us to determine a single effective radius for each target based on exclusion and screening effects from the core electrons. By using this method, r_0 for the positron-argon system is deduced to be $1.7a_0$ from r_0 for electron-argon system ($r_0 = 2.6$ for s-, 2.1 for p-wave, and 2.6 for the higher partial waves) which yields the best experimentally determined scattering length ($a_s = -1.45a_0$).

The differential cross sections are calculated from the formula given by Nesbet [3] using the calculated partial wave phase shifts, δ_ℓ up to $\ell = 5$:

---

*Physics Dept., Univ. of Missouri at Kansas City, Kansas City, MO 64110, USA

$$\frac{d\sigma}{d\Omega} = |\int(\theta)|^2,\qquad(2)$$

where

$$\int(\theta) = \frac{1}{k} \sum_{\ell=0}^{5} (2\ell + 1)e^{i\delta\ell} \sin \delta_\ell P_\ell(\cos\theta)$$

$$+ \pi\alpha_d k \left[\frac{1}{3} - \frac{1}{2}\sin\frac{\theta}{2} - \sum_{\ell=1}^{5} \frac{P_\ell(\cos\theta)}{(2\ell + 3)(2\ell - 1)}\right]\qquad(3)$$

The results are shown in Figs. 1-3.

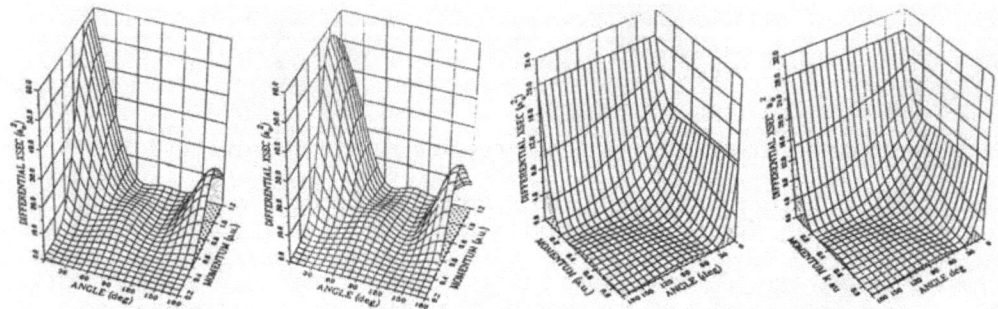

Fig.1 Differential cross sections for electron scattering from argon. Left: present calculation Right: calculated from the experimentally determined phase shifts by Williams [4].

Fig.2 Differential cross sections for positron scattering from argon. Left: present calculation Right: calculated from the phase shifts given by McEachran et al. [5].

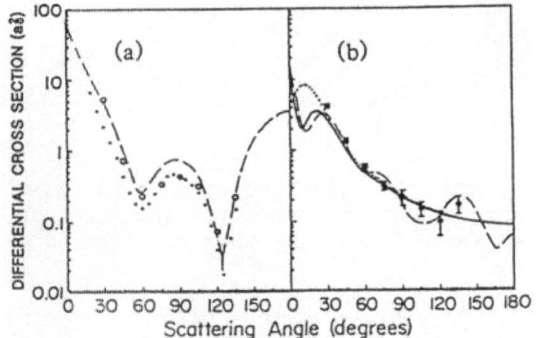

Fig.3 Differential cross sections for electron(a) and positron(b) scattering from argon at 100 eV taken from Ref.[1]. Present results are indicated by dashed lines.

REFERENCES

1. G. M. A. Hyder, M. S. Dababneh, Y.-F. Hsieh, W. E. Kauppila, Phys. Rev. Lett., 57, 2252 (1986).
2. H. Nakanishi and D. M. Schrader, Phys. Rev. A, 34, 1810 and 1823 (1986).
3. R. K. Nesbet, Phys. Rev. A, 20, 58 (1979).
4. J. F. Williams, J. Phys. B, 12, 265 (1979).
5. R. P. McEachran, A. G. Ryman, and A. D. Stauffer, J. Phys. B, 12, 1031 (1979).

IONIZATION OF POSITRONIUM BY ELECTRONS

G. Peach

Department of Physics and Astronomy, University College

Gower Street, London WC1E 6BT, UK

A computer program originally developed to calculate cross sections for
the ionization of neutral atoms by electrons and heavy particle impact,
has been adapted to calculate the Born cross section for the ionization
of positronium by electrons. The cross section can be written as

$$Q(k_i) = \left\{ \int \frac{k_i}{k_f} \, |f_B|^2 \, dk \, d\hat{k}_f \right\}_{av}$$

where f_B is the Born scattering amplitude and 'av' denotes an average
over all the degenerate sublevels of the initial atomic state. Atomic
units are used throughout and Mk_i, Mk_f and μk are the momenta of the
incident, scattered and ejected electrons respectively. The main difference between this application of the Born approximation and the original
one arises from the fact that the centre of mass (CM) of the positronium
atom is at the point with position vector ρ, where 2ρ gives the position
of the bound electron relative to the positron. The reduced mass of
positronium is μ, and the momenta Mk_i and Mk_f are measured relative to
point CM where $M = 2/3$ and $\mu = 1/2$ in units of the electron mass.
    The Born approximation for ionization is then given by

$$f_B = -\frac{4i}{K^2} \int \varphi_i(\rho) \sin(K.\rho) \, \psi(-k,\rho) \, d\rho \; ,$$

where $K = M(k_i - k_f)$ and $\varphi_i(\rho)$ and $\psi(-k,\rho)$ are wave functions for the
initial and final states of the electron that is ejected from the target.
These wave functions are normalised according to

$$\int |\varphi_i(\rho)|^2 \, d\rho = 1 \; ; \quad \int \psi^*(k,\rho) \, \psi(k',\rho) \, d\rho = \delta(k-k') \; ,$$

and $\psi(-k,\rho)$ is the complex conjugate of a free wave function containing
an incoming scattered wave. Conservation of energy gives

$$M(k_i^2 - k_f^2) = \mu\left(\frac{1}{n^2} + k^2\right)$$

where n is the principal quantum number of the initial bound state i.
    The Born cross section for the ionization of positronium is
presented in the table below. The 'half-range' version of the Born
approximation has been used in which the energy of the scattered electron

is always greater than or equal to that of the ejected electron. A more detailed account of this work will be published in Journal of Physics B.

Table

E(eV)[†]	6.8029	7.6330	8.5644	10.782	13.574	17.088	21.513	27.083
$Q(\pi a_0^2)$	0.0000	0.7364	1.7929	3.7254	4.9815	5.5473	5.5881	5.2902

E(eV)	34.095	42.923	54.037	68.029	85.644	107.82	135.74	170.88
$Q(\pi a_0^2)$	4.8066	4.2463	3.6781	3.1412	2.6555	2.2278	1.8584	1.5373

E(eV)	215.13	270.83	340.95	429.23	540.37	680.29	1078.2	1708.8
$Q(\pi a_0^2)$	1.2666	1.0420	0.8560	0.7024	0.5757	0.4713	0.3151	0.2099

[†]E is the energy of the incident electron relative to the centre of mass of the positronium atom.

POSITRONIUM FORMATION AT CONDENSED ATOMIC SEMILAYERS OF KRYPTON ON CARBON

P. Rice-Evans and K.U. Rao

Department of Physics, Royal Holloway & Bedford New College

University of London, Egham, Surrey TW20 OEX, England

As the temperature of a surface in the presence of a gas is lowered, the gas begins to condense. A monolayer develops first, and then a bi-layer, trilayer, etc. Physisorption is determined by van der Waals forces and fluid submonolayers will be described approximately by the ideal two dimensional gas model with the Boltzmann approximation [1]. The relation between the coverage (n molecules m^{-2}) and the pressure (P) is

$$P = \frac{nkT}{\lambda} \exp(-\varepsilon_o/kT)$$

where $\lambda = h/(2\pi mkT)^{\frac{1}{2}}$, ε_o is the binding energy of the atom or molecule and n is the coverage.

To maximise the probability of a slow positron interacting with a mono-layer surface we have exploited the properties of exfoliated graphite. The form grafoil has a layered structure, the plane of the leaves coinciding with the basal plane of the carbon, and it offers a very large specific surface area (~ 10 m^2 gm^{-1}). We created a sandwich specimen with ^{22}NaCl deposited centrally. The positrons would inevitably pass through many leaves in slowing down, eventually to appear at a surface with either epithermal or thermal energies. They would then have the chance of interacting with any physisorbed layer.

Our experiment consisted of cooling the specimen in the presence of krypton, maintained at 2.7 Torr as a consequence of the reservoir being held at 77K. A germanium detector measured the $2\gamma/3\gamma$ ratio as a function of temperature, although for statistical reasons in Fig. 1. we actually plot the area of the 511 keV photo peak (A).

Below 140 K we see a sharp decline in A which we ascribe to an enhanced 3γ emission and concomitant diminution of 2γ. This is the signal for ortho-positronium. The graph peaks at about 117 K and steeply returns to its original value (which also corresponds to the value of A in the presence of vacuum.

This positronium must be associated with the development of the mono-layer. Notwithstanding our earlier conclusions for nitrogen [2], on the basis of recent measurements on methane [3], we now think the peak in Ps production corresponds to a monolayer coverage of 50%. At this coverage and temperature the krypton monolayer would be in a fluid state [4]. It is

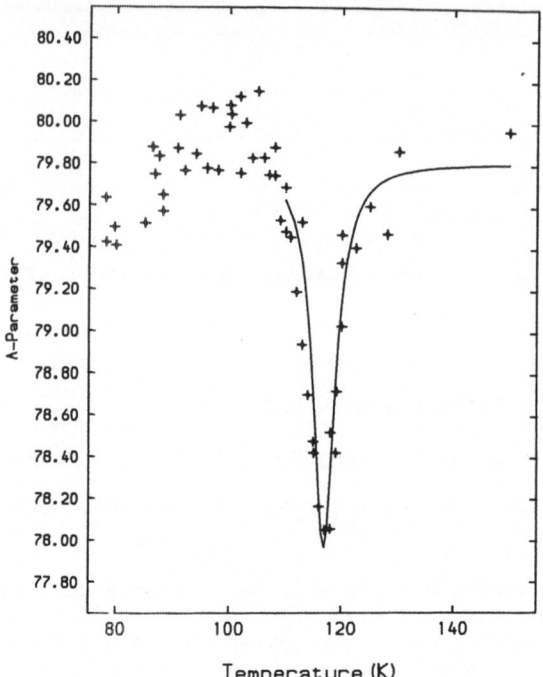

Fig. 1. The variation of the intensity of the 511 keV photo peak,
taken to be indicative of the 2γ/3γ ratio, and hence the
formation of ortho-positronium, as a function of temperature,
and therefore of adlayer coverage.

not easy to speculate why Ps formation is a maximum at 50%, although as Jean
et al [5] have noted, this corresponds with maximum surface irregularity.
We intend to examine the process in more detail in the near future.

REFERENCES

[1] Dash J.G., Films on Solid Surfaces, Academic Press, New York (1975).

[2] Rice-Evans, P., Mousavi-Madani, M., Rao, K.U., Britton, D.T., and
     Cowan, B.P., Phys. Rev. B34:6117 (1986).

[3] Rice-Evans, P. and Rao, K.U., Proc. Int. Symp. on Positron Annihilation
     Studies of Fluids, Arlington, Texas, June (1987) ed. S.C. Sharma
     (in press).

[4] Caflisch, R.G., Nihat Berken, A. and Kardar, M., Phys. Rev. B31:4527
     (1985).

[5] Jean, Y.C., Yu. C and Zhou, D.M., Phys. Rev. B32:4313 (1985).

POSITRON–Na SCATTERING USING CCA

K.P. Sarkar, Madhumita Basu and A.S. Ghosh

Indian Association for the Cultivation of Science

Theoretical Physics Department, Calcutta-32, India

The Detroit group have started experiments on positron-alkali atom scattering, and they have reported total cross sections for e^+-K scattering[1]. They are going to report total cross sections for other alkali atoms in the very near future. The Calcutta group (Ghosh and his colla-borators[2-5]) has investigated e^+-Li scattering in detail. Guha and Ghosh[2] and Majumdar and Ghosh[3] have reported the positronium formation cross section using the adiabatic coupled-static and distorted wave methods. Basu and Ghosh[4] have predicted the ionization cross section for e^+-Li scattering using the distorted wave and First Born approximations. Khan, Dutta and Ghosh[5] have performed a close coupling calculation retaining five (2s, 2p, 3s, 3p, 3d) eigenstates for e^+-Li scattering. They have reported the results for elastic as well as inelastic processes. The findings of the Calcutta group show that inelastic processes dominate the scattering process. The Detroit group has also reached a similar conclusion.

Elastic and excitation cross sections in e^+-Na scattering have been calculated using the close-coupling method at low incident energies In our coupling scheme we have included five (3s, 3p, 4s, 3d, 4p) target eigenstates. The Na atom is a highly polarizable target. Inclusion of five eigenstates accounts for the polarizability reliably. In our calculations we have assumed that only the valence electron is excited. The wave function of the Na ground state is taken from Clementi and Roetti[6]. Excited state wave functions are obtained from Kundu et al[7] and Kundu and Mukherjee[8]. The excited state wave function has got 12 basis sets. We have neglected the effect of positronium formation. We have solved the coupled differential equations using the code developed by Khan et al[5].

REFERENCES

1. T.S. Stein, W.E. Kauppila, in Electronic and Atomic Collisions eds. D.C. Lorentz et al (Ameterdam Elsevier, p. 105 (1986).

2. S. Guha and A.S. Ghosh, Phys. Rev. A 23:743 (1981).

3. P.S. Majumdar and A.S. Ghosh, Phys. Rev. A34:4433 (1986).

4. M. Basu and A.S. Ghosh, J. Phys. B 19:1249 (1986).

5. P. Khan, S.K. Dutta and A.S. Ghosh, J. Phys. B. (in press).

6. E. Clementi and C. Roetti, At. Data Nucl. Data Tables 14:187 (1979)

7. B. Kundu, D. Ray and P.K. Mukherjee, Phys. Rev. A34:62 (1986).

8. B. Kundu and P.K. Mukherjee, Phys. Rev. A35: (1987).

ELASTIC DIFFERENTIAL POSITRON SCATTERING FROM ARGON

Arnim Schwab, Petra Höner, Wilhelm Raith,
and Günther Sinapius

Fakultät für Physik, Universität Bielefeld
D-4800 Bielefeld, Federal Republic of Germany

Up to now the only experimental data on elastic differential scattering
of positrons from a crossed beam experiment are those of Hyder et al. (1986)
for Ar in the energy range from 100 eV to 300 eV. There are two theoretical
papers on this subject of Nahar and Wadehra (1986) from 3 eV to 300 eV and
of McEachran and Stauffer (1985) from 10 eV to 250 eV. At low energies and
small angles these theories differ significantly. We are performing measurements in the energy and angular range where these calculations do not agree.

The apparatus is shown schematically in Fig. 1. The slow positrons are
produced by a ^{22}Na-source (8 mCi) with a tungsten-mesh moderator and are
electrostatically guided through the scattering region. The unscattered
beam is monitored by a channel-electron multiplier mounted off-axis. The
diameter of the e^+-beam is limited by a 4 mm-aperture in front of the scattering region. At 30 eV the intensity is roughly 6000 e^+/sec.

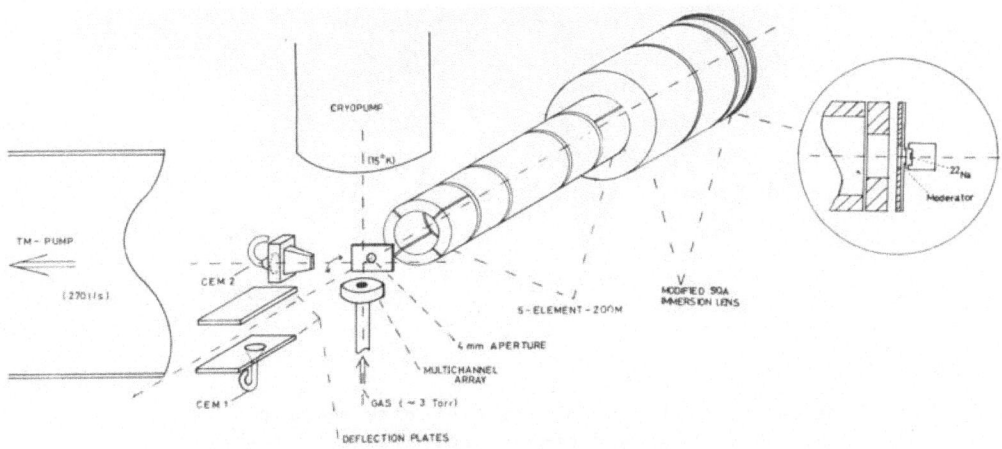

Fig. 1. Lay-out of the Bielefeld crossed-beam experiment

To form an atomic beam, the target gas flows through a glass-capillary array with an open area of 4 mm diameter. The density of the beam is about 10^{13} atoms/cm^3. Most of the gas beam is pumped by a cryopump. Elastically scattered positrons are counted by a rotatable CEM; inelastically scattered ones are rejected by a retarding potential. Due to the low positron beam intensity the signal is only about 0.1 - 0.4/sec whereas γ-rays from the source and from annihilating positrons cause a background of around 5/sec. Long-time computer-controlled data taking is employed to reduce the statistical errors.

In order to reduce the effect of long term instabilities of the apparatus the angular range is scanned every day. At each angle four different types of measurements are performed.
  a) gas beam on, CEM 2 open for scattered positrons
  b) gas beam on, CEM 2 positively biased
  c) gas beam off, CEM 2 open for scattered positrons
  c) gas beam off, CEM 2 positively biased.

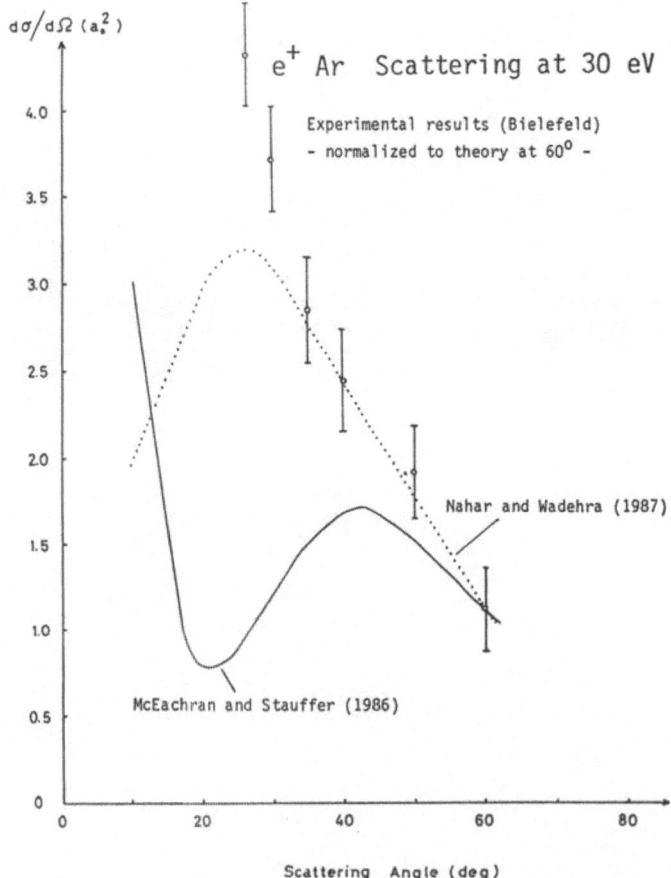

Fig. 2. Relative differential elastic cross section
        for 30 eV e$^+$Ar scattering, normalized to theory
        at 60°

When the atom beam is turned off the gas flows through a bypass. Relative cross sections are obtained from (a-b) - (c-d). The second expression accounts for positrons scattered from residual gas outside the atomic beam. Terms b and d correct for background (Ps, fast e^+). Our first results (Fig. 2) show an angular dependence which disagrees with both theoretical predictions.

Hyder, G.M.A., Dababneh, M.S., Hsieh, Y.-F., Kauppila, W.E., Kwan, C.K., Mahdavi-Hezaveh, M., and Stein, T.S., 1986, Positron Differential Elastic-Scattering Cross Section Measurements for Argon, Phys. Rev. Lett. 57, 18:2252-2255.

McEachran, R.P., and Stauffer, A.D., 1985, Differential Cross-Sections for Positron Noble Gas Collisions, in: "Proceedings of the third International Workshop on Positron(Electron)-Gas Scattering" (Detroit), Kauppila, W.E., Stein, T.S., Wadehra, J.M., ed., World Scientific, Singapore.

Nahar, S.N., and Wadehra, J.M., 1986, Elastic Scattering of Positrons and Electrons by Argon, submitted to Phys. Rev. A.

ANGULAR DIFFERENTIAL CROSS SECTIONS AND MODERATION EFFECTS

Masayoshi Senba

TRIUMF and Department of Chemistry
University of British Columbia
Vancouver, B.C., Canada, V6T 2A3

It has been found that the elastic collisions of both muons and muonium play essential roles in determining the muonium fraction and the spin polarization, and that the muonium fraction is especially sensitive to a very subtle balance between the charge exchange and elastic processes near the charge exchange thresholds (Senba, 1987; Fleming and Senba,1987). In the present work, we discuss how the angular distribution of elastic collisions affects the slowing down of a light particle in gases.

In the case where a light projectile of mass m is scattered by a stationary target atom of mass M, the average energy loss of the projectile per collision of a type k can be expressed (Senba, 1987) by,

$$dE/dN \ = \ \ln(f_k(E))[\ E + I_k(E)/(1-f_k(E)) \ ] \tag{1}$$

where $I_k(E)$ is essentially the endothermicity of the collision, and $f_k(E)$ is the angular average of a kinematic factor weighted by the angular dependence of the collision cross section,

$$f_k(E)= \int d\Omega \left[\frac{2m^2}{(M+m)^2} \cos\theta \left\{ \cos\theta + \sqrt{\cos\theta + \frac{M^2-m^2}{m^2} - \frac{M(M+m)I(E)}{m^2 E}} \right\} \right.$$

$$\left. + \frac{M-m}{M+m} \right] [d\sigma_k(E,\theta)/d\Omega]/\sigma_k(E) \tag{2}$$

and $\sigma_k(E)= \int d\Omega d\sigma(E,\theta)/d\Omega$ is the total cross section of this process. If the scattering is isotropic, $f_k(E)$ takes an energy-independent value, $f=(M^2-m^2/3)/(M+m)^2$. For elastic collisions $(I_k(E)=0)$, dE/dN takes a particularly simple form

$$dE/dN \ = \ E \ \ln(f(E)) \tag{3}$$

where subscripts k's are suppressed for the elastic process.

Based on the theoretical calculation of the angular differential elastic cross section for the $e^+ +$ Ar system (Nahar and Wadehra, 1987), the f(E) for the elastic process of this system has been calculated as a function of positron energy between 3eV and 300eV. Figure 1 shows the energy loss per collision (solid circles) together with the corresponding quantity based on the isotropic scattering assumption (solid line). It has been found that the quantity dE/dN is much smaller than its isotropic values, because of the forward-peaked nature of the scattering. Since dE/dt=nV(E)σ(E)dE/dN, where n and V(E) are the number density of the gas atoms and the relative velocity, respectively, the smaller dE/dN implies that the slowing down time is correspondingly longer.

The slowing down efficiency for a given mechanism k can be characterized by a figure of merit, F_K, defined as,

$$F_K = dE/dN \; \sigma_K(E) = \ln(f_K(E))[E+I_K(E)/(1-f_K(E))]\sigma_K(E)$$

$$= [\; E(1-f_K(E)) + I_K(E) \;]\sigma_K(E) \qquad (4)$$

Since $E(1-f_K(E))$ becomes smaller than $I_K(E)$ at low energies, the quantity F for an inelastic process is approximately given by $F=I_K(E)\sigma_K(E)$ towards the end of charge exchange. Therefore, the figures of merit for inelastic processes are relatively independent of the details of the angular dependence of the processes. For the elastic process, on the other hand, F is given by $F=E \ln(f(E))\sigma(E)$ and it should be noted that the cross sections for elastic collisions can be much larger than those for inelastic processes at low energies. Since the neutral fraction is determined by the subtle competition between the elastic and inelastic (charge exchange) processes at low energies, the neutral fraction depends on the details of the angular differential cross section as well as the total cross section of the elastic process.

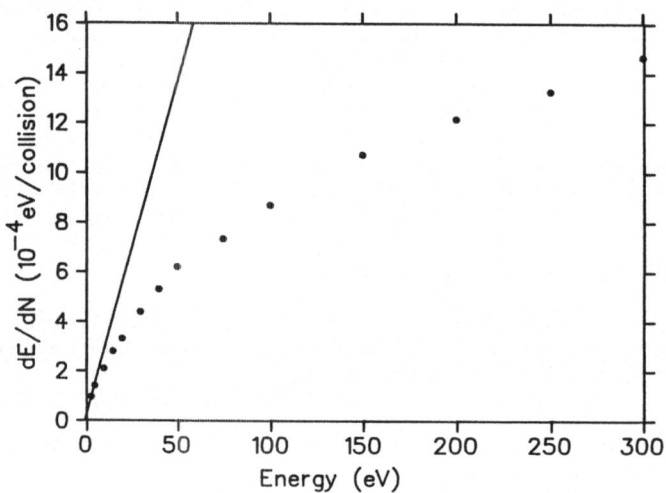

Figure 1.   dE/dN for the elastic positron-Ar scattering.
The solid line assumes an isotropic scattering.
The solid circles are calculated from the work
by Nahar and Wadehra (1987).

References

Fleming D.G., and Senba M., 1987, (this workshop).
Nahar S.N., and Wadehra J.M., 1987, Phys. Rev., A35:2051.
Senba M., 1987, (in preparation).

POSITRONIUM FORMATION IN THE EXCITED STATES (2s,2p) IN POSITRON HYDROGEN ATOM COLLISIONS

N.C. Sil, S. Tripathi and C. Sinha

Theoretical Physics Department, Indian Association for the Cultivation of Science, Jadavpur, Cal-32.

Differential and total positronium (Ps) formation cross sections in the excited 2s and 2p states in e^+-H collisions are studied in the framework of the Glauber eikonal approximation which satisfies the proper boundary conditions and takes a consistent account of the projectile-nucleus potential term (both in the phase and in the interaction). The post form of the amplitude for Ps formation in the Glauber eikonal approximation is given by

$$T_{if}^{post} = -\frac{\mu_f}{2\pi} \int e^{-i\vec{k}_f \cdot \vec{S}} \Phi_{2S,2p}^*(r_{12}) \left(\frac{1}{r_1} - \frac{1}{r_2}\right) \Phi_{1s}(r_2)$$

$$x \; e^{i\vec{k}_i \cdot \vec{r}_1} \left(\frac{r_{12}-z_{12}}{r_1-z_1}\right)^{-1} d\vec{r}_1 \; d\vec{r}_2 \qquad (1)$$

where $\eta = \frac{1}{v_i}$; $\vec{S} = \frac{\vec{r}_1 + \vec{r}_2}{2}$; $\vec{r}_{12} = \vec{r}_1 - \vec{r}_2$; z_1 and z_{12} are the z components of the respective vectors, v_i is the projectile velocity. Φ_i and Φ_f are the bound state wave functions for the hydrogen atom (1S) and positronium (2s or 2p) respectively; N_i and N_f are normalization constants. K_i, k_f are the initial and final momenta. We have used atomic units throughout the work, in which μ, the reduced mass of the final system comes out to be 2.

To deal with the eikonal phase terms occurring in eqn.(1), we use, as in our previous work[1], the following contour integral representation

$$y^{-i\eta} = -\frac{1}{2i \, \sin(\pi i \eta) \, \Gamma(i\eta)} \int_c (-\lambda)^{i\eta \, -1} e^{-\lambda y} \, d\lambda \qquad (2)$$

By virtue of this representation we have been able to reduce the six dimensional integral in (1) to a single dimensional one containing Gauss hypergeometric functions. The resulting one dimensional integral is evaluated numerically by Gaussian quadrature methods.

Here we present some preliminary cross section results for positronium formation to the 2s state in the present eikonal approximation as well as the corresponding Born results.

Incident psoitron energy (eV)	Present cross section (πa_o^2)	Born (πa_o^2)
50	0.688 (-1)	0.656 (-1)
80	0.144 (-1)	0.155 (-1)
100	0.608 (-2)	0.688 (-2)

REFERENCE

1.   C. Sinha, S. Tripathi and N.C. Sil, Phys. Rev. A34:1026 (1986).

IONISATION OF ATOMIC HYDROGEN BY POSITRON IMPACT

Gottfried Spicher, Andreas Gläsker, Wilhelm Raith,
Günther Sinapius, and Wolfgang Sperber

Fakultät für Physik, Universität Bielefeld
D-4800 Bielefeld, Federal Republic of Germany

THE APPARATUS

We have built a crossed beam apparatus to measure the relative ionisation cross sections of atomic hydrogen by positron impact. A schematic
diagram is given in Fig. 1.

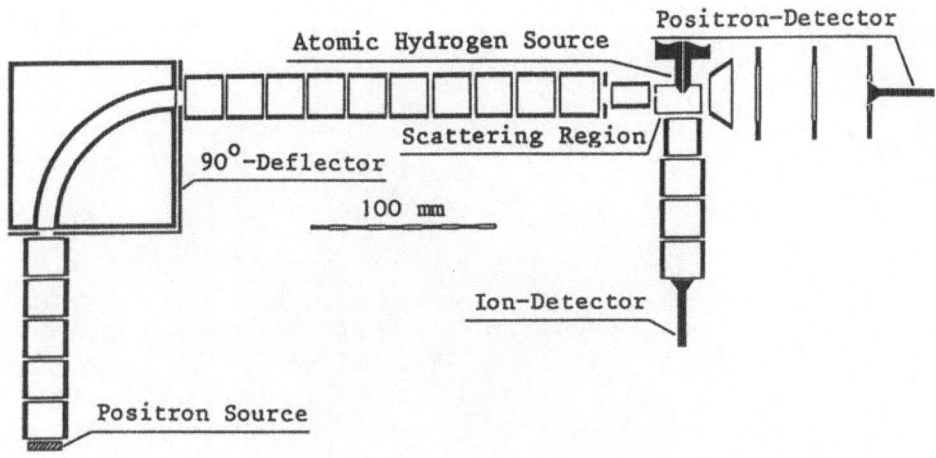

Fig. 1. Schematic diagram of apparatus

Fast positrons from a ^{22}Na source are moderated on tungsten meshes.
The low energy positrons are accelerated, pass the 90° spherical deflector
and are transported into the scattering region. The atomic hydrogen beam
is supplied by a Slevin atomic hydrogen source[1]. Ions produced by positron
beam and focussed onto a Channel Electron Multiplier (CEM). The primary
positron beam and the forward scattered positrons are detected by another
CEM.

Because of the H_2-molecules in the atomic beam and the residual gas
both H^+ and H_2^+ ions are produced:

Process	Cross section	Threshold
$e^+ + H \rightarrow Ps + H^+$	$\sigma_{Ps}(H)$	6.8 eV
$e^+ + H \rightarrow e^+ + e^- + H^+$	$\sigma^+_{ion}(H)$	13.6 eV
$e^+ + H_2 \rightarrow Ps + H_2^+$	$\sigma_{Ps}(H_2)$	8.6 eV
$e^+ + H_2 \rightarrow e^+ + e^- + H_2^+$	$\sigma^+_{ion}(H_2)$	15.4 eV
$e^+ + H_2 \rightarrow Ps + H + H^+$	$\sigma_{Ps}(H_2,H)$	11.1; 20.2 eV
$e^+ + H_2 \rightarrow e^+ + e^- + H + H^+$	$\sigma^+_{ion}(H_2,H)$	17.9; 27.0 eV

DATA TAKING

Time-of-flight (TOF) spectroscopy is applied to distinguish the different processes and to obtain relative cross sections from threshold up to
$\approx$ 1000 eV. Ionisation without positronium formation leads to time correlated
signals on both detectors, N_1 and N_2, which originate from H^+ and H_2^+. Measurements are performed with the discharge in the hydrogen source turned on
and off. At a given positron intensity cross sections are obtained from
the ion counts N^{on} and N^{off} and the time correlated signals N_1^{on}, N_2^{on} and
N_1^{off}, N_2^{off}:

$$N_1^{on} = K^* \sigma^+_{ion}(H) \, d_1^{on} + K^* \sigma^+_{ion}(H_2,H) \, d_2^{on}$$

$$N_1^{off} = K^* \sigma^+_{ion}(H_2,H) \, d_2^{off}$$

$$N_2^{on} = K^* \sigma^+_{ion}(H_2) \, d_2^{on}$$

$$N_2^{off} = K^* \sigma^+_{ion}(H_2) \, d_2^{off}$$

$$N^{on} = K \, [\sigma_{Ps}(H) + \sigma^+_{ion}(H)] \, d_1^{on}$$
$$+ K \, [\sigma_{Ps}(H_2) + \sigma^+_{ion}(H_2) + \sigma_{Ps}(H_2,H) + \sigma^+_{ion}(H_2,H)] \, d_2^{on}$$

$$N^{off} = K \, [\sigma_{Ps}(H_2) + \sigma^+_{ion}(H_2) + \sigma_{Ps}(H_2,H) + \sigma^+_{ion}(H_2,H)] \, d_2^{off}$$

The constants K and K^* account for the detection probabilities for
ions and for correlated positron-ion pairs. The density for atomic and molecular hydrogen with the discharge turned on and off is given by $d_{1/2}^{on/off}$.
If the densities and detection probabilities are kept constant the
relative ionisation cross sections can easily be deduced.

All equations are also valid (with $\sigma_{Ps} = 0$) for the ionisation of hydrogen by electron impact. By switching the polarity of the optical ele-
ments for the primary beam transport we can measure $\sigma^-_{ion}(H)$, $\sigma^-_{ion}(H_2)$ and
σ^-_{21} as a function of energy. Comparison of our electron data to literature
values[2,3] will yield the constants $d_{1/2}^{on/off}$, K and K^* and hence a normalisation for the positron data. If measurements are extended to the energy
range where electron and positron impact ionisation cross sections converge, the convergence can also be utilized for normalisation.

REFERENCES

L. J. Slevin, W. Stirling, Radio frequency atomic hydrogen beam source,
    Rev. Sci. Instr. 52:1780 (1981); source manufact. by Leisk
    Engineering Ltd.

2. W.L. Fite, R.T. Brackman, Collisions of electrons with hydrogen atoms:
    I. Ionisation, Phys. Rev. 122:1141 (1958).

3. E.W. Rothe, L.L. Marion, R.H. Neynaber, S.M. Trujillo, Electron impact
    ionisation of atom hydrogen and atomic oxygen, Phys. Rev. 125:582
    (1962).

CROSS SECTIONS FOR e^+ + H^- → Ps(nℓ) + H(1s) IN FOCK-TANI REPRESENTATION

Jack C. Straton and Richard J. Drachman

National Aeronautics and Space Administration

Goddard Space Flight Center, Code 681, Greenbelt, MD 20771

Reliable cross sections for the various possible positronium formation processes are essential for an accurate theoretical value for the width of the .511 MeV annihilation line found in the region of the galactic center,[1] solar flares[2] and in planetary nebulae.[3] In transition regions of planetary nebulae the concentration of the negative hydrogen ion should be large enough [4] that the reaction

$$e^+ + H^- → Ps(nℓ) + H(1s) \tag{1}$$

should make an important contribution to the line width. And because there is no threshold for this reaction, it may be the only competitor to direct-annihilation[5] at energies below the 6.8 eV threshold for electron capture from neutral hydrogen even in regions where the H^- density is low.

The present calculation utilizes Fock-Tani representation,[6] which treats composite species exactly and includes reactants, intermediate states and products symmetrically within a single Hamiltonian. The first-order Fock-Tani matrix elements contain higher-order effects than do those arising from the first Born approximation since each term in the Fock-Tani potential contains corrections orthogonalizing unbound particles to their respective bound species (three species in this case). And because each term in the potential is smaller, the overall convergence should be improved. Ojha et al[7] have calculated the first-order Fock-Tani cross sections for charge transfer in proton-hydrogen collisions and have obtained good agreement with experiment[8] for differential angles within 1 mrad of the forward direction at 25, 60 and 125 keV and for total cross sections at energies greater than 10 keV.

Choudhury et al.[9] have calculated cross sections for (1) in the first-order Coulomb-Born approximation for energies 20 to 500 eV. Since it is expected that a first-order Born calculation will not be reliable at energies in the range 0 to 6.8 eV, the present work will compare the reliability of the post form of the first-order Fock-Tani T-matrix. Because Choudhury found a post-prior discrepancy arising from their use of an approximate H wave function, an additional level of refinement would come from using the generalized Fock-Tani transformation[10] which is explicitly post/prior-symmetrical at each order (a symmetry of the exact T-matric in any representation).

Because the creation operators for the various species (anti-) commute, the same decomposition of the Fock-Tani Hamiltonian is used for initial and final states so that the asymptotic part contains the net Coulomb force on the positron. Therefore, the Lippmann–Schwinger equations give the (orthogonalized) Coulomb-Born T-matrix as the lowest-order approximation without resorting to a distorted wave formalism.[9] Also, since the Fock-Tani potential has an occupation number form, each of the 31 terms corresponds to a unique and immediately identifiable physical process (scattering, ionization, recombination, and rearrangement of the various species). The lowest-order Fock-Tani T-matrix for (1) is a sum of ($\sqrt{2}$ times) the (post) Coulomb-Born contribution,[9]

$$
T^1_{CB} = \int d\bar{x}d\bar{y}d\bar{y}' \phi^*_{n\ell A}(\bar{x}\bar{y}') u^*_{1sE}(\bar{y}) \left\{ \frac{1}{x} - \frac{1}{y'} - \frac{1}{|\bar{x}-\bar{y}|} + \frac{1}{|\bar{y}-\bar{y}'|} \right\} \psi_B(\bar{y}'\bar{y}) \chi^+_C(\bar{k}_i,\bar{x}) , \quad (2)
$$

the (post) direct-orthogonalization correction,

$$
T^1_{DO} = - \int d\bar{x}d\bar{y}d\bar{y}'d\bar{y}_2 d\bar{x}_2 \phi^*_{n\ell A}(\bar{x}_2\bar{y}_2) u^*_{1sE}(\bar{y}) \left\{ \frac{1}{x_2} - \frac{1}{y_2} - \frac{1}{|\bar{x}_2-\bar{y}|} + \frac{1}{|\bar{y}_2-\bar{y}|} \right\}
$$

$$
\times \; \Delta_A(\bar{x}_2\bar{y}_2;\bar{x}\bar{y}') \psi_B(\bar{y}'\bar{y}) \chi^+_C(\bar{k}_i,\bar{x}) , \quad (3)
$$

and the (post) exchange-orthogonalization correction

$$
T^1_{EO} = \int d\bar{x}d\bar{y}d\bar{y}'d\bar{x}_1 d\bar{y}_1 \phi^*_{n\ell A}(\bar{x}_1\bar{y}_1) u^*_{1sE}(\bar{y}) \left\{ \frac{1/2}{|\bar{x}-\bar{y}_1|} - \frac{1}{|\bar{x}_1-\bar{y}|} + \frac{1}{|\bar{y}-\bar{y}_1|} - \frac{1/2}{|\bar{y}'-\bar{y}_1|} \right.
$$

$$
\left. + \overleftarrow{H}(\bar{x}_1\bar{y}_1) + \frac{1}{2}\overrightarrow{H}(\bar{x}\bar{y}') \right\} \Delta_A(\bar{x}_1\bar{y};\bar{x}\bar{y}') \psi_B(\bar{y}'\bar{y}_1) \chi^+_C(\bar{k}_i,\bar{x}) , \quad (4)
$$

where $H(\bar{x}\bar{y}) = T(\bar{x}) + T(\bar{y}) + 1/x - 1/y - 1/|\bar{x}-\bar{y}|$. The wave functions, ϕ_A, include the center of mass motion so that the bound-state kernel is

$$
\Delta(\bar{x}\bar{y};\bar{x}'\bar{y}') = \sum_\gamma \phi_{\gamma A}(\bar{x}\bar{y}) \phi^*_{\gamma A}(\bar{x}'\bar{y}') = \delta(\bar{R}-\bar{R}') \sum_\gamma u_{\lambda A}(\bar{r}) u^*_{\lambda A}(\bar{r}') , \quad (5)
$$

where $\bar{R}$ is the center-of-mass coordinate and $\bar{r}$ is the relative coordinate. Note that (3) and (4) may be rewritten in terms of the set of continuum wave functions that are the orthogonal complement to the bound-state kernel,

$$
\Lambda_A(\bar{x}\bar{y};\bar{x}'\bar{y}') = \delta(\bar{x}-\bar{x}') \delta(\bar{y}-\bar{y}') - \Delta_A(\bar{x}\bar{y};\bar{x}'\bar{y}')
$$

$$
= \delta(\bar{R}-\bar{R}') \int d\bar{k} \, \chi_A(\bar{k},\bar{r}) \chi^*_A(\bar{k},\bar{r}') , \quad (6)
$$

from which, by orthogonality of the u's and χ's, the internuclear term can be shown to (approximately) cancel between the CB and DO terms if the positron is replaced by a proton, just as in proton-hydrogen charge exchange.[11] In the present case the first and second and the third and fourth terms of (3) cancel if nℓ and λ (in the sum of (5) are of the same parity so that the nℓ = λ = 1s exchange terms in (4) may not be negligible.

REFERENCES

1.   M. Leventhal, et al. Astrophys. J. 225:L11 (1978).
2.   E.L. Chupp et al. Nature 241:333 (1973).
3.   M Leventhal et al. Nature 266:696 (1977).
4.   J.H. Black, Astrophys. J. 222:125 (1978).

5. B.L. Brown et al. Phys. Rev. A33:2281 (1986).
6. M.D. Girardeau, Phys. Rev. A26:217 (1982).
7. P.C. Ojha et al. Phys. Rev. A33:112 (1986).
8. P.J. Martin et al Phys. Rev. A23:3357 (1981), and G.W. McClure,
    Phys. Rev. 148:47 (1966).
9. Choudhury et al. Phys. Rev. A33:2358 (1986).
10. M.D. Girardeau (private communication).
11. J.C. Straton, Phys. Rev. A34:3725 (1987).

ELASTIC SCATTERING OF POSITRONS FROM ARGON

J. M. Wadehra and Sultana N. Nahar

Department of Physics and Astronomy, Wayne State University

Detroit, Michigan 48202, USA

Differential and integrated cross sections for the elastic scattering of low- and intermediate-energy (3 - 300 eV) positrons and electrons by argon atoms are calculated using partial wave method. Model potentials are used to represent the interactions between positrons or electrons and argon atoms. For each impact energy, the phase shifts of the lower partial waves are obtained exactly by numerical integration of the radial part of the Schrodinger equation. The Born approximation is used to obtain the contribution of the higher partial waves to the scattering amplitude.

The model potential for positron-argon interaction contains the static potential of the target atom and a Buckingham type polarization potential with an adjustable parameter d. The electron-argon interaction is represented by the target static potential (with proper sign), the Buckingham type polarization potential with parameter d and an exchange potential. The value of the parameter d, which depends on the projectile impact energy, is determined by fitting the calculated electron-argon scattering cross sections (i.e., differential, integrated and momentum transfer cross sections) and the phase shifts with the measured values of the same for a particular energy. Then the same value of d is used for the calculation of cross sections for positron scattering from argon at the same impact energy. When normalized at 90°, the relative values of the differential cross sections for the elastic scattering of positrons from argon at impact energies of 100, 200 and 300 eV measured by Hyder et al.[1] agree well with the present calculations as shown in Fig. 1 on the next page. Presently the group of Kauppila and Stein[2] is making measurements for positron scattering from argon at lower impact energies. Their preliminary results are showing encouraging agreement with the present calculations.

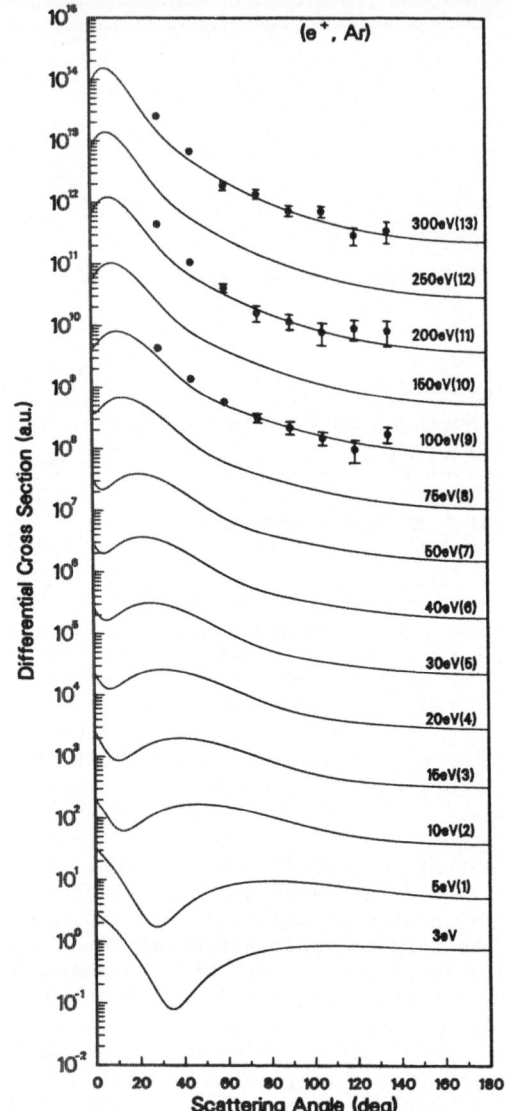

Fig. 1. Differential cross sections for the elastic
scattering of positrons by argon at various impact energies.
Solid lines are the present theoretical curves. The number
in parenthesis following an energy value indicates the power
of ten by which the cross section values are multiplied. The
experimental values are from Ref. 1.

ACKNOWLEDGEMENT

Support of NSF and AFOSR is gratefully acknowledged.

RFERENCES

1. G.M.A. Hyder, M.S. Dababneh, Y.-F. Hsieh, W.E. Kauppila, C.K. Kwan,
   M. Mahdavi-Hezaveh and T.S.Stein, Phys. Rev. Lett. 57, 2252 (1986).
2. W.E. Kauppila and T.S. Stein (private communication).

POSITRONIUM FORMATION FROM ATOMIC HYDROGEN

J. M. Wadehra and Sultana N. Nahar

Department of Physics and Astronomy, Wayne State University

Detroit, Michigan 48202, USA

The first Born approximation and the distorted wave Born
approximation are used to calculate the cross sections for positronium
(Ps) formation in all bound states by the impact of intermediate energy
(20 - 500 eV) positrons on atomic hydrogen. Differential and integrated
cross sections for the formation of Ps(1s), Ps(2s), Ps($2p_0$) and
Ps($2p_1$) are calculated individually and the $1/n^3$ behavior (n being the
principal quantum number) for charge transfer cross sections is used for
$n \geqslant 3$ to obtain the total cross sections for positronium formation. The
formation of Ps in s-state is evaluated using formulation of the
distorted wave Born approximation similar to that described in Ref. 1.
All calculations are carried out using the prior form of the
interaction. The p-state wave functions of Ps, unlike spherically
symmetric s-state wave functions, are angle dependent and introduce
complexity in the calculations of capture cross sections. The complexity
is reduced by expressing the angle dependent part of the wave function
in terms of an exponential factor. It is observed in the present
calculations that the cross section for Ps formation in n = 1 state
dominates significantly over that for n = 2 state. No experimental
values of cross sections for Ps formation from atomic hydrogen are
available at present. The present results for the formation of Ps(1s)
compare[2] favorably with some of the other theoretical investigations.
The features of the present differential cross section curves for Ps
formation showing a large maximum in the forward direction followed by a
minimum also agree well with works of other investigators. The total
cross sections for the formation of Ps in all bound states at various
impact energies are shown in Fig. 1 on the next page.

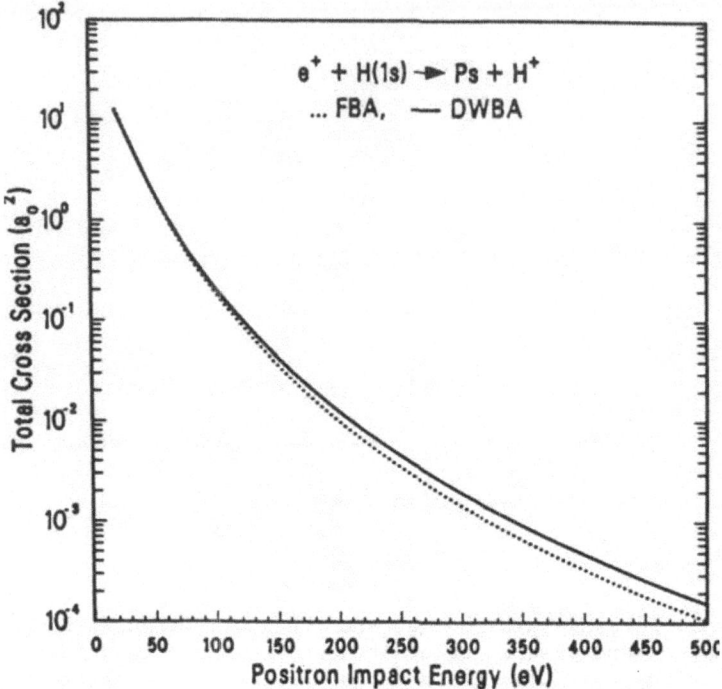

Fig. 1. Total integrated cross sections for positronium formation from atomic hydrogen at various positron impact energies.

ACKNOWLEDGEMENT

Support of NSF and AFOSR is gratefully acknowledged.

RERERENCES

1. R. Shakeshaft and J.M. Wadehra, Phys. Rev. A22, 968 (1980); Sultana N. Nahar and J.M. Wadehra, ibid, A35, XXXX (1987).
2. Sultana N. Nahar, Ph.D. Thesis, Wayne State University (1987).

POSITRON–ATOMIC HYDROGEN SCATTERING

AT MEDIUM TO HIGH ENERGIES

H.R.J. Walters

Department of Applied Mathematics
The Queen's University
Belfast BT7 1NN, U.K.

Calculations of 1s → 1s, 1s → 2s and 1s → 2p transitions have been made for the incident energy range 54.4 to 300 eV. The approximation is similar to that used by van Wyngaarden and Walters[1] in the corresponding electron case, namely, multi-pseudostate close-coupling supplemented by a second Born term to take account of target states of angular symmetries not included in the pseudostate set. To be specific, the scattering amplitude is approximated by

$$f(PS) + f(SBA, AM \geqslant 3) \tag{1}$$

where $f(PS)$ is the close-coupled pseudostate amplitude calculated using states only of s-, p- and d-type, and $f(SBA, AM \geqslant 3)$ is the plane wave second Born term resulting from coupling to target states with angular momenta (AM) greater than or equal to 3. We have considered two sets of pseudostates, those of Fon et al[2], and an improved set introduced by van Wyngaarden and Walters[1]. The Fon et al set consists of three eigenstates, 1s, 2s, 2p, and six pseudostates, $\overline{3s}$, $\overline{4s}$, $\overline{3p}$, $\overline{4p}$, $\overline{3d}$, $\overline{4d}$. The improved set also has the same three eigenstates but includes eighteen pseudostates, i.e., $\overline{3s}$ to $\overline{8s}$, $\overline{3p}$ to $\overline{8p}$, $\overline{3d}$ to $\overline{8d}$, although, following van Wyngaarden and Walters[1], only pseudostates corresponding to open channels are retained in the close-coupling calculation at any given incident energy.

As a sample of the results we show in figure 1 differential cross sections for 1s → 1s, 1s → 2s and 1s → 2p scattering at 100 eV calculated according to (1) using the improved pseudostates. These results are compared with a coupled-channel optical model calculation of Bransden et al[3] and with the unitarised eikonal-Born series (UEBS) approximations of Byron et al[4,5]. All three approximations are in fair agreement at forward angles but significant differences appear at large angles. Also shown are the corresponding electron scattering cross sections of van Wyngaarden and Walters[1] which lie systematically below the present positron results for 1s → 2s and 1s → 2p excitation but consistently above for elastic scattering.

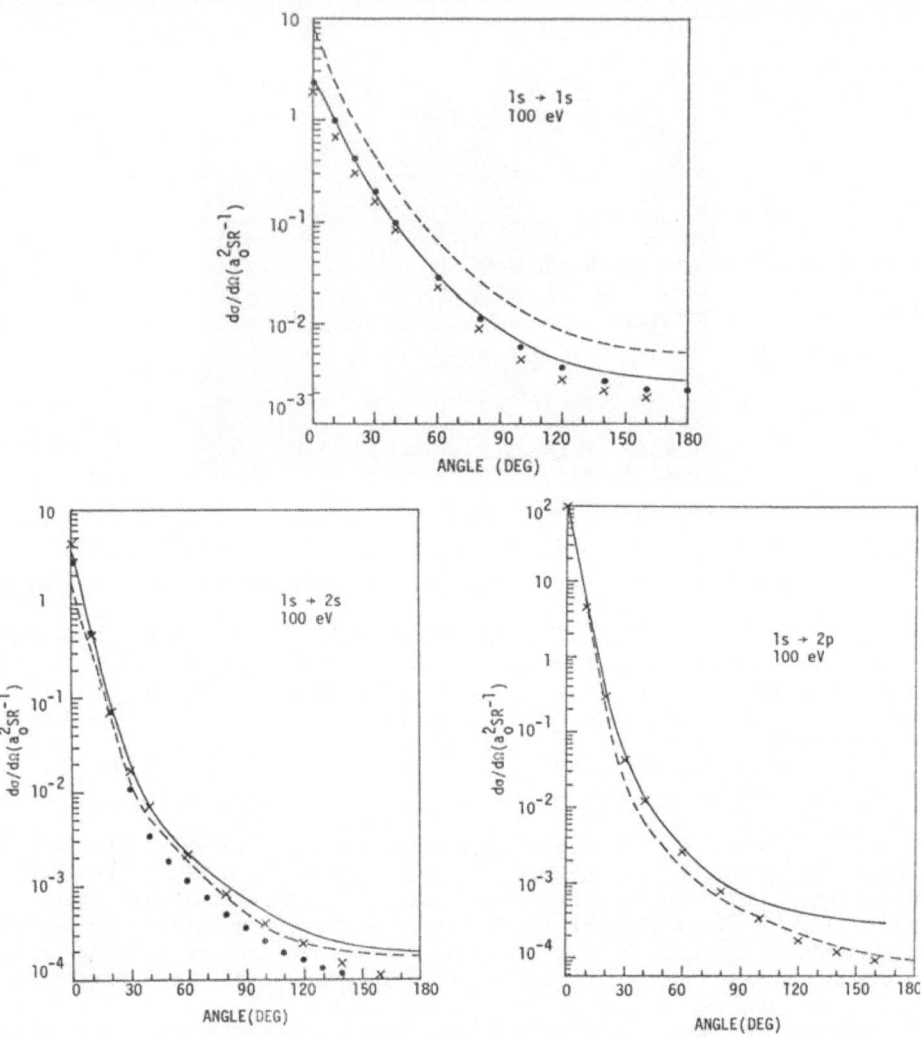

Fig. 1. Differential cross sections for positron scattering by atomic
hydrogen at 100 eV. Notation: ———, present results; X, optical
model results of Bransden et al[3]; ●, UEBS cross sections of
Byron et al[4],[5]; – – – –, corresponding electron scattering
cross sections of van Wyngaarden and Walters[1].

REFERENCES

1. W.L. van Wyngaarden and H.R.J. Walters, J.Phys.B. 19:929 (1986).
2. W.C. Fon, K.A. Berrington, P.G. Burke and A.E. Kingston, J.Phys.B.
   14:1041 (1981).
3. B.H. Bransden, I.E. McCarthy and A.T. Stelbovics J.Phys.B. 18:823
   (1985).
4. F.W. Byron Jr., C.J. Joachain and R.M. Potvliege, J.Phys.B. 14:L609
   (1981).
5. F.W. Byron Jr., C.J. Joachain and R.M. Potvliege, J.Phys.B. 18:1637
   (1985).

LOW ENERGY POSITRON TRANSMISSION MEASUREMENT FROM THIN SINGLE-CRYSTAL

W FOIL

N. Zafar[a], F.M. Jacobsen[b], J. Chevallier[b], M. Charlton[a]
and G. Laricchia[a]
a) Department of Physics and Astronomy, University College
London, Gower Street, London WC1E 6BT, U.K.
b) Institute of Physics, Aarhus University, DK-8000 Aarhus-C,
Denmark

Thin single-crystal metal foils have been proposed and reported to be successful as transmission mode moderators by several authors, e.g. Mills and Wilson[1] with Al, Cu and Si, Schultz et al[2] with Ni and Chen et al[3] with W, all as remoderators. Lynn et al[4] have used W as a primary moderator. W appears to be a good choice for this use namely because of its high density and its high efficiency in backscattering mode; therefore a study has been made into the transmission efficiency of thin, single-crystal W(100) and W(110) foils in the range 1,000 to 18,000 Å annealed in poor vacuum ($\simeq 10^{-2}$ torr) and tested in vacuum of approximately 10^{-6} torr.

The foils were epitaxially grown and the thickness determined to an accuracy of ± 10%. They were placed in the centre of an "oven" configuration and heated resistively, first at a few hundred degrees for 1 minute, then flashed between 1800 and 2400°C, as measured by an optical pyrometer and subsequently allowed to cool quickly. The foils were transferred in air to the testing system and placed within 1 - 2 mm of the source. The detection system was the usual configuration of retarding field analyser and a channeltron with a magnetic guidance field of $\simeq$ 100 gauss.

The yields were seen to increase with the number of heating cycles for all the foils until they plateaued off at efficiencies of the order of 10^{-4}. The highest yield was obtained from a 2000 Å sample flashed at 2000-2200° C after undergoing 5 heating cycles; this was 170 slow e^+, from an extended 64 μCi Na^{22} source.

The factors that must be taken into account before comparing values of efficiency from theory and experiment are namely the number of grids in the path of the e^+, the backscattering from the source mount, the detection efficiency of the channeltron and the solid angle effect. Taking the backscattering to be 50% and the grid transmission as 80% an efficiency of 2.5 × 10^{-4} was calculated.

To obtain an estimate of the number of fast e^+ incident on the foil the source was divided into ring elements and the percentage of e^+ emitted into the solid angle extended by the foil was calculated for each element. This gave an efficiency of 4.6 × 10^{-4}. The detection efficiency of the channeltron was not corrected for in this instance.

The value obtained is about an order of magnitude lower than that predicted by theory ($\simeq 10^{-3}$)[5]. This suggests that, perhaps, the annealing technique may not be adequate to remove all the defects or the contamination although Lynn et al[4] have obtained similar efficiencies from a different and more involved method using O_2 to remove the C impurities in the foil and testing in UHV conditions.

This study has shown that relatively good yields can be obtained from thin single-crystal W foils annealed in poor vacuum and used in non-UHV conditions. Further studies will involve extending the technique by attempting to O_2 treat the foils in low vacuum and also to compare the yields from different face orientations.

REFERENCES

1. A.P. Mills, Jr. and R.J. Wilson, Phys. Rev. A 26:490(1982).
2. P.J. Schultz, E.M. Gullikson and A.P. Mills, Jr., Phys. Rev. B 34:442 (1986).
3. D.M. Chen, K.G. Lynn, R. Pareja and B. Nielsen, Phys. Rev. B 31:4123 (1985).
4. K.G. Lynn, B. Nielsen and J.H. Quaterman, Appl. Phys. Lett. 47:239 (1985).
5. A. Vehanen and J. Makinen, Appl. Phys. A 36:97 (1985).

INDEX